Crop Science and Technology

Crop Science and Technology

Edited by **Corey Aiken**

R CALLISTO
REFERENCE

New York

Published by Callisto Reference,
106 Park Avenue, Suite 200,
New York, NY 10016, USA
www.callistoreference.com

Crop Science and Technology
Edited by Corey Aiken

International Standard Book Number: 978-1-63239-631-0 (Hardback)

Contents

Preface

This book discusses the fundamentals as well as modern approaches of crop science and technology. It elucidates the key concepts and other innovative aspects related to this field. Crop Science deals with assortment and cultivation of crops. It also incorporates development and analysis of the production methods, techniques for improving quality, methods to reduce usage of pesticides and other harmful chemicals. Most of the topics introduced in this book cover new techniques and applications in the field of crop science. It will prove to be a valuable asset for readers who are interested in this field. It aims to serve as a resource guide for students and experts alike and contribute to the growth of this area.

Significant researches are present in this book. Intensive efforts have been employed by authors to make this book an outstanding discourse. This book contains the enlightening chapters which have been written on the basis of significant researches done by the experts.

Finally, I would also like to thank all the members involved in this book for being a team and meeting all the deadlines for the submission of their respective works. I would also like to thank my friends and family for being supportive in my efforts.

Editor

The use of organic biostimulants in hot pepper plants to help low input sustainable agriculture

Andrea Ertani*, Paolo Sambo, Carlo Nicoletto, Silvia Santagata, Michela Schiavon and Serenella Nardi

Abstract

Background: World demand for agricultural products is increasing. New insights are required in order to achieve sufficient and sustainable yields to meet global food request. Chemical fertilizers have been studied for almost 200 years, and it is unlikely that they could be improved. However, to produce food for a growing world population, various methods to increase the efficiency of chemical fertilizers are investigated. One approach to increasing crop productivity is the development of environment-friendly organic products named biostimulants which stimulate plant growth by enhancing the efficiency of chemical fertilizers. Most studies have tested these products in short-term experiments, but little information is available on their effect on plants at the maturity stage of growth. On this account, this paper focuses on the effects of two biostimulants, red grape skin extract (RG) and alfalfa hydrolyzate (AH), throughout the entire plant development.

Results: The findings obtained in the present investigation demonstrate the effectiveness of RG and AH in improving growth and the nutritional value of peppers. Specifically, the two biostimulants increased the phenol concentration, antioxidant activity, and ascorbic acid concentration in fruits, as well as the capsaicin concentration in plants. Differences in effectiveness between RG and AH were likely related to the characteristics of the starting matrixes as well as to the industrial processes used for their production. The efficiency of RG and AH in promoting plant growth and yield could also be due to their content in indole-3-acetic acid (IAA), isopentenyladenosine (IPA), phenols, and amino acids.

Conclusions: In the light of these results, the application of biostimulants could be considered as a good strategy for obtaining high yields of nutritionally valuable vegetables with lower environmental impact.

Keywords: Peppers; Biostimulants; Endogenous hormones; Phenols; Antioxidants

Background

The climate changes occurring in recent years have affected the farming, imposing constraints and objectives frequently ignored in the past. The environmental question and quality of the products affect farmers' choices in different ways, depending on the specific vulnerability of the environment where the crops are cultivated [1]. In this context, crop-derived food safety and nutritional value have become important issues worldwide [2]. For example, there is increasing interest in studying and quantifying the antioxidant and anti-inflammatory constituents of plants in terms of their potential health benefits [3,4].

To increase the content of nutritional constituents in plants, many approaches have been studied to increase plant nutrients capture and yield such as genetic selection, which has included allele selection, gene and genome duplication, and new genotypes creation. However, despite the advantages that these techniques offer, some of them may also pose potential problems for food safety and require special attention in order to ensure consumer health protection [5,6].

On this account, the use of biostimulants in agricultural practices is proposed as a safe tool to enhance the nutritional properties of food crops. Biostimulants are recognized as environment-friendly compounds with beneficial effects on plants [7,8]. In particular, they decrease the use of mineral fertilizers by increasing the amount of micro- and macro-nutrients taken up by plants, positively influencing root morphology and plant growth [9,10]. They also display hormone-

---

* Correspondence: andrea.ertani@unipd.it
Department of Agronomy, Animals, Food, Natural Resources and
Environment - DAFNAE, University of Padua, Viale dell'Università 16, 35020,
Legnaro, Padova, Italy

like activity and influence plant metabolism by interacting with biochemical processes and physiological mechanisms, such as glycolysis and nitrogen assimilation [11,12-13].

The mechanisms behind the physiological and biochemical effects of biostimulants on plants are often unknown, because of the heterogeneous nature of the raw materials used for their production. Furthermore, these effects are often the result of many components that may work synergistically in different ways.

Recent studies suggest that active molecules contained in biostimulants can promote nitrogen assimilation through stimulation of the activity and transcription of N assimilation and Krebs' cycle enzymes [7-11]. Furthermore, the induction of the metabolic pathway associated with the synthesis of phenylpropanoids in plants treated with biostimulants may explain why these products can help plants to overcome stress situations [14,15]. In particular, a protein hydrolyzate-based fertilizer obtained from alfalfa hydrolyzate plants (AH) was proved to help maize plants to overcome salinity stress through stimulation of enzymes functioning in nitrogen metabolism, enhancement of phenylalanine ammonia-lyase (PAL) activity and transcription, and increase of flavonoid synthesis [16]. PAL is an important enzyme that catalyzes the first committed step in the biosynthesis of phenolics by converting phenylalanine to trans-cinnamic acid and tyrosine to p-coumaric acid, opening the way to the secondary metabolism. The biostimulant activity of AH was related to the presence of triacontanol (TRIA) and indole-3-acetic acid (IAA), two important regulators of plant metabolism [15].

In recent studies, the application of products with biostimulant action to pepper plants was found to exert positive effects on plant growth and yield without loss of fruit quality [13,17,18]. Pepper is an important agricultural crop due to the nutritional value of its fruits, which are an excellent source of a wide array of phytochemicals with well-known antioxidant properties. The main antioxidant compounds include carotenoids, capsaicinoids, and phenolic compounds, particularly flavonoids, quercetin, and luteolin [19,20].

Most of studies testing biostimulants analyze their effects on seed germination and plants growth in short-term experiments, but little information is available on their effects at flowering and maturity stages of plants. On this account, the purpose of this investigation was to study the effects of two biostimulants on plant growth, phenological development, and some metabolic parameters of pepper plants (total phenols, ascorbic acid, capsaicin, dihydrocapsaicin), throughout the development from seedling to fruit.

Methods

Characterization of the products

The two products used in this study (red grape skin extract (RG) and AH) were manufactured by ILSA (Arzignano-Vicenza, Italy). AH was produced by a fully controlled enzymatic hydrolysis using vegetal material from alfalfa (*Medicago sativa* L.) plants. The elemental composition, organic matter content, and physical properties of AH were previously reported by Schiavon et al. [7]. Specifically, the water percentage in AH was 43.5% (v/v), and the amount of organic matter was low 37.6% (w/v). The content of ash and the electrical conductivity (ECw) were both high, with values of 18.9% (w/v) and 16 dS m-1, respectively. The percentage of inorganic nitrogen in the form of ammonia and nitrates was low (0.38% and 0.03 w/w, respectively), whereas the total amount of free amino acids was high up to 1.916% (w/w) and correlated with the free α-amino nitrogen (α-NH2-N). The hydrolytic process employed for EM production was effective in the weight-average molecular weight (MW) reduction, as confirmed by data on EM hydrolysis degree (DH) and MW.

RG was obtained from the skin of red grape wine via cool extraction according to the method of Machado [21]. RG aqueous extract was prepared by extracting 10 g of dried ground plant material with 50 mL of deionized water in a shaker for 2 h at room temperature. The extract was then filtered with cellulosic membrane filters at 0.8 μm (Membra-Fil° Whatman Brand, Whatman, Milan, Italy). The pH was determined in water (3:50 w/v). Total organic carbon (TOC) was measured using an element analyzer (varioMACRO CNS, Hanau, Germany).

Total phenols in RG extracts were determined according to Arnaldos et al. [22]. Specifically, RG (1 mL) was maintained in ice with pure methanol (1:1, v/v) for 30 min and centrifuged at 5,000g for 30 min at 4°C. One mL of 2% Na_2CO_3 and 75 μL of Folin-Ciocalteau reagent (Sigma-Aldrich, St. Louis, MO, USA) were added to 100 μL of phenolic extract. After 15 min of incubation at 25°C in the dark, the absorbance at 725 nm was measured. Gallic acid was used as standard according to Meenakshi et al. [23].

For reducing sugars determination, a sample of RG product was dried for 48 h at 80°C, ground in liquid nitrogen, and then 100 mg were extracted with 2.5 mL 0.1 N H_2SO_4. Samples were incubated in a heating block for 40 min at 60°C and then centrifuged at 6,000g for 10 min at 4°C. After filtration (0.2 μm, Membra-Fil° Whatman Brand, Whatman, Milan, Italy), the supernatants were analyzed by HPLC (Perkin Elmer 410, Perkin Elmer, Waltham, MA, USA). The soluble sugars were separated through a Biorad Aminex 87 C column (300 × 7.8 mm; Bio-Rad Laboratories, Inc.,

Hercules, CA, USA) using H_2O as eluent at a flow rate of 0.6 mL min^{-1}.

Indole-3-acetic acid (IAA) in the RG was determined using an enzyme-linked immuno-sorbent assay (ELISA) standardized with methylated IAA (Phytodetek-IAA, Sigma, St. Louis, MO, USA). The ELISA test utilized a monoclonal antibody to IAA and was sensitive in the 0.05-to-100 pmol range. The tracer and standard solutions were prepared following the manufacturer's instructions, and absorbances were read at 405 nm with a microplate reader (Bio-Rad Laboratories, Inc., Hercules, CA, USA).

Isopentenyladenosine (IPA) in RG and AH was determined by ELISA, a Phytodetek-IPA with an anti-IPA monoclonal antibody was used (Sigma-Aldrich, St. Louis, MO, USA). The competitive antibody-binding method was adopted to measure the IPA concentration. IPA labeled with alkaline phosphatase (tracer) was added with the sample to antibody coated microwells. A competitive binding reaction was set up between a constant amount of the IPA tracer, a limited amount of the antibody and the unknown sample containing IPA (Sigma-Aldrich, St. Louis, MO, USA). One hundred μL of standard IPA concentration or serial dilutions of RG and AH and 100 μL diluted tracer were added to each well. For the standard curve, progressions of 100, 50, 20, 5, 1, 0.1, and 0.02 pmol IPA 100 μL^{-1} were used, whereas for RG and AH, the progressions were 20, 10, 7.5, 5.0, 3.5, and 2.5 μg C 100 μL^{-1}. After incubation at 4°C for 3 h, the wells were decanted and any unbound tracer was washed out by adding 200 μL of washing solution before adding 200 μL of substrate solution for colorimetric detection. After 60 min at 37°C, the optical density (OD) was read at 405 nm using a microplate reader (Bio-Rad Laboratories, Inc., Hercules, CA, USA).

Experimental design and growth conditions

The experimental trial was derived from the factorial combination of three types of treatment: no biostimulant, a cool extract from RG, and an AH. Two doses (50 and 100 mL L^{-1} for RG, and 25 and 50 mL L^{-1} for AH) of biostimulants were applied to two randomized blocks of pots, each consisting of 50 pots (ten pots per treatment to test biostimulants at flowering stage and ten pots per treatment to test biostimulants at maturity stage). The two doses for each biostimulant were in accordance with previous tests showing that doses in the 1-to-200 mL L^{-1} range were those that more interfered with pepper metabolism. Each pot was filled with 2 L perlite/vermiculite mixture. At the beginning of the trial, 7-day-old seedlings of chili pepper (*Capsicum chinense* L. *cv.* Fuoco della Prateria) were transplanted at a density of one plant per pot. Plants were grown until maturity in a tunnel maintained at 25°C/15°C day/night,

receiving natural light. Plants were treated twice, i.e., at the second and fourth week after transplanting, by spraying each one with 4.5 mL of RG or AH on the leaves. All treatments were irrigated with a half-strength Hoagland's nutrient solution. Plants from 50 pots were sampled at flowering (4 weeks after transplanting), and plants from the remaining 50 pots at maturity (6 weeks after transplanting), harvesting leaves and fruits separately.

Analysis of sugars in plant material

Leaf and fruit samples (2 g) were homogenized in water (20 mL) with an Ultra Turrax T25 (IKA, Staufen, Germany) at 13,500 rpm until uniform consistency. Samples were filtered with filter paper (589 Schleicher) and further filtered through cellulose acetate syringe filters (0.45 μm). The analysis of the extracts was performed using an HPLC apparatus (Jasco X.LC system, Jasco Co., Tokyo, Japan) consisting of a model PU-2080 pump, a model RI-2031 refractive index detector, a model AS-2055 autosampler and a model CO-2060 column oven. ChromNAV Chromatography Data System was used as software. Sugars were separated on a Hyper-Rez XP Carbohydrate Ca^{++} analytical column (7.7 × 300 mm, ThermoScientific, Waltham, MA, USA) operating at 80°C. Isocratic elution was effected using water at a flow rate of 0.6 mL/min. The peaks were identified by comparing the retention time with those of standard compounds. To calculate the concentrations in the extract, a calibration curve was drawn for four solutions of known concentration in water.

Determination of total phenols in plant material

The concentration of total phenols was determined by the Folin-Ciocalteu (FC) assay with gallic acid as calibration standard, using a Shimadzu UV-1800 spectrophotometer (Shimadzu Corp., Columbia, MD, USA). The FC assay was performed by pipetting 200 μL of plant extract (obtained as described above for sugars analysis) into a 10-mL PP tube. This operation was followed by addition of 1 mL of Folin-Ciocalteu's reagent. The mixture was vortexed for 20 to 30 s. Eight hundred microliters of sodium carbonate solution (20% *w/v*) was added to the mixture 5 min after the addition of the FC reagent. This was recorded as time zero; the mixture was then vortexed for 20 to 30 s after addition of sodium carbonate. After 2 h at room temperature, the absorbance of the colored reaction product was measured at 765 nm. The total phenols concentration in the extracts was calculated from a standard calibration curve obtained with different concentrations of gallic acid, ranging from 0 to 600 μg mL^{-1}. Results were expressed as milligrams of gallic acid equivalent per kilogram of fresh weight [24].

Determination of total antioxidant activity by ferric reducing antioxidant power

The assay was based on the methodology of Benzie and Strain [25]. Ten grams of plant material was homogenized in 20 mL of HPLC grade methanol using an Ultra-Turrax tissue homogenizer (Takmar, Cincinnati, OH, USA) at moderate speed (setting of 60) for 30 s. The ferric-reducing antioxidant power (FRAP) reagent was freshly prepared, containing 1 mM 2,4,6-tripyridyl-2-triazine (TPTZ) and 2 mM ferric chloride in 0.25 M sodium acetate buffer at pH 3.6. One hundred microliters of the methanol extract was added to 1,900 μL of FRAP reagent and accurately mixed. After leaving the mixture at 20°C for 4 min, the absorbance was determined at 593 nm. Calibration was against a standard curve (0 to 1,200 μg mL^{-1} ferrous ion) obtained by the addition of freshly prepared ammonium ferrous sulfate. FRAP values were calculated as microgram per milliliter ferrous ion (ferric-reducing power) and are presented as milligram per kilogram of Fe^{2+E} (ferrous ion equivalent).

Extraction and determination of ascorbic acid

After harvest, the samples were immediately stored at −80°C before analysis. Five-gram samples were homogenized until uniform consistency in a meta-phosphoric acid and acetic acid solution. This solution was used for the ascorbic acid extraction after quantitative reduction of 2,6-dichlorophenolindophenol dyestuff by ascorbic acid and extraction of the excess dyestuff using xylene. The excess of ascorbic acid was measured at 500 nm in a Shimadzu UV 160A spectrophotometer (Shimadzu Corp., Columbia, MD, USA) and compared with a vitamin C reference standard (ISO/6557-2-1984 method).

Capsaicin and dihydrocapsaicin determination

Fresh leaves and green and red peppers (2 g) were extracted with 20 mL of acetone, followed by homogenization with an Ultra-Turrax T 25 (IKA, Staufen, Germany) for 30 s at 17,500 rpm. The extract was filtered with filter paper and then through regenerated cellulose syringe filters (0.45 μm). The analysis of the extracts was performed using an HPLC system (X-LC Jasco Co., Tokyo, Japan) equipped with a DAD detector (MD-2015, Jasco Co., Tokyo, Japan) and autosampler (AS-2055, Jasco Co., Tokyo, Japan). Samples (20 μL injection volume) were separated on a Tracer Extrasil ODS-2 (250 × 45 mm, 5 μm, Teknokroma, Barcelona, Spain) HPLC column. The mobile phase consisted of two solvents: water (A) and methanol (B) (50:50, v/v). Isocratic elution for 10 min was used, followed by gradient elution 50% to 90% B for 10 min. The flow rate was 1 mL/min and column temperature was 25°C. Detection was set at 278 nm. To calculate the concentrations in the extract,

a calibration curve was drawn for four solutions of known concentration in acetone.

Data analysis

The data represent the means of measurements from five plants per treatment, each representing one biological replicate. Analysis of variance (ANOVA) was performed using the SPSS for Windows software, version 18.0 (SPSS, Chicago, IL, USA) and was followed by pairwise *post hoc* analyses (Student-Newman-Keuls test) to determine which means differed significantly at $p < 0.05$.

Results

Characterization of biostimulants and hormones quantification

The chemical characteristics of RG and AH are reported in Table 1. The values of pH were acid in both cases: 2.9 for RG and 5.9 for AH. Total carbon (TC) was 1.23% for RG and 18.8 for AH. The total sugars content was 5,700 mg/L for RG and 4,642 mg/L for AH. The level of total phenols in RG and AH was appreciable: 970 and 2,576 mg/L, respectively. Hormone determination via ELISA assay evidenced an IAA concentration of 2.92 nmol mg carbon^{-1} and an IPA concentration of 0.073 nmol mg carbon^{-1} in RG; 18.46 nmol mg carbon^{-1} for IAA and 0.055 nmol mg carbon^{-1} for IPA in AH.

Effects of RG and AH on growth parameters

To establish the stimulatory effects of RG and AH on growth, the fresh weight of leaves and peppers was measured (Figures 1A,B). At the first sampling time, RG and AH increased the leaf biomass at both tested rates. More specifically, RG at the lower and AH at the higher rate enhanced leaf weight (+68% and +165%) in comparison to untreated plants. The weight of green peppers showed a similar trend as the leaves, but both RG and AH rates, reduced the biomass of red peppers.

At the second sampling time (Figure 1B), the opposite trend of leaf and pepper biomass production in relation to the application of biostimulants was observed: RG and AH addition weakly stimulated the leaf biomass, decreased green pepper weight, and significantly increased the red pepper weight.

Table 1 Chemical properties and hormone content of red grape (RG) and alfalfa hydrolyzate (AH) products

Property	Unit	RG	AH
pH		2.9	5.9
Total carbon (TC)	%	1.23	18.8
Total sugars	mg L^{-1}	5,700	4,642
Total phenols	mg L^{-1}	970	2,576
Indole-3-acetic acid (IAA)	nmol mg carbon^{-1}	2.92	18.46
Isopentenyladenosine (IPA)	nmol mg carbon^{-1}	0.073	0.055

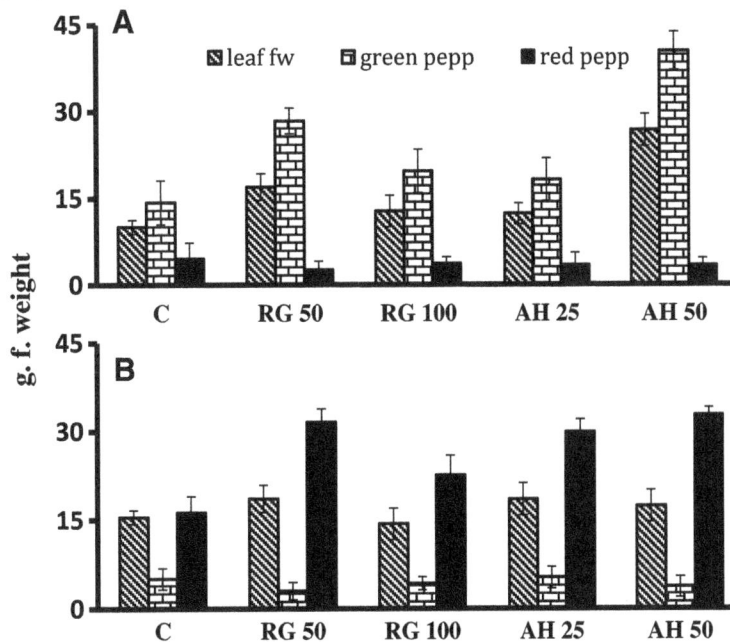

Figure 1 Effect of spray-solutions with red grape skin extract (RG) or alfalfa hydrolyzate (AH). On fresh weight of leaf and green and red fruits of hot pepper. **(A)** first harvest (flowering); **(B)** second harvest (maturity); error bars represent ± standard deviation ($n = 5$).

Table 2 reports the number of peppers at different maturity stages. Application of RG and AH at both tested doses increased the number of peppers in comparison to the untreated plants at flowering stage. Specifically, the amount of green peppers increased (+104% and +175%) when RG and AH were furnished at the lower and at the higher rate, respectively. The number of red peppers was slightly enhanced after biostimulants treatment. On the contrary, at maturity stage (Table 2), both RG dosages

Table 2 Effect of RG and AH on green and red pepper number at flowering and maturity

Treatment	mL L^{-1}	Flowering	%	Maturity	%
Green pepper number					
Control	-	18.2 ± 1.2c	100	4.3 ± 0.3c	100
RG	50	37.1 ± 1.7b	204	5.1 ± 0.4b	119
RG	100	21.5 ± 1.1c	110	5.0 ± 0.1b	116
AH	25	20.2 ± 0.8c	111	6.0 ± 0.4a	140
AH	50	50.1 ± 6.1a	275	4.2 ± 0.6c	98
Red pepper number					
Control	-	4.3 ± 0.3a	100	18.3 ± 0.9c	100
RG	50	2.1 ± 0.1b	49	32.5 ± 1.6b	178
RG	100	4.3 ± 0.3a	100	30.1 ± 1.5b	164
AH	25	4.2 ± 0.5a	98	36.0 ± 1.8a	197
AH	50	4.1 ± 0.1a	95	37.0 ± 1.8a	202

Data represent the means of measurements using five plants per treatment. Different letters in the same column indicate significant differences between treatments ($p < 0.05$) according to Student-Newman-Keuls test.

increased the number of green and red peppers with increments ranging from +19% to +78% in comparison to the control plants. The application of AH to plants induced the greatest effect on red peppers, regardless of the tested dose.

Effects of RG and AH on reducing sugars concentration

The effects of RG and AH on pepper plants were first investigated by determining the concentration of reducing sugars (Table 3). At the flowering phase, there was an increase in sugar concentration in leaves of pepper plants sprayed with RG, at both doses. More specifically, the increase was by about 111% at the lower and by about 75% at the higher RG dose. With respect to green peppers, the concentration of sugars was enhanced by the two treatments at the maturity stage. In particular, RG increased the sugars concentration by 71% and 354% at the lower and higher rates, respectively. At the same stage, the AH addition increased the reducing sugars with increments of 77% and +50% in red peppers of plants treated with the lower and higher rates, respectively.

Total phenols, FRAP, and ascorbic acid concentration

The samples were analyzed for total phenolic compounds, which are mainly responsible for the antioxidant activity of plant extracts. Different concentrations of total phenols were found in plants in response to RG and AH addition (Table 4). Their amount was significantly enhanced in leaves of plants treated with AH, at

Table 3 Effect of RG and AH on reducing sugar content in pepper plants at flowering

Treatment		Reducing sugars (glucose+fructose)			
		µg	(%)	µg	(%)
		Flowering		Maturity	
Leaf					
C	-	4.19 ± 1.12c	100	19.37 ± 1.05a	100
RG	50	8.83 ± 1.78b	211	16.61 ± 0.89b	86
RG	100	7.34 ± 0.54b	175	17.46 ± 1.13ab	90
AH	25	7.00 ± 1.82b	167	18.39 ± 1.23ab	95
AH	50	14.89 ± 2.45a	355	16.35 ± 0.97b	84
Green peppers					
C	-	17.56 ± 2.33b	100	2.64 ± 0.23c	100
RG	50	28.77 ± 2.59a	164	4.52 ± 0.36b	171
RG	100	14.28 ± 1.24b	81	11.99 ± 0.99a	454
AH	25	10.82 ± 0.78c	62	9.47 ± 0.85a	359
AH	50	32.89 ± 3.01c	187	5.66 ± 0.68b	214
Red peppers					
C	-	23.87 ± 3.48a	100	67.12 ± 4.66d	100
RG	50	13.84 ± 1.55c	58	123.80 ± 7.48a	184
RG	100	19.14 ± 2.95b	80	84.09 ± 5.51c	125
AH	25	17.68 ± 2.66b	74	118.86 ± 6.36a	177
AH	50	17.48 ± 2.59b	73	100.83 ± 5.88b	150

Data represent the means of measurements using five plants per treatment. Different letters in the same column indicate significant differences between treatments ($p < 0.05$) according to Student-Newman-Keuls test. Data are expressed as µg reducing sugars per leaf or pepper dry weight.

both tested rates. The increments ranged from +5% in leaves of plants sprayed with RG at the lower rate to +58% in those supplied with AH at the higher rate, in comparison to the untreated plants. In both red and green peppers, the amount of total phenols increased after RG treatment at the higher dose and after application of AH at both tested doses. More specifically, the highest values were observed in red (+55%) and green peppers (+54%) of plants sprayed with RG at the higher dose.

An increase in FRAP was found in leaves of pepper plants sprayed with AH at both tested rates, with values ranging from +22% to +24% compared to the controls (Table 4). The stimulation of FRAP in green peppers by AH was much more pronounced (+61% at the lower and +33% at the higher rate). On the contrary, the effect of biostimulants on red peppers led to a weak increment of FRAP. The application of biostimulants caused a slight increase in the ascorbic acid concentration, with the highest values observed in leaves and green peppers of plants sprayed with AH at the lower dose (+18% and +20%, respectively).

Overall, the content of phenolics, ascorbic acid, and FRAP were more enhanced at the second sampling time

corresponding to plant maturity (Table 5). The greatest increments of total phenols were observed in leaves (+45%) and green peppers (+140%) of plants treated with both AH doses. The FRAP in peppers was increased after AH treatment, with values ranging from +59% in red peppers at the higher rate to +27% in green peppers at the lower one. The ascorbic acid concentration was enhanced in leaves of plants supplied with RG at the lower (+17%) and higher (+24%) dose, as well as after treatment with the lower dose of AH (+37%).

With respect to red peppers, the greatest increase of ascorbic acid (+45%) was evident in plants supplied with RG at the higher dose, while in green pepper both AH and RG at all the tested doses increased ascorbic acid concentration. Specifically, the greater increments were observed in plants treated with RG at the higher rate (+106%) and with AH at the lower rate (+153%).

Capsaicin and dihydrocapsaicin determination

Capsaicin was not detected in leaves of pepper plants (Table 6). In red peppers, RG at the lower rate determined an enhancement of capsaicin content (+24%) in comparison to untreated plants, while AH increased the amount of this compound at both tested doses: +30% and +37% at the lower and higher rate, respectively. On the contrary, in green peppers, biostimulants did not stimulate the synthesis of capsaicin (Table 6). The dihydrocapsaicin concentration was enhanced in leaves of RG- and AH-treated plants, with increments ranging from +117% in plants supplied with RG at the lower rate to +98% in plants sprayed with AH at the same dose. With respect to the concentration of this metabolite in red and green peppers, there was only an increase of dihydrocapsaicin in the case of red peppers on plants with AH added at the higher dose (+46%). At the second sampling time, neither capsaicin nor dihydrocapsaicin were present in leaves, while their concentration was increased in red peppers by the two biostimulants at both rates, with increments ranging from +162% to +341%. RG and AH treatments increased these two metabolites to a lesser extent in green peppers.

Discussion

A number of researches have described the positive effects of biostimulants on plant growth and physiology. In particular, previous studies reported the improvement of pepper plant biomass and yield by either natural biostimulants [26] or humic substances from composted sludge [13]. Accordingly, RG and AH application to plants resulted in increased leaf biomass and weight of green peppers at the flowering stage, as well as in increased red pepper number and growth at maturity. The stimulation of plant growth by RG and AH was possibly due to their content in IAA and IPA, as both hormones

Table 4 Effect of biostimulants on total phenols, FRAP and ascorbic acid in pepper plants at flowering

Treat	mL L^{-1}	Total phenols		FRAP		Ascorbic acid	
		mg GAE kg^{-1} fw	(%)	mg Fe^{2+}E kg^{-1} fw	(%)	mg kg^{-1} fw	(%)
Leaf							
C	-	1,102.1 ± 60.3c	100	3,042.5 ± 121.1b	100	1,857.1 ± 83.1ab	100
RG	50	1,161.4 ± 53.1c	105	2,907.3 ± 116.4b	96	1,709.2 ± 73.1b	92
RG	100	1,364.4 ± 65.3b	124	3,211.1 ± 130.1b	106	1,962.2 ± 38.4a	106
AH	25	1,686.1 ± 73.1a	153	3,697.1 ± 190.1a	122	2,190.4 ± 93.1a	118
AH	50	1,745.1 ± 80.2a	158	3,781.3 ± 173.3a	124	1,906.3 ± 87.3a	103
Red peppers							
C	-	1,081.1 ± 51.1c	100	3,090.1 ± 152.3b	100	1,740.2 ± 81.1a	100
RG	50	1,080.2 ± 38.1c	100	3,178.3 ± 120.3b	103	1,677.3 ± 63.2b	96
RG	100	1,678.1 ± 80.7a	155	4,510.5 ± 220.1a	146	1,754.3 ± 75.1ab	101
AH	25	1,317.3 ± 60.1b	122	3,135.2 ± 130.4b	101	1,905.0 ± 91.0a	110
AH	50	1,138.3 ± 60.7c	105	3,280.2 ± 133.0b	106	1,877.1 ± 83.1a	108
Green peppers							
C	-	3,621.4 ± 153.2b	100	1,078.0 ± 38.2c	100	1,502.6 ± 63.1b	100
RG	50	2,797.2 ± 110.2c	77	942.4 ± 40.2d	87	1,687.4 ± 80.2a	112
RG	100	5,590.2 ± 216.2a	154	1,584.1 ± 83.8a	147	1,639.4 ± 53.2a	109
AH	25	5,559.1 ± 230.8a	154	1,733.1 ± 73.1a	161	1,806.0 ± 63.2a	120
AH	50	5,439.4 ± 250.4a	150	1,432.3 ± 53.2b	133	1,697.0 ± 80.1a	112

Data represent the means of measurements using five plants per treatment. Different letters in the same column indicate significant differences between treatments ($p < 0.05$) according to Student-Newman-Keuls test.

are known to play a pivotal role in plant development and initiation of fruit ripening. In support of this, previous work showed that physiological active concentrations of IAA contained in high molecular-weight humic substances [14] and in an alfalfa-based biostimulant EM [7] accounted for the enhancement of maize biomass production.

The more pronounced growth of pepper plants treated with RG and AH was consistent with the reduction in concentration of soluble sugars in leaves at maturity, which could be partially ascribed to the higher consumption of these compounds in the respiratory process. Indeed, biostimulants can induce the activity and gene expression of several enzymes involved in the tricarboxylic acid cycle (TCA), as previously observed in maize plants treated with an alfalfa based-hydrolizate [7]. Additionally, soluble sugars in leaves decreased because of their translocation into the fruits at plant maturity. Sugars provide energy for fruit development [27] and represent a major determinant of fruit quality. Therefore, changes in their composition and content lead to easily perceptible alterations in fruit flavor, which is an important, non-visual attribute for customers [28,29]. In this respect, the higher content of reducing sugars in peppers of plants treated with RG and AH compared to peppers of untreated plants indicated that both biostimulants can efficiently improve the fruit quality.

The increased accumulation of sugars in peppers appeared to be strongly associated to the level of ascorbic acid. Pepper is a good source of this antioxidant compound, which is notoriously involved in the photosynthetic process and in the synthesis of ethylene, gibberellins and anthocyanins, as well as in the control of cell growth and in the reduction of oxidative stress. Research performed on the ascorbate deficient mutant of *Arabidopsis thaliana* provided evidence that the ascorbic acid biosynthetic pathway starts from glucose, thus underlining a positive relationship between ascorbic acid and sugars concentrations in plants. In agreement with these findings, in the current investigation, the amount of ascorbic acid was enhanced in green peppers by RG and AH mainly at the maturity stage of plants, as in the case of soluble sugars. Increased levels of ascorbic acid were also observed in tomato plants treated with a seaweed-based biostimulant [30].

A recent study showed that agro-industrial residues and an alfalfa-based hydrolizate EM could function as biostimulants in agriculture as they increased maize plant yield and resistance to stress conditions through the enhancement of phenol production [15,16]. In particular, these biostimulants were able to induce the activity of the PAL enzyme, which is a notorious key regulator of phenol compound biosynthesis [16]. Possibly, RG and AH enhanced

Table 5 Effect of biostimulants on total phenols, FRAP and ascorbic acid in pepper plants at maturity

Treatment	mL L^{-1}	Total phenols		FRAP		Ascorbic acid	
		mg GAE kg^{-1} fw	(%)	mg Fe^{2+}E kg^{-1} fw	(%)	mg kg^{-1} fw	(%)
Leaf							
C	-	1,318 ± 55.7c	100	3,620.4 ± 62.3c	100	528.3 ± 42.3c	100
RG	50	1,784 ± 80.3ab	135	3,740.2 ± 72.2b	103	618.4 ± 56.3d	117
RG	100	1,649 ± 60.2b	125	3,634.2 ± 48.4c	100	652.7 ± 25.2b	124
AH	25	1,906 ± 62.84a	145	3,890.0 ± 53.3a	107	723.3 ± 38.3a	137
AH	50	1,894 ± 82.2a	144	3,990.1 ± 25.8a	110	525.4 ± 52.2c	100
Red peppers							
C	-	1,196.4 ± 38.7c	100	4,218.1 ± 113.2c	100	900.3 ± 76.7c	100
RG	50	1,425.5 ± 82.5b	119	4,610.1 ± 216.1c	109	1,084.2 ± 112.1b	120
RG	100	1,684.4 ± 32.8a	141	4,748.3 ± 221.3c	112	1,307.5 ± 103.0a	145
AH	25	1,732.3 ± 83.1a	145	5,770.0 ± 286.3b	136	1,045.1 ± 128.0b	116
AH	50	1,667.1 ± 77.0a	139	6,721.0 ± 283.3a	159	1,185.1 ± 176.3ab	131
Green peppers							
C	-	2,364.4 ± 101.2d	100	12,010.7 ± 421.6c	100	498.5 ± 56.2d	100
RG	50	3,447.4 ± 128.3c	146	12,749.5 ± 526.3c	106	705.3 ± 123.0c	142
RG	100	4,320.5 ± 222.1b	183	13,645.8 ± 628.3bc	113	1,029.5 ± 101.81b	206
AH	25	5,670.0 ± 271.5a	240	15,260.1 ± 623.5a	127	1,263.2 ± 112.21a	253
AH	50	5,678.0 ± 215.4a	240	14,539.1 ± 618.6a	121	1,021.1 ± 116.07b	205

Data represent the means of measurements using five plants per treatment. Different letters in the same column indicate significant differences between treatments ($p < 0.05$) according to Student-Newman-Keuls test.

the phenol content of peppers, especially at the maturity stage, by acting on PAL enzyme, too. This increase in phenolic compounds was likely responsible for the higher values of FRAP activity measured in peppers, as supported by a number of authors [31,32].

Other additional positive effects of AH and RG in pepper included the increased number of red fruits and stimulation of capsaicinoid production. In the first case, the higher number of fruits indicated an acceleration of the phenological development in plants supplied with the two biostimulants. This kind of 'priming' can be very important in countries with a temperate climate, where the cultivation period of peppers is relatively short and early cold temperatures in autumn can have a negative effect on flower and fruit development, as well as on fruit quality [33]. Priming may also result in economic and environmental benefits, since it can contribute to a reduction in the number of pesticide applications needed for the control of pepper pests and diseases [17].

The induction of capsaicinoid synthesis was particularly remarkable at the maturity stage of plants, probably because of the cumulative effects of the treatments. Capsaicin is one of the most important pepper metabolites as it is responsible for the fruit pungency. The increase of this compound in treated plants is of practical interest and has great relevance in medicine because capsaicin possesses analgesic and anti-inflammatory activities [34].

Although the RG and AH induced similar responses in pepper plants, the degree of each response could vary depending on the type of biostimulant. Differences in effectiveness between RG and AH could be related to the characteristics of the starting matrixes and/or to the industrial processes employed for their production. The cool-extraction method used to produce RG was had the advantage of preserving the organoleptic properties of the raw materials [21], while the hydrolytic process used for AH production could cause chemical changes in thermolabile bioactive compounds [35]. Moreover, the effectiveness of RG and AH could be in part due to their content in IAA and IPA and in part to the presence of compounds with known biological activity, such as phenols.

Conclusions

Several studies investigate the effects of biostimulants originated from different sources on the morphological and physiological parameters of plants. The novelty of this research is the attention paid to the effects of RG and AH on the nutritional features of fruits in consideration of the increased demand for functional foods by consumers. The obtained results demonstrated the effectiveness of RG and

Table 6 Effect of biostimulants on capsaicin and dihydrocapsaicin concentration in pepper plants at flowering and maturity

Treatment	mL L^{-1}	Capsaicin (μg g^{-1} dw) Flowering	(%)	Dihydrocapsaicin (μg g^{-1} dw)	(%)	Capsaicin (μg g^{-1} dw) Maturity	(%)	Dihydrocapsaicin (μg g^{-1} dw)	(%)
Leaf									
C	-	-	-	0.91 ± 0.02b	100	-	-	-	-
RG	50	-	-	1.98 ± 0.10a	217	-	-	-	-
RG	100	-	-	1.80 ± 0.08a	198	-	-	-	-
AH	25	-	-	1.80 ± 0.06a	198	-	-	-	-
AH	50	-	-	0.76 ± 0.021c	84	-	-	-	-
Red peppers									
C	-	155.4 ± 7.2c	100	42.8 ± 1.18b	100	48.3 ± 3.8d	100	17.2 ± 0.6c	100
RG	50	192.0 ± 8.3b	124	34.7 ± 1.53c	80	321.4 ± 12.1a	668	70.2 ± 12.5a	394
RG	100	124.6 ± 5.8d	81	26.3 ± 1.12d	61	261.0 ± 61.6b	544	50.3 ± 11.4b	281
AH	25	201.1 ± 6.2a	130	38.1 ± 1.76b	91	220.1 ± 52.5c	459	46.3 ± 12.0b	262
AH	50	211.2 ± 7.8a	137	62.7 ± 3.83a	146	335.0 ± 72.1a	696	78.2 ± 14.1a	441
Green peppers									
C	-	179.3 ± 18.3a	100	46.35 ± 11.2a	100	89.0 ± 13.2c	100	27.2 ± 1.9c	100
RG	50	183.1 ± 17.5a	102	40.14 ± 11.7b	87	162.1 ± 25.7a	182	28.3 ± 1.2c	106
RG	100	152.2 ± 15.4b	85	31.53 ± 11.8c	68	157.2 ± 25.3a	176	54.5 ± 12.8a	199
AH	25	153.3 ± 15.8b	86	38.74 ± 11.5b	84	98.2 ± 12.8c	111	25.4 ± 13.1c	93
AH	50	160.2 ± 13.2b	89	47.27 ± 12.16a	102	117.1 ± 16.2b	132	43.6 ± 15.5b	161

Data represent the means of measurements using five plants per treatment. Different letters in the same column indicate significant differences between treatments ($p < 0.05$) according to Student-Newman-Keuls test.

AH in improving the nutritional value of peppers by increasing their content in phytochemicals with nutritional value.

Competing interests

The authors declare that they have no competing interests.

Authors' contributions

AE carried out the laboratory studies, drafted the manuscript, and made the corrections. CN and SS carried out the laboratory studies. PS participated in the drafted the manuscript. MS helped to draft the manuscript. SN conceived of the study, participated in its design and coordination, and helped to draft the manuscript. All authors read and approved the final manuscript.

Acknowledgement

This research was funded by a grant from the ILSA company Arzignano, Vicenza, to Padua University.

References

1. Giardini L (2004) Productivity and sustainability of different cropping systems. Patron (Ed.) BO, pp. 1–359
2. Taylor AW, Coveney J, Ward PR, Dal Grande E, Mamerow L, Henderson J, Meyer SB (2012) The Australian food and trust survey: demographic indicators associated with food safety and quality concerns. Food Control 25:476–483
3. Menichini F, Tundis R, Bonesi M, Loizzo MR, Conforti F, Statti G, De Cindio B, Houghton PJ (2009) The influence of fruit ripening on the phytochemical content and biological activity of Capsicum chinensis Jaqc. cv habanero. Food Chem 114:553–560
4. Mueller M, Hobiger S, Jungbauer A (2010) Anti-inflammatory activity of extracts from fruits, herbs and spices. Food Chem 122:987–996
5. Ahmad P, Ashraf M, Younis M, Hu X, Kumar A, Akram NA, Al-Qurainy F (2012) Role of transgenic plants in agriculture and biopharming. Biotech Adv 30:524–540
6. Sherlock R, Morrey JD (2002) Ethical issues in biotechnology. Rowman and Littlefield. Publishers, Inc. Lanham, MD, Lanham MD
7. Schiavon M, Ertani A, Nardi S (2008) Effects of an alfalfa protein hydrolysate on the gene expression and activity of enzymes of TCA cycle and N metabolism in Zea mays L. J Agric Food Chem 56:11800–11808
8. Muscolo A, Sidari M, Nardi S (2013) Humic substance: relationship between structure and activity. Deeper information suggests univocal findings. J Geochem Explor 129:57–63
9. Nardi S, Carletti P, Pizzeghello D, Muscolo A (2009) Biological activities of humic substances. In: Senesi N, Xing B, Huang PM (ed) Biophysico-chemical processes involving natural nonliving organic matter in environmental systems. PART I. Fundamentals and impact of mineral-organic-biota interactions on the formation, transformation, turnover, and storage of natural nonliving organic matter (NOM). John Wiley and Sons, Hoboken, NJ, pp 301–335
10. Ertani A, Nardi S, Altissimo A (2013) Review: long-term research activity on the biostimulant properties of natural origin compounds. Acta Hort 1009:181–188
11. Ertani A, Cavani L, Pizzeghello D, Brandellero E, Altissimo A, Ciavatta C, Nardi S (2009) Biostimulant activity of two protein hydrolysates on the growth and nitrogen metabolism in maize seedlings. J Plant Nutr Soil Sci 172:237–244
12. Vaccaro S, Muscolo A, Pizzeghello D, Spaccini R, Piccolo A, Nardi S (2009) Effect of a compost and its water-soluble fractions on key enzymes of nitrogen metabolism in maize seedlings. J Agric Food Chem 57:11267–11276
13. Azcona I, Pascual I, Aguirreolea J, Fuentes M, Garcia-Mina JM, Sanchez-Diaz M (2011) Growth and development of pepper are affected by humic substances derived from composted sludge. J Plant Nutr Soil Sci 174:916–924

14. Schiavon M, Pizzeghello D, Muscolo A, Vaccaro S, Francioso O, Nardi S (2010) High molecular size humic substances enhance phenylpropanoid metabolism in maize (Zea mays L.). J Chem Ecol 36:662–669

15. Ertani A, Schiavon M, Muscolo A, Nardi S (2013) Alfalfa plant-derived biostimulant stimulates short-term growth of salt stressed Zea mays L. plants. Plant Soil 364:145–158

16. Ertani A, Schiavon M, Altissimo A, Franceschi C, Nardi S (2011) Phenol-containing organic substances stimulate phenylpropanoid metabolism in Zea mays L. J Plant Nutr Soil Sci 3:496–503

17. Pascual I, Azcona I, Morales F, Aguirreolea J, Sanchez-Diaz M (2009) Growth, yield and physiology of verticillium-inoculated pepper plants treated with ATAD and composted sewage sludge. Plant Soil 319:291–306

18. Pascual I, Azcona I, Aguirreolea J, Morales F, Corpas FJ, Palma JM, Rellán-Álvarez R, Sánchez-Díaz M (2010) Growth, yield and fruit quality of pepper plants amended with two sanitized sewage sludges. J Agric Food Chemi 58:6951–6959

19. Deli J, Matus Z, Tóth G (2000) Carotenoid composition in the fruits of asparagus officinalis. J Agric Food Chem 48:2793–2796

20. Howard LR, Talcott ST, Brenes CH, Villalon B (2000) Changes in phytochemical and antioxidant activity of selected pepper cultivars (Capsicum species) as influenced by maturity. J Agric Food Chem 48:1713–1720

21. Machado S (2007) Allelopathic potential of various plant species on downy brome: implications for weed control in wheat production. Agron J 99:127–132

22. Arnaldos TL, Muñoz R, Ferrer MA, Calderón AA (2001) Changes in phenol content during strawberry (Fragaria ananassa, cv. Chandler) callus culture. Physiol Plant 113:315–322

23. Meenakshi S, Manicka GD, Tamilmozhi S, Arumugam M, Balasubramanian T (2009) Total flavonoid in vitro antioxidant activity of two seaweeds of Rameshwaram. Global J Pharmacol 3:59–62

24. Nicoletto C, Santagata S, Bona S, Sambo P (2013) Influece of cut number on qualitative traits in different cultivars of sweet basil. Ind Crops Prod 44:465–472

25. Benzie IFF, Strain JJ (1996) The ferric reducing ability of plasma (FRAP) as a measure of "Antioxidant Power": The FRAP essay. Anal. Biochem 239:70-76

26. Paradikovic N, Vinkovic T, Vrcek IV, Zuntar I, Bojic M, Medic-Saric M (2011) Effect of natural biostimulants on yield and nutritional quality: an example of sweet yellow pepper (Capsicum annuum L.) plants. J Sci Food Agric 91:2146–2152

27. Ruan YL (2012) Signaling role of sucrose metabolism in development. Mol Plant 5(4):1–3

28. Awad MA, de Jager A (2002) Influences of air and controlled atmosphere storage on the concentration of potentially healthful phenolics in apples and other fruits. Postharv Biol Technol 27:53–58

29. Giovannoni JJ (2004) Genetic regulation of fruit development and ripening. Plant Cell 16:170–S18

30. Zodape ST, Gupta A, Bhandari SC (2011) Foliar application of seaweed as biostimulant for enhancement of yield and quality of tomato (Lycopersicon esculentum Mill.). J Sci Ind Res 70:215–219

31. Oboh G, Ademosun AO (2012) Characterization of the antioxidant properties of phenolic extracts from some citrus peels. J Food Sci Tech 49:729–736

32. Serrano M, Zapata PJ, Castillo S, Guillén F, Martínez-Romero D, Valero D (2010) Antioxidant and nutritive constituents during sweet pepper development and ripening are enhanced by nitrophenolate treatments. Food Chem 118:497–503

33. Polowick PL, Sawhney VD (1985) Temperature effects on male fertility and flower and fruit development in Capsicum annuum L. Scientia Horticulture 25:117–127

34. Govindarajan VS, Sathyanarayana MN (1991) Capsicum production, technology, chemistry and quality. Part V. Impact on physiology, pharmacology, nutrition and metabolism: structure, pungency, pain and desensitization sequences. Crit Rev Food Sci Nutr 29:435–473

35. Buchanan BB, Wilhelm G, Russell LJ (2003) Biochemistry & molecular biology of plants. American Society of Plant Physiologists Publisher, Rockville, Maryland, pp 1044–1100

2

Separation of acid-soluble constituents of soil humic acids by dissolution in alkaline urea solution and precipitation with acid

Masakazu Aoyama

Abstract

Background: Humic substances are considered to be composed of relatively small, heterogeneous molecules bound by weak linkages. The dissociation of acid-soluble constituents from soil humic acids (HAs) during the preparative polyacrylamide gel electrophoresis in the presence of concentrated urea has been previously demonstrated. Moreover, the dissociation of acid-soluble constituents has been attributed to the action of concentrated urea. The aim of this study was to investigate the effects of concentrated urea on the dissociation of acid-soluble constituents of soil HAs.

Results: Three types of soil HAs were solubilized in 0.1 M NaOH containing 7 M urea and precipitated after 16 h by acidifying the samples to pH 1.0. The acid-soluble constituents were separated from the dark-colored precipitates by concentrated urea treatment and accounted for 16–45 % of the total organic carbon in HAs. Approximately half of the acid-soluble constituents was recovered in the DAX-8-adsorbed fraction. The humification degree of the DAX-8-adsorbed fraction was considerably lower than that of the corresponding unfractionated HA. In contrast, the humification degree of the precipitated fraction increased due to the separation of acid-soluble constituents. The molecular sizes of the DAX-8-adsorbed and DAX-8-non-adsorbed fractions, estimated by high-performance size exclusion chromatography, were similar and smaller than the precipitated fraction. Three-dimensional excitation-emission matrix fluorescence spectroscopy revealed that the acid-soluble constituents exhibited fluorescence similar to that of fulvic acid (FA), added to which the DAX-8-non-adsorbed fraction exhibited protein-like fluorescence. Diffuse reflectance infrared Fourier transform spectroscopy showed that the DAX-8-adsorbed fraction contained proteinous moieties and the DAX-8-non-adsorbed fraction was rich in proteinous and polysaccharide moieties.

Conclusions: The present findings suggest that soil HAs are formed by the molecular associations between dark-colored acid-insoluble constituents, FA-like acid-soluble constituents, protein-like constituents, and polysaccharides bound by weak linkages.

Keywords: Acid-soluble constituents; Concentrated urea; Degree of humification; Fluorescence; Humic acid

Background

Soil organic matter is a complex, heterogeneous mixture resulting from the decomposition of plants, animals, and microorganisms in the soil environment. Soil humic acids (HAs) are usually extracted from soil using an alkaline solution and precipitated by acidification; therefore, these are considered to be heterogeneous in composition. Recent research has suggested that humic substances are formed due to the association between heterogeneous, relatively small molecules derived as a result of the degradation and decomposition of biological material. It is considered that the constituent molecules are bound by weak linkages, such as hydrogen-bonding and hydrophobic interactions [1–3].

Our previous study on two-dimensional (2-D) electrophoresis of HAs in the presence of 7 M urea showed that the HAs were separated into their constituents with different charge characteristics and molecular sizes [4]. In that study, the 7 M urea was used to facilitate the dissociation of HA constituents by disrupting the hydrogen-

Correspondence: aoyamam@hirosaki-u.ac.jp
Faculty of Agriculture and Life Science, Hirosaki University, Hirosaki 036-8561, Japan

bonding and hydrophobic interactions [1, 5, 6]. However, we observed a low recovery of organic carbon in the precipitates after acidification to pH 1.0, indicating the dissolution of a significant quantity of HAs in the acid solution upon being subjected to electrophoresis [4].

We further subjected the HAs to polyacrylamide gel electrophoresis (PAGE) in the presence of 7 M urea, using a preparative electrophoresis system [7]. Acidification of the electrophoretic fractions resulted in the separation of acid-soluble constituents from the dark-colored precipitates. A part of the acid-soluble constituents could be recovered by adsorption onto DAX-8 resin. The degree of humification was higher in the precipitates and lower in the acid-soluble DAX-8-adsorbed constituents, when compared to the corresponding whole HA. Thus, our previous studies have demonstrated that acid-soluble HA constituents can be dissociated by electrophoresis in the presence of concentrated urea.

Concentrated urea enhances the solvation capacity of alkaline solution, leading to the use of an alkaline concentrated urea solution in the extraction of humin from soils [8]. Concentrated urea could also be used in the fractionation of HAs into constituents displaying different chemical properties, using size exclusion chromatography [1, 9, 10]. Furthermore, soil HA constituents with different fluorescent properties were separated by ultrafiltration in the presence of concentrated urea [11, 12]. The acid-soluble constituents can be dissociated by acidification after dissolution in alkaline medium [13]. However, the abovementioned findings suggested that the dissociation of acid-soluble constituents from soil HAs was attributable to the action of concentrated urea.

The purpose of this study was to investigate the effect of concentrated urea on the dissociation of acid-soluble constituents from soil HAs. Three soil HA samples were solubilized in 0.1 M NaOH containing 7 M urea and precipitated by acidification to pH 1.0. The supernatant solutions were further separated into the DAX-8-adsorbed and DAX-8-non-adsorbed fractions. The fractionated constituents were characterized by degree of humification, high-performance size exclusion chromatography (HPSEC) with UV and fluorescence detections, three-dimensional excitation-emission matrix (3-D EEM) fluorescence spectroscopy, and diffuse reflectance infrared Fourier transform (DRIFT) spectroscopy.

Materials and methods
Preparation of HAs
The HAs used in this study were prepared from a Fluvisol (Fujisaki HA) and an Andosol (Takizawa HA) and purchased from the Japanese Humic Substances Society (Dando HA). The Fujisaki and Takizawa HAs were extracted twice from the soil samples containing 500 mg of organic carbon with 150 mL of 0.1 M NaOH under N_2 by intermittent shaking

for 24 h and were separated by acidification to pH 1.0 with HCl. The HA precipitates were obtained by centrifugation at 10,000×g for 30 min and dissolved in 150 mL of 0.1 M NaOH containing 0.3 M KCl, then precipitated again by acidification to pH 1.0 with HCl. The dissolution-precipitation cycle was repeated twice. The resultant precipitates were suspended in a 0.1 M HCl/0.3 M HF solution and shaken overnight. This step was repeated three times to minimize the ash content. The suspension was dialyzed against ultrapure water and freeze-dried. The Dando HA is the standard HA provided by the Japanese Humic Substances Society, and the details were reported elsewhere [14]. The elemental compositions of the HA samples used are presented in Table 1.

Fractionation with an alkaline concentrated urea solution
Forty milligrams of the HA samples was dissolved in 40 mL of 0.1 M NaOH containing 7 M urea and left standing for 16 h at 25 °C under N_2, then acidified with HCl to reach a pH of 1.0. The dark-colored precipitates were obtained by centrifugation at 10,000×g for 10 min and washed four times with 0.01 M HCl and once with ultrapure water. The resultant precipitates were freeze-dried and designated as the precipitated fraction. The supernatant and the washings were combined and passed through a column containing DAX-8 resin (Supelco, Bellefonte, PA, USA). The column was washed with 0.1 M HCl followed by ultrapure water. The pass-through solution of the DAX-8 column and the washings were combined and dialyzed in a Spectra/Por 7 membrane tubing (nominal molecular weight cutoff of 1000 Da; Spectra/Por, Rancho Dominguez, CA, USA) against ultrapure water, and the dialysate was designated as the DAX-8-non-adsorbed fraction (>1 kDa). The fraction was concentrated to 10 mL using a rotary evaporator at 40 °C. The concentrated solution was divided into five 2-mL portions and freeze-dried. The constituents adsorbed onto the DAX-8 resin were eluted with 0.1 M NaOH and passed through a cation exchange resin (H^+ form) column. This was designated as the DAX-8-adsorbed fraction. The fraction was concentrated to 10 mL using a rotary evaporator at 40 °C, divided into five 2-mL portions, and freeze-dried. As a control treatment, 40 mg of the HA samples was dissolved in 40 mL of 0.1 M NaOH (without urea) and treated in the same manner, with the exception that the DAX-8-

Table 1 Elemental composition of the humic acid samples used

Humic acid	Ash[a] (g kg^{-1})	Elemental composition[b] (g kg^{-1})				
		C	H	N	S	O
Fujisaki	9.6	530	51.7	56.2	7.4	355
Dando	6.7	530	52.5	44.9	2.9	369
Takizawa	6.7	568	41.0	39.5	2.5	349

[a]On a moisture-free basis
[b]On a moisture and ash-free basis

non-adsorbed portion was used only for the analysis of organic carbon concentration without dialyzing and freeze-drying.

Determination of organic carbon

For the precipitated fraction, 1 mg of the freeze-dried sample was dissolved in 800 µL of 0.1 M NaOH and a part of the solution was mixed with four times the volume of 0.067 M potassium dihydrogen phosphate [15], vortexed, and then allowed to stand at room temperature overnight to remove inorganic carbon. The total organic carbon concentration of the mixture was determined using a total organic carbon analyzer (TOC-V_E; Shimadzu Co., Ltd., Kyoto, Japan). For the DAX-8-adsorbed and DAX-8-non-adsorbed fractions, a divided portion of the fractions was dissolved in 800 µL of 0.1 M NaOH and determined the concentration of total organic carbon in the same manner as described above.

Evaluation of humification degree

The absorbances of the precipitated and DAX-8-adsorbed fractions used for the analysis of total organic carbon content at 400 and 600 nm were determined using a V-630 spectrophotometer (JASCO Corp., Tokyo, Japan). The degree of humification of the precipitated and DAX-8-adsorbed fractions was evaluated based on the A_{600}/C and $\log(A_{400}/A_{600})$ values, where A_{400}, A_{600}, and C denoted the absorbance at 400 and 600 nm and the carbon concentration (mg mL^{-1}), respectively [15].

HPSEC

The molecular size distribution was estimated by HPSEC as described in our previous study [4]. For HPSEC, the freeze-dried samples were dissolved in 1 mL of 0.1 M NaOH and then neutralized by passing through a cation-exchange cartridge (H$^+$ form; Dionex OnGuard II H, Thermo Fisher Scientific K.K., Yokohama, Japan). The solution was adjusted to the same composition as the mobile phase and then filtered through a 0.45-µm membrane filter. A 100-µL portion of the sample solution was injected. The chromatograms were monitored by UV absorption using a photodiode array detector (MD-2018, JASCO Corp., Tokyo, Japan) and by fluorescence at excitation and emission wavelengths of 460 and 520 nm, respectively, using a fluorescence detector (FP-920, JASCO Corp., Tokyo, Japan). To estimate the molecular weight (MW), the column was calibrated using polyethylene glycols as the MW standards [7].

3-D EEM fluorescence spectroscopy

The freeze-dried sample was dissolved in 1 mL of 0.1 M NaOH and neutralized by passing through a cation-exchange cartridge (H$^+$ form; Dionex OnGuard II H, Thermo Fisher Scientific K.K., Yokohama, Japan). The

solution was diluted with ultrapure water and added with 0.05 M phosphate buffer (pH 8.0) to a final concentration of 5 mg organic carbon L^{-1} in 0.01 M phosphate buffer. The 3-D EEM fluorescence spectra were recorded in a 1-cm quartz cell using a FP-8300 scanning spectrofluorometer (JASCO Corp., Tokyo, Japan) equipped with an automatic higher order diffraction cut filter. The spectra were recorded over the excitation and emission wavelength ranges of 200–550 nm and 250–600 nm, respectively, and then corrected for instrumental bias according to the manufacturer's method. Inner-filter effect was corrected using the following equation [16, 17]:

$$\text{Em}_{real} = \text{Em}_{obs} \times 10^{b \times (Aex + Aem)}$$

where Em_{obs} was the observed fluorescence intensity, Em_{real} denoted the fluorescence in the absence of self-absorption, b was 0.5 cm, and the path length to the center of the cell for both excitation and emission, A_{ex} and A_{em}, denoted the absorbance at excitation and emission wavelengths, respectively. The Raman scatter effect was minimized by subtracting EEM spectrum of 0.01 M phosphate buffer (pH 8.0). The relative fluorescence intensity was expressed as the quinine sulfate unit (QSU) using the fluorescence intensity of a quinine sulfate solution (0.01 mg L^{-1} in 0.05 M H$_2$SO$_4$) at the excitation/emission wavelengths ($\lambda_{ex}/\lambda_{em}$) = 350/450 nm.

DRIFT spectroscopy

DRIFT spectra of the fractions were recorded using an FT/IR-4100 spectrometer (JASCO Corp., Tokyo, Japan) equipped with a DR-81 diffuse reflectance accessory (JASCO Corp., Tokyo, Japan). The freeze-dried samples were thoroughly mixed with 50–100 times the amount of potassium bromide (FT-IR grade; Wako Pure Chemical Industries, Ltd., Osaka, Japan) in an agate mortar and pestle and placed in an aluminum sample cup. Spectra were collected from 4000 to 800 cm^{-1} and averaged over 100 scans and then transformed into Kubelka-Munk units [18]. The resolution was set at 4 cm^{-1}. To identify the principal bands that contribute to the more complex band resulting from overlapping features, Fourier self-deconvolution (FSD) was performed for the wavenumber region between 1800 and 800 cm^{-1}, using the JASCO FT-IR software provided with the spectrometer.

Results and discussion
Distribution of HA constituents among the fractions

The dissolution of HAs in 0.1 M NaOH alone and subsequent acidification resulted in the dissociation of acid-soluble constituents from the dark-colored precipitates (Fig. 1). The proportion of the organic carbon, recovered as precipitates, was 84 % for the Fujisaki HA sample, 84 % for the Dando HA sample, and 92 % for the Takizawa HA

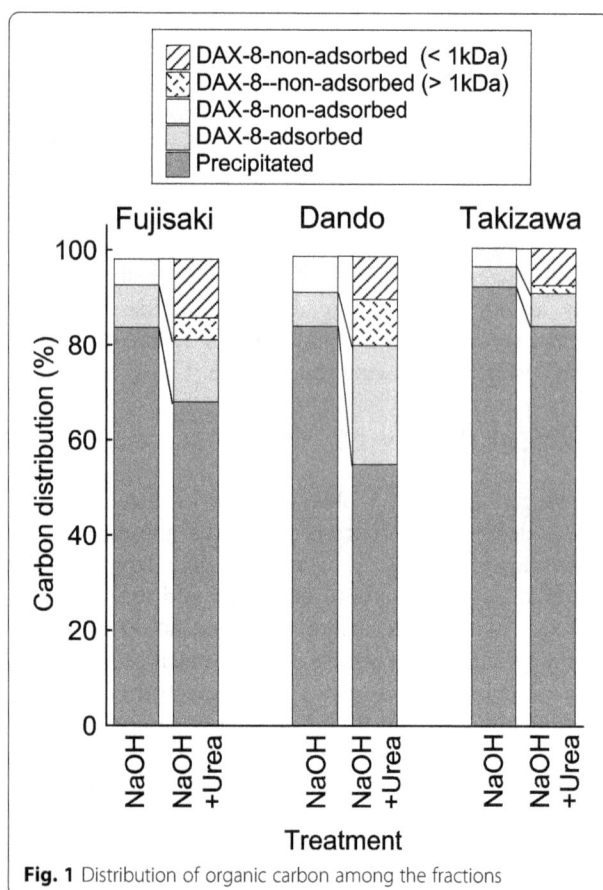

Fig. 1 Distribution of organic carbon among the fractions

sample. Among the acid-soluble constituents, the brown-colored constituents were recovered by adsorption onto the DAX-8 resin (DAX-8-adsorbed fraction). The organic carbon in the DAX-8-adsorbed fraction was approximately half of that in the acid-soluble constituents. Dissolution in alkaline concentrated urea enhanced the dissociation of acid-soluble constituents. The precipitated fraction accounted for 68 % of the total organic carbon in the Fujisaki HA sample, 54 % in the Dando HA sample, and 84 % in the Takizawa HA sample. The proportion of the brown-colored DAX-8-adsorbed fraction was significantly higher than when treated with 0.1 M NaOH. The DAX-8-non-adsorbed fraction was nearly colorless, and the proportion of the total organic carbon in this fraction was also increased by treatment with concentrated urea. However, a substantial part was lost during the dialysis (Fig. 1). The proportion of organic carbon in the unrecovered part was estimated to be 12, 9, and 8 % for the Fujisaki HA, Dando HA, and Takizawa HA samples, respectively.

Our previous study has demonstrated that acid-soluble constituents were separated from HAs during preparative PAGE in the presence of concentrated urea [7]. This study further revealed that the dissociation of acid-soluble constituents from HAs occurred when the HAs were

dissolved in alkaline concentrated urea solution, followed by the acidification of the solution. Moreover, it was confirmed that the brown-colored acid-soluble constituents, as well as the nearly colorless acid-soluble constituents, were dissociated from HAs by concentrated urea treatment. Concentrated urea is considered to disrupt the hydrogen-bonding and hydrophobic interactions [1, 12, 19]. Therefore, the dissociation of HA constituents observed in this study can be attributed to the disruption of the hydrogen-bonding and hydrophobic interactions triggered by concentrated urea.

Humification degree of the fractions

In order to evaluate the humification degree of the precipitated and DAX-8-adsorbed fractions, A_{600}/C and $\log(A_{400}/A_{600})$ values were used (Fig. 2). An increase in the A_{600}/C value and a decrease in the $\log(A_{400}/A_{600})$ value were known to indicate an increase in the degree of humification of HA [15]. The degree of humification used here is synonymous with the degree of darkening [15, 20]. Kumada [20] classified HAs into four types according to their optical properties: type A HAs are the most humified, type B HAs are the intermediates between type A and type Rp, type Rp HAs are the least humified, and type P HAs are moderately humified, as indicated in Fig. 2.

The DAX-8-adsorbed fraction exhibited a lower A_{600}/C value and a higher $\log(A_{400}/A_{600})$ value compared to the corresponding whole HA sample irrespective of treatment with concentrated urea. The reverse was true for the precipitated fraction. These results indicated that the acid-soluble constituents of HA were characterized by a low degree of humification and that their dissociation from HA resulted in an increase in the humification degree of precipitated constituents. These findings were in agreement with our previous study, where preparative PAGE of HAs was carried out in the presence of concentrated urea [7]. The degree of humification of the precipitated fraction was significantly higher when treated with alkaline concentrated urea. This was attributed to the higher dissociation of acid-soluble constituents in the presence of concentrated urea.

Molecular size distribution of the fractions

The molecular size distributions were estimated using HPSEC with UV detection at 280 nm (Fig. 3). The molecular size distributions of the whole samples of Fujisaki and Dando HAs were similar but differed largely from that of Takizawa HA. An intense peak was observed at the void volume (V_0), and a broad peak was eluted in the MW region of 2–20 kDa for the Fujisaki and Dando HA whole samples. In contrast, for the Takizawa HA, a broad peak was eluted at a MW of 2 kDa. Thus, the

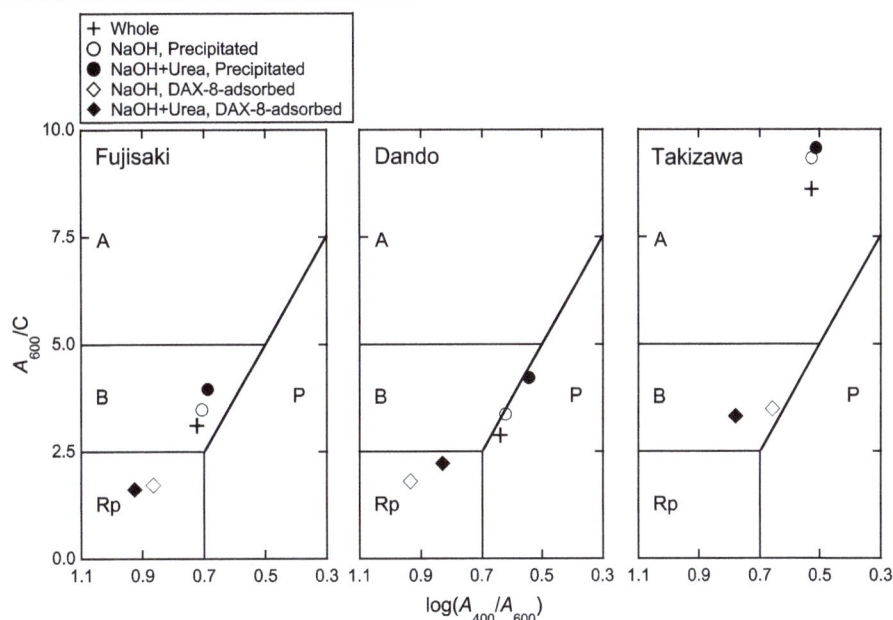

Fig. 2 log(A_{400}/A_{600}) versus A_{600}/C diagram of precipitated and DAX-8-adsorbed fractions. The symbols (A, B, P, and Rp) in the figure indicate the types of HAs in Kumada's classification system [20]. The degree of humification is higher in the order A > B or P > Rp

Takizawa HA consisted mainly of relatively small molecular size constituents compared to the Fujisaki and Dando HAs.

For control treatment, the molecular size distribution of the precipitated fraction was similar to that of the corresponding whole HA, while the peak of DAX-8-adsorbed fraction was observed at a MW of 2 kDa irrespective of the HAs used. The molecular size of the DAX-8-adsorbed fraction was similar to those of fulvic acids (FAs) [21]. Higher molecular size distributions were obtained from the precipitated fractions treated with alkaline urea compared to those having received the treatment with 0.1 NaOH alone. This was attributed

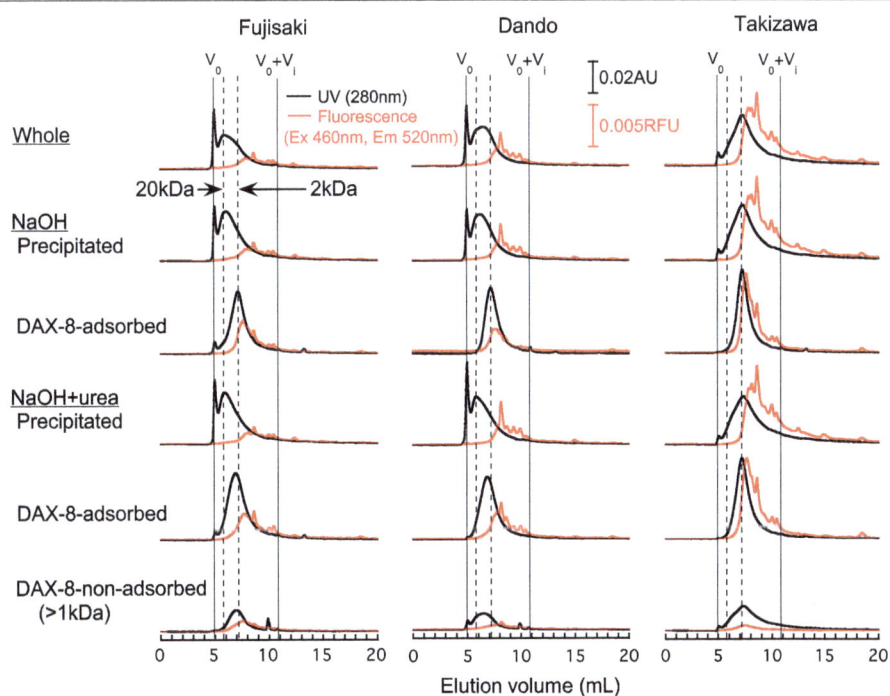

Fig. 3 Size exclusion chromatograms (UV detection at 280 nm; fluorescence detection at excitation 460 nm and emission 520 nm) of whole humic acid samples and their fractions normalized to the concentration of 100 mg carbon L^{-1}. V_0 void volume, $V_0 + V_i$ total effective column volume

to the higher dissociation of DAX-8-adsorbed and DAX-8-non-adsorbed fractions when treated with concentrated urea. The molecular size of the DAX-8-adsorbed fraction was similar to (in the case of Takizawa HA) or larger than (Fujisaki and Dando HAs) those observed for the control treatment. The DAX-8-non-adsorbed fraction (>1 kDa) showed a similar molecular size to that observed for the DAX-8-adsorbed fraction; however, the intensity of the peak was observed to be relatively lower. The molecular sizes of the acid-soluble constituents dissociated by concentrated urea treatment were similar to those dissociated by preparative PAGE in the presence of concentrated urea [7].

When detected by fluorescence at $\lambda_{ex}/\lambda_{em} = 460/520$ nm, the peaks for whole HAs were eluted over a wide range of elution volumes (Fig. 3), as observed in previous studies [21, 22]. However, the intensities of the fluorescence-detected peaks varied with the HA samples (lowest in the Fujisaki HA and highest in the Takizawa HA).

Fluorescence-detected peaks were observed for all the fractions (Fig. 3), with the peak intensity being relatively low in the DAX-8-non-adsorbed fraction. This indicated that the fluorescent substances were mainly partitioned into both the precipitated and DAX-8-adsorbed fractions. The fluorescence-detected peaks of the DAX-8-adsorbed and DAX-8-non-adsorbed fractions were eluted in advance of that of the precipitated fraction. Therefore, the elution profile of the precipitated fraction lacked the largest molecular size components, compared to the corresponding whole HA. This indicates that the molecular sizes of the fluorescent substances in the DAX-8-adsorbed and DAX-8-non-adsorbed fractions were higher than those in the precipitated fractions.

Fluorescent properties of the fractions
3-D EEM fluorescence spectroscopy was utilized to investigate the fluorescent properties of whole HAs and their fractions. The 3-D EEM contour plots are shown in Fig. 4, and the fluorescence maxima and their relative intensities have been summarized in Table 2.

The fluorescence peaks at $\lambda_{ex}/\lambda_{em} = 265-275/505-540$ nm (H1) and 430-460/510-540 nm (H4) were observed for all the whole HA samples, with the relative fluorescence intensities varying with each HA. In addition, the fluorescence peak was observed at $\lambda_{ex}/\lambda_{em} = 360-365/505$ nm (H3) for the Fujisaki and Dando HA samples and at $\lambda_{ex}/\lambda_{em} = 310/510$ nm (H2) for the Dando HA sample. The relative fluorescence intensities (QSU) were significantly high in the Takizawa HA compared to the other HAs. This is in agreement with the results of HPSEC with fluorescence detection. Our previous studies [4, 7, 21–23] showed that the fluorescent substances were considerably more in Andosol HAs than in HAs prepared from the other types of soils.

For the control treatment, the fluorescence maxima and relative intensities of the precipitated fractions were nearly identical to those observed for the whole HA samples. In contrast, the positions of fluorescence peaks of the DAX-8-adsorbed fraction (F1–3) did not coincide with those of the precipitated fraction and the whole HA, with the excitation and emission wavelengths being shorter for the former compared to the latter. The positions of the fluorescence maxima resembled those of FAs [24]. However, the fluorescence maxima and relative fluorescence intensities of the DAX-8-adsorbed fraction were similar between the different HA samples.

For the treatment with concentrated urea, the fluorescence maxima and relative intensities of the precipitated fraction were observed to be identical to those of the whole HA and the precipitated fraction of the control treatment. The fluorescence maxima and relative fluorescence intensities of the DAX-8-adsorbed fraction were similar to those observed in the DAX-8-adsorbed fraction from the control treatment. However, a fluorescence peak at $\lambda_{ex}/\lambda_{em} = 275/315$ nm (P) was observed for the DAX-8-adsorbed fraction of the Dando HA sample, which was attributed to the fluorescence of protein-like substances [25, 26].

The position of the major fluorescence maximum of DAX-8-non-adsorbed fraction varied for each HA sample: it was observed at $\lambda_{ex}/\lambda_{em} = 270/465$ nm (F2) for the Fujisaki HA sample, at 215/430 nm (F1) for the Dando HA sample, and at 225/440 nm (F1) for the Takizawa HA sample. All spectra were associated with a secondary maximum (P) at $\lambda_{ex}/\lambda_{em} = 280/330$ nm (Fujisaki HA) or 270/330 nm (Dando and Takizawa HAs). These secondary maxima were attributed to the fluorescence of protein-like substances [25, 26].

Richard et al. [12] fractionated a soil HA sample by ultrafiltration, in the presence of 7 M urea. They reported that the fraction with molecular size 0.5–1 kDa exhibited an emission maxima at a shorter wavelength compared to the fractions with molecular size >1 kDa. This indicated that the relatively smaller HA constituents with an emission maximum at a shorter wavelength were dissociated in the presence of concentrated urea, an observation that was confirmed by the present results.

Infrared spectroscopy of the fractions
Figure 5 shows the DRIFT spectra of the whole HAs and their fractions. The spectrum of the precipitated fraction was similar to that of the corresponding whole HA, irrespective of treatment with concentrated urea. In contrast, the spectra of the DAX-8-adsorbed and DAX-8-non-adsorbed fractions were different from their whole samples and precipitated fractions. Absorption peaks at 2850 and 2920 cm^{-1}, assigned to the symmetric and asymmetric methylene stretching bands in aliphatic chains

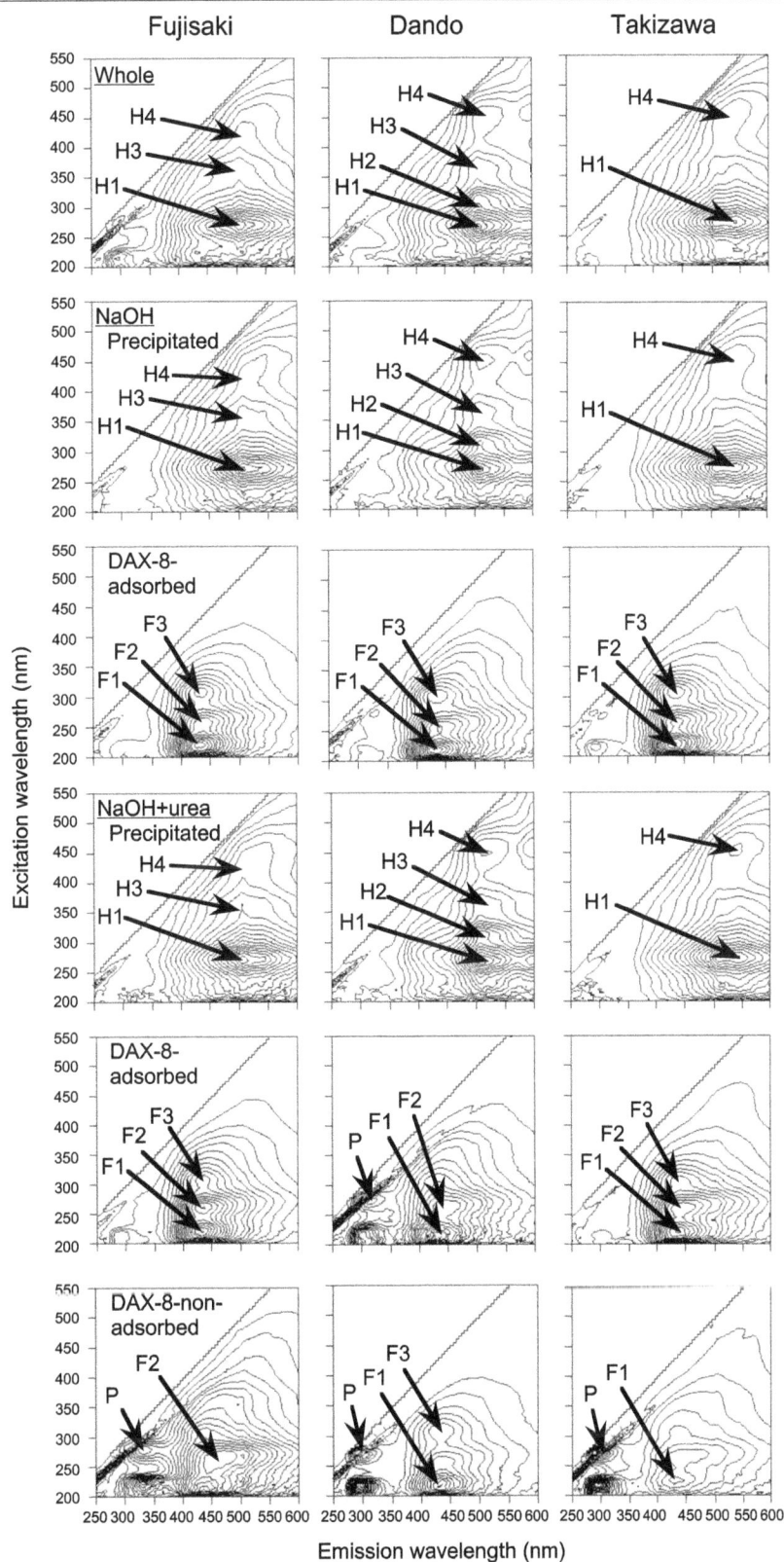

Fig. 4 Three-dimensional excitation-emission matrix (3-D EEM) fluorescence spectra of whole humic acid samples and their fractions. Emission intensities were normalized on the maximum. *Arrows* indicate the positions of fluorescence peaks. *H1–4* humic acid-specific fluorescence peaks, *F1–3* fulvic acid-like fluorescence peaks [24], *P* protein-like fluorescence peaks [25, 26]

Table 2 Positions and relative intensities of the fluorescence peaks

Fraction	Peak[a]	Fujisaki		Dando		Takizawa	
		$\lambda_{ex}/\lambda_{em}$[b]	QSU	$\lambda_{ex}/\lambda_{em}$[b]	QSU	$\lambda_{ex}/\lambda_{em}$[b]	QSU
Whole	H1	270/505	65	265/505	54	275/540	134
	H2	–	–	310/510	36	–	–
	H3	365/505	27	360/505	23	–	–
	H4	430/510	18	460/515	16	450/540	48
NaOH							
Precipitated	H1	270/505	54	270/510	56	275/545	180
	H2	–	–	310/510	38	–	–
	H3	365/505	22	365/505	23	–	–
	H4	430/505	16	450/510	18	450/535	57
DAX-8-adsorbed	F1	220/430	107	220/430	100	220/430	104
	F2	255/435	77	255/440	71	255/435	80
	F3	310/435	57	310/435	55	310/435	59
NaOH + urea							
Precipitated	H1	270/505	48	270/510	69	275/545	181
	H2	–	–	310/510	48	–	–
	H3	360/505	21	365/505	27	–	–
	H4	430/505	16	450/510	21	450/540	59
DAX-8-adsorbed	F1	220/435	105	210/435	96	220/435	154
	F2	260/435	89	260/440	72	260/440	127
	F3	305/435	58	–	–	305/440	81
	P	–	–	275/315	32	–	–
DAX-8-non-adsorbed (>1 kDa)	F1	–	–	215/430	30	225/440	29
	F2	270/465	31	–	–	–	–
	F3	–	–	310/435	15	–	–
	P	280/330	14	270/310	12	270/310	26

QSU quinine sulfate unit
[a]Indicated in Fig. 4
[b]Excitation/emission wavelengths (nm)

[27], respectively, were observed in most of the spectra. The peaks were intense in the spectra obtained for the Fujisaki and Dando HAs, especially in the precipitated and DAX-8-non-adsorbed fractions. Absorption bands in the wavenumber region between 800 and 1800 cm^{-1} were observed to be overlapping. Therefore, the FSD was applied in order to enhance the resolution of the absorption bands (Fig. 6).

For the whole HAs, the peak observed at 1720 cm^{-1}, assigned to the C=O stretching of carboxyl group, was more intense in the Takizawa HA. On the other hand, the peaks at around 1670 and 1540 cm^{-1}, attributed to the amide I and amide II bands of proteinous moieties [28], and at 1510 cm^{-1} (vibrations of aromatic moieties in lignin) [29–31], 1410 cm^{-1} (the symmetric carboxylate stretching) [28], and 1080 and 1030 cm^{-1}, attributed to C–H stretching of polysaccharide moieties [28], were observed to be more intense in the Fujisaki and Dando HA samples.

The control and concentrated urea-treated precipitated fractions exhibited similar spectra to those displayed by the corresponding whole HA. In contrast, the control-treated DAX-8-adsorbed fraction showed a similar spectrum, irrespective of the type of HA. The spectrum was characterized by a more intense peak representing the carboxyl group at 1720 cm^{-1} and a less intense peak at 1600 cm^{-1}, attributed to the C=C stretching in aromatic rings [28] compared to the whole HAs. These spectral characteristics were similar to that observed for FA [27, 32]. The DAX-8-adsorbed fraction from the concentrated urea treatment displayed spectra with more intense peaks at 1670 and 1540 cm^{-1} due to the amide I and amide II bands, compared to those shown by the DAX-8-adsorbed fractions from the control treatment. The spectra of DAX-8-non-adsorbed fraction were characterized by prominent amide peaks at 1670 and 1540 cm^{-1} and by intense polysaccharide peaks at 1080 and 1030 cm^{-1}. The presence of proteinous

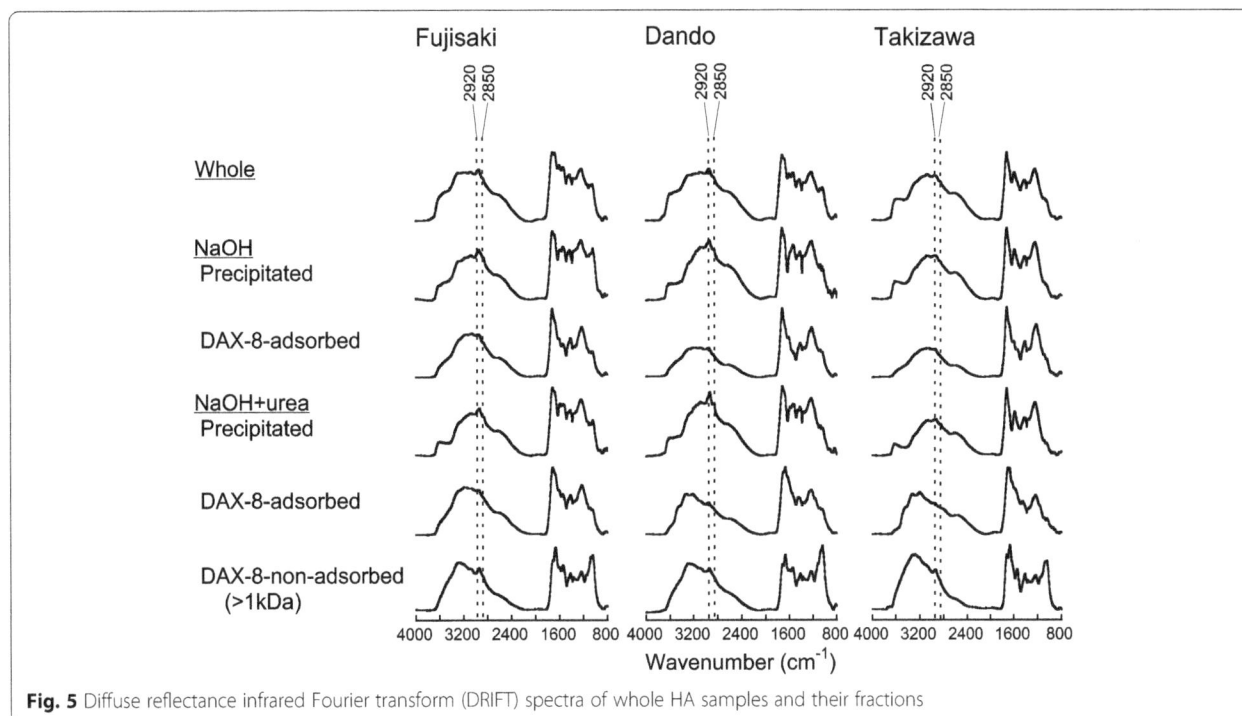

Fig. 5 Diffuse reflectance infrared Fourier transform (DRIFT) spectra of whole HA samples and their fractions

moieties in the acid-soluble constituents is in agreement with the results of 3-D EEM fluorescence spectroscopy.

Conclusions

The acid-soluble constituents of soil HAs obtained by treatment with concentrated urea were characterized by

lower degrees of humification and smaller molecular sizes and displayed FA-like fluorescence. These features indicate that the acid-soluble constituents of soil HAs dissociated by concentrated urea treatment expressed similar properties to FAs. This was supported by the results of infrared spectroscopy. Infrared and fluorescence spectroscopies

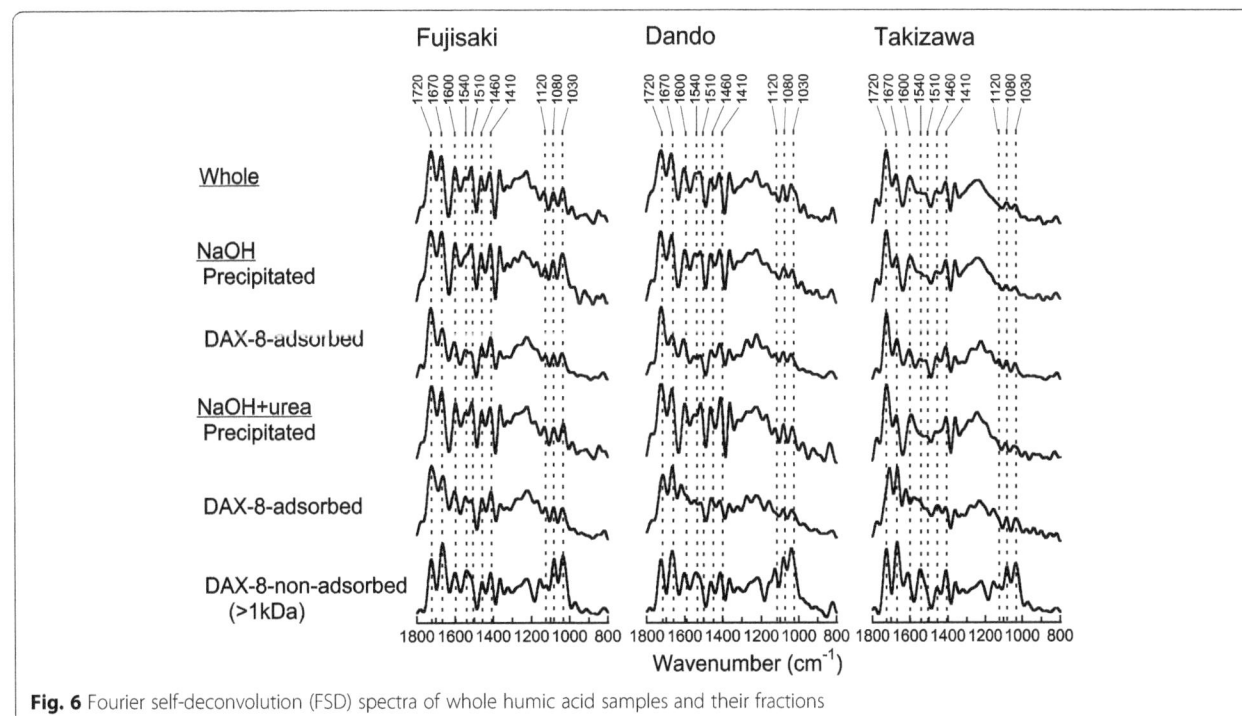

Fig. 6 Fourier self-deconvolution (FSD) spectra of whole humic acid samples and their fractions

revealed that the acid-soluble constituents contained proteinous moieties and the DAX-8-non-adsorbed fraction was rich in polysaccharide moieties. In contrast, the humification degree of the precipitated fraction increased due to the separation of acid-soluble constituents. The precipitated fraction still contained smaller molecular size fluorescent substances. The present findings suggest that soil HAs are composed of dark-colored acid-insoluble constituents, FA-like acid-soluble constituents, protein-like constituents, and polysaccharides bound by weak linkages.

Abbreviations

3-D EEM: three-dimensional excitation-emission matrix; DRIFT: diffuse reflectance infrared Fourier transform; FA: fulvic acid; FSD: Fourier self-deconvolution; HA: humic acid; HPSEC: high-performance size exclusion chromatography; MW: molecular weight; PAGE: polyacrylamide gel electrophoresis.

Competing interests

The author declares that he has no competing interests.

Acknowledgements

This study was supported by Grant-in-Aid for Scientific Research (No. 26450072) from the Japan Society for the Promotion of Science (JSPS).

References

1. Piccolo A (2001) The supramolecular structure of humic substances. Soil Sci 166:810–832
2. Simpson AJ, Kingery WL, Hayes MHB, Spraul M, Humpfer E, Dvortsak P, Kerssebaum R, Godejohann M, Hofmann M (2002) Molecular structures and associations of humic substances in the terrestrial environment. Naturwissenschaften 89:84–88
3. Sutton R, Sposit G (2005) Molecular structure in soil humic substances: the new view. Environ Sci Technol 39:9009–9015
4. Karim S, Okuyama Y, Aoyama M (2013) Separation and characterization of the constituents of compost and soil humic acids by two-dimensional electrophoresis. Soil Sci Plant Nutr 57:130–141
5. Trubetskoj OA, Kudryavceva LY, Shirshova LT (1991) Characterization of soil humic matter by polyacrylamide gel electrophoresis in the presence of denaturating agents. Soil Biol Biochem 23:1179–1181
6. Trubetskoj OA, Trubetskaya OE, Khomutova TE (1992) Isolation, purification and some physico-chemical properties of soil humic substances fractions obtained by polyacrylamide gel electrophoresis. Soil Biol Biochem 24:893–896
7. Karim S, Aoyama M (2013) Fractionation of the constituents of soil humic acids by preparative polyacrylamide gel electrophoresis in the presence of concentrated urea. Soil Sci Plant Nutr 59:827–839
8. Song G, Hayes MHB, Novotny EH, Simpson AJ (2011) Isolation and fractionation of soil humin materials using alkaline urea and dimethylsulphoxide plus sulphuric acid. Naturwissenschaften 98:7–13
9. Francioso O, Montecchio D, Gioacchini P, Cavani L, Ciavatta C, Trubetskoj O, Trubetskaya O (2009) Structural differences of Chernozem soil humic acids SEC–PAGE fractions revealed by thermal (TG–DTA) and spectroscopic (DRIFT) analyses. Geoderma 152:264–268
10. Trubetskoj OA, Hatcher PG, Trubetskaya OE (2010) ¹H-NMR and ¹³C-NMR spectroscopy of chernozem soil humic acid fractionated by combined size-exclusion chromatography and electrophoresis. Chem Ecol 26:315–325
11. Trubetskaya OE, Shaloiko LA, Demin DV, Marchenkov VV, Proskuryakov II, Coelho C, Trubetskoj OA (2011) Combining electrophoresis with detection under ultraviolet light and multiple ultrafiltration for isolation of humic fluorescence fractions. Anal Chim Acta 690:263–268
12. Richard C, Coelho C, Guyot G, Shaloiko L, Trubetskoj O, Trubetskaya O (2011) Fluorescence properties of the <5 kDa molecular size fractions of a soil humic acid. Geoderma 163:24–29
13. Baglieri A, Vindrola D, Gennari M, Negre M (2014) Chemical and spectroscopic characterization of insoluble and soluble humic acid fractions at different pH values. Chem Biol Technol Agric 1:1–11
14. Watanabe A, Maie N, Hepburn A, McPhail DB, Abe T, Ikeya K, Ishida Y, Ohtani H (2004) Chemical characterization of Japanese Humic Substances Society standard soil humic and fulvic acids by spectroscopic and degradative analyses. Humic Subs Res 1:18–28
15. Ikeya K, Watanabe A (2003) Direct expression of an index for the degree of humification of humic acids using organic carbon concentration. Soil Sci Plant Nutr 49:47–53
16. Ohno T (2002) Fluorescence inner-filtering correction for determining the humification index of dissolved organic matter. Environ Sci Technol 36:742–746
17. Childers JW, Palmer RA (1986) A comparison of photoacoustic and diffuse reflectance detection in FTIR spectrometry. Am Lab 18:22–38
18. Fery-Forgues S, Lavabre D (1999) Are fluorescence quantum yields so tricky to measure? A demonstration using familiar stationery products. J Chem Educ 76:1260
19. Trubetskoj OA, Trubetskaya OE, Afanas'eva GV, Reznikova OI, Saiz-Jimenez C (1997) Polyacrylamide gel electrophoresis of soil humic acid fractionated by size-exclusion chromatography and ultrafiltration. J Chromatogr A 767:285–292
20. Kumada K (1987) Chemistry of soil organic matter. Japan Scientific Societies Press-Elsevier, Tokyo
21. Aoyama M (2001) Do humic substances exhibit fluorescence? In: Swift RS, Spark KM (ed) Understanding and managing organic matter in soils, sediments, and waters. International Humic Substances Society, Inc, St Paul, pp 125–131
22. Aoyama M (1999) Chromatographic separation of fluorescent substances from humic acids. In: Davies G, Ghabbour EA (ed) Understanding humic substances: advanced methods. Properties and applications. The Royal Society of Chemistry, Cambridge, pp 179–189
23. Aoyama M, Watanabe A, Nagao S (2000) Characterization of the 'fluorescent fraction' of soil humic acids. In: Ghabbour EA, Davies G (ed) Humic substances: versatile components of plant. Soil and Water. The Royal Society of Chemistry, Cambridge, pp 125–133
24. Alberts JJ, Takács M (2004) Total luminescence spectra of IHSS standard and reference fulvic acids, humic acids and natural organic matter: comparison of aquatic and terrestrial source terms. Org Geochem 35:243–256
25. Coble PG (1996) Characterization of marine and terrestrial DOM in seawater using excitation-emission matrix spectroscopy. Mar Chem 51:325–346
26. Yamashita Y, Tanoue E (2003) Chemical characterization of protein-like fluorophores in DOM in relation to aromatic amino acids. Mar Chem 82:255–271
27. Ding G, Amarasiriwardena D, Herbert S, Novak J, Xing B (2000) Effect of cover crop systems on the characteristics of soil humic substances. In: Ghabbour EA, Davies G (ed) Humic substances: versatile components of plant, soil and water. The Royal Society of Chemistry, Cambridge, pp 53–61
28. D'Orazio V, Senesi N (2009) Spectroscopic properties of humic acids isolated from the rhizosphere and bulk soil compartments and fractionated by size-exclusion chromatography. Soil Biol Biochem 41:1775–1781
29. Zaccheo P, Cabassi G, Ricca G, Crippa L (2002) Decomposition of organic residues in soil: experimental technique and spectroscopic approach. Org Geochem 33:327–345
30. Boeriu CG, Bravo D, Gosselink RJA, van Dam JEG (2004) Characterisation of structure-dependent functional properties of lignin with infrared spectroscopy. Ind Crops Prod 20:205–218
31. Ferrari E, Francioso O, Nardi S, Saladini M, Ferro ND (2011) DRIFT and HR MAS NMR characterization of humic substances from a soil treated with different organic and mineral fertilizers. J Mol Struc 998:216–224
32. Baes AU, Bloom PR (1989) Diffuse reflectance and transmission Fourier transform infrared (DRIFT) spectroscopy of humic and fulvic acids. Soil Sci Soc Am J 53:695–700

3

Indigenous rhizobia population influences the effectiveness of *Rhizobium* inoculation and need of inorganic N for common bean (*Phaseolus vulgaris* L.) production in eastern Ethiopia

Anteneh Argaw[1*] and Angaw Tsigie[2]

Abstract

Background: Supplement with inorganic N application is essential to improve the common bean production in sub-Saharan Africa. However, the influence of indigenous rhizobial population on the inorganic N requirement with *Rhizobium* inoculation to secure sustainable way of common bean production system is not well known. The effect of different rates of N application either alone or in combination with *Rhizobium* inoculation on the nodulation, yield and yield traits of common bean cultivated in soils with different rhizobial population were conducted.

Methods: Twelve treatments were produced by factorially combined six levels of N fertilizer (0, 20, 40, 60, 80 and 100 kg N ha^{-1}) and two *Rhizobium* inoculation treatments (inoculated and uninoculated). The treatments were laid out in randomized completely block design and all treatments were replicated three times.

Result: Regardless of soil types, nodule number and nodule dry weight decreased with increasing rates of N application. 20 kg N ha^{-1} both alone and in combination with *Rhizobium* inoculation resulted in the largest nodulation in all soil types. The largest nodulation were induced in soil with large rhizobial population. *Rhizobium* inoculation significantly ($P < 0.05$) improved yield and yield traits of common bean. Moreover, our result revealed that the largest values of investigated traits were observed in inoculated treatment, as compared to the corresponding N rates of uninoculated treatments. The 20, 100 and 40 kg N ha^{-1} treatments resulted in significantly greater plant total tissue N at soil types with small, medium and large rhizobial population, respectively, as compared to unfertilized control. The highest total biomass yield (TBY) and grain yield (GY) at soil types with small and medium rhizobial population were obtained by the 100 kg N ha^{-1} treatment in combination with *Rhizobium* inoculation, while 20 and 40 kg N ha^{-1} applications produced the greatest TBY and GY, respectively, in soil with large rhizobial population.

Conclusion: These results indicate that N requirement is varied based on rhizobial population and effectiveness of native rhizobia in N$_2$ fixation.

Keywords: Common bean (*Phaseolus vulgaris* L.), Ethiopia, Indigenous rhizobia, *Rhizobium leguminosarum* bv. Phaseoli

*Correspondence: antenehargaw@gmail.com
[1] College of Agriculture and Environmental Sciences, School of Natural
Resources Management and Environmental Sciences, Haramaya
University, Dire Dawa, Ethiopia
Full list of author information is available at the end of the article

Background

Common bean (*Phaseolus vulgaris* L.) is one of most important food legume cultivated on greater than 4 million ha, providing the main source of dietary protein and carbohydrate for eastern and central African peoples, and containing about 20–25 % of the protein [1]. In Ethiopia, common bean is one of the major grain legumes, with its production centered in small farmers' fields where the soil fertility is depleted. Furthermore, the use of nitrogen (N) fertilizer is limited and average yields are low, usually less than 1 ton ha^{-1} [2]. In contrast, some studies indicated that up to 4600 kg ha^{-1} seed yield of common bean were obtained from research managed experimental plots in Ethiopia [3–5]. Low soil fertility, especially low N content, is the most important limiting nutrient status for common bean production in the tropics, including Ethiopia [6, 7].

Due to high price of N mineral fertilizers, their use by subsistent farmers in sub-Saharan Africa (SSA) to increase crop production has been limited. This condition has therefore necessitated an approach to crop production that emphasizes biological N$_2$ fixation (BNF). However, common bean is considered as a poor nitrogen fixing plant in comparison to other grain legumes due to its promiscuous nature, i.e., it forms symbiosis with many rhizobia species [8]. On top of this, several findings revealed that common bean generally responds poorly to *Rhizobium* inoculation under field conditions [9–11]. The reason for the failure to respond inoculation is believed to be due to a high and inefficient population of native common bean nodulating rhizobia in soil [9, 12, 13].

It is indicated that N$_2$ fixation in common bean can be increased through highly efficient *Rhizobium* inoculation [14]. Moreover, elite *Rhizobium* isolates improved the productivity of common bean, although in soil with high rhizobial nodulating population [15–17]. Asad et al. [15] found that the efficient *Rhizobium* inoculation did not fulfill the N needs of common bean. In most cases, common bean is able to fix up to 50 kg N ha^{-1} [18], which is less than 50 % of the plant N requirement [19]. Therefore, common bean requires mineral N application to achieve substantial yields under the current cropping system in SSA. However, N mineral recommended rates >40 kg N ha^{-1} suppress nodulation and N$_2$ fixation [10, 20, 21]. On the other hand, the application of a small amount of fertilizer N (<30 kg N ha^{-1}) enhanced nodulation, but grain yield improvement was not satisfactory [10, 22]. Higher stimulation of plant growth, N$_2$ fixation and grain yield of common bean have been recorded at low levels of N fertilizer applied with *Rhizobium* inoculation [17, 23–25]. Therefore, inoculation trials must emphasize not only the benefits of common bean inoculation, but also the combination of that practice with N

fertilization, in order to achieve a decrease in mineral N input, whilst still obtaining maximum yields. However, the amount of N required in conjunction with elite *Rhizobium* inoculation to get maximum yield of common bean in soil with different rhizobial population is unknown. Hence, the objective of this study was to investigate the effect of naturalized common bean rhizobia population on the N rates of applied when alone or in combination with *Rhizobium* inoculation on nodulation, yield and yield traits of common bean in major growing areas of eastern Ethiopia.

Methods

Study areas

Field experiments were conducted on four locations of Eastern Ethiopia having different indigenous rhizobia nodulating common bean in 2012 cropping season. The experimental sites were located in the Hirna [N09°13.157″ and E041°06.488″ at an altitude of 1808 m above sea level (m.a.s.l.)], Fedis (N09°06.941″ and E042°04.835″ at an altitude of 1669 m.a.s.l.) Babillae (N09°13.234″ and E042°19.407″ 1669 m.a.s.l.) and Haramaya (N09°24.954″ and E042°02.037″ at an altitude of 2020 m.a.s.l.) agricultural research centers. The soils had not been inoculated before with rhizobia isolates nodulating common bean. Before sowing, soil samples were taken from 0–20 cm depth to determine baseline soil properties. Soil samples were air-dried, crushed, and passed through a 2-mm sieve prior to physical and chemical analysis. Details of physical and chemical characteristics of the soil of experimental sites are given in Table 1.

Sources of seeds and *Rhizobium* strain

A common bean var. Dursitu was supplied by Lowland Pulses Research Project, Haramaya University, Ethiopia. Variety was selected based on their yield, their maturity time, and its better performance in eastern Ethiopia. Strain of *Rhizobium leguminosarum* bv. Phaseoli (HU*Pv*R-16) was obtained from Biofertilizer Research and Production Project, Haramaya University (Haramaya, Ethiopia). This strain was selected because it was previously found efficient while tested in this region on two improved varieties of common bean under laboratory and greenhouse conditions [26].

Inoculums preparation

Agar slope of HU*Pv*R-16 strain was obtained from Soil Microbiology Research Laboratory, Haramaya University, Ethiopia. For purification, this isolate was preliminarily cultured in YEMA medium (10 g mannitol, 1 g yeast-extract, 1 g KH$_2$PO$_4$, 0.1 g NaCl, and 0.2 g MgSO$_4$·7H$_2$O per liter, pH 6.8) and incubated at 28 °C for 5 days. The pure colony of the isolate was later transferred to YEM broth

Table 1 Soil analysis of experimental sites before sowing

Soil properties	Hirna soil	Babillae soil	Haramaya soil	Fedis soil
pH in H_2O	7.25	6.66	7.84	7.76
EC (mS/cm)	0.06	0.04	0.14	0.06
Organic carbon (%)	1.65	0.56	1.96	1.32
Total nitrogen (%)	0.16	0.06	0.12	0.12
Available P (mg kg^{-1})	27.11	2.22	9.94	1.78
Ca (cmol(+)kg^{-1})	39.88	4.18	31	23.12
Mg (cmol(+)kg^{-1})	9.00	3.5	8.7	12.87
Na (cmol(+)kg^{-1})	0.14	0.15	0.33	0.12
K (cmol(+)kg^{-1})	0.80	0.34	0.14	1.09
CEC (cmol(+)kg^{-1})	40.03	6.59	25.98	32.22
Zn (mg kg^{-1})	0.95	0.26	0.11	0.10
B (mg kg^{-1})	0.83	ND	0.15	0.75
NH_4-N (mg kg^{-1})	33.77	25.57	–	20.10
NO_3-N (mg kg^{-1})	33.74	27.98	–	27.75
Clay (g kg^{-1})	49	18	33	36
Silt (g kg^{-1})	39	6	18	45
Sand (g kg^{-1})	12	79	49	19
Textural class	Clay	Sandy loam	Sandy clay loam	Silty clay loam
Number of indigenous rhizobia of common bean g^{-1} soil	1.1×10^4	<10	2.8×10^3	2.5×10^2

medium and incubated at 28 °C for 5 days with gentle shaking at 120 rpm in shaker incubator. By this procedure, cell density in the culture was estimated by measuring optical density (540 nm) to determine whether the *Rhizobium* culture reached the middle or late logarithmic phase. *Rhizobium* inoculant was prepared by mixing 30 g of sterilized decomposed filter-mud with 15 ml of broth culture containing HU*PvR*-16 strain in polyethylene bags. After incubating the inoculated filter-mud for 2 weeks at 28 °C, the count of the *Rhizobium* was reached 1×10^9 g^{-1} of inoculant. Populations of rhizobia in the inoculants were determined by duplicate plate counts (Vincent, 1970).

Enumeration of indigenous rhizobia nodulating common bean

The initial indigenous rhizobia population was determined by the plant infection technique, using inoculation of serially diluted soil on germinated common bean seedling for nodulation assessment following the method of Brockwell et al. [27]. This experiment was conducted under controlled condition in growth chamber. The most probable number (MPN) was calculated from the most likely number, using the MPN tables of Vincent [28]. The rhizobial population that nodulated common bean in all study sites are indicated in Table 1.

Experimental design

Field trials on three soil types which had different rhizobial population nodulating common bean were conducted in order to investigate the effect of indigenous rhizobial population on the effect of N rates when applied alone or in combination with *Rhizobium* inoculation. The treatment effects were evaluated via determination of nodulation, yield and yield traits of common bean. The experimental design was a split plot in randomized complete block design (RCBD) with three replications. Main plot treatments consisted of six levels of inorganic N: 0, 20, 40, 60, 80 and 100 kg N ha^{-1}. Two *Rhizobium* inoculations (inoculated and uninoculated) were assigned as subplot treatments. Nitrogen fertilizer in each level was divided into two equal parts: (1) the first part of the N (20 kg N ha^{-1}) was applied along the furrow by hand and incorporated before planting time, and (2) the remaining parts were applied at flowering stages (R_3-stage).

The area was moldboard-plowed and disked before planting. The sizes of the main and subplot were 3×5 m^2 and 3×2 m^2, respectively. There were five rows per subplot and the spacing was 40 cm between rows, 10 cm between plants, 1 m between subplots and 1.5 between main plots. Disinfected seeds of common bean were sown after they were moistened with a 20 % solution of sucrose and then inoculated (7 g inoculant per kg seed) with *Rhizobium*. Inoculated seeds were hand planted on July 7, 2012. Phosphorus (P) was uniformly applied at planting at rate of 20 kg P ha^{-1} as triple superphosphate. Two seeds were sown per hill. After germination, the plants were thinned to one seedling per hill to obtain about 30 plants per row.

Weeds were controlled over the growth period with hand hoeing. At late flowering and early pod setting stage (R_3 stage), five plants from central rows were randomly chosen and harvested to record number of nodule plant^{-1} (NN), nodule dry weight plant^{-1} (NDW) and shoot dry weight plant^{-1} (SDW). Shoots of the plants were dried and later ground to pass a 0.5 cm sieve. Total N determinations were done by the Kjeldahl method of Bremner [29]. At physiological maturity stage on October 30, 2012, yield and yield traits of common bean were recorded. Number of pods plant^{-1} (NPP), number of seeds plant^{-1} (NSP), 100 seed weight, total biomass yield (TBY) and grain yield (at 13 % moisture content) (GY) were determined.

Data analysis

Data were submitted to analysis of variance using SAS version 9.1. Statistically significant differences between means were also determined by the LSD test. The bar graphs were constructed using Microsoft excel version 10.

Result and discussion

Nodulation

The soil samples from four experimental sites were aseptically collected to enumerate the rhizobia population that nodulated common bean by using the plant infection method. The result of this experiment revealed that the rhizobia of common bean varied from <10 to 10^5 g^{-1} of soil. Based on this rhizobial population, the experimental sites were grouped into three soil types. Accordingly, Babillae soil had <100 rhizobia of common bean g^{-1} soil. Rhizobia population >1000 g^{-1} of soil was found in Haramaya and Hirna soils. While the rhizobial population in Fedis soil was between 100 and 1000 g^{-1} soil. Therefore, Babillae, Fedis, and Haramaya and Hirna soils were categorized into low, medium and high rhizobia containing soil types, as it has been previously described by Howieson and Ballard [30]. None of common bean cultivating history in Babillae and Fedis sites could be attributed to lower rhizobial population nodulating common bean [31]. Continuous cultivation of the host plant could increase rhizobial population at Haramaya and Hirna soils [32].

Different rates of N application either alone or in combination with *Rhizobium leguminosarum* bv. Phaseoli inoculation significantly ($P < 0.05$) affected common bean nodulation in all study sites (Table 2). In all soil types, the NN and NDW were significantly decreased with increasing rates of N application either alone or in combination with inoculation. Other authors have found similar trends of nodulation along N rates with different

Table 2 Nodulation status and shoot dry weight of common bean var. Dursitu along different rates of N application with and without inoculation of *Rhizobium leguminosarum* bv. *Phaseoli* at selected areas of eastern Ethiopia

Treatments	NN			NDW			SDW		
	Soil type 1	Soil type 2	Soil type 3	Soil type 1	Soil type 2	Soil type 3	Soil type 1	Soil type 2	Soil type 3
Control	28.33d	71.67bcd	163.00abcd	0.1217cd	0.1920bc	0.4407b	36.23cd	31.37e	49.92b
20 kg N ha^{-1}	88.67c	146.67a	206.00a	0.1880b	0.4097a	0.3399bc	35.30cd	38.03cde	58.13ab
40 kg N ha^{-1}	65.00c	106.67b	149.67abcde	0.1263cd	0.3963a	0.2299bc	41.17bc	48.57bcd	60.77ab
60 kg N ha^{-1}	18.67d	45.00def	91.33cde	0.0437d	0.1727bc	0.1300c	49.97ab	53.23abc	63.60ab
80 kg N ha^{-1}	31.00d	45.00def	93.00cde	0.1033cd	0.1610bcd	0.1621c	47.53ab	55.70ab	63.67ab
100 kg N ha^{-1}	22.33d	20.00f	68.00e	0.0193d	0.0160e	0.2127bc	44.83abc	56.33ab	76.47a
Rhizobium sp.	82.33c	90.67bc	186.67ab	0.3063ab	0.2533b	0.7184a	30.77d	39.00cde	59.07ab
Rhizobium sp. + 20 kg N ha^{-1}	161.67a	64.33cde	178.67abc	0.3900a	0.2010bc	0.7111a	42.63bc	36.40de	70.50ab
Rhizobium sp. + 40 kg N ha^{-1}	124.33b	46.67def	111.50bcde	0.3207ab	0.0897cde	0.3337bc	52.93a	64.90a	78.03a
Rhizobium sp. + 60 kg N ha^{-1}	26.00d	27.33ef	71.50de	0.1530cd	0.0516de	0.1596c	47.47ab	52.83abc	75.68a
Rhizobium sp. + 80 kg N ha^{-1}	17.33d	20.00f	170.00abc	0.0287d	0.0340e	0.3238bc	50.77ab	51.70abcd	76.38a
Rhizobium sp. + 100 kg N ha^{-1}	19.67d	22.33f	63.17e	0.0210d	0.0279e	0.2124bc	43.93abcd	51.97abcd	66.92ab
Mean	57.11	58.86	129.38	0.1518	0.1671	0.3312	43.63	48.35	66.59
F value	66.17***	29.26***	7.05***	21.55***	33.36***	12.78***	11.46***	9.80***	3.71**
LSD	30.06	37.15	93.59	0.1389	0.119	0.27	10.22	16.20	22.32
CV (%)	17.88	21.44	36.85	31.07	24.19	41.54	7.95	11.38	17.07

Means in the same column followed by the same letter are not significantly different at the 5 % probability level by Tukey's test

NS non-significant, NN nodule number, NDW nodule dry weight, SDW shoot dry weight

* Significant at 0.05; ** significant at 0.01; *** significant at 0.001

legume pulses [21, 33, 34]. In soil with low rhizobial population, Rhizobium inoculation alone produced significantly higher NN and NDW than the uninoculated treatments. In contrast, the data indicated the non-significant increases of NN and NDW due to inoculation in soil types with medium and high rhizobial population. Similarly, the effect of inoculation on nodulation of common bean varied due to different indigenous rhizobial population, as previously observed by Asad et al. [15]. In soil having low rhizobial population, Rhizobium inoculated with 20 and 40 kg N ha^{-1} resulted in significantly higher NN than the remaining N treatments. Moreover, Rhizobium inoculated with 20 kg N ha^{-1} produced the highest NDW. The overall effect of inoculation on NN and NDW in this soil type was higher than for the corresponding N rates of applied without inoculation (Fig. 1a, b), thus showing the competitive advantage of inoculated Rhizobium over the indigenous rhizobial population. Similarly, low rate of N with inoculation in soil with low indigenous rhizobial population resulted in an enhancement of nodulation [24]. In contrast, N application as low as 15 kg N ha^{-1} and applied at sowing, suppressed nodule dry weight of common bean [25]. Furthermore, N applied at planting had beneficial effect on nodulation at late growth stage [35].

The soil type with medium rhizobial population responded to the 20 kg N ha^{-1} with significantly higher NN and NDW than the other treatments, followed in the order by that obtained from 40 kg N ha^{-1}. On top of this, it was also observed a slight reduction of NN and NDW due to inoculation as compared to the uninoculated treatment (Fig. 1a, b). Similar result was previously observed by Msumali and Kipe-Nolt, [36], when indigenous rhizobia may have been more potent than inoculated Rhizobium strain. On the other hand, for soil having high rhizobial population, the 20 kg N ha^{-1} treatment alone gave the highest NN, although without significant difference from that obtained with either Rhizobium inoculation alone or 40 kg N ha^{-1} alone or inoculation applied with 20 kg N ha^{-1}. Moreover, the highest NDW were obtained from Rhizobium inoculation applied either alone or in combination with 20 kg N ha^{-1}. In general, it was observed a slight increment of NN and a remarkable improvement of NDW due to inoculation, as compared to the uninoculated treatments with corresponding rates of N (Fig. 1a, b). Similar to this finding, a significant improvement of nodulation due to inoculation in soil with >1000 rhizobial population was observed by Hungria et al. [17]. Inoculation together with a 20 kg N ha^{-1} treatment resulted in the highest NDW at soil with low and high rhizobial population. While the 20 kg N ha^{-1} treatment alone gave the highest NDW in soil with medium rhizobial population. The highest NN produced

in soil with low, medium and high rhizobial population were 161.67, 146.67 and 206.00, respectively. These various NN in different soil types could be attributed to different number of rhizobia present in these three soil types. Similarly, Patrick and Lowther [37] found that the size of rhizobial population affects nodulation. Previous work confirmed that the largest the soil native rhizobial population, the greater is the nodulation [38]. The lowest NN and NDW produced in soil having relatively lower rhizobial population was previously observed by Chemining'wa and Vessey [21].

Shoot dry weight

The experimental treatments (different rates of N application solely and in combination with Rhizobium inoculation) resulted in significant variation of SDW (Table 2). In all soil types, a significant improvement of SDW was observed with increasing rates of N applied either alone or in combination with Rhizobium inoculation. In soil with low and medium rhizobial population, Rhizobium inoculation together with the 40 kg N ha^{-1} treatment, in which the highest nodulation was produced, resulted in significantly higher SDW than those produced at N rates lower than 40 kg N ha^{-1} in both inoculation treatments. This indicates the importance of higher nodulation for SDW production of common bean. In contrast, the high N rate application decreased nodulation, but an increase of above ground biomass production was observed [21]. In soil having high rhizobial population, the non-significant difference in SDW among N treatments, excluding the control, was observed. This implies that the sole N reserves in soil with lower N rate of application could have been sufficient for maximum shoot biomass production of common bean at R_3 stage. The highest SDW obtained from soils having low, medium and high rhizobial population were 52.93, 64.90 and 78.03 g, respectively. All these shoot biomass were obtained from Rhizobium inoculated with 40 kg N ha^{-1}. It has been shown that number of native rhizobia had a detrimental impact on productivity of above ground dry biomass [39]. Denton et al. [40] found increased shoot biomass production with increased Rhizobium inoculation rate. In all soil types, the lowest SDW were produced for the control treatments (uninoculated and unfertilized). This suggests that the N is the major limiting nutrient for common bean production in all study sites.

Number of pods per plant

The treatments of this experiment affected significantly ($P < 0.05$) the NPP in all soil types (Table 3). In soil having low rhizobial population, 20 kg N ha^{-1} alone gave the highest NPP. Inoculation with 60 kg N ha^{-1} resulted in significantly higher NPP than those produced

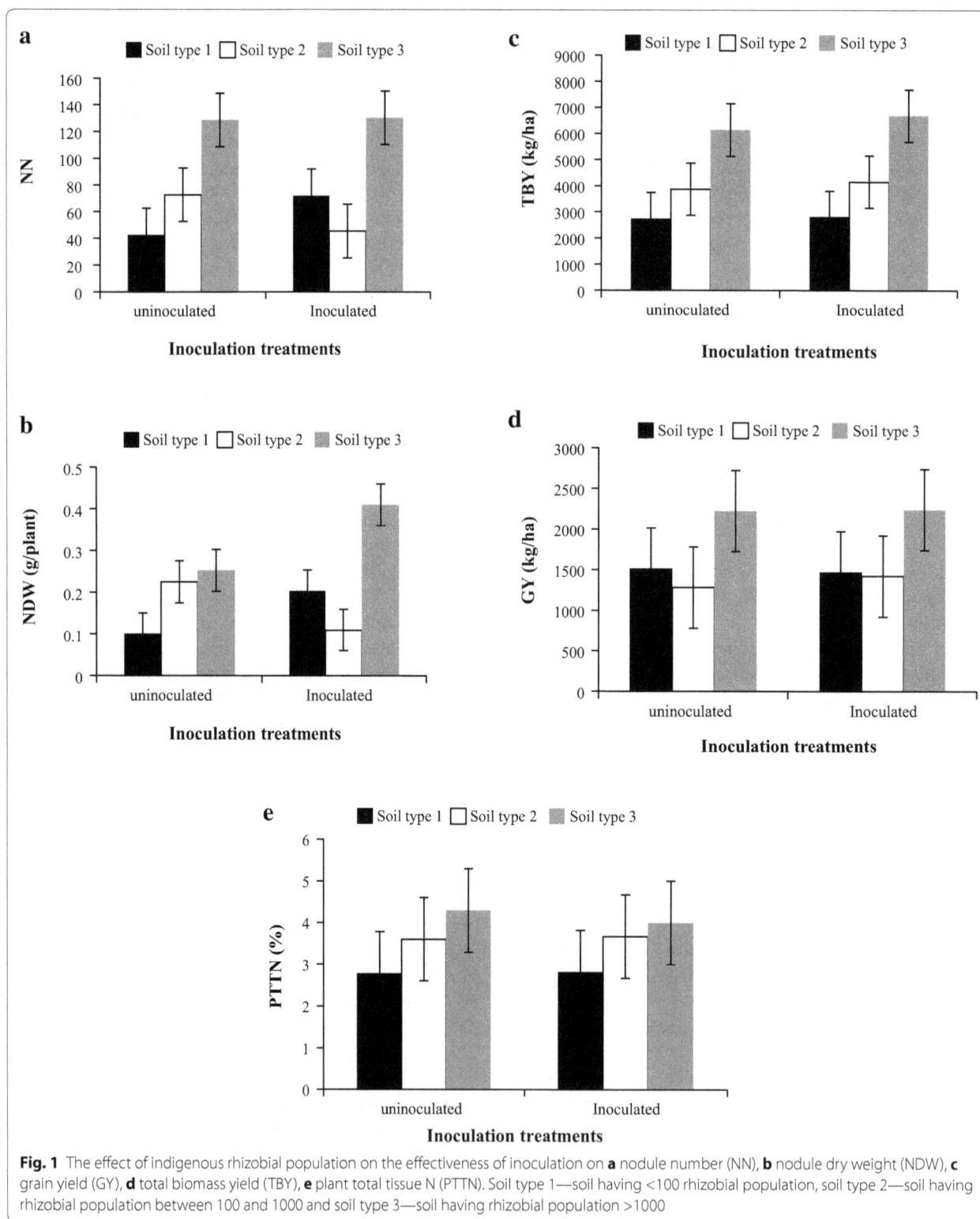

Fig. 1 The effect of indigenous rhizobial population on the effectiveness of inoculation on **a** nodule number (NN), **b** nodule dry weight (NDW), **c** grain yield (GY), **d** total biomass yield (TBY), **e** plant total tissue N (PTTN). Soil type 1—soil having <100 rhizobial population, soil type 2—soil having rhizobial population between 100 and 1000 and soil type 3—soil having rhizobial population >1000

for the control and inoculation alone, in soils having medium and high rhizobial population. Similarly, a significant improvement of NPP ranging from 20.2 at the

control to 24.15 at N treated plants was obtained from faba bean [41]. The highest NPP produced in soils having low, medium and high were 18.63, 24.99 and 31.05,

Table 3 Number of pods per plant, number of seeds per pod and 100 seed weight of common bean var. Dursitu along different rates of N application with and without inoculation of *Rhizobium leguminosarum* bv. *Phaseoli* at selected areas of eastern Ethiopia

Treatments	NPP			NSP			100 seed weight		
	Soil type 1	Soil type 2	Soil type 3	Soil type 1	Soil type 2	Soil type 3	Soil type 1	Soil type 2	Soil type 3
Control	12.55bc	10.33d	20.89c	4.61b	5.40bc	5.82b	19.77c	20.90cd	19.33a
20 kg N ha^{-1}	18.63d	18.66ab	20.83c	6.78a	6.07abc	6.62ab	21.30ab	22.80abc	19.18a
40 kg N ha^{-1}	17.11ab	23.22ab	23.00bc	6.55a	6.30ab	6.70ab	21.37ab	22.07bc	18.92a
60 kg N ha^{-1}	16.00abc	18.00abc	25.22abc	6.44a	6.97a	6.92ab	21.17abc	22.13abc	19.42a
80 kg N ha^{-1}	14.89abc	22.00ab	25.83abc	6.99a	6.40ab	7.13a	21.20abc	22.90ab	19.42a
100 kg N ha^{-1}	16.44ab	22.22ab	25.78abc	6.66a	6.63a	6.70ab	21.57a	22.90ab	19.02a
Rhizobium sp.	10.78c	11.22 cd	23.66bc	5.33ab	5.50bc	6.50ab	20.00bc	19.90d	19.28a
Rhizobium sp. + 20 kg N ha^{-1}	13.66abc	16.44bcd	29.33ab	6.89a	5.20c	6.65ab	21.07abc	22.57abc	18.87a
Rhizobium sp. + 40 kg N ha^{-1}	16.00abc	22.00ab	29.11ab	6.66a	6.30ab	6.85ab	21.40ab	23.03ab	18.97a
Rhizobium sp. + 60 kg N ha^{-1}	16.22abc	24.99a	31.05a	6.22ab	6.30ab	6.35ab	22.17a	22.20abc	18.83a
Rhizobium sp. + 80 kg N ha^{-1}	16.44ab	22.11ab	28.89ab	6.99a	6.40ab	6.72ab	21.47ab	21.37bcd	18.87a
Rhizobium sp. + 100 kg N ha^{-1}	15.22abc	22.22ab	27.05abc	6.89a	6.00abc	6.38ab	20.97abc	24.07a	19.10a
Mean	15.33	19.45	25.89	6.42	6.12	6.61	21.12	22.24	19.10
F value	3.93**	11.06***	5.86***	4.68***	6.87***	2.11*	5.09***	8.41***	0.21 ns
LSD	5.48	7.20	6.66	1.72	1.02	1.10	1.49	1.94	2.30
CV (%)	12.13	12.58	13.10	9.11	5.64	8.48	2.38	2.96	6.12

Means in the same column followed by the same letter are not significantly different at the 5 % probability level by Tukey's test

NS non-significant, NPP number of pods per plant, NSP number of seeds per pod

* Significant at 0.05; ** significant at 0.01; *** significant at 0.001

respectively, confirming the positive effect of indigenous rhizobial population on NPP.

Number of seeds per pod

The data revealed significant variation of NSP due to the treatments (Table 3). In all soil types, NSP increased with increasing rates of N application solely and in combination with *Rhizobium* inoculation. Previous study reported that N nutrition increased the seeds per pod [42, 43]. In soil having low rhizobial population, significantly higher NSP (6.99) was recorded at 80 kg N ha^{-1} and inoculation in conjunction with 80 kg N ha^{-1} as compared to the control. In soils having medium rhizobial population, 60 kg N ha^{-1} resulted in significantly higher NSP than those produced at N rates below 20 kg N ha^{-1} in both inoculation treatments. 80 kg N ha^{-1} resulted in the highest NSP in soil having high rhizobial population. Non-significant differences in NSP obtained from different N rates, excluding control were observed for both inoculation treatments, in soils with low and high rhizobial population.

100 seed weight

The 100 seed weight of common bean was significantly ($P < 0.05$) varied due to the treatments, except those observed in soil having high rhizobial population, in

which this trait exhibited a non-significant difference (Table 3). The non-significant difference in common bean seed size was previously observed by Mulas et al. [44]. Similar to this, 100 seed weight of soybean was not improved by either inoculation or N fertilizer application [45]. In soil having low rhizobial population, significantly higher 100 seed weight was obtained from 100 kg N ha^{-1} and *Rhizobium* inoculated with 60 kg N ha^{-1}, as compared to the control treatment. The sole 100 kg N ha^{-1} addition and in combination with *Rhizobium* inoculation gave significantly higher 100 seed weight compared to the control, in soil having medium rhizobial population. This result is in agreement with that obtained previously by El Hardi and Elsheikh [46] who found that inoculation and N application significantly improved 100 seed weight of chickpea over the control. N application increased 100 seed weight by 7.4 % over those produced for the uninoculated treatment [41].

Total biomass yield

The significant variation of TBY due to treatments was observed at $P \leq 0.05$ (Table 4). In all soil types, *Rhizobium* inoculated with 100 kg N ha^{-1} resulted in significantly higher TBY than those produced in the control treatment and inoculation alone. This result supports the findings of Mulas et al. [44] that inoculation and

Table 4 Total biomass yield, grain yield and total plant tissue N of common bean var. Dursitu along different rates of N application with and without inoculation of *Rhizobium leguminosarum* bv. *Phaseoli* at selected areas of eastern Ethiopia

Treatments	TBY			GY			PTTN		
	Soil type 1	Soil type 2	Soil type 3	Soil type 1	Soil type 2	Soil type 3	Soil type 1	Soil type 2	Soil type 3
Control	2055.6ef	2077.8c	4917.6c	1025.65f	1082.13cd	1946.05bc	2.3800de	2.5200c	4.0983ab
20 kg N ha^{-1}	2277.8def	4018.5ab	5609.5bc	1254.35e	1270.19bc	2121.48abc	3.2433a	3.5533b	4.5300ab
40 kg N ha^{-1}	2631.5cd	4333.3a	6351.4abc	1568.52c	1307.13bc	2262.27ab	2.8533abcd	3.7800ab	4.6533a
60 kg N ha^{-1}	3131.5abc	4055.6a	6481.5abc	1727.63bc	1289.44bc	2207.04abc	2.5567bcde	3.6933ab	4.0767ab
80 kg N ha^{-1}	3082.4bc	4388.9a	6721.8ab	1674.26bc	1280.46bc	2380.60a	2.5167cde	3.8600ab	4.2900ab
100 kg N ha^{-1}	3260.8ab	4374.1a	6742.8ab	1830.09b	1454.26ab	2416.44a	3.1567abc	4.2000a	4.1200ab
Rhizobium sp.	1787.0f	3133.3b	5472.6bc	996.39f	1005.37d	1941.67c	2.1600e	3.3833b	4.0517ab
Rhizobium sp. + 20 kg N ha^{-1}	2148.1def	4087.0a	7405.6a	1302.22de	1258.70bc	2169.17abc	3.1800ab	3.3633b	4.1867ab
Rhizobium sp. + 40 kg N ha^{-1}	2498.1de	4249.2a	6668.1ab	1211.48ef	1569.72a	2381.16a	2.6567abcde	3.6800ab	4.0800ab
Rhizobium sp. + 60 kg N ha^{-1}	3375.4ab	4277.8a	6851.6ab	1697.59bc	1461.85ab	2366.06a	2.7000abcde	3.8200ab	4.0033ab
Rhizobium sp. + 80 kg N ha^{-1}	3421.4ab	4462.2a	6486.1abc	1511.57cd	1571.57a	2240.34abc	3.2000ab	3.9200ab	3.9467ab
Rhizobium sp. + 100 kg N ha^{-1}	3648.1a	4740.7a	7222.2a	2089.54a	1653.89a	2329.40a	3.0133abcd	3.8767ab	3.7583b
Mean	2776.48	4016.54	6410.89	1490.78	1350.39	2230.14	2.8014	3.6375	4.1496
F value	32.12***	16.30***	4.95***	57.11***	19.75***	5.99***	8.10***	14.64***	1.73 ns
LSD	557.86	913.75	15,380.6	226.81	226.53	318.67	0.6467	0.5621	0.8943
CV (%)	6.82	7.73	12.56	5.17	5.70	7.28	7.84	5.25	10.98

Means in the same column followed by the same letter are not significantly different at the 5 % probability level by Tukey's test

NS non-significant, *TBY* total biomass yield, *GY* grain yield, *PTTN* plant total tissue N

* Significant at 0.05; ** significant at 0.01; *** significant at 0.001

inorganic N application enhanced the common bean production in all soil type regardless of the indigenous rhizobial population. Significant enhancement of above ground biomass production by 22 % due to *Rhizobium* inoculated with inorganic N over the uninoculated treatment was also observed by Chemining'wa and Vessey [21]. In soil having medium rhizobial population, it was observed a non-significant difference in TBY produced at 20 kg N ha^{-1} and beyond rates of N, with both inoculation treatments. This could have confirmed the presence of effective common bean–rhizobia symbiosis at low inorganic N [47] and thus satisfy the N need for boost the biomass production. The control treatment, inoculation alone, and 20 kg N ha^{-1} alone gave significantly lower TBY than those produced in other treatments in soil having high rhizobial population. In soil having low rhizobial population, statistically lower TBY was produced for control treatment, and for those with 20 and 40 kg N ha^{-1} either alone or in combination with *Rhizobium* inoculation, as compared to those produced in other treatments. This may indicate that the native rhizobia in this site are capable but not effective to fix N$_2$ to satisfy the N requirement of common bean. The presence of higher rhizobial population is not an indicator of the symbiotic effectiveness between rhizobia and N derived from the atmosphere [31]. In all soil types, the present study revealed that inoculation slightly increased the TBY as compared to those obtained from the corresponding N treatments without *Rhizobium* inoculation (Fig. 1c). The highest TBY produced in soils having low, medium and high rhizobial population were 3648.1, 4740.7 and 7222.2 kg ha^{-1}, respectively. The highest biomass in soil having high rhizobial population was previously confirmed by Furseth et al. [48]. These authors found that yield of soybean positively correlated with soil indigenous rhizobial population across environments.

Grain yield

The GY of common bean revealed significant variation due to treatments at $P \leq 0.05$ (Table 4). In soil having low rhizobial population, increasing rates of N application increased GY production, though the highest nodulation and PTTN were recorded at 20 kg N ha^{-1}. The 100 kg N ha^{-1} addition in conjunction with inoculation also gave significantly higher GY than those produced in other treatments. Similarly, inhibition effect of higher N application on nodulation and nitrogenase enzyme without affecting grain yield was previously determined by Rai [23]. Our result may be confirm that the previous findings, although N mineral reduced the nodulation and N$_2$ fixation and yields of common bean can be improved by increasing N availability [49]. da Silveira et al. [47] also found that common bean responded well up to 200 kg N ha^{-1}. Asad et al. [15] observed that nodulation improvement due to inoculation had not significantly enhanced plant biomass production. This present study also confirms that the common bean–rhizobia symbiosis was not satisfactory to fulfill the N needs as it has been previously observed by Fesonko et al. [50]. Contrary to this, Denton et al. [40] showed that improvement of shoot N increased the grain yield of faba bean by 1 Mg ha^{-1} in soil having low rhizobial population. This indicates that biological N$_2$ fixation alone is not a sufficient N requirement of common bean to get the local attainable yield in the prevailing environmental condition.

Inorganic N at 100 kg N ha^{-1} with *Rhizobium* inoculation gave the highest GY at soil having medium rhizobial population as it has been observed in PTTN. This may indicate that N derived from indigenous and inoculated rhizobia alone did not satisfy the N requirement of common bean. Voisin et al. [51] observed that high N requirement at the seed setting stage was supported by both N$_2$ fixation and mineral N supply. Although N fertilizer enhanced the plant N uptake, the common bean production did not improve [25]. The present study indicates a non-significant difference in GY obtained from inoculation when either coupled to 40, 60, 80 and 100 kg N ha^{-1} treatment, or sole application of 100 kg N ha^{-1} addition. This may indicate the inoculated *Rhizobium* satisfies the N requirement of common bean with relatively low inorganic N application.

In soil having high rhizobial population, the 40 kg N ha^{-1} applied in conjunction with *Rhizobium* resulted in significantly higher GY than those produced at the control and inoculation alone. Similarly, a previous finding recommended N application to maximize common bean yield in soil having high indigenous rhizobial population [31]. In this soil type, it was recorded a non-significant difference in GY at different N rates of application, excluding control. Besides this, treatment with low N rate applied together with both inoculation treatments, produced statistically similar GY with those found at higher N rates application. The synergetic effect between low rates of N application and *Rhizobium* inoculation on common bean production in soil having rhizobial population >1000 was previously confirmed by Hungria et al. [17]. These authors found the highest seed yield from treatments with *Rhizobium* inoculated together with 15 kg N ha^{-1} applied at planting and further 15 kg N ha^{-1} addition at early flowering stages. Due to the fact that biological N$_2$ fixation is not active at early stage of common bean, a starter dose of N application is required to enhance plant growth and eventually improve the grain yield production [35, 52, 53]. Nitrogen application in addition to starter N reduced nodulation and failed to increase the common bean yield [24]. Vargas et al. [31] found that increasing rates of N decreased the number of nodules from inoculated cells, whereas it increased that by indigenous rhizobia.

The overall effect of inoculation on GY was slightly decreased in soil having low rhizobial population but slightly improved in soil having medium and high rhizobial population, as compared to that obtained at corresponding rates of N application without inoculation (Fig. 1d). This result supports the finding of Hungria et al. [54] who showed that inoculation increased the yield of common bean as compared to the control treatment. A previous study also reported that an inoculation response was observed in soil having rhizobial population between 300 and 1000 g^{-1} soil [55, 56]. The highest GY produced for soil types having low, medium and high rhizobial population were 2089.54, 1653.89 and 2381.16 kg ha^{-1}, respectively. These GY are comparable with those previously produced at 100 kg N ha^{-1} N, for in which 2000 kg ha^{-1} was reported [47]. At least 29 % faba bean yield advantage was provided for soil having higher rhizobial population over that yield obtained from low rhizobial population, as determined by Sorwle and Mytton [41].

Plant total tissue N

The PTTN was significantly varied due to different treatments at $P \leq 0.05$ (Table 4). In soil having low and high rhizobial population, N application beyond 20 kg N ha^{-1} with both inoculation treatments resulted in a slight decrease of PTTN. 20 kg N ha^{-1} application resulted in the highest PTTN in soil having low rhizobial population. The lower PTTN at higher N application could be attributed to the negative effect of inorganic N on fixed N through the inhibition of the nitrogenase activity [20, 57].

In soil having medium rhizobial population, progressively larger rates of N application increased the PTTN, though the highest NN and NDW was produced at 40 kg N ha^{-1}. The 100 kg N ha^{-1} treatment gave significantly higher PTTN than those obtained from either the control treatment or that with 20 kg N ha^{-1} applied alone or in combination with *Rhizobium* inoculation. This could be due to low effectiveness of indigenous rhizobia of common bean in N$_2$ fixation. Asad et al. [15] found the significant improvement of plant total N accumulation of common bean was due to N application rather than to inoculation. This was attributed to the fact that common bean rarely derives more than 50 % of its N from symbiotic N$_2$ fixation [58].

In soil type having high rhizobial population, 40 kg N ha^{-1} alone resulted in significantly higher PTTN than that obtained from inoculation with the 100 kg N ha^{-1} addition. Shutsrirung et al. [59] demonstrated that inoculation may not be effective in soil having high effective native rhizobial population. In top of this, PTTN was also decreased with increasing rates of N application following both inoculation treatments. This indicates that the native rhizobia in this site might have been more effective in N_2 fixation than the inoculated stain. Furthermore, inoculation slightly increased the PTTN in soil having low and medium rhizobial population whereas it showed a decreasing trend in soil with high rhizobial population, in comparison to PTTN found for corresponding N rates without inoculation (Fig. 1e). The highest PTTN was observed at low, medium and high rhizobial population and reached 3.24, 4.20 and 4.65 %, respectively. The highest PTTN could be attributed to a higher nodulation induced by the high rhizobial population and, thus, an enhanced N_2 fixation [38, 48]. Due to synergetic effect of a starter N application on symbiotic N_2 fixation, the highest accumulated N in plants was recorded for the 20 kg N ha^{-1} application in soil having high rhizobial population [24].

Correlation analysis

The significant correlation among investigated traits of common bean was observed in all soil types at $P < 0.05$ (Tables 5, 6, 7). The amount of NN produced at soil with low rhizobial nodulating population inversely correlated with TBY ($r = -0.6754$; $P \leq 0.05$) and with GY ($r = -0.5629$; $P \leq 0.05$) (Table 5). In this soil type, a negative and strong correlation between NDW with TBY ($r = -0.7563$; $P \leq 0.01$), GY ($r = -0.6953$; $P \leq 0.05$) was observed. NN had an inverse and strong correlation with both SDW ($r = -0.5958$; $P \leq 0.05$) and GY ($r = -0.6090$; $P \leq 0.05$) in soil with medium rhizobial population (Table 6). Conversely, NDW and GY revealed an inverse and significant correlation ($r = -0.6356$; $P \leq 0.05$). In soil with high rhizobial population, the correlation analyses

between the NN and both TBY ($r = 0.5599$; $P \leq 0.05$) and GY ($r = 0.7538$; $P \leq 0.01$) were positive (Table 7), whereas NDW had an inverse and significant correlation with GY ($r = -0.6917$; $P \leq 0.05$). In general, this finding indicates that N mineral fertilizer rather than symbiotic N_2 fixation had a determinant effect on common bean yield. This result supports the results of the other authors [60], in which NDW and N fixation showed negative correlation with yield related parameters. La Favre and Eaglesham [61] also observed an inverse and significant relationship between the starting N application and both NN and NDW. In contrast, Aggarwal [62], who conducted experiment in Malawi, found positive and significant correlations between NN and GY for common bean under inorganic N treatment. Mothapo et al. [38] also reported that plant biomass production correlated positively with NDW when inoculation alone was applied.

In all soil types, also SDW of common bean had a significant correlation with NPP, NSP, 100 seed weight, TBY and GY. However, NPP and NSP had the non-significant association with SDW with low and high rhizobial population (Tables 5, 6, 7). In particular, in soil with low rhizobial population, all listed growth traits had positive and significant correlation with SDW at $P \leq 0.05$ (Table 5). SDW had also a positive and significant ($P \leq 0.05$) correlation with 100 seed weight ($r = 0.5087$) and TBY ($r = 0.6996$), and even highly significant ($P \leq 0.01$) with NPP ($r = 0.7719$), NSP ($r = 0.7751$) and GY ($r = 0.7270$). In the same soil, a positive and strong correlation was also observed between SDW and either NPP ($r = 0.8726$; $P \leq 0.001$), or TBY ($r = 7378$; $P \leq 0.01$) or GY ($r = 0.7496$; $P \leq 0.01$). Highly significant relationships among SDW, shoot N content and seed yield are previously shown by Pereira and Bliss [63].

Table 5 Correlation among the investigated traits of common bean treated different rates of N with and without inoculation in soil having rhizobial population nodulating common bean <100 g^{-1} soil

Traits	NN	NDW	SDW	NPP	NSP	100 SW	TBY	GY	HI
NN									
NDW	0.93***								
SDW	−0.19 ns	−0.28 ns							
NPP	−0.15 ns	−0.39 ns	0.44 ns						
NSP	0.11 ns	−0.14 ns	0.59*	0.66*					
100 SW	−0.08 ns	−0.21 ns	0.67*	0.78**	0.74**				
TBY	−0.68*	−0.76**	0.69*	0.50*	0.56*	0.65*			
GY	−0.56*	−0.70*	0.52*	0.46 ns	0.59*	0.58*	0.91***		
HI	0.13 ns	−0.04 ns	0.08 ns	−0.05 ns	0.30 ns	0.08 ns	0.09 ns	−0.00 ns	
PTTN	0.13 ns	−0.18 ns	0.23 ns	0.66*	0.72*	0.56*	0.36 ns	0.38 ns	0.52*

Ns non-significant, *NN-Nodule* number per plant, *NDW* nodule dry weight per plant (g plant^{-1}), *SDW* shoot dry weight (g plant^{-1}), *NPP* number of pods per plant, *NSP* number of seeds per pod, *100 SW* 100 seed weight (g), *TBY* total biomass yield (kg ha^{-1}), *GY* grain yield (kg ha^{-1}), *HI* harvest index

* Significant at 0.05; ** highly significant at 0.01; *** very highly significant at 0.001

Table 6 Correlation among the investigated traits of common bean treated different rates of N with and without inoculation in soil having rhizobial population nodulating common bean between 100 and 1000 g^{-1} soil

Traits	NN	NDW	SDW	NPP	NSP	SW	TBY	GY	HI
NN									
NDW	0.96***								
SDW	−0.60*	−0.55*							
NPP	−0.40 ns	−0.34 ns	0.77**						
NSP	−0.35 ns	−0.26 ns	0.78**	0.66*					
SW	−0.26 ns	−0.26 ns	0.51*	0.66*	0.34 ns				
TBY	−0.36 ns	−0.29 ns	0.70*	0.88***	0.57*	0.71**			
GY	−0.61*	−0.64*	0.73**	0.81**	0.49 ns	0.69*	0.77**		
HI	−0.25 ns	−0.29 ns	0.19 ns	0.26 ns	0.16 ns	0.46 ns	0.17 ns	0.50*	
PTTN	−0.42 ns	−0.36 ns	0.77**	0.84**	0.72**	0.52*	0.91***	0.69*	0.03 ns

ns non-significant, *NN-Nodule* number per plant, *NDW* nodule dry weight per plant (g plant^{-1}), *SDW* shoot dry weight (g plant^{-1}), *NPP* number of pods per plant, *NSP* number of seeds per pod, *100 SW* 100 seed weight (g), *TBY* total biomass yield (kg ha^{-1}), *GY* grain yield (kg ha^{-1}), *HI* harvest index

* Significant at 0.05; ** highly significant at 0.01; *** very highly significant at 0.001

Table 7 Correlation among the investigated traits of common bean treated different rates of N with and without inoculation in soil having rhizobial population nodulating common bean >1000 g^{-1} soil

Traits	NN	NDW	SDW	NPP	NSP	SW	TBY	GY	HI
NN									
NDW	0.72**								
SDW	0.45 ns	−0.23 ns							
NPP	0.45 ns	−0.10 ns	0.87***						
NSP	0.18 ns	−0.29 ns	0.41 ns	0.27 ns					
SW	0.00 ns	−0.07 ns	−0.68*	−0.62*	0.05 ns				
TBY	0.56*	−0.27 ns	0.74**	0.79**	0.49 ns	−0.50 ns			
GY	0.75**	−0.69*	0.75**	0.61*	0.55*	−0.39 ns	0.77**		
HI	0.33 ns	0.34 ns	0.08 ns	−0.15 ns	−0.28 ns	0.06 ns	−0.49 ns	−0.25 ns	
PTTN	0.44 ns	−0.02 ns	0.38 ns	−0.53*	0.27 ns	0.03 ns	−0.26 ns	−0.09 ns	0.11 ns

Ns non-significant, *NN-Nodule* number per plant, *NDW* nodule dry weight per plant (g plant^{-1}), *SDW* shoot dry weight (g plant^{-1}), *NPP* number of pods per plant, *NSP* number of seeds per pod, *100 SW* 100 seed weight (g), *TBY* total biomass yield (kg ha^{-1}), *GY* grain yield (kg ha^{-1}), *HI* harvest index

* Significant at 0.05; ** highly significant at 0.01; *** very highly significant at 0.001

In sites with soil having a low rhizobial population, GY of common bean correlated positively with SDW ($r = 0.5152$; $P \leq 0.05$), NSP ($r = 0.5907$; $P \leq 0.05$), 100 seed weight ($r = 0.5791$; $P \leq 0.05$) and TBY ($r = 0.9121$; $P \leq 0.001$) (Table 5). Similarly, strong and positive correlation was observed between GY with SDW ($r = 0.7271$; $P \leq 0.01$), NPP ($r = 0.8098$; $P \leq 0.01$), 100 seed weight ($r = 0.6939$; $P \leq 0.05$) and TBY ($r = 0.7713$; $P \leq 0.01$) in experimental soil with medium rhizobial population (Table 6). In soil showing a high rhizobial population, we noted a positive correlation between GY and SDW ($r = 0.7496$; $P \leq 0.01$), NPP ($r = 0.6067$; $P \leq 0.050$), NSP ($r = 0.5474$; $P \leq 0.05$) and TBY ($r = 0.7712$; $P \leq 0.01$) (Table 7). Similarly, Bayuelo-Jiménez et al. [64] had previously indicated a positive and significant correlation among yield and yield components of common bean.

The correlation analysis indicated a positive relationship between PTTN and NPP ($r = 0.6644$; $P \leq 0.05$), NSP ($r = 0.7195$; $P \leq 0.05$) and 100 seed weight ($r = 0.5562$; $P < 0.05$) in experimental sites showing soil with low rhizobial population (Table 5). In those with medium rhizobial population, a positive and strong correlation between PTTN and SDW ($r = 0.7665$; $P \leq 0.01$), NPP ($r = 8386$; $P \leq 0.01$), NSP ($r = 0.7154$; $P \leq 0.01$), 100 seed weight ($r = 0.5206$; $P \leq 0.05$), TBY ($r = 0.9114$; $P \leq 0.001$) and GY ($r = 0.6941$; $P \leq 0.05$) (Table 6). These results agree with the previous studies, which reported the significant correlation between plant N accumulation with seed yield, seed weight and total biomass [45, 50]. Ruiz-Díez et al. [45] also demonstrated that plant N accumulation was the most suitable trait for the selection of highly effective and highly competitive *Rhizobium* isolates. In contrast, in soil with high rhizobial population, PTTN was only significantly but inversely correlated with NPP ($r = -0.5296$; $P \leq 0.05$), thus, indicating the negative impact of mineral N treatment on soil productivity, despite the plant N accumulation was noted (Table 7). The present study also indicated a non-significant relationship between PTTN and both NN and NDW in all soil types. Similar result was previously observed by da Silva et al. [24] who reported that the correlation between fixed N and nodulation was decreased with increasing rates of N application. In contrast to this, Tsia et al. [25] found a positive and significant relationship between both NN and NDW and N in shoots at 45 days after emergence.

Conclusion

The results of this study indicate the significant effect of N treatments on the nodulation, yield and yield traits of common bean in the major growing areas of eastern Ethiopia. Our experiments indicated the need of inoculation of *Rhizobium* beside that of a starting N dose in soil having <100 rhizobial population g^{-1} of soil. An N requirement to reach the highest yield of common bean was not affected by the native rhizobial population. The effect of *Rhizobium* inoculation on nodulation of common bean appears to depend on the type of soil. Our data also indicated that the amount of mineral N required for maximum seed yield is dependent on the original soil rhizobial population. Moreover, the indigenous rhizobial population that nodulates common bean also affects the bean production. Further research on effectiveness of the combination between indigenous rhizobia and inoculated rhizobial population would be recommended.

Authors' contributions

Both of us participated equally starting from the development of the research idea, writing proposal and competing research grant and development of this manuscript. But I participated more in the management and collection of data from the field experiment, which is why I am the first author of this manuscript. Both authors read and approved the final manuscript.

Author details

[1] College of Agriculture and Environmental Sciences, School of Natural Resources Management and Environmental Sciences, Haramaya University, Dire Dawa, Ethiopia. [2] Ethiopian Institute of Agricultural Research, Holleta Agricultural Research Center, Holleta, Ethiopia.

Acknowledgements

This work was carried out with the support of Ethiopia Institute of Agricultural Research under National Biofertilizer Development Project hosted at Holleta Agricultural Research Center, Soil microbiology research group. The author also thanks Berhanu Mengistu, Dejene Ayenew and Girmaye Mekonnen for managing the field experiments.

Competing interests

The authors declare that they have no competing interests.

References

1. Broughton WJ, Hernandez G, Blair M, Beebe S, Gepts P, Vanderleyden J. Beans (*Phaseolus* spp.)—model food legumes. Plant Soil. 2003;252:55–128.

2. Abebe G, Mazengia W, Nikus O. Soil fertility and crop management research of lowland food legumes in the rift valley. In Proceedings of the workshop on food and forage legumes. Edited by Ali K, Ahmed S, Beniwal S, Kenneni G, Malhotra RS, Makkouk K, Halila MH. International Center for Agriculture Research in the Dry Areas (ICARDA), Aleppo, Syria, 2006. pp. 157–166.

3. IAR. Melkassa Agricultural Research Center Progress Report for the period 1995/96. Melkassa: Institute of Agricultural Research (IAR); 1997.

4. IAR. Melkassa Agrictural Research Center Progress Report for the period 1997/98. Melkassa: Institute of Agricultural Research (IAR); 1998.

5. Assefa T, Assefa H, Kimani P: Development of improved Haricot bean germplasm for the Mid-and Low-altitude sub-humid agro-ecologies of Ethiopia. In Proceedings of the Workshop on food and forage legumes. Edited by Ali K, Ahmed S, Beniwal S, Kenneni G, Malhotra RS, Makkouk K, Halila MH. International Center for Agriculture Research in the Dry Areas (ICARDA), Aleppo, Syria; 2006. pp. 87–94.

6. Jansa J, Bationo A, Frossard E, Rao IM. Options for improving plant nutrition to increase common bean productivity in Africa. In: Bationo A, editor. Fighting poverty in Sub-Saharan Africa: the multiple roles of legumes in integrated soil fertility management. Dordrecht: Springer Science + Business Media B.V.; 2011. p. 201–40.

7. Lunze L, Abang MM, Buruchara R, Ugen MA, Nabahungu NL, Rachier GO, Ngongo M, Rao I. Integrated Soil Fertility Management in Bean Based Cropping Systems of Eastern, Central and Southern Africa. In: Soil fertility improvement and integrated nutrient management—a global perspective. Edited by Whalen J. InTech; 2012. pp. 239–272 **(ISBN: 978-953-307-945-5)**.

8. Hardarson G, Bliss FA, Cigales-Riveri MR, Henson RA, Kipe-Nolt JA, Longeri L, Manrique A, Pena-Cabriales JJ, Pereira PAA, Sanabria CA, Tsai SM. Genotypic variation in biological nitrogen fixation by common bean. Plant Soil. 1993;152:59–70.

9. Graham PH, Apolitano C, Ferrera R, Halliday J, Lepiz R, Menéndez D, Ríos R, Saito SMT, Viteri S. The International Bean Inoculation Trail (IBIT). Results for the 1978–1979 Trial. In: Graham PH, Harris S, editors. Biological Nitrogen Fixation Technology for Tropical Agriculture. CIAT, Cali; 1981.

10. Graham PH. Some problems of nodulation and symbiotic fixation in *Phaseolus vulgaris* L.: a review. Field Crops Res. 1981;4:93–112.

11. Buttery BR, Park SJ, Findlay WI. Growth and yield of white bean (*Phaseolus vulgaris* L.) in response to nitrogen, phosphorus and potassium fertilizer and to inoculation with *Rhizobium*. Can. J Plant Sci. 1987;67:425–32.

12. Rodriguez-Navarro DN, Buendia AM, Camacho M, Lukas MM, Santamatria C. Characterization of *Rhizobium* spp. bean isolates from South West Spain. Soil Biol Biochem. 2000;32:1601–13.

13. Brockwell J, Bottomley PJ. Recent advances in inoculant technology and prospects for the future. Soil Biol Biochem. 1995;27:683–97.

14. Giller KE, Cadisch G. Future benefits from biological nitrogen fixation: an ecological approach to agriculture. Plant Soil. 1995;174:255–77.

15. Asad S, Malik KA, Hafeez FY. Competition between inoculated and indigenous *Rhizobium/Bradyrhizobium* spp. strains for nodulation of grain and fodder legumes in Pakistan. Biol Fertil Soils. 1991;12:107–11.

16. Mostasso L, Mostasso FL, Vargas MAT, Hungria M. Selection of bean (*Phaseolus vulgaris*) rhizobial strains for the Brazilian Cerrados. Field Crops Res. 2002;73:121–32.

17. Hungria M, Campo RJ, Mendes IC. Benefits of inoculation of the common bean (*Phaseolus vulgaris*) crop with efficient and competitive *Rhizobium tropici* strains. Biol Fertil Soils. 2003;39:88–93.

18. Bliss FA. Utilizing the potential for increased nitrogen fixation in common bean. Plant Soil. 1993;152:157–60.

19. van Kessel C, Hartely C. Agricultural management of grain legumes: has it led to an increase in nitrogen fixation. Field Crops Res. 2000;65:165–81.

20. Saxena AK, Rathi SK, Tilak KVBR. Selection and evaluation of nitrate-tolerant strains of *Rhizobium leguminosarum* biovar *viceae* specific to the lentil. Biol Fertil Soils. 1996;22:126–30.

21. Cheminingwa GN, Vessey JK. The abundance and efficacy of *Rhizobium leguminosarum* bv'. *viciae* in cultivated soils of the eastern Canadian prairie. Soil Biol Biochem. 2006;38:294–302.

22. Santalla M, Amurrio JM, Rodiño AP, de Ron AM. Variation in traits affecting nodulation of common bean under intercropping with maize and sole cropping. Euphytica. 2001;122:243–55.

23. Rai R. Effect of nitrogen levels and Rhizobium strains on symbiotic N_2 fixation and grain yield of Phaseolus vulgaris L. genotypes in normal and saline-sodic soils. Biol Fertil Soils. 1992;14:293–9.

24. da Silva PM, Tsai SM, Bonetti R. Response to inoculation and N fertilization for increased yield and biological nitrogen fixation of common bean (Phaseolus vulgaris L.). Plant Soil. 1993;152:123–30.

25. Tsai SM, DA Silva PM, Cabezas WL, Bonetti R. Variability in nitrogen fixation of common bean (Phaseolus vulgaris L) intercropped with maize. Plant Soil. 1993;152:93–101.

26. Argaw A: Symbiotic and phenotypic characterization of rhizobia nodulating common bean (Phaseolus vulgaris L.) from Eastern Ethiopia. MSC thesis, Addis Ababa University, Addis Ababa, Ethiopia; 2007.

27. Brockwell J. Accuracy of a plant-infection technique for counting populations of Rhizobium trifolii. Appl Microbiol. 1963;2:377–83.

28. Vincent JM: A manual for the practical study of root-nodule bacteria. (IBP handbook no 15) Blackwell, Oxford; 1970.

29. Bremner JM: Inorganic forms of nitrogen. In: Methods of soil analysis. Edited by Black C.A. et al. Part 2, 2nd ed., agron. Monogr. 9. ASA and SSSA, Madison, WI; 1965.

30. Howieson JG, Ballard RA. Optimising the symbiosis in stressful and competitive environments in southern Australia—some contemporary thoughts. Soil Biol Biochem. 2004;36:1261–73.

31. Vargas MAT, Mendes IC, Hungria M. Response of field-grown bean (Phaseolus vulgaris L.) to Rhizobium inoculation and nitrogen fertilization in two Cerrados soils. Biol Fertil Soils. 2000;32:228–33.

32. Vlassak K, Vanderleyden J, Franco AA. Competition and persistence of Rhizobium etli in tropical soil during successive bean (Phaseolus vulgaris L.) cultures. Biol Fertil Soils. 1996;21:61–8.

33. Leidi EO, Rodriguez Navarro DN. Nitrogen and phosphorus availability limit N fixation in bean. New Phytol. 2000;147:337–46.

34. Li YY, Yu CB, Cheng X, Li CJ, Sun HJ, Zhang FS, Lambers H, Li L. Intercropping alleviates the inhibitory effect of N fertilization on nodulation and symbiotic N fixation of faba bean. Plant Soil. 2009;323:295–308.

35. Eaglesham AR, Hassouna S, Seegers R. Nitrogen fixation by cowpea and soybean. Agron J. 1983;75:61–6.

36. Msumali GP, Kipe-Nolt JA. The usefulness of bean nodulation data relative to other symbiotic parameters in judging success of Rhizobium inoculation. Biol Agric Hortic. 2002;19:355–64.

37. Patrick HN, Lowther WL. Influence of the number of rhizobia on the nodulation and establishment of Trifolium ambiguum. Soil Biol Biochem. 1995;27:717–20.

38. Mothapo NV, Grossman JM, Sooksa-nguan T, Maul J, Bräuer SL, Shi W. Cropping history affects nodulation and symbiotic efficiency of distinct hairy vetch (Vicia villosa Roth.) genotypes with resident soil rhizobia. Biol Fertil Soils. 2013;49:871–9.

39. Ballard RA, Charman N, McInnes A, Davidson JA. Size, symbiotic effectiveness and genetic diversity of field pea rhizobia (Rhizobium leguminosarum bv. viciae) populations in South Australian soils. Soil Biol Biochem. 2004;36:1347–55.

40. Denton MD, Pearce DJ, Peoples MB. Nitrogen contributions from faba bean (Vicia faba L.) reliant on soil rhizobia or inoculation. Plant Soil. 2013;365:363–74.

41. Sorwli FK, Mytton LR. Nitrogen limitations to field bean productivity: a comparison of combined nitrogen applications with Rhizobium inoculation. Plant Soil. 1986;94:267–75.

42. Deshwal VK, Dubey RC, Maheshwari DK. Isolation of plant growth-promoting Bradyrhizobium (Arachis) sp. with biocontrol potential against Macrophomina phaseolina causing charcoal rot of peanut. Curr Sci. 2003;84:443–8.

43. Mazen MM, El-Batanony NH, Abd El-Monium MM, Massoud ON. Cultural filtrate of Rhizobium spp. and arbuscular mycorrhiza are potential biological control agents against root rot fungal diseases of faba bean. Glob J Biotechnol Biochem. 2008;3:32–41.

44. Mulas D, García-Fraile P, Carro L, Ramírez-Bahena MH, Velázquez E, González-Andrés F. Distribution and efficiency of Rhizobium leguminosarum strains nodulating Phaseolus vulgaris in Northern Spanish soils: selection of native strains that replace conventional N fertilization. Soil Biol Biochem. 2011;43:2283–93.

45. Ruiz-Díez B, Fajardo S, Fernández-Pascual M. Selection of rhizobia from agronomic legumes grown in semiarid soils to be employed as bioinoculants. Agron J. 2012;104:550–9.

46. El Hardi EA, Elsheikh EAE. Effect of Rhizobium inoculation and nitrogen fertilization on yield and protein content of six chickpea (Cicer arietinum L.) cultivars in marginal soils under irrigation. Nutr Cycl Agroecosys. 1999;54:57–63.

47. da Silveira PM, Braz AJBP, Kliemann HJ, et al. Nitrogen fertilization of common bean grown under no-tillage system after several cover crops. Pesqui Agropecu Bras. 2005;40:377–81.

48. Furseth BJ, Conley SP, Ané JM. Soybean response to soil rhizobia and seed-applied rhizobia inoculants in Wisconsin. Crop Sci. 2012;52:339–44.

49. Herridge DF. Contributions of fixed nitrogen and soil nitrate to the nitrogen economy of irrigated soybean. Soil Biol Biochem. 1988;20:711–7.

50. Fesenko AN, Provorov NA, Orlova IE, Orlov VP, Simarov BV. Selection of Rhizobium leguminosarum by. viceae strains for inoculation of Pisum sativum L. cultivars: analysis of symbiotic efficiency and nodulation competitiveness. Plant Soil. 1995;172:189–98.

51. Voisin A, Salon C, Munier-Jolain NG, Ney B. Quantitative effects of soil nitrate, growth potential and phenology on symbiotic nitrogen fixation of pea (Pisum sativum L.). Plant Soil. 2002;243:31–42.

52. Sanginga N, Thottappilly G, Dashiell K. Effectiveness of rhizobia nodulating recent promiscuous soyabean selections in the moist savanna of Nigeria. Soil Biol Biochem. 2000;32:127–33.

53. Gan YT, Warkentin TD, McDonald CL, Zentner RP, Vandenberg A. Seed yield and yield stability of chickpea in response to cropping systems and soil fertility in Northern Latitudes. Agron J. 2009;101:1113–22.

54. Hungria M, Andrade DS, Chueire LMO, Probanza A, GuttierrezMaero FJ, Megias M. Isolation and characterization of new efficient and competitive bean (Phaseolus vulgaris L.) rhizobia from Brazil. Soil Biol Biochem. 2000;32:1515–28.

55. Herridge DF. Inoculation technology for legumes. In: Dilworth MJ, James EK, Sprent JI, Newton WE (eds). Leguminous nitrogen-fixing symbioses. Dordrecht: Kluwer; 2008.

56. Thies JE, Singleton PW, Bohlool B. Influence of the size of indigenous rhizobial populations on establishment and symbiotic performance of introduced rhizobia on field grown legumes. Appl Environ Microb. 1991;57:19–28.

57. Sanginga N, Wirkom LE, Okogun A, Akobundu IO, Carsky RJ, Tian G. Nodulation and estimation of symbiotic nitrogen fixation by herbaceous and shrub legumes in Guinea savanna in Nigeria. Biol Fertil Soils. 1996;23:441–8.

58. Wortmann CS. Nutrient dynamics in a climbing bean and sorghum crop rotation in the Central Africa highlands. Nutr Cycl Agroecosys. 2001;61:267–72.

59. Shutsrirung A, Sutigoolabud P, Santasup C, Senoo K, Tajima S, Hisamatsu M, Bhromsiri A. Symbiotic efficiency and compatibility of native rhizobia in northern Thailand with different soybean cultivars. Soil Sci Plant Nutr. 2002;48:491–9.

60. Nleya T, Walley FL, Vandenberg A. Response of determinate and indeterminate common bean genotypes to rhizobium inoculant in a short season rainfed production system in the canadian prairie. J Plant Nutr. 2009;32:44–57.

61. LA Favre AK, Eaglesham ARJ. The effects of a high level of N, applied at planting, on nodulation of soybean (Glycine max (L.) Merr.) by diverse strains of Bradyrhizobium. Plant Soil. 1987;102:267–70.

62. Aggarwal VD. An overview of SADC/CIAT bean activities in Malawi. Bean Res. 1994;8:15–31.

63. Pereira PAA, Bliss FA. Selection of common bean (Phaseolus vulgaris L.) for N_2 fixation at different levels of available phosphorus under field and environmentally-controlled conditions. Plant Soil. 1989;115:75–82.

64. Bayuelo-Jiménez JS, Peña-Valdivia CB, Rogelio Aguirre RJ. Yield components of samples of two wild Mexican common bean (Phaseolus vulgaris L.) populations grown under cultivation. S Afr J Plant Soil. 1999;16:197–203.

4

Hygienic characteristics of radishes grown in soil contaminated with *Stenotrophomonas maltophilia*

Alessandro Miceli, Alessandra Martorana, Giancarlo Moschetti and Luca Settanni[*]

Abstract

Background: *Stenotrophomonas maltophilia* is a plant growth-promoter. This bacterium is also implicated in human diseases. Thus, after the use of this bacterium in agriculture, the safety of the final products has to be verified. Due to the ubiquitous presence of *S. maltophilia* in soil, in this study a massive contamination was simulated to evaluate the growth and safety of *Raphanus sativus* L..

Results: Different inoculums and soil treatment conditions were tested. Soils were analysed weekly and the radishes at harvest for their microbial loads and presence/persistence of *S. maltophilia* LMG 6606. The concentration of the bacterium added in the different trials decreased during the first week, but increased thereafter and determined a significant increase of growth parameters of radishes.

Conclusions: The addition of *S. maltophilia* LMG 6606 to non-autoclaved soil enhanced the productivity of radishes. The bacterium did not internalize in the hypocotyls, but colonized the external surface ensuring the safety of the products. Thus, a sanitizing bath of hypocotyls before consumption is necessary.

Keywords: Hygienic safety, Microbial internalization, Plant growth, *Raphanus sativus* L., *Stenotrophomonas maltophilia*

Background

Soil fertility is a complex concept that involves many interacting parameters. Cultivated plants may suffer nutritional stresses when the amount or availability of soil nutrients is lower than that required for sustaining metabolic processes in each growth stage [1]. Thus, restoring of nutrients and enhancing their availability by improving soil characteristics and efficiency of plants, are the main objectives of the modern agriculture. Due to the increasing sensitivity to environmental and economic issues, researchers and consumers are more and more aware of the impact of agriculture on the environment. Lowering the use of chemical inputs and the search for alternative ways to improve a more sustainable agriculture is a current challenge [2].

Sustainable agricultural production systems may be obtained by enhancing the uptake efficiency of nutrient of plants, that might be achieved through a better comprehension of the role of plant–microbe–soil interaction and the association of soil microorganisms with roots in the rhizosphere [3]. Within soil microbiota, plant growth-promoting (PGP) microorganisms constitute a heterogeneous group of bacteria and fungi that have gained particular importance for their stimulating effects. They are able to promote directly plant metabolism by nutritional and/or hormonal ways [4–9] as well as indirectly through the production of antimicrobial compounds, the reduction of iron available to phytopathogens, the synthesis of fungal cell wall-lysing enzymes, the competition with detrimental microorganisms for colonization sites on the roots, and the induced systemic resistance [5, 10].

The most studied PGP rhizobacteria belong to Gram-negative genera [11]. Among these, *Stenotrophomonas maltophilia* is an important species and represents one of the four dominant bacterial species in the rhizosphere of European cereal cultivations [12]. *S. maltophilia* is a Gammaproteobacterium distributed worldwide [13] and

---

*Correspondence: luca.settanni@unipa.it
Department of Agricultural and Forest Science, University of Palermo,
Viale delle Scienze 4, 90128 Palermo, Italy

typically found in soil where it plays a defining role in the nitrogen and sulphur cycles [14–16] and it often dominates the microbial communities detected outside and inside the plants [13]. However, *S. maltophilia* is a known human pathogen [17]. It may be responsible for respiratory tract infections, bacteremia, biliary sepsis, infections of the bones and joints, urinary tract and soft tissues, endophthalmitis, eye infections, endocarditis and meningitis [18].

In the last years, the change in lifestyle and the consequent need of consuming ready-to-eat foods determined the increase of fresh-cut vegetable use. Since these products do not undergo any treatment before consumption, their hygienic safety is of paramount importance. Among fresh vegetables, the request of radish is on the increase [19] due to its positive effect on the consumer's health [20–22].

Radish (*Raphanus sativus* L.) is an important vegetable of the *Brassicaceae* family, grown and consumed all over the world, due to its wide adaptation, high yield, and high nutritional content. It is normally consumed in salads and comes in a variety of forms and skin colors. Most known varieties are round and red-skinned. The most popular part for eating is the napiform hypocotyl, although the entire plant is edible and the tops can be used as a leaf vegetable. Radishes are eaten raw or cooked or processed by pickling, canning or drying [23]. Radish is not only a vegetable crop but also an important source of medicinal compounds. In fact, it is used by people with different gastrointestinal, biliary, hepatic, urinary and respiratory disorders, and in cardiovascular diseases such as hypertension [24]. An important quality characteristic of radish is the total antioxidant activity due to molecules, such as ascorbic acid and phenol, mainly phenolic acids, with free radical scavenging activity [25, 26]. The protective role of these molecules in the prevention of human degenerative diseases has been widely demonstrated by many studies [27–29]. Moreover, as other Crucifers, radish contains many other compounds as glucosinolates that are associated with cancer protection [30].

The interaction between *S. maltophilia* and *R. sativus* has been recently approached in a floating cultivation system [31]. However, their reciprocal effects in soil and the safety aspects related to the safety of the radishes have not been evaluated yet. In the present study high levels of *S. maltophilia* were added to soil, simulating a massive environmental contamination, to: investigate the survival of *S. maltophilia* in soil during the whole crop cycle of *R. sativus*; evaluate the effects of this bacterium on plants; monitor its transfer to the radishes; and determine their viability in the radishes ready for consumption.

Methods

Microbial strain, plant seeds and experimental plan

Stenotrophomonas maltophilia LMG 6606, a strain originating from rhyzosphere as reported in the strain details provided by the Belgian Co-ordinated Collection of Micro-organisms (BCCM/LMG), was propagated in Nutrient Broth (NB) (Oxoid, Milan, Italy) at 28 °C for 24 h. Seeds of radish (*R. sativus* L.) cultivar Saxa three were purchased from Blumen (Piacenza, Italy).

The experimental plan included eight different conditions for the growth of radish plants: ASS, autoclaved soil inoculated with *S. maltophilia* LMG 6606; AS, autoclaved soil added with Ringer's solution (Sigma-Aldrich, Milan, Italy); nASS, non-autoclaved soil inoculated with *S. maltophilia* LMG 6606; nAS, non-autoclaved soil added with Ringer's solution; ASSwS, autoclaved soil weekly inoculated with *S. maltophilia* LMG 6606; ASwR, autoclaved soil weekly added with Ringer's solution; nASSwS, non-autoclaved soil weekly inoculated with *S. maltophilia* LMG 6606; nASwR, non-autoclaved soil weekly added with Ringer's solution. Four replicate pots were produced for each trial. Two independent experiments were performed in two consecutive weeks.

Inoculation of soil and plant development

The pots used in this work were 13.5 cm × 13.5 cm × 16 cm and were filled with 2.5 L of commercial soil SER CA-V7 (Vigorplant Italia srl, Piacenza, Italy). This soil is a mixture of slightly or fully decomposed raised bog peat (pH 6.0) fertilized with 800 g m^{-3} of a mineral fertilizer (NPK 12-11-18). The pots were previously treated with a NaClO solution (5 % *v/v*) for 24 h. A part of the bulk soil, placed in autoclave bags, was autoclaved twice (in two consecutive days) for 70 min at 120 °C [32]. The fresh inoculums of *S. maltophilia* LMG 6606 was prepared after overnight development as reported by Settanni et al. [33]. The cell suspension was added to the autoclaved and non-autoclaved soil in the ratio 1:10 (*v/v*), vigorously mixed with sterile spoons to obtain the homogenous distribution of the bacterial inocula and transferred into the pots (trials ASS, nASS, ASSwS and nASSwS). The trials AS, nAS, ASwR and nASwR were prepared with Ringer's solution. Seeds of radish were sown in five dibblings for each pot. Pots were watered from below (sub-irrigation) with sterile water, in order to avoid further microbial contamination to the soil and kept at 25 °C in a climatic chamber till seed germination. From the third day, the trials ASSwS and nASSwS were weekly added with 300 mL of *S. maltophilia* LMG 6606 cell suspension (concentration at about 10^8 CFU mL^{-1}), while the trials ASwR and nASwR were added with the same volume of Ringer's solution.

After seed emergence, only five plants per pot were left to grow (274 plants m^{-2}). The pots were transferred in an unheated plastic greenhouse and received the same volume of water that varied daily according to environmental conditions and plant needs until harvesting.

Microbiological analyses

The sampling for microbiological analysis included soil during plant growth, aseptically collected as described by Settanni et al. [33] at T_0 and at 7-day intervals, and hypocotyls at harvest, aseptically collected as reported by Settanni et al. [31]: four hypocotyls were collected from each replicate of all trials, two hypocotyls for the direct microbial count and two for the internal *S. maltophilia* LMG 6606 detection.

Soil samples (10 g) were diluted (1:10) with sodium pyrophosphate (0.16 % *w/v*) solution in sterile flasks under agitation (10 min at 150 rpm). Radishes (approximately 10 g) were first subjected to the removal of the soil adhering to the surface as described by Brandl et al. [34] and then homogenised in Ringer's solution by a stomacher (BagMixer® 400, Interscience, Saint Nom, France) at the maximum speed for 2 min. The decimal serial dilutions of both soil and radishes continued in Ringer's solution. Total mesophilic count (TMC) were determined on Plate Count Agar (PCA) (Oxoid), incubated aerobically at 30 ˚C for 72 h; presumptive *S. maltophilia* were enumerated on vancomycin–imipenem–amphotericin B (VIA) agar [35], incubated aerobically at 30 ˚C for 48 h. Plate counts were performed in duplicate.

Data from bacterial counts were averaged and converted to log CFU g^{-1} dry weight (dw) for soil samples and to log CFU g^{-1} for radishes. Moisture of soil (5 g) was obtained after drying (24 h at 105 ± 1 ˚C) in an oven and weighting the residual.

Recognition of *S. maltophilia* LMG 6606 and evaluation of the internalization

Approximately ten colonies from the two highest dilutions of sample suspensions at each analysis were picked up from VIA agar plates based on their morphology (colour, edge, surface and elevation) and cultured in NB overnight at 30 ˚C. The cultures were sub-cultivated onto NA and stored in glycerol (20 %, *v/v*) stocks at –80 ˚C. The isolates and *S. maltophilia* LMG 6606 were analysed by randomly amplified polymorphic DNA-PCR (RAPD-PCR) as described by Settanni et al. [31].

The internal presence of *S. maltophilia* LMG 6606 in radishes was investigated as reported by Settanni et al. [31]. Briefly, radishes of each trials were collected, superficially sterilized and then transferred in stomacher bags, added with VIA broth (final ratio 1:10) and homogenized as reported above.

Analyses of plants

Plants were harvested 4 weeks after sowing and washed accurately with tap water. After air drying, leaves, hypocotyls and roots were separated and the number of leaves and their area, the root elongation and the radish diameter were recorded on four plants for each replicate of each trial. Leaf area of each plant was calculated by digital image analysis. Leaves were scanned (Epson Perfection 4180 Photo, Seiko Epson Corp. Japan) with 350 dpi of resolution and the images were saved in TIFF format. The images were analysed with the ImageJ 1.46r software (National Institutes Health, Bethesda, MD, USA).

Fresh and dry biomass of the different plant parts were calculated by weighting before and after they were oven dried to a constant weight at 80 ˚C.

Color of radishes was measured on two points of four hypocotyls from each replicates of all trials, using a colorimeter (Chroma Meter CR-400C, Minolta, Osaka, Japan). The Hunter scale parameters were determined: *L* (lightness, ranging from 0 to 100, from black to white), *a* (positive values indicating redness and negative values, greenness) and *b* (positive values indicate yellowness and negative values, blueness).

The firmness of radishes was determined using a digital penetrometer (mod. 53205, TR Snc. Italy) equipped with a flat 6 mm diameter stainless steel cylinder probe. Four hypocotyls were punched and the mean peak force was calculated in Newton.

Radishes (50 g) were homogenized in 50 mL of distilled water and homogenate centrifuged (5000 rpm, 15 min); the supernatant was taken for analysis of soluble solids content (SSC), ascorbic acid and N–NO$_3^-$. SSC was measured using a digital refractometer (MTD-045nD, Three-In-One Enterprises Co. Ltd. Taiwan). Ascorbic acid and nitrate content were measured by the Reflectoquant test strips and a RQflex hand-held reflectometer (Merck, Darmstadt, Germany) [procedures described in Art. 1.16971.0001 and 1.16981.0001 by Merck (http://www.merckmillipore.com/chemicals)].

Water use efficiency (WUE) was calculated as the ratio between total dry weight of plant and total amount of water supplied.

Statistical analyses

The study was carried out in a completely randomized design. To determine the effects of microorganisms and time on soil microbial load, a two-way ANOVA was carried out. A one way ANOVA was performed for other data. When a significant F value was detected, Tukey–Kramer's multiple range test was used to determine differences among microbial populations and plant parameters of the different trials (significance level $P < 0.05$).

Principal components analysis was employed to investigate any underlying relationship among the different trials based on the agronomic and quality parameters of radish plants at harvest. The input matrix for the analysis consisted of leaf number, leaf fresh weight, leaf dry matter, leaf area, root length, root fresh weight, root dry matter, radish fresh weight, radish dry matter, radish diameter, firmness, L^*, a^*, b^*, SSC, nitrate, ascorbic acid, plant WUE. For the selection of the optimum number of principal components (PCs), factors with eigenvalues greater than 1.0 were retained. In addition the plot of the PCs enabled the investigation of correlations between the variables of the input data set. To this end, the initial variables were projected into the subspace defined by the reduced number of PCs (first and second components) and correlated variables were identified. In the current approach, the Principal components analysis was implemented with SPSS version 14.0 (SPSS Inc. Chicago, IL, USA).

Results

Microbiological analyses

The microbiological counts of soil during the growth of radishes are reported in Table 1. The interaction microorganisms x time resulted significant for both PCA and VIA media ($P < 0.001$). The autoclaved soil un-inoculated showed a TMC of 3.53 Log CFU gdw^{-1} and, surprisingly, the bacterial load estimated on VIA was at almost 2 Log CFU gdw^{-1}, whereas the non-autoclaved un-inoculated soil showed levels of ca. 7 Log CFU gdw^{-1} on both PCA and VIA media. Soon after inoculation, the concentration of TMC and *S. maltophilia* LMG 6606 or stenotrophomonads were above 10^8 CFU gdw^{-1}. In general, TMC was at higher levels than the microbial developments detected on VIA, but the trials nASS, nASSwS, nAS and nASwR showed an opposite behaviour at 7 day. During radish growth, the highest microbial concentrations were observed for the trial AS and ASwR, whose TMC, at the second week, were 9.58 and 9.57 Log CFU gdw^{-1}, respectively. Lower concentrations were estimated for the trials nASS, nASSwS, nAS and nASwR, for which TMC was below 8 Log CFU gdw^{-1} for the entire period of observation. The concentrations on VIA evaluated for the trials weekly added with *S. maltophilia* LMG 6606 (ASSwS and nASSwS) showed trends almost comparable to those displayed by the corresponding trials not subjected to the weekly additions (ASS and nASS), but the levels estimated for the second were slightly lower.

Table 1 Microbial counts (Log CFU gdw^{-1}) in soil as function of time and trials

Trials[A]	Days				
	0	7	14	21	28
PCA[B]					
AS	3.53 ± 0.37i	7.61 ± 0.66eh	9.58 ± 0.24a	9.16 ± 0.11ab	8.98 ± 0.28ac
ASS	8.21 ± 0.25cf	8.06 ± 0.39dg	8.96 ± 0.30ac	9.03 ± 0.07ac	9.10 ± 0.37ab
ASSwS	8.21 ± 0.25cf	9.11 ± 0.15ab	8.74 ± 0.20ad	9.20 ± 0.31ab	8.85 ± 0.34ad
ASwR	3.53 ± 0.37i	7.29 ± 0.60gh	9.57 ± 0.21a	9.17 ± 0.11ab	9.06 ± 0.22ab
nAS	6.88 ± 0.20h	7.24 ± 0.08gh	7.11 ± 0.15h	7.41 ± 0.35fh	7.65 ± 0.15eh
nASS	8.45 ± 0.34be	7.26 ± 0.10gh	7.52 ± 0.08fh	7.35 ± 0.17gh	7.52 ± 0.18fh
nASSwS	8.45 ± 0.34be	7.30 ± 0.11gh	7.50 ± 0.15fh	7.48 ± 0.08fh	7.57 ± 0.13fh
nASwR	6.88 ± 0.20h	7.04 ± 0.02h	7.09 ± 0.09h	7.25 ± 0.13gh	7.34 ± 0.18gh
VIA[B]					
AS	2.12 ± 0.10p	7.77 ± 0.09ci	8.67 ± 0.04ac	6.91 ± 0.18io	6.61 ± 0.7lo
ASS	8.28 ± 0.02ae	7.19 ± 0.48gm	8.15 ± 0.25af	7.67 ± 0.15dk	8.01 ± 0.44ag
ASSwS	8.28 ± 0.20ae	7.72 ± 0.21ci	8.81 ± 0.08ab	7.18 ± 0.15gn	7.88 ± 0.32bh
ASwR	2.12 ± 0.35p	7.03 ± 0.56ho	8.85 ± 0.30a	7.25 ± 0.29fl	6.71 ± 0.14ko
nAS	6.85 ± 0.10io	7.17 ± 0.08go	6.90 ± 0.04io	6.23 ± 0.26no	6.90 ± 0.36io
nASS	8.35 ± 0.02ad	7.76 ± 0.24ci	7.49 ± 0.23dl	6.22 ± 0.29o	6.68 ± 0.09lo
nASSwS	8.35 ± 0.20ad	7.49 ± 0.11dl	7.35 ± 0.42el	6.97 ± 0.27ho	7.05 ± 0.12ho
nASwR	6.85 ± 0.35io	7.47 ± 0.11dl	7.07 ± 0.32go	6.60 ± 0.36lo	6.27 ± 0.08mo

Data represent the mean of four replicates of two independent experiments. For each media, data followed by the same letter are not significantly different according to Tukey–Kramer's multiple range test at $P < 0.05$

[A] Trials: *ASS* autoclaved soil inoculated with *S. maltophilia* LMG 6606, *AS* autoclaved soil, *nASS* non-autoclaved soil inoculated with *S. maltophilia* LMG 6606, *nAS* non-autoclaved soil, *ASSwS* autoclaved soil weekly inoculated with *S. maltophilia* LMG 6606, *ASwR* autoclaved soil weekly added with Ringer's solution, *nASSwS* non-autoclaved soil weekly inoculated with *S. maltophilia* LMG 6606, *nASwR* non-autoclaved soil weekly added with Ringer's solution

[B] Media: *PCA* plate count agar, *VIA* vancomycin–imipenem–amphotericin B agar

The trials not inoculated with *S. maltophilia* LMG 6606 (AS and nAS) showed levels of count on VIA particularly high. At harvest, the hypocotyls were analysed for TMC and stenotrophomonad concentrations and no statistical significant differences were registered for the microbial concentrations of the radishes cultivated in the eight different conditions (results not shown). TMC were in the range 7.07–7.63 Log CFU g^{-1}, while the counts detected on VIA were at least 1 Log cycle lower for each trial followed.

Monitoring of *S. maltophilia* LMG 6606

All ten presumptive stenotrophomonad isolates collected from VIA medium at the highest dilutions of soil samples were characterized at strain level by RAPD-PCR analysis. DNA from the pure culture of *S. maltophilia* LMG 6606 was used for strain recognition. The comparison of the polymorphic profiles is shown in Fig. 1. *S. maltophilia* LMG 6606 dominated the microbial community found on VIA medium for the trials ASS, nASS, ASSwS and nASSwS until the 28th day of experimentation. Despite the high counts detected on VIA for the trials AS, nAS, ASwR and nASwR, no colony showed a RAPD profile superimposable to that of *S. maltophilia* LMG 6606 excluding, at least at the highest dilutions of soil samples, a cross-contamination among inoculated and un-inoculated trials. One main RAPD pattern was recognised for the stenotrophomonad isolates from the un-inoculated autoclaved soil (AS and ASwR) trials and another main profile for the isolates from the un-inoculated non-autoclaved soil (nAS and nASwR) trials.

The same procedure was applied on the stenotrophomonad isolates collected from VIA medium at the highest dilutions of radishes and it produced the same results (not shown) registered for soil: *S. maltophilia* LMG 6606 dominated the microbial community of radishes for the trials ASS, nASS, ASSwS and nASSwS, whereas no colonies collected from the trials AS, nAS, ASwR and nASwR shared the same *S. maltophilia* LMG 6606 RAPD profile.

The last data could not indicate whether the bacterium added to soil was adherent to the radish surface or in the inner part. Thus, at harvest, the radishes were also specifically investigated for the internal presence of *S. maltophilia* LMG 6606. The enrichment cultures obtained in VIA broth after incubation of the homogenized surface sterilized hypocotyls were streaked onto the corresponding agar medium. No colonies developed for the trials AS, nASS, nAS, ASwR and nASwR (results not shown) indicating that no stenotrophomonads internalized in these conditions. On the contrary, a development was observed for the trials ASS, ASSwS and nASSwS. These colonies, characterized by the same appearance (colour, morphology, edge, surface and elevation), were randomly collected and subjected to the RAPD analysis as reported above; all cultures shared the same profile of *S. maltophilia* LMG 6606 (results not shown) demonstrating its internalization. Interestingly, the only inoculated trial that did not show the internal presence of *S. maltophilia* LMG 6606 in radishes, at harvest, was nASS. To confirm the last data, the enrichment cultures from nASS were also streaked onto Nutrient Agar. Seventy-nine colonies were isolated and analysed by RAPD-PCR (results not shown), but none of them shared the same profile of *S. maltophilia* LMG 6606.

Plant growth

Seed germination and seedling emergence occurred after 3 days from sowing in non-autoclaved soil and 1 day later in autoclaved soil. Inoculation of soil with *S. maltophilia* LMG 6606 had no effect on seed germination.

During the cultivation period, average maximum and minimum temperatures inside the greenhouse ranged between 29.5 and 18.4 °C, respectively. Soil temperature did not greatly differ from air temperature (29.2 and 17.8 °C, respectively).

Plant growth was significantly influenced by soil treatment (Table 2). After 28 days from sowing, some differences were registered for the development of plants. The above-ground part showed only little differences in the number of leaves plant^{-1} among the eight trials, but leaf fresh weight of nASS and nASSwS was higher (14.4 and 17.6 g plant^{-1}, respectively) than that found for the other trials (10.5 g plant^{-1} on average). No differences were found in dry matter percentage as function of treatments, while leaf dimensions were influenced by soil autoclaving as well as by inoculation with *S. maltophilia* LMG 6606. The lowest leaf area was recorded for the plants of trials AS and ASwR (141.5 cm^2 on average) that differed significantly from that of the plants of trials ASS, ASSwS, nAS and nASwR (193.3 cm^2 on average). The non-autoclaved

(See figure on next page.)

Fig. 1 Monitoring of *S. maltophilia* LMG 6606 during the growth cycle of radish performed by RAPD-PCR profile comparison. Trials: *ASS* autoclaved soil inoculated with *S. maltophilia* LMG 6606, *ASSwS* autoclaved soil weekly inoculated with *S. maltophilia* LMG 6606, *nASS* non-autoclaved soil inoculated with *S. maltophilia* LMG 6606, *nASSwS* non-autoclaved soil weekly inoculated with *S. maltophilia* LMG 6606, *AS* autoclaved soil, *nAS* non-autoclaved soil, *ASwR* autoclaved soil weekly added with Ringer's solution, *nASwR* non-autoclaved soil weekly added with Ringer's solution. *Lanes: M* marker (GeneRuler 100 bp Plus DNA ladder, M·Medical Srl, Milan, Italy); *1*, *S. maltophilia* LMG 6606; *2–11*, colonies randomly collected from the highest dilutions of soil samples on VIA agar from each trial

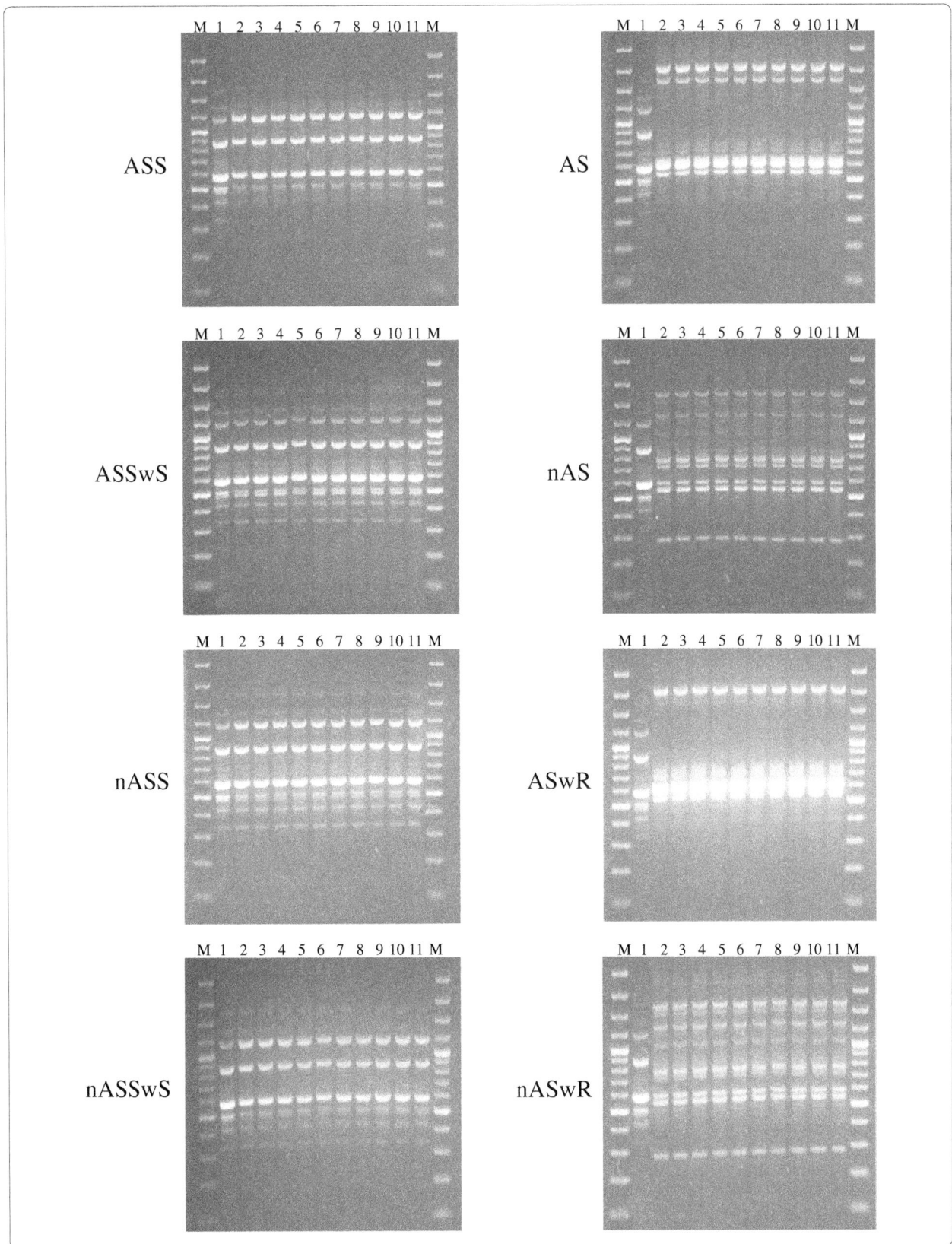

Table 2 Agronomic and quality parameters of radish plants at harvest

Parameters	Trials							
	ASS	ASSwS	AS	ASwR	nASS	nASSwS	nAS	nASwR
Leaves								
Number (n. plant^{-1})	5.0 ± 0.3a	4.8 ± 0.2a	4.8 ± 0.2a	4.7 ± 0.2a	4.6 ± 0.2ab	4.6 ± 0.2ab	4.5 ± 0.3ab	4.1 ± 0.2b
Fresh weight (g)	10.5 ± 0.6c	10.7 ± 0.5c	10.3 ± 0.4c	10.8 ± 0.5c	14.4 ± 0.8b	17.6 ± 0.9a	10.3 ± 0.5c	10.3 ± 0.6c
Dry matter (%)	6.5 ± 0.3a	7.1 ± 0.4a	7.1 ± 0.5a	7.1 ± 0.3a	7.3 ± 0.4a	7.2 ± 0.4a	6.8 ± 0.3a	6.7 ± 0.4a
Leaf area (cm^2)	186.4 ± 8.7b	196.4 ± 9.3b	138.3 ± 6.9c	144.8 ± 7.2c	297.0 ± 16.9a	311.5 ± 14.6a	198.9 ± 9.7b	191.6 ± 9.8b
Roots								
Length (cm)	31.5 ± 1.6ab	33.6 ± 1.8a	28.1 ± 1.4ab	29.1 ± 1.5ac	26.3 ± 1.5bc	23.9 ± 1.1c	27.6 ± 1.4ac	27.8 ± 1.6ac
Fresh weight (g)	1.06 ± 0.05a	1.04 ± 0.04a	0.91 ± 0.05a	0.8 ± 0.04a	0.49 ± 0.02b	0.50 ± 0.03b	0.53 ± 0.04b	0.40 ± 0.03b
Dry matter (%)	6.2 ± 0.3b	6.2 ± 0.4b	7.2 ± 0.4b	7.3 ± 0.4b	7.7 ± 0.4b	8.1 ± 0.5b	12.8 ± 0.6a	11.7 ± 0.5a
Radish								
Fresh weight (g)	12.2 ± 0.7b	10.0 ± 0.6b	10.1 ± 0.5b	12.0 ± 0.6b	25.0 ± 1.4a	23.9 ± 1.2a	15.3 ± 0.8b	14.8 ± 0.9b
Dry matter (%)	3.9 ± 0.2a	3.9 ± 0.3a	3.9 ± 0.4a	3.8 ± 0.2a	4.1 ± 0.3a	4.3 ± 0.5a	4.3 ± 0.2a	4.5 ± 0.4a
Diameter (mm)	24.9 ± 1.2b	23.4 ± 1.2b	23.1 ± 1.3b	24.6 ± 1.2b	34.0 ± 1.7a	31.8 ± 1.6a	25.5 ± 1.3b	26.6 ± 1.5b
Firmness (N)	39.1 ± 2.1a	38.6 ± .0a	36.2 ± 1.8a	37.1 ± 1.9a	30.3 ± 1.7b	29.9 ± 1.5b	31.2 ± 1.6b	30.2 ± 1.5b
L*	32.9 ± 1.7c	34.0 ± 1.6c	35.2 ± 1.8ac	34.2 ± 1.7bc	36.8 ± 1.9ac	38.7 ± 1.8ab	39.6 ± 2.0a	39.0 ± 2.0a
a*	45.8 ± 2.3ab	46.0 ± 2.4ab	42.5 ± 2.1b	43.5 ± 2.2b	47.9 ± 2.6a	50.3 ± 2.5a	49.4 ± 2.9a	49.4 ± 2.5a
b*	20.6 ± 1.0ab	20.7 ± 1.2ab	19.2 ± 1.0b	18.8 ± 0.9b	21.0 ± 1.1ab	23.2 ± 1.2ab	25.1 ± 1.3a	25.0 ± 1.4a
SSC (Brix)	2.2 ± 0.1b	2.2 ± 0.1b	2.8 ± 0.1a	2.6 ± 0.2ab	2.3 ± 0.1b	2.3 ± 0.2b	2.3 ± 0.1b	2.3 ± 0.1b
N-NO$_3^-$ (mg kg^{-1} FW)	644.6 ± 33.2a	580.3 ± 29.8a	583.7 ± 29.2a	563.7 ± 28.2a	522.7 ± 25.1a	543.4 ± 27.8a	567.0 ± 28.4a	599.2 ± 30.2a
Ascorbic acid (mg kg^{-1} FW)	205.0 ± 10.3ab	197.0 ± 9.9ab	220.0 ± 11.0a	206.0 ± 10.3ab	182.0 ± 9.1ab	166.0 ± 8.3ab	132.0 ± 6.6b	192.0 ± 9.6ab
Plant WUE (g DM kg^{-1} H$_2$O)	2.03 ± 0.11b	2.03 ± 0.10b	2.0 ± 0.11b	2.14 ± 0.12b	3.51 ± 0.18a	3.88 ± 0.16a	2.39 ± 0.12b	2.33 ± 0.13b

Data within a row followed by the same letter are not significantly different according to Tukey–Kramer's multiple range test at $P < 0.05$

Percentages were subjected to angular transformation prior to perform statistical analysis

soils inoculated with *S. maltophilia* LMG 6606, subjected or not to the weekly additions, determined the greatest total leaf area which was about 300 cm^2.

The roots of radish plants had a lower fresh weight in the trials with non-autoclaved soil than autoclaved soil trials, but dry matter content was higher only for nAS and nASwR. Root elongation was influenced by inoculation with *S. maltophilia* LMG 6606. However, this inoculums determined a decrease of root length in non-autoclaved soil trials, while an opposite behaviour was observed for the autoclaved soil trials.

After 28 day from sowing, radish hypocotyls reached commercial maturity. The highest average fresh weight was recorded in nASS and nASSwS (25.0 and 23.9 g, respectively), and was almost twice those reached by the hypocotyls of the other trials. A positive correlation was found between radish fresh weight and size. The biggest hypocotyls developed in non-autoclaved inoculated soils which overcame the diameter of 30 mm. No significant difference was found for dry matter percentage among the eight trials.

Colour modifications of radish were evaluated in terms of L^*, a^* and b^* values. Radishes grown in non-autoclaved soil had a darker colour especially for ASS and ASSwS trials ($L^* = 33.5$ on average), while trials nAS and nASwR determined the greatest values of L^* (39.6 and 39.0, respectively). The inoculation with *S. maltophilia* LMG 6606 did not influence neither the redness (a^*) nor the yellowness (b^*) of radish colour. The trials AS and ASwR were characterised by the lowest redness and differed significantly from non autoclaved trials. The only significant differences for yellowness were noticed between radishes from trials AS and ASwR (19.0 on average) and trials nAS and nASwR (25.0 on average).

Soil autoclaving determined an increase of radish firmness both in inoculated and un-inoculated trials (37.7 N on average), while a lower value was registered for those grown in non-autoclaved soils (30.4 N on average).

Soluble solid contents were almost comparable among the different trials, with the exception of AS that was significantly higher. Also nitrate and ascorbic acid content showed similar values among trials with no significant differences between inoculated and un-inoculated trials.

Stenotrophomonas maltophilia LMG 6606 increased significantly WUE for producing dry matter (DM) when inoculated in non-autoclaved soil trials: nASSwS and nASS ranged from 3.51 to 3.88 g DM kg^{-1} H$_2$O, respectively, against 2.15 g DM kg^{-1} H$_2$O on average for the other trials.

Multivariate data analysis

The results of the principal components analysis showed four principal components (PCs) with eigenvalues higher than 1.00 (Table 3), accounting for 56.5, 15.4, 7.7 and 6.34 % of the total variance, respectively. This indicated that the initial 18 variables could be expressed as a linear combination of four PCs explaining 85.9 % of the total variance. PC1 was mainly related to leaf fresh weight, leaf dry matter, leaf area, root length, root fresh weight, root dry matter, radish fresh weight, radish dry matter, diameter, firmness, ascorbic acid, plant WUE and color components (L^*, a^*, b^*); PC2 was related to leaf number and b^*; PC3 was related to SSC, and finally PC4 to nitrate content (Table 3). The projection of the original variables on the plane of the two first PCs could clearly illustrate such relationship as shown in the plot of loadings (Fig. 2a). The discrimination of the various trials can be visualized in the plot of scores (Fig. 2b) where three clusters could be clearly distinguished. The trials with autoclaved soil were close each other and located mainly in the negative part of *F1* axis; they were clearly separated from the trials with autoclaved soil that were located in the positive part of *F1* axis. Among these trials nASSwS and nASwR were clearly separated from nAS and nASS. The trial nASS also showed the greatest score for Factor 1.

Combining the information from the plot of loadings and scores, it can be inferred that nAS and especially nASS influenced positively fresh and dry matter of leaves and radishes, colour and firmness of hypocotyls and plant water use efficiency.

Table 3 Correlation of variables to the factors of the PCA analysis based on factor loadings

Variable	Factor 1	Factor 2	Factor 3	Factor 4
Leaf number	−0.236	*0.716*	0.242	−0.210
Leaf fresh weight	*0.809*	0.427	0.037	−0.065
Leaf dry matter	*0.781*	0.422	−0.070	0.082
Leaf area	*0.870*	0.310	0.072	0.182
Root length	*−0.773*	0.174	0.210	0.457
Root fresh weight	*−0.772*	0.452	0.206	0.017
Root dry matter	*−0.836*	−0.115	0.259	−0.137
Radish fresh weight	*0.940*	0.173	−0.075	−0.006
Radish dry matter	*0.956*	0.107	−0.063	−0.050
Radish diameter	*−0.841*	0.430	0.087	0.080
Firmness	*0.858*	0.242	−0.002	0.073
L^*	*0.794*	−0.518	0.088	−0.128
a^*	*0.771*	−0.398	0.328	0.027
b^*	0.506	*−0.726*	0.397	0.004
SSC	−0.282	−0.072	*−0.842*	−0.338
Nitrate	−0.020	−0.229	−0.392	*0.817*
Ascorbic acid	*−0.721*	−0.411	−0.079	−0.110
Plant WUE	*0.939*	0.283	−0.066	0.011

Values in italic within the same factor indicate the variable with the largest correlation

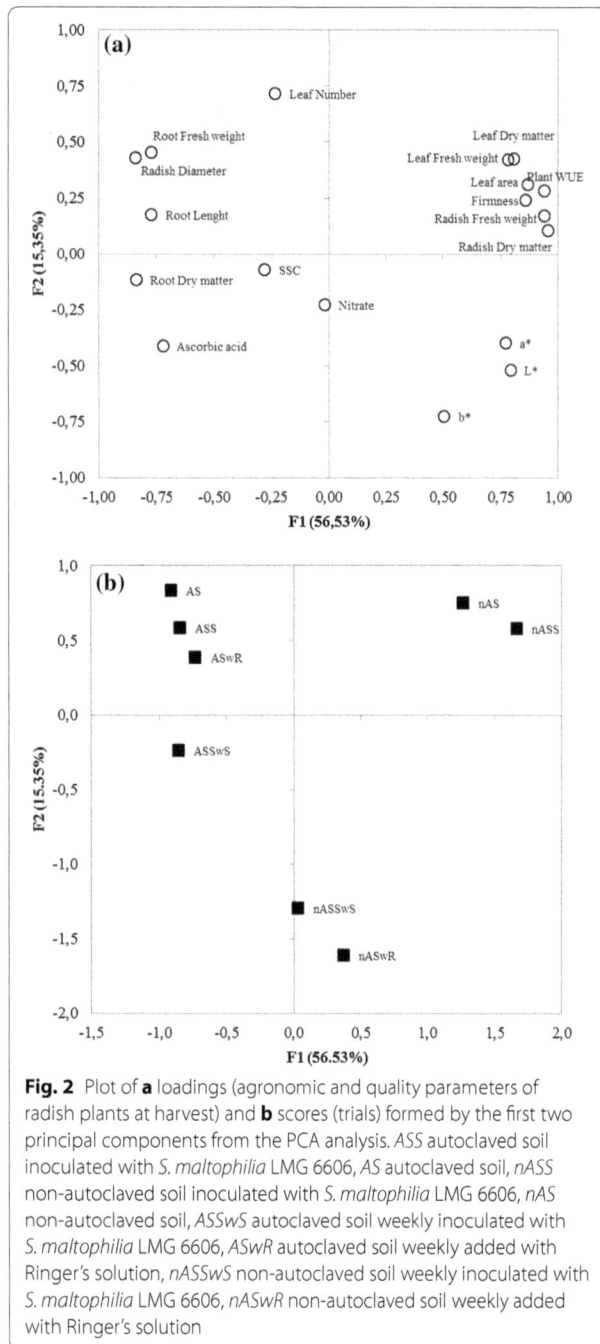

Fig. 2 Plot of **a** loadings (agronomic and quality parameters of radish plants at harvest) and **b** scores (trials) formed by the first two principal components from the PCA analysis. *ASS* autoclaved soil inoculated with *S. maltophilia* LMG 6606, *AS* autoclaved soil, *nASS* non-autoclaved soil inoculated with *S. maltophilia* LMG 6606, *nAS* non-autoclaved soil, *ASSwS* autoclaved soil weekly inoculated with *S. maltophilia* LMG 6606, *ASwR* autoclaved soil weekly added with Ringer's solution, *nASSwS* non-autoclaved soil weekly inoculated with *S. maltophilia* LMG 6606, *nASwR* non-autoclaved soil weekly added with Ringer's solution

Discussion

The positive effects of *S. maltophilia* on the plant growth has been reported for several species [36–39]. However, *S. maltophilia* is also reported to be a human pathogen. For this reason, its association with vegetables, especially when eaten raw, is detrimental for the final quality of these products. The aim of this work was to modify the microbial composition of soil with the addition of *S. maltophilia* LMG 6606 in order to simulate a massive

contamination and to assess its effects on the growth of *R. sativus*.

The experimentation was carried out with soil amended with inorganic fertilizer; organic amendments were not included to avoid the transfer of consistent concentrations of microorganisms [33] commonly present in compost or manure. The trials were carried out in a greenhouse during the spring season, with an average air temperature of 24.0 and 23.5 °C reached in the soil. In this conditions, *S. maltophilia* LMG 6606 found a temperature range compatible with its growth.

Stenotrophomonas maltophilia may increase the plant growth not only by direct production of growth promoting substances but also indirectly by interacting with the native root microflora [40]. Kwok et al. [41] reported that the biocontrol ability of *S. maltophilia* might be enhanced in presence/combination with other soil microorganisms. For these reasons, our experimental design included autoclaved and non-autoclaved soils, in order to exclude or include, respectively, the interaction between *S. maltophilia* LMG 6606 and the resident soil microbiota. Plate counts specific for the added bacterium were performed on VIA medium that, to our knowledge, is still considered to be highly selective for the isolation of *S. maltophilia* [35]. The non-autoclaved un-inoculated soils showed an initial load of 6.92 Log CFU gdw^{-1} on VIA. Thus, for the trials carried out with non-autoclaved un-inoculated soil (trials nAS and nASwR), these populations were referred to as stenotrophomonads. This bacterial group was detected also in the soil soon after autoclaving, although at very low levels (2.13 CFU gdw^{-1}). Thus, the common protocol applied to sterilize soil (two autoclaving cycles at 121 °C for 70 min at 24 h interval) was not enough, in this study, to destroy completely the bacterial component and, as a consequence, the microbial interaction could not be totally excluded in the autoclaved soil trials, at T_0.

The trials artificially contaminated with more than 10^8 CFU gdw^{-1} of *S. maltophilia* LMG 6606 showed a decrease of these inocula during the first week of radish growth, but their number increased thereafter. De Boer et al. [42] registered an opposite trend during the first week of observation of sands inoculated with stenotrophomonads, but the stimulating effect was imputed to the development of fungi. However, the different behaviour of this bacterial population might be due to the different conditions tested in the two works; the soil amended with inorganic fertilizer used in our work and the sand used by De Boer et al. [42] represented two distinct ecosystems, characterized by differences in pH, water activity and oxidation–reduction potential. Furthermore, the differences can also depend on the strain-specific characteristics. A continuous decrease of *S.*

maltophilia inoculated at 10^7 CFU gdw^{-1} in different soil types with different management regimes at 28 ℃ was reported by Messiha et al. [43] who studied the antagonistic effects of *S. maltophilia* against *Ralstonia solanacearum*. Those authors stated that the reasons for the differences in survival of *S. maltophilia* in the various soils were not clear, but supposed a direct effect of nitrate and ammonium contents and pH. The increase in concentration of *S. maltophilia* LMG 6606, observed in this study from the second week, might be due to the production of methionine by *R. sativus* roots. *S. maltophilia* requires methionine [44] and colonizes mainly the rhizosphere of cruciferous plants which produce high concentrations of sulphur-containing compounds [45].

The persistence of *S. maltophilia* LMG 6606 was monitored by plate count and polymorphic profile by RAPD analysis, after isolation. The isolates collected from VIA agar at the highest dilutions of soils were analyzed and compared with the pure strain. The direct comparison of RAPD patterns allowed to confirm that *S. maltophilia* LMG 6606 dominated the stenotrophomonad group of soil in all the inoculated trials during the entire growth cycle of radishes. This approach was previously successfully applied on the recognition and monitoring of other food/spoilage bacteria tested in similar experiments [31, 33].

In our study, the strain *S. maltophilia* LMG 6606 determined a significant increase of the values of several parameters (leaf area, fresh and dry matter yield, radish size and WUE) indicating the growth-promotion of radish, even though the plants grown in presence of *S. maltophilia* LMG 6606 behaved differently in autoclaved or non-autoclaved soil. Plants exhibited a significantly greater growth in the latter condition. This might be explained with the fact that soil treatments, especially steaming or autoclaving, can result in soil toxicity [46].

Dry matter yield of control trials was similar to those obtained by other authors [47]. The increase in yield was significant only for non-autoclaved soil and ranged from 47 to 67 % for nASS and nASSwS, respectively. The growth increase may vary greatly depending on crop and PGP rhizobacteria strains used [48]. Antoun et al. [47] reported an increase of 15 % in the dry matter yield of radish using the strain *Bradyrhizobium japonicum* Tal 629.

The results of the principal components analysis showed four principal components (PCs). Combining the information from the plot of loadings and scores, it can be concluded that soil autoclaving negatively influenced growth and quality parameters, thus their use need further investigation. Furthermore, principal components analysis was able to differentiate the trials with respect to

agronomic and quality trait of radish. Hence, soil autoclaving should be avoided, while the inoculation of non-autoclaved soil with *S. maltophilia* LMG 6606 only once before plant establishment was positively related with some growth parameter.

Radishes were microbiologically investigated at harvest. TMC and stenotrophomonad concentrations were not statistically different for all trials. Thus, inoculated trials produced radishes characterized by the same levels of microbial contamination of those obtained from control trials. The levels of TMC were in the range 10^6–10^7 CFU g^{-1} and were superimposable to those reported for radishes grown in hydroponic systems [31].

The isolates collected from the highest dilutions of hypocotyls confirmed that *S. maltophilia* LMG 6606 dominated the microbial community of radishes for all inoculated trials as assessed by RAPD analysis. However, to retrieve the exact location of this bacterium on the radishes, inside or outside the hypocotyls, they were surface sterilized and subjected to an enrichment in VIA broth. This procedure, developed specifically for radishes in a previous work [31], indicated that *S. maltophilia* LMG 6606 internalized in all inoculated trials except that carried out in non-autoclaved soil not subjected to the weekly addition (nASS).

Conclusions

The addition of *S. maltophilia* LMG 6606 to the non-autoclaved soil amended with inorganic fertilizer enhanced the productivity of radishes grown in greenhouse. *S. maltophilia* LMG 6606 colonized the external surface of the hypocotyls, but it did not internalize. Thus, a massive contamination of soil with *S. maltophilia* determines a growth promotion of *R. sativus*, but due to the hygienic implication of this bacterium, a sanitizing bath of hypocotyls before eating is mandatory. Works are being prepared to better evaluate the biocontrol activity of *S. maltophilia* in presence of other soil microorganisms in specific combinations.

Authors' contributions

AMi provided financial support of the experiment and conducted greenhouse experiment and agronomic and quality determinations, general conception and coordination of the experiments, interpretation of results and manuscript writing. AMa carried out the microbiological and molecular genetic analyses. GM provided general conception and interpretation of results and manuscript writing. LS provided financial support of the experiment, offered expert advice during the conduction of the assays, interpretation of results and manuscript writing. All authors read and approved the final manuscript.

Competing interests

The authors declare that they have no competing interests.

References

1. Rengel Z, Marschner P. Nutrient availability and management in the rhizosphere: exploiting genotypic differences. New Phytol. 2005;168:305–12.
2. Smit B, Pilifosova O, Burton I, Challenger B, Huq S, Klein RJT, Yohe G. Adaptation to climate change in the context of sustainable development and equity. In: McCarthy JJ, Canziani OF, Leary NA, Dokken DJ, White KS, editors. Climate change 2001: impacts, adaptation and vulnerability. Contribution of working group ii to the third assessment report of the intergovernmental panel on climate change. Cambridge: Cambridge University Press; 2001. p. 877–912.
3. Richardson AE, Barea JM, McNeil AM, Prigent-Combaret C. Acquisition of phosphorus and nitrogen in the rhizosphere and plant growth promotion by microorganisms. Plant Soil. 2009;321:305–39.
4. Biswas JC, Ladha JK, Dazzo FB. Rhizobia inoculation improves nutrient uptake and growth of lowland rice. Soil Sci Soc Am J. 2000;64:1644–50.
5. Dobbelaere S, Vanderleyden J, Okon Y. Plant growth-promoting effects of diazotrophs in the rhizosphere. Crit Rev Plant Sci. 2003;22:107–49.
6. Glick BR. The enhancement of plant growth by free-living bacteria. Can J Microbiol. 1995;41:109–17.
7. Glick BR, Liu C, Ghosh S, Dumbroff EB. Early development of canola seedling in the presence of the plant growth-promoting rhizobacterium *Pseudomonas putida* GR12-2. Soil Biol Biochem. 1997;29:1233–9.
8. Richardson AE. Prospects for using soil microorganisms to improve the acquisition of phosphorus by plants. Aust J Plant Physiol. 2001;28:897–906.
9. van Loon LC, Glick GR. Increased plant fitness by rhizobacteria. In: Sandermann H, editor. Molecular Ecotoxicology of Plants, vol. 170. Berlin: Springer-Verlag; 2004. p. 177–205.
10. Kloepper JW, Lifshitz R, Zablotowicz RM. Free-living bacterial inocula for enhancing crop productivity. Trends Biotechnol. 1989;7:39–43.
11. Kloepper JW. Plant growth-promoting rhizobacteria as biological control agents. In: Metting Jr FB, editor. Soil microbial ecology-applications in agricultural and environmental management. New York: Marcel Dekker; 1993. p. 255–74.
12. Lambert B, Joos H. Fundamental aspects of rhizobacterial plant growth promotion research. Trends Biotechnol. 1989;7:215–9.
13. Denton M, Kerr KG. Microbiological and clinical aspects of infection associated with *Stenotrophomonas maltophilia*. Clin Microbiol Rev. 1998;11:57–80.
14. Banerjee M, Yesmin L. Sulfur-oxidizing plant growth promoting rhizobacteria for enhanced canola performance. 2002. US Patent 07491535.
15. Berg G, Marten P, Ballin G. *Stenotrophomonas maltophilia* in the rhizosphere of oilseed rape—occurrence, characterization and interaction with phytopathogenic fungi. Microbiol Res. 1996;151:19–27.
16. Park M, Kim C, Yang J, Lee H, Shin W, Kim S, Sa T. Isolation and characterization of diazotrophic growth promoting bacteria from rhizosphere of agricultural crops of Korea. Microbiol Res. 2005;160:127–33.
17. Lockhart SR, Abramson MA, Beekmann SE, Gallagher G, Riedel S, Diekema DJ, Quinn JP, Doern GV. Antimicrobial resistance among Gram-negative bacilli causing infections in intensive care unit patients in the United States between 1993 and 2004. J Clin Microbiol. 2007;45:3352–9.
18. Brooke JS. *Stenotrophomonas maltophilia*: an emerging global opportunistic pathogen. Clin Microb Rev. 2012;25:2–41.
19. Salerno A, Pierandrei F, Rea E, Colla G, Rouphael Y, Saccardo F. Floating system cultivation of radish (*Raphanus sativus* L.): production and quality. Acta Hort. 2005;69:87–92.
20. Giusti MM, Wrostald RE. Characterization of red radish anthocyanins. J Food Sci. 1996;61:322–6.
21. Jing P, Zhao SJ, Ruan SY, Xie ZH, Dong Y, Yu LL. Anthocyanin and glucosinolate occurrences in the roots of Chinese red radish (*Raphanus sativus* L.), and their stability to heat and pH. Food Chem. 2012;133:1569–76.
22. Lu ZL, Liu LW, Li XY, Gong YQ, Hou XL, Zhu XW, Yang JL, Wang LZ. Analysis and evaluation of nutritional quality in Chinese radish (*Raphanus sativus* L.). Agr Sci China. 2008;7:823–30.
23. Curtis IS. The noble radish: past, present and future. Trends Plant Sci. 2003;8:305–7.
24. Watt JM, Breyer-Brandwijk MG. The medicinal and poisonous plants of southern and eastern Africa. Edinburgh and London: E. & S. Livingstone; 1962.
25. Sgherri C, Cosi E, Navarri-Izzo F. Phenols and antioxidative status of *Raphanus sativus* grown in excess copper. Physiolia Plantarum. 2003;118:21–8.
26. Takaya Y, Kondo Y, Furukawa T, Niwa M. Antioxidant constituent of radish sprout (kiware-daikon), *Raphanus sativus* L. J Agric Food Chem. 2003;51:8061–6.
27. Basu TK, Temple NJ, Garg ML. Antioxidants in human health and disease. Wallingford: CABI Publishing; 1999.
28. Ghiselli A, D'Amicis A, Giocosa A. The antioxidant potential of Mediterranean diet. Eur J Cancer Prev. 1997;6:15–9.
29. Meyer AS, Yi OS, Pearson DA, Waterhouse AI, Frankel EN. Inhibition of human low-density lipoprotein oxidation in relation to composition of phenolic antioxidants in grapes. J Agric Food Chem. 1997;45:1638–43.
30. Fahey JW, Zalcmann AT, Talalay T. The chemical diversity and distribution of glucosinolates and isothiocyanates among plants. Phytochemistry. 2001;56:5–51.
31. Settanni L, Miceli A, Francesca N, Cruciata M, Moschetti G. Microbiological investigation of *Raphanus sativus* L. grown hydroponically in nutrient solutions contaminated with spoilage and pathogenic bacteria. Int J Food Microbiol. 2013;160:344–52.
32. Marques APGC, Pires C, Moreira H, Rangel AOSS, Castro PML. Assessment of the plant growth promotion abilities of six bacterial isolates using *Zea mays* as indicator plant. Soil Biol Biochem. 2010;42:1229–35.
33. Settanni L, Miceli A, Francesca N, Moschetti G. Investigation of the hygienic safety of aromatic plants cultivated in soil contaminated with *Listeria monocytogenes*. Food Control. 2012;26:213–9.
34. Brandl MT, Haxo AF, Bates AH, Mandrell RE. Comparison of survival of *Campylobacter jejuni* in the phyllosphere with that in the rhizosphere of spinach and radish plants. Appl Environ Microbiol. 2004;70:1182–9.
35. Kerr KG, Denton M, Todd N, Corps CM, Kumari P, Hawkey PM. A new selective differential medium for isolation of *Stenotrophomonas maltophilia*. Eur J Clin Microbiol Infect Dis. 1996;15:607–10.
36. Idris A, Labus Chagne N, Korsten L. Efficacy of rhizobacteria for growth promotion in sorghum under greenhouse conditions and selected modes of action studies. J Agric Sci. 2009;147:17–30.
37. Sturz AV, Matheson BG, Arsenault W, Kimpniski J, Christie BR. Weeds as a source of plant growth promoting rhizobacteria in agricultural soils. Can J Microbiol. 2001;47:1013–24.
38. de Freitas JR, Banerjee MR, Germida JJ. Phosphate-solubilizing rhizobacteria enhance the growth and yield but not phosphorus uptake of canola (*Brassica napus* L.). Biol Fertil Soils. 1997;24:358–64.
39. Fages J, Arsac JF. Sunflower inoculation with *Azospirillum* and other plant growth promoting rhizobacteria. Plant Soil. 1991;137:87–90.
40. Kloepper JW, Schroth MN. Relationship of in vitro antibiosis of plant growth-promoting rhizobacteria to plant growth and the displacement of root microflora. Phytopathology. 1981;71:1020–4.
41. Kwok OCH, Fahy PC, Hoitink HAJ, Kuter GA. Interactions between bacteria and *Trichoderma hamatum* in suppression of *Rhizoctonia* damping-off in bark compost media. Phytopathology. 1987;77:1206–12.
42. De Boer W, Gunnewiek PJAK, Kowalchuk GA, Van Veen JA. Growth of chitinolytic dune soil beta-subclass *Proteobacteria* in response to invading fungal hyphae. Appl Environ Microbiol. 2001;67:3358–62.
43. Messiha NAS, van Diepeningen AD, Farag NS, Abdallah SA, Janse JD, van Bruggen AHC. *Stenotrophomonas maltophilia*: a new potential biocontrol agent of *Ralstonia solanacearum*, causal agent of potato brown rot. Eur J Plant Pathol. 2007;118:211–25.
44. Ikemoto S, Suzuki K, Kaneko T, Komagata K. Characterization of strains of *Pseudomonas maltophilia* which do not require methionine. Int J Syst Bacteriol. 1980;30:437–47.
45. Debette J, Blondeau R. Presence de *Pseudomonas maltophilia* dans la rhizosphere de quelque plantes cultivee. Can J Microbiol. 1980;26:460–3.
46. Warcup JH. Chemical and biological aspects of soil sterilization. Soils Fertil. 1957;20:1–5.
47. Antoun H, Beauchamp CJ, Goussard N, Chabot R, Lalande R. Potential of *Rhizobium* and *Bradyrhizobium* species as plant growth promoting rhizobacteria on non-legumes: effect on radishes (*Raphanus sativus* L.). Plant Soil. 1998;204:57–67.
48. Lucy M, Reed E, Glick BR. Applications of free living plant growth-promoting rhizobacteria. Antonie Van Leeuwenhoek. 2004;86:1–25.

5

Plant growth-promoting effects of rhizospheric and endophytic bacteria associated with different tomato cultivars and new tomato hybrids

Gennaro Roberto Abbamondi[1,2][*] iD, Giuseppina Tommonaro[1], Nele Weyens[2], Sofie Thijs[2], Wouter Sillen[2], Panagiotis Gkorezis[2], Carmine Iodice[1], Wesley de Melo Rangel[3], Barbara Nicolaus[1] and Jaco Vangronsveld[2]

Abstract

Background: Conventional agriculture relies on chemical pesticides and fertilizers, which can degrade ecosystems. A reduction of these harmful practices is required, replacing (or integrating) them with more eco-friendly approaches, such as microbial inoculation. Tomato is an important agricultural product, with a high content of bioactive compounds (folate, ascorbate, polyphenols, and carotenoids). The focus of this research was to investigate the plant growth-promoting (PGP) abilities of bacterial strains isolated from different tomato cultivars, with the aim to develop systems to improve plant health and crop productivity based on microbial inoculation.

Methods: A pool of different tomato cultivars already available on the market and new tomato hybrids were selected based on their nutritional quality (high content of biologically active compounds). A total of 23 strains were isolated from tomato roots (11 rhizospheric strains and 12 root endophytes). The cultivable isolates were analyzed for a number of different PGP traits: organic acids (OA), indole acetic acid (IAA), ACC deaminase, and siderophore production. The effects of microbial inoculation on root growth of *Arabidopsis thaliana* were also evaluated using a Vertical Agar Plate assay.

Results: A high percentage of the isolated strains tested positive for the following PGP traits: 73 % were able to produce OA, 89 % IAA, 83 % ACC deaminase, and 87 % siderophores. The most striking result were remarkable increases in the formation of root hairs for most of the inoculated plants. This effect was obvious for all *A. thaliana* seedlings inoculated with the isolated endophytes, and for the 50 % of the seedlings inoculated with the rhizospheric strains.

Conclusions: A better knowledge of the plant growth-promotion activity of these strains can provide an important contribution to increase environmental sustainability in agriculture.

Background

Feeding an increasing number of people is one of the major challenges of the twenty-first century [1, 2]. Established strategies to enhance crop productivity are the use of chemical fertilizers, manures, and pesticides. However, these approaches often have a negative impact on the environment: leaching of nitrate into

groundwater, surface run-off of phosphorus and nitrogen, and eutrophication of aquatic ecosystems [3]. The growing interest in environmental sustainability has led to considerable efforts to minimize the use of chemical fertilizers and pesticides, replacing (or integrating) these conventional approaches with more eco-friendly methods, such as the application of beneficial microorganisms [4–6]. Plants form mutually beneficial associations with microbes [7]. These associations play essential roles in agricultural and food safety, and contribute to the environmental equilibrium [8]. A clear distinction should

*Correspondence: g.r.abbamondi@icb.cnr.it
[2] Hasselt University, Environmental Biology, Centre for Environmental Sciences, Agoralaan, building D, 3590 Diepenbeek, Belgium
Full list of author information is available at the end of the article

be drawn between bacteria residing in the rhizosphere or phyllosphere (the aerial habitat influenced by plants) and bacteria living inside the plant, the so-called endophytes. Endophytic bacteria reside in specific tissues of the plant (such as root cortex or xylem) and develop a close association with the plant, with exchange of nutrients, enzymes (lipase, catalase, oxidase, etc.), functional agents (siderophores, biosurfactants, etc.), and also "signals" [9, 10]. Endophytes colonize their plant host tissues in which they persist without exerting the negative effects of a pathogen (disruption of respiration, photosynthesis, translocation of nutrients, transpiration, etc.). On the contrary, the presence of these endophytic bacteria in the host plant leads to beneficial effects on its health and/or growth. Plant growth-promoting bacteria promote plant health and growth via three mechanisms: phytostimulation, biofertilization, and biocontrol [11].

Phytostimulators enhance plant growth in a direct way, usually by the production of phytohormones (auxins, cytokinins, gibberellins) [12]. The production of plant hormones such as indole-3-acetic acid (IAA) is classified as "direct" promotion. The synthesis of 1-aminocyclopropane-1-carboxylate (ACC) deaminase can be included in the same group: ACC deaminase cleaves ACC, the immediate precursor of ethylene, and thereby reduces its biosynthesis; ethylene inhibits growth of roots and shoots; therefore lower levels of this plant hormone lead to plant growth promotion.

Biofertilizing strains can fix nitrogen or increase the availability of phosphorus and iron.

Biocontrol agents protect plants from infections by phytopathogens (through competition for nutrients, induced systemic resistance, production of antimicrobial secondary metabolites). Siderophore production is considered an indirect biocontrol trait; iron chelation limits the amount of trace metals available to potential plant pathogens [13].

Tomato represents one of the most important agricultural products. It can be consumed not only as raw fruit, but also transformed by industrial processes (pulped, canned, incorporated in sauce). It is known to be rich in several bioactive compounds (folate, ascorbate, polyphenols, carotenoids) [14–16].

Recent studies underline the importance of the microbial metabolism and host interaction for growth, quality, and health of edible and medicinal plants [17–19]. The bacterial community composition of tomato plants and the influence of inoculation of rhizobacteria on plant health is also a topic of increasing scientific interest [20–23]. A recent study showed that *Trichoderma*-enriched biofertilizer enhances production and nutritional quality of tomato and minimizes NPK fertilizer use [24]. However, the unraveling of the correlation between the

in vitro detection of Plant Growth-Promoting (PGP) traits in bacterial isolates and a real contribution to the host plant health in vivo is still lacking. Moreover, in this study we not only investigated the PGP characteristics of bacteria isolated from roots of tomato cultivars already available on the market, but we also focused on new tomato hybrids of industrial interest. These new tomato hybrids, both fresh fruits and transformed products, could have a high potential for the agro-food industry because of their nutritional quality (high content of biologically active compounds) [14–16]. A screening of PGP traits of the isolates was performed to detect ACC deaminase, organic acids (OA), indole acetic acid (IAA), and siderophore (SID) production. The effects of microbial inoculation on root development were evaluated using a standardized test system with *Arabidopsis thaliana* grown on Vertical Agar Plates (VAPs).

Methods
Isolation of bacteria from different tomato cultivars and new tomato hybrids
Isolation of bacteria from the rhizosphere
In 2012, eleven bacterial strains, colonizing the rhizosphere of tomato plants, were collected from an experimental field situated in the Campania Region (southern Italy). The soil had originated from limestone and tuff which have been covered with a thick layer of volcanic material. High levels of oxygen and organic matter and a high availability of potassium contribute to the high fertility of the soil (see Additional file 1). Tomato roots were sampled at the late stage of maturation, placed in 50 mL tube with sterile phosphate-buffered saline (PBS; $NaH_2PO_4 \cdot H_2O$, 6.33 g L^{-1}; $Na_2HPO_4 \cdot 7H_2O$, 16.5 g L^{-1}; pH 7.4), and stored at 4 °C until isolation. Samples were rinsed several times in PBS in order to remove adhering soil particles. Root cuttings were rubbed against a sterile 200-μm nylon mesh to retain roots particles and large cell debris. A sucrose solution (9 ml; 4 % w/v) was added to the filtrates obtained from 1 g of roots. Serial 10-fold dilutions of the suspensions in the same sucrose solution were prepared, and 100 μl aliquots were spread on modified Luria–Bertani (LB) plates containing 4 g L^{-1} of NaCl instead of 10 g L^{-1}. After 7 days incubation at 30 °C, single colonies were selected and further purified on the basis of morphological differences. Subsequently, all strains were grown aerobically in modified Luria–Bertani (LB) medium (NaCl 0.4 % w/v).

Isolation of endophytic bacteria from roots
In 2013, 12 strains were isolated from tomato plant roots from the above-mentioned experimental field, but this time a specific protocol was applied to isolate endophytic bacteria. Fresh roots were collected in sterile

50 ml Falcon tubes with 25 ml P-buffer (per L: 6.33 g of $NaH_2PO_4 \cdot H_2O$; 16.5 g of $Na_2HPO_4 \cdot 7H_2O$; 200 µl Tween 40). Samples were washed several times with P-buffer until the roots were free of all attached soil particles; then distilled H_2O was used to remove foam (Tween 40). Sterilization was performed using a 5 % sodium hypochlorite solution for 5 min, and then the roots were rinsed 5 times in sufficient volumes of new sterile water. After rinsing, the samples were cut in smaller fragments with a sterile razor blade and crushed with sterile mortar and pestle in sterile 10 mM $MgSO_4$-solution (1 g roots in 10 ml). Serial 10-fold dilutions of the suspensions were prepared, and 100 µl aliquots were streaked on Luria–Bertani (LB) plates. To verify the surface sterility of the roots, 100 µl of the last rinsing water was plated on LB medium.

Bacterial strain and growth media

The eleven rhizospheric strains and the twelve root endophytic strains were grown aerobically at 30 °C while shaking (120 rpm) in modified (NaCl 0.4 % w/v) and standard (NaCl 1.0 % w/v) Luria–Bertani (LB) medium, respectively. The pH of medium was 7.0–7.2. Inocula for the assays to detect plant growth-promoting (PGP) traits were prepared as described below. 10 µl of the glycerol-stocked bacterial isolates stored at −80 °C were grown in 96-well plates containing 1 mL of liquid LB medium for 48 h at 30 °C with shaking. The cultures were centrifuged at 4000 rpm for 20 min at room temperature and then the supernatant was discharged. Cell pellets were washed twice with 1 mL of 10 mM $MgSO_4$ and re-suspended in 650 µL of $MgSO_4$.

Genotypical analysis of the isolates

Genomic DNA of isolated bacteria was extracted using the DNeasy® 96 Blood and Tissue Kit (Qiagen). The qualitative and quantitative analysis of the extracted DNA was performed using the Nanodrop ND-1000 Spectrophotometer (Isogen Life Sciences). An aliquot (1 µl) of the extracted DNA of each bacterial strain was used in a PCR reaction without further purification to amplify a fragment of the 16S ribosomal DNA coding region. Two primers, the universal 1392R (5′-ACGGGCGG TGTGTGTRC-3′) and the bacteria-specific 27F (5′-AGAGTTTGATCMTGGCTCAG-3′), were used for the amplification. PCR products were sequenced by Macrogen Europe (Amsterdam, The Netherlands). Sequencing data were processed using the DNATools application developed by Sam Achten and Lenny Jorissen (Hasselt University). The Smith–Waterman algorithm was chosen for local sequence alignment and the consensus was obtained by comparing the quality of bases. The obtained sequences were compared with reference strains in the National Center for Biotechnology Information (NCBI) database.

Plant growth-promoting (PGP) traits of selected bacterial isolates

ACC-deaminase activity

Bacterial isolates with the ability to produce ACC deaminase were identified according to the method developed by Belimov et al. (2005) that was slightly modified [25, 26]. 250 µl of each last bacterial suspension in 10 mM $MgSO_4$ (see *Bacterial strain and growth media*) was added to 1.2 mL of salts minimal medium (SMN) containing 5 mM ACC as a sole source of N. The cultures were incubated at 30 °C for 72 h with shaking (150 rpm), then harvested by centrifugation at 4000 rpm for 20 min at room temperature. The supernatant was discharged and the pellets were re-suspended in 100 µl of 0.1 M Tris–HCl buffer (pH = 8.5); cells were disrupted by the addition of 3 µl toluene followed by vigorous vortexing for 10 min. Subsequently, 10 µL of 0.5 M ACC and 100 µL of 0.1 M Tris–HCl buffer (pH = 8.5) were added and followed by shaking for 10 min. The microplates were incubated for 30 min at 30 °C with shaking (150 rpm). Then, 690 µl of 0.56 N HCl and 150 µl of 0.2 % 2, 4-dinitrophenylhydrazine reagent (in 2 N HCl) were added to the cell suspensions. The plates were kept at 30 °C for 30 min, supplemented with 1 ml of 2 N NaOH. Well plates without ACC were used as negative controls. The change of color from yellow to brown was considered as positive.

IAA production

The ability of the strains to produce indole acetic acid (IAA) was tested using the method of Gordon and Weber (1951) slightly modified [27, 28]. Briefly, 50 µl of each last bacterial suspension in 10 mM $MgSO_4$ (see *Bacterial strain and growth media*) was inoculated in 96-well plates containing 1 mL 1/10 diluted 869-rich medium complemented with 50 mg mL^{-1} tryptophan and incubated at 30 °C for 4 days at 150 rpm in the dark. Afterward, bacterial cultures were centrifuged at 4000 rpm for 20 min, and 0.5 mL of the supernatant was added to 1 mL of Salkowski reagent (98 mL 35 % $HClO_4$, 2 mL 0.5 M $FeCl_3$) and gently vortexed. After 20 min, the development of a pink color was considered as positive (bacteria producing IAA).

Organic acids production

For the identification of the strains able to produce various organic acids, a protocol developed by Cunningham and Kuiack [29] was used. 50 µl of each last bacterial suspension in 10 mM $MgSO_4$ (see *Bacterial strain and growth media*) was inoculated in 800 µl of Sucrose Tryptone medium (ST) containing 20 g L^{-1} sucrose and 5 g L^{-1} tryptone. ST medium was supplemented with 10 mL of trace elements solution of the following composition: $NaMoO_4$ 20 mg L^{-1}, H_3BO_3

20 mg L^{-1}, CuSO$_4$·5H$_2$O 20 mg L^{-1}; FeCl$_3$ 100 mg L^{-1}; MnCl$_2$·4H$_2$O 20 mg L^{-1}; ZnCl$_2$ 280 mg L^{-1}. After 15 min, the color change of the alizarine red pH indicator from red (pH > 6) to yellow (pH ~ 5 or below) was considered as positive.

Siderophore production

The ability of the isolates to produce siderophores was evaluated through a qualitative test in liquid 284 medium (that stimulates siderophore production) with chrome azurol S (CAS) shuttle solution [30]. Briefly, 50 µl of each last bacterial suspension in 10 mM MgSO$_4$ (see *Bacterial strain and growth media*) was inoculated in 96-well plates containing 800 µl of 284 medium prepared at three different iron concentrations: without iron, with 0.25 µM, and 3 µM Fe(III) citrate. The microplates were incubated at 30 °C at 150 rpm for 5 days. After incubation, 100 µl of the blue Chromium-Azurol S (CAS) reagent was added. Plates were kept at room temperature for 4 h, and the change of color from blue to orange/yellow was considered as positive.

Evaluation of the effects of microbial inoculation on root growth

A Vertical Agar Plate (VAP) assay was performed to evaluate the effects of microbial inoculation on root development of *A. thaliana* seedlings [31, 32].

Growth medium and growth conditions

The basic medium was based on a 50-fold dilution of Gamborg's B5 medium and contained KNO$_3$ 0.5 mM, MgSO$_4$·7H$_2$O 0.02 mM, CaCl$_2$·6H$_2$O 0.02 mM, NaH$_2$PO$_4$ 0.022 mM, MnSO$_4$·H$_2$O 0.94 µM, (NH$_4$)$_2$SO$_4$ 0.02 mM as macronutrients, KI 90 nM, H$_3$BO$_3$ 0.97 nM, ZnSO$_4$·7H$_2$O 0.14 nM, CuSO$_4$·5H$_2$O 2 nM, Na$_2$MoO$_4$·2H$_2$O 20.6 nM, CoSO$_4$·H$_2$O 2.6 nM as micronutrients, FeCl$_3$ 3.6 µM, 2-[*N*-Morpholino]ethanesulfonic acid 2.56 mM MES (Sigma), pH = 5.7. Only for the plates used for germination, this basic medium was complemented with 5 g L^{-1} sucrose. *A. thaliana* seeds were surface sterilized for 1 min in 0.1 % Cl-solution, followed by five washes with sterile deionized water. The seeds were rinsed 4 times for 5 min in a larger volume of sterile deionized water and then dried in a laminar air flow. Sterilized seeds were stored for 3 days at 4 °C in the dark, then transferred with a sterile toothpick onto 12 × 12 cm transparent plates containing 40 mL of solid medium (1 % agar nr. 4 for plant tissue culture) complemented with sucrose. Plates were incubated vertically in a growth chamber at 22 °C, with a 16-h light/8-h dark regime and a light intensity of 150 µmol m^{-2} s^{-1}. After 6 days, plantlets growing on the surface of the agar were transferred to fresh growth media.

Vertical Agar Plate (VAP) assay

In order to estimate the effects of bacterial inoculation on root growth, a Vertical Agar Plate (VAP) assay was performed. 5 ml of 48 h culture of each strain was centrifuged (4000 rpm, 15 min) and brought to OD$_{660}$ = 0.5 using 10 mM MgSO$_4$·7H$_2$O, then 400 µl of this inoculum was transferred to fresh growth medium and equally distributed over the surface of the plate. For the "control," 400 µl 10 mM MgSO$_4$·7H$_2$O was used.

Arabidopsis thaliana seedlings were prepared as described previously and transferred to the VAPs (five plants per plate, in triplicate). Plates were incubated vertically in a growth chamber at 22 °C, with a 16-h light/8-h dark regime and a light intensity of 150 µmol m^{-2} s^{-1}.

Analysis of root growth

Root growth in the vertical agar plates was monitored daily for 8 days, and the plates were then scanned at 300 dpi (Canon CanoScan 4400F). Primary root length and total lateral root length were calculated after analysis of scanned images using the RootNav image analysis tool (University of Nottingham) [33]. 2D pictures of the roots were also created with the same software in order to graphically compare the root system architectures. Root hair visibility of scanned images was improved using the GNU Image Manipulation Program (GIMP).

Statistical analysis

All statistical analyses were performed using the 3.1.1 version of R (The R Foundation for Statistical Computing, Vienna, Austria). Since for all the data groups considered, normality and homoscedasticity assumptions were met, parametric tests were used in all analyses. Student's *t* test was used for comparing two averages, while ANOVA with post hoc Tukey's Honest Significant Difference correction for multiple correction was applied when considering more than two averages.

Results

Genotypic characterization

After the isolation, all morphologically different bacteria were purified, but only 2/3 of the strains showed to be cultivable in the subsequent steps; hence, 16S rDNA-based identification was performed on 15 strains (6 rhizospheric and 9 endophytic strains).

Among the 15 strains that were genotypically characterized, five belong to the genus *Agrobacterium*, and others to the genera *Microbacterium* (2), *Bacillus* (2), *Rhizobium* (2), *Ensifer* (1), *Chryseobacterium* (1), *Pseudomonas* (1), and *Rhodococcus* (1) (Table 1).

Table 1 Tomato cultivars—genotypic characterization

	Sample no.	Genotypic characterization			Cultivar
		Phylum	Family	Genus	
Rhizospheric strains	[1]	Proteobacteria	Rizobiaceae	Ensifer sp.	Plus Licopene
	[3]	Bacteroidetes	Flavobacteriaceae	Chryseobacterium sp.	
	[7]	Firmicutes	Bacillaceae	Bacillus sp.	
	[4]	Firmicutes	Bacillaceae	Bacillus sp.	Indigo Perù
	[6]	n.p.	n.p.	n.p.	
	[5]	n.p.	n.p.	n.p.	Super San Marzano
	[2]	Actinobacteria	Microbateriaceae	Microbacterium sp.	
	[8]	Actinobacteria	Microbateriaceae	Microbacterium sp.	
Root endophytes	[9]	Proteobacteria	Pseudomonadaceae	Pseudomonas sp.	San Marzano x Black Tomato hybrid
	[10]	n.p.	n.p.	n.p.	
	[11]	Proteobacteria	Rhizobiaceae	Rhizobium sp.	
	[12]	Actinomycetales	Nocordiaceae	Rhodococcus sp.	
	[13]	Proteobacteria	Rhizobiaceae	Agrobacterium sp.	
	[14]	Proteobacteria	Rhizobiaceae	Agrobacterium sp.	Black Tomato
	[15]	Proteobacteria	Rhizobiaceae	Agrobacterium sp.	
	[16]	Proteobacteria	Rhizobiaceae	Rhizobium sp.	
	[17]	Proteobacteria	Rhizobiaceae	Agrobacterium sp.	San Marzano
	[18]	Proteobacteria	Rhizobiaceae	Agrobacterium sp.	

Genotypic characterization of cultivable strains isolated from tomato roots (6 rhizospheric strains, 9 root endophytes). *n.p.* not performed, due to cultivation problems with standard media (LB with standard NaCl concentration and modified, NaCl 0.4 % w/v)

Screening of the plant growth-promoting (PGP) traits

All isolates were tested for their in vitro putative PGP traits. Specific tests were performed in order to detect ACC deaminase, organic acids (OA), indole acetic acid (IAA), and siderophore (SID) production. Results indicate a high percentage of strains showing PGP traits; in particular, 73 % were able to produce organic acids (OA), 89 % IAA, 83 % ACC deaminase, and 87 % siderophores (Table 2).

It is remarkable that for most root endophytes sidero-phore production was detected. Also among the rhizos-phere strains a high percentage of siderophore producers was found; in fact, only the two strains belonging to genus *Microbacterium* tested negative for this feature.

For the other traits that were investigated (ACC deaminase, organic acids (OA), and indole acetic acid (IAA) production) also high percentages of positives were detected. All the isolates showed capability to exhibit at least one of the three PGP traits, and most of the strains tested positive for at least two of the three traits. *Agrobacterium* strains tested positive for all features.

Vertical Agar Plate (VAP) assay

After inoculation of most of the strains, the length of the main root was comparable to the control (not inoculated seedlings) or even shorter (Table 3, see Additional file 2). More specifically, concerning the rhizosphere bacteria,

inoculation of the two *Microbacterium* sp. (numbers 2 and 8) and the two *Bacillus* sp. (numbers 4 and 7) strains resulted in a root length similar (±5 %) to the control; inoculation of strains 5, 6 and *Chryseobacterium* sp. (number 3) led to significant inhibitions (p value <0.05) of the primary root length, while only for *Ensifer* sp. (number 1) a trend of primary root length increase was observed (no significant difference). Concerning the root endophytes, only seedlings inoculated with strains *Rhizobium* sp. (number 11) and *Agrobacterium* sp. (number 13) showed root lengths comparable to the control (±5 %), while for the other endophytic strains shorter primary roots were found. More specifically, for strain number 10, *Pseudomonas* sp. (number 9), *Agrobacterium* sp. (number 15), and *Rhizobium* sp. (number 16) we registered significant differences (p value <0.05) compared to the control; for *Agrobacterium* sp. (number 18) a relevant but not significant difference ($p = 0.053$) was observed, and for the other strains (*Rhodococcus* sp. n. 12 and *Agrobacterium* sp. n. 14 and n. 17) there was a trend of lower values without being significantly different from the control.

Similar results were observed for the lateral roots; in general, the total lateral root length of inoculated plants was lower than that of the control plants (not inoculated seedlings), in some cases comparable to it (Table 4). More specifically, the total lateral root lengths of seedlings

Table 2 Plant growth-promoting traits—siderophore production ability

	Strain no.	Genotypic characterization	OA[a]	IAA[a]	ACC[a]	Sid[b] - Fe(III) citrate		
						No	0.25 μM	3 μM
Rhizospheric strains	[1]	*Ensifer* sp.	+	++	-	++	++	+/-
	[2]	*Microbacterium* sp.	+	+	+/-	-	-	-
	[3]	*Chryseobacterium* sp.	-	+	-	+	+	-
	[4]	*Bacillus* sp.	-	-	+/-	+	+	+/-
	[5]	n.p.	n.p.	++	++	n.p.	n.p.	n.p.
	[6]	n.p.	n.p.	++	++	n.p.	n.p.	n.p.
	[7]	*Bacillus* sp.	+++	-	++	+	+	+/-
	[8]	*Microbacterium* sp.	++	+	-	-	-	-
Root endophytes	[9]	*Pseudomonas* sp.	-	+	+++	+++	+++	+
	[10]	n.p.	n.p.	++	++	n.p.	n.p.	n.p.
	[11]	*Rhizobium* sp.	+++	++	++	+++	+++	++
	[12]	*Rhodococcus* sp.	-	+	++	+++	+++	+
	[13]	*Agrobacterium* sp.	+++	+++	++	+++	+++	++
	[14]	*Agrobacterium* sp.	++	+++	++	+++	+++	++
	[15]	*Agrobacterium* sp.	+	+++	++	+++	+++	++
	[16]	*Rhizobium* sp.	+++	+++	+	+++	+++	+
	[17]	*Agrobacterium* sp.	+++	+++	+++	+++	+++	++
	[18]	*Agrobacterium* sp.	+++	+++	+++	+++	+++	++

[a] Plant growth-promoting (PGP) traits of selected bacterial isolates: ACC deaminase (ACC), organic acids (OA), and Indole acetic acid (IAA) production. *n.p.* not performed, due to cultivation problems with standard media (LB with standard NaCl concentration and modified, NaCl 0.4 % w/v)

[b] Evaluation of siderophore (Sid) production ability of selected bacterial isolates. Evaluation of the positivity to the tests, from negative (−) to the highest activity (+++): −, ±, +, ++, +++

inoculated with the rhizosphere strains *Ensifer* sp. (number 1) and *Chryseobacterium* sp. (number 3) were comparable (±5 %) to the control; for strains 5 and 6 and *Bacillus* sp. (number 4) a decreasing trend was observed but only for strain 8 this decrease was relevant ($p = 0.054$), while for *Bacillus* sp. (number 7) and the two *Microbacterium* sp. (numbers 2 and 8) trends to increased total lateral root length were registered. All plants inoculated with the root endophytes showed inhibitions of the total lateral root length. Only for strain number 10, statistical analysis showed a significant difference, but relevant differences were observed for *Agrobacterium* sp. (numbers 15 and 18) and *Rhizobium* sp. (number 16); p values were, respectively, 0.069, 0.052, and 0.072.

The most striking effects of bacterial inoculation on root development were strong increase in root hairs formation for most of the inoculated seedlings. Increases of root hairs formation were observed for plants inoculated with the rhizospheric strains 5, 6, *Ensifer* sp. (number 1) and *Chryseobacterium* sp. (number 3), and for all seedlings inoculated with endophytes (Fig. 1).

Discussion

In this study, we report the isolation of a total of 23 strains from the roots of tomato plants. In particular, 11

strains were isolated from the rhizosphere of 4 tomato cultivars (commercial names: Plus Licopene, San Marzano giallo, Indigo Perù, and Super San Marzano), while 12 endophytes were isolated from 3 different cultivars (San Marzano, Black Tomato, and a new hybrid created by cross-pollination between these two species).

The highest efficiencies (in terms of root hair development and OA, IAA, ACC deaminase, and siderophore production) were registered for endophytic strains. In comparison to the rhizospheric ones, endophytes interact more closely with their host because of their localization in the plant tissues [34]. Such a close link leads to highly evolved mutualistic interactions, from which both plants and bacteria benefit. Endophytes represent a rich source of bioactive compounds that can positively influence plant growth through a number of different mechanisms [35]. Moreover, bacterial endophytes are competing with phytopathogens due to the fact that they colonize the same ecological niches; due to this, they often develop abilities that protect plants from infections (biocontrol agents) [36]. For these reasons, we expected to find higher efficiency in terms of PGP traits for endophytes compared to the strains isolated from the tomato rhizosphere.

This difference between the rhizosphere strains and the endophytes was also highlighted by the statistical

Table 3 VAP test—primary root length

	Sample	Primary root length (cm)	Stdev	p-value comparing to the control
Rhizospheric strains	Control	6.94	0.46	
	1	7.90	0.96	0.2
	2	7.11	0.43	1.0
	3	5.20	0.37	0.001 ***
	4	6.88	0.32	1.0
	5	5.03	0.75	0.001 ***
	6	4.28	0.47	0.001 ***
	7	7.12	0.58	1.0
	8	6.83	0.47	1.0
Root endo-phytes	Control	7.51	0.81	
	9	6.19	0.22	0.048 *
	10	3.77	0.85	0.001 ***
	11	7.60	0.54	1.0
	12	6.84	0.37	0.8
	13	7.60	0.53	1.0
	14	6.98	0.76	0.95
	15	4.69	0.64	0.001 ***
	16	6.02	0.52	0.02 *
	17	6.33	0.70	0.1
	18	6.20	0.50	0.053 +

Primary root length measurement after 8 days growth. Control: *Arabidopsis thaliana* VAPs not inoculated; *Stdev* standard deviation

Statistical analysis: significant differences *0.01 < p < 0.05, **0.001 < p < 0.01, ***p < 0.001; not significant but relevant differences +0.05 < p < 0.1

Table 4 VAP test—total lateral root length

	Sample	Total lateral root length (cm)	Stdev	p-value comparing to the control
Rhizospheric strains	Control	12.50	3.38	
	1	12.83	5.08	1.0
	2	14.20	1.89	1.0
	3	12.16	2.51	1.0
	4	10.07	2.45	0.94
	5	9.95	2.10	0.92
	6	6.12	1.84	0.054 +
	7	15.44	4.21	0.84
	8	15.65	2.66	0.79
Root endophytes	Control	12.03	3.58	
	9	8.15	2.03	0.67
	10	4.55	3.55	0.018 *
	11	8.97	2.59	0.89
	12	7.40	1.69	0.42
	13	8.23	2.47	0.7
	14	8.89	3.19	0.88
	15	5.59	2.20	0.069 +
	16	5.63	3.23	0.072 +
	17	10.73	5.37	1.0
	18	5.36	2.71	0.052 +

Total lateral root length measurement after 8 days growth. Control: *Arabidopsis thaliana* VAPs not inoculated; *Stdev* standard deviation

Statistical analysis: significant differences *0.01 < p < 0.05, **0.001 < p < 0.01, ***p < 0.001; not significant but relevant differences +0.05 < p < 0.1

analysis of the outcomes of the VAP test. A comparison, separately for endophytes and rhizosphere strains, between inoculated and non-inoculated plants showed statistically significant effects of inoculation on primary root length (p value = 0.03) and lateral root length (p value = 0.005) for the endophytes, but not for the rhizosphere strains (p values = 0.28 and 0.82, respectively).

This dissimilarity between rhizosphere strains and endophytes in general was once more retrieved in the form of a statistically significant difference between rhizosphere strains and endophytes in general for their effects on lateral root length (p value = 0.0001), although this was not observed for primary root length (p value = 0.8). This discrepancy between lateral and primary root length suggests a significant interaction between parameter (primary or lateral root length) and bacterial origin (endosphere or rhizosphere), which was confirmed by an ANOVA.

The obtained results also suggest a causal relationship between indole acetic acid (IAA) production capacity and the stimulation of root hair development in *A. thaliana*. IAA is the most studied auxin; in fact, auxin and IAA are often used as interchangeable terms

[37]. It was suggested that the quantity of IAA produced by *Sphingomonas* sp. (11.23 ± 0. 93 µM/ml), an endophyte isolated from the leaves of *Tephrosia apollinea*, should be linked to the increases of surface area and root length, the loss of cell wall and the release of exudates in tomato (*Solanum lycopersicum* L.) plants [38]. It is known that root hair initiation and elongation are among the best characterized auxin-associated phenotypes [39–41]. High levels of IAA promote the formation of lateral roots and induce a decrease in primary root length and increases in root hair development [42]. In our work, all isolates that increased root hair development in *A. thaliana* in the VAP assay, also showed positive results in the IAA production test. Based on this, we may speculate that, for the investigated strains, IAA synthesis plays an essential role in the regulation of this PGP trait.

Conclusions

The improved capacity to acquire nutrients and water due to the increase of the surface area of the plant's root system induced by inoculation of most of the isolates, makes them potentially exploitable to enhance crop

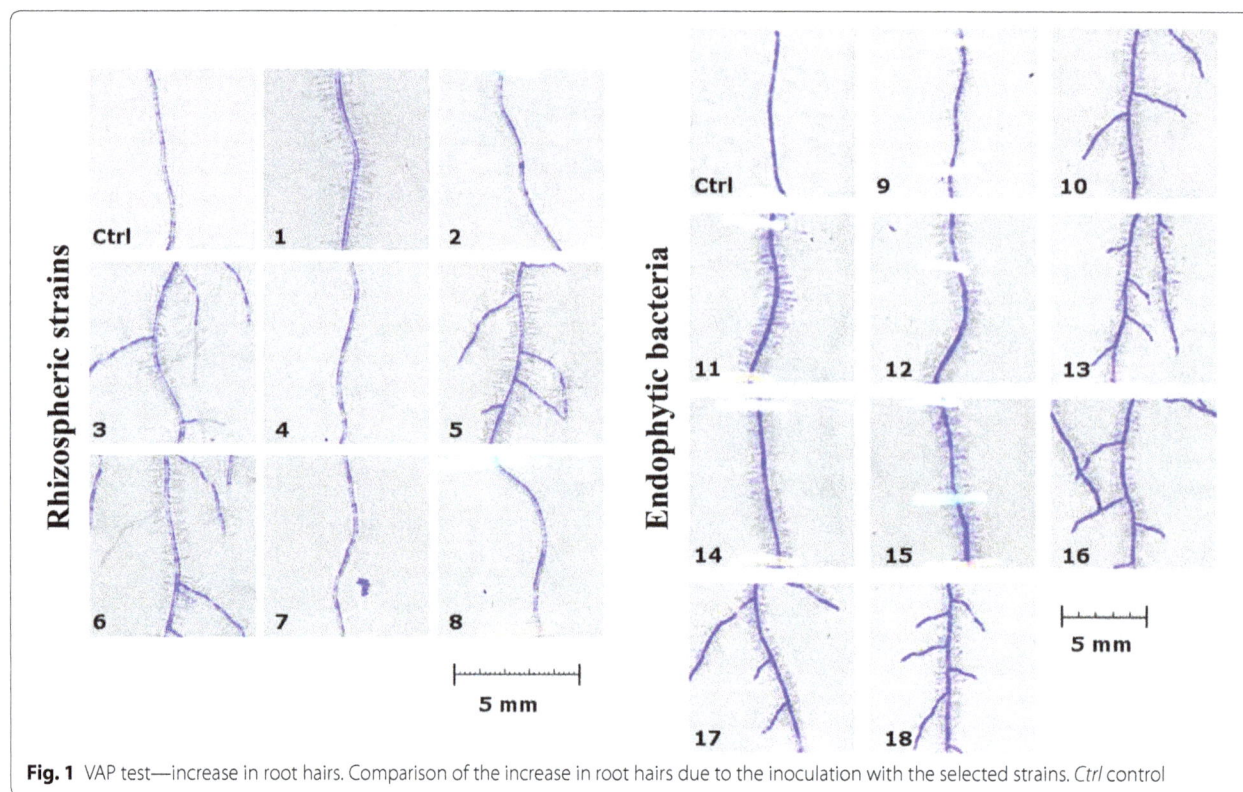

Fig. 1 VAP test—increase in root hairs. Comparison of the increase in root hairs due to the inoculation with the selected strains. *Ctrl* control

productivity. However, the in vitro detection of PGP traits in natural endophytic bacteria does not implicate a general and predictable improvement of growth and health of all types of host plants in vivo [43], and therefore further investigations are needed to assess if the strains that we isolated from the different tomato cultivars effectively induce beneficial effects in vivo in tomato and also in other plant species.

Moreover, important processes of many beneficial rhizosphere-colonizing microorganisms, including some PGP traits (i.e., siderophore production), are regulated by Quorum Sensing (QS) [44]. QS is the mechanism that describes the coordinated gene expression that microbial communities show at high cell density [45]. Further analyses are in progress to investigate QS activity in the isolated strains.

Authors' contributions

GRA, JV, GT, and BN designed the study, participated in its coordination, and helped to draft the manuscript. GRA performed the laboratory studies, wrote the manuscript, and made the corrections. WdMR contributed to carry out the PGP bioassays. WS participated in data acquisition and statistical analysis. ST, NW, and CI contributed in experimental designing. All authors read and approved the final manuscript.

Author details

[1] CNR, National Research Council of Italy, Institute of Biomolecular Chemistry, Via Campi Flegrei, 34, 80078 Pozzuoli, NA, Italy. [2] Hasselt University, Environmental Biology, Centre for Environmental Sciences, Agoralaan, building D, 3590 Diepenbeek, Belgium. [3] Laboratory of Biology, Microbiology and Soil Biological Processes, Soil Science Department, Federal University of Lavras, PO box 3037, 37200-000 Lavras, Minas Gerais, Brazil.

Acknowledgements

This work was supported by a dedicated grant from the Italian Ministry of Economy and Finance to the National Research Council for the project "Innovazione e Sviluppo del Mezzogiorno, Conoscenze Integrate per Sostenibilita' ed Innovazione del Made in Italy Agroalimentare, Legge n. 191/2009" and the UHasselt Methusalem project 08M03VGRJ. Funding was also received from the COST Action (FA1103, COST-STSM-ECOST-STSM-FA1103-090914-049031) through a Short-Term Scientific Mission for GRA to Hasselt University. The authors would like to thank Bram Beckers and Marijke Gielen (Hasselt University) for assistance with the Genotypical analysis and the Vertical Agar Plate (VAP) assay, and for comments that greatly improved the manuscript.

Competing interests

The authors declare that they have no competing interests.

References

1. Berg G. Plant-microbe interactions promoting plant growth and health: perspectives for controlled use of microorganisms in agriculture. Appl Microbiol Biotechnol. 2009;84(1):11–8.
2. Adesemoye AO, Kloepper JW. Plant-microbes interactions in enhanced fertilizer-use efficiency. Appl Microbiol Biotechnol. 2009;85(1):1–12.
3. Camargo JA, Alonso A. Ecological and toxicological effects of inorganic nitrogen pollution in aquatic ecosystems: a global assessment. Environ Int. 2006;32(6):831–49.
4. Farrar K, Bryant D, Cope-Selby N. Understanding and engineering beneficial plant-microbe interactions: plant growth promotion in energy crops. Plant Biotechnol J. 2014;12(9):1193–206.
5. Hardoim PR, van Overbeek LS, van Elsas JD. Properties of bacterial endophytes and their proposed role in plant growth. Trends Microbiol. 2008;16(10):463–71.
6. Enya J, Shinohara H, Yoshida S, Negishi TTH, Suyama K, Tsushima S. Culturable leaf-associated bacteria on tomato plants and their potential as biological control agents. Microb Ecol. 2007;53(4):524–36.
7. Lugtenberg B, Kamilova F. Plant-growth-promoting rhizobacteria. Annu Rev Microbiol. 2009;63:541–56.
8. Mendes R, Garbeva P, Raaijmakers JM. The rhizosphere microbiome: significance of plant beneficial, plant pathogenic, and human pathogenic microorganisms. FEMS Microbiol Rev. 2013;37(5):634–63.
9. Parsek MR, Greenberg EP. Acyl-homoserine lactone quorum sensing in Gram-negative bacteria: a signaling mechanism involved in associations with higher organisms. Proc Natl Acad Sci USA. 2000;97(16):8789–93.
10. Hardoim PR, van Overbeek LS, Berg G, Pirttilä AM, Compant S, Campisano A, Döring M, Sessitsch A. The hidden world within plants: ecological and evolutionary considerations for defining functioning of microbial endophytes. Microbiol Mol Biol Rev. 2015;79(3):293–320.
11. Bloemberg GV, Lugtenberg BJJ. Molecular basis of plant growth promotion and biocontrol by rhizobacteria. Curr Opin Plant Biol. 2001;4(4):343–50.
12. Weyens N, van der Lelie D, Taghavi S, Vangronsveld J. Phytoremediation: plant-endophyte partnerships take the challenge Phytoremediation: plant-endophyte partnerships take the challenge. Curr Opin Biotechnol. 2009;20(2):248–54.
13. Jin CW, Ye YQ, Zheng SJ. An underground tale: contribution of microbial activity to plant iron acquisition via ecological processes. Ann Bot. 2014;113(1):7–18.
14. Tommonaro G, De Prisco R, Abbamondi GR, Nicolaus B. Bioactivity of tomato hybrid powder: antioxidant compounds and their biological activities. J Med Food. 2013;16(4):351–6.
15. Tommonaro G, de Prisco R, Abbamondi GR, Marzocco S, Saturnino C, Poli A, Nicolaus B. Evaluation of antioxidant properties, total phenolic content, and biological activities of new tomato hybrids of industrial interest. J Med Food. 2012;15(5):483–9.
16. Tommonaro G, Caporale A, De Martino L, Popolo A, De Prisco R, Nicolaus B, Abbamondi GR, Saturnino C. Antioxidant and cytotoxic activities investigation of tomato seed extracts. Nat Prod Res. 2014;28(10):764–8.
17. Koberl M, Schmidt R, Ramadan EM, Bauer R, Berg G. The microbiome of medicinal plants: diversity and importance for plant growth, quality and health. Front Microbiol. 2013;4:400.
18. Palaniyandi SA, Yang SH, Zhang L, Suh J-W. Effects of actinobacteria on plant disease suppression and growth promotion. Appl Microbiol Biotechnol. 2013;97(22):9621–36.
19. Garcia-Fraile P, Carro L, Robledo M, Ramirez-Bahena M-H, Flores-Felix J-D, Teresa Fernandez M, Mateos PF, Rivas R, Mariano Igual J, Martinez-Molina E, et al. Rhizobium promotes non-legumes growth and quality in several production steps: towards a biofertilization of edible raw vegetables healthy for humans. Plos One. 2012;7(5):e38122.
20. Botta AL, Santacecilia A, Ercole C, Cacchio P, Del Gallo M. In vitro and in vivo inoculation of four endophytic bacteria on Lycopersicon esculentum. New Biotechnol. 2013;30(6):666–74.
21. Yim W, Seshadri S, Kim K, Lee G, Sa T. Ethylene emission and PR protein synthesis in ACC deaminase producing Methylobacterium spp. inoculated tomato plants (Lycopersicon esculentum Mill.) challenged with Ralstonia solanacearum under greenhouse conditions. Plant Physiol Biochem. 2013;67:95–104.
22. Palaniyandi SA, Damodharan K, Yang SH, Suh JW. Streptomyces sp. strain PGPA39 alleviates salt stress and promotes growth of 'Micro Tom' tomato plants. J Appl Microbiol. 2014;117(3):766–73.
23. Goudjal Y, Toumatia O, Yekkour A, Sabaou N, Mathieu F, Zitouni A. Biocontrol of Rhizoctonia solani damping-off and promotion of tomato plant growth by endophytic actinomycetes isolated from native plants of Algerian Sahara. Microbiol Res. 2014;169(1):59–65.
24. Molla AH, Md. Haque M, Md. Haque A, Ilias GNM. Trichoderma-enriched biofertilizer enhances production and nutritional quality of tomato (Lycopersicon esculentum Mill.) and minimizes NPK fertilizer use. Agric Res 2012, 1(3):265–72.
25. Belimov AA, Hontzeas N, Safronova VI, Demchinskaya SV, Piluzza G, Bullitta S, Glick BR. Cadmium-tolerant plant growth-promoting bacteria associated with the roots of Indian mustard (Brassica juncea L. Czern.). Soil Biol Biochem. 2005;37(2):241–50.
26. Truyens S, Jambon I, Croes S, Janssen J, Weyens N, Mench M, Carleer R, Cuypers A, Vangronsveld J. The effect of long-term cd and ni exposure on seed endophytes of Agrostis capillaris and their potential application in phytoremediation of metal-contaminated soils. Int J Phytorem. 2014;16(7–8):643–59.
27. Gordon SA, Weber RP. Colorimetric estimation of indoleacetic acid. Plant Physiol. 1951;26(1):192–5.
28. Patten CL, Glick BR. Role of Pseudomonas putida indoleacetic acid in development of the host plant root system. Appl Environ Microbiol. 2002;68(8):3795–801.
29. Cunningham JE, Kuiack C. Production of citric and oxalic acids and solubilization of calcium-phosphate by Penicillium bilaii. Appl Environ Microbiol. 1992;58(5):1451–8.
30. Schwyn B, Neilands JB. Universal chemical assay for the detection and determination of siderophores. Anal Biochem. 1987;160(1):47–56.
31. Zhang HM, Forde BG. An Arabidopsis MADS box gene that controls nutrient-induced changes in root architecture. Science. 1998;279(5349):407–9.
32. Remans T, Nacry P, Pervent M, Girin T, Tillard P, Lepetit M, Gojon A. A central role for the nitrate transporter NRT2.1 in the integrated morphological and physiological responses of the root system to nitrogen limitation in Arabidopsis. Plant Physiol. 2006;140(3):909–21.
33. Pound MP, French AP, Atkinson JA, Wells DM, Bennett MJ, Pridmore T. RootNav: navigating images of complex root architectures. Plant Physiol. 2013;162(4):1802–14.
34. Weyens N, Beckers B, Schellingen K, Ceulemans R, Croes S, Janssen J, Haenen S, Witters N, Vangronsveld J. Plant-associated bacteria and their role in the success or failure of metal phytoextraction projects: first observations of a field-related experiment. Microb Biotechnol. 2013;6(3):288–99.
35. Ryan RP, Germaine K, Franks A, Ryan DJ, Dowling DN. Bacterial endophytes: recent developments and applications. FEMS Microbiol Lett. 2008;278(1):1–9.
36. Gupta P, Puniya B, Barun S, Asthana M, Kumar A. Isolation and characterization of endophytes from different plants: effects on growth of Pennisetum typhoides. Biosci, Biotechnol Res Asia. 2014;11(1):223–34.
37. Gamalero E, Glick BR. Mechanisms used by plant growth-promoting bacteria. In: Maheshwari DK, editor. Bacteria in agrobiology: plant nutrient management. Berlin, Heidelberg: Springer; 2011. p. 17–46.
38. Khan AL, Waqas M, Kang S-M, Al-Harrasi A, Hussain J, Al-Rawahi A, Al-Khiziri S, Ullah I, Ali L, Jung H-Y, et al. Bacterial endophyte Sphingomonas sp LK11 produces gibberellins and IAA and promotes tomato plant growth. J Microbiol. 2014;52(8):689–95.
39. Overvoorde P, Fukaki H, Beeckman T. Auxin control of root development. Cold Spring Harb Perspect Biol 2010;2(6):a001537.
40. Taghavi S, Garafola C, Monchy S, Newman L, Hoffman A, Weyens N, Barac T, Vangronsveld J, van der Lelie D. Genome survey and characterization of endophytic bacteria exhibiting a beneficial effect on growth and development of poplar trees. Appl Environ Microbiol. 2009;75(3):748–57.
41. Cho HT, Cosgrove DJ. Regulation of root hair initiation and expansin gene expression in Arabidopsis. Plant Cell. 2002;14(12):3237–53.
42. Vacheron J, Desbrosses G, Bouffaud M-L, Touraine B, Moenne-Loccoz Y, Muller D, Legendre L, Wisniewski-Dye F, Prigent-Combaret C. Plant growth-promoting rhizobacteria and root system functioning. Front Plant Sci 2013;4:356.

43. Long HH, Schmidt DD, Baldwin IT. Native bacterial endophytes promote host growth in a species-specific manner; phytohormone manipulations do not result in common growth responses. Plos One. 2008;3(7):e2702.

44. Ortiz-Castro R, Contreras-Cornejo HA, Macias-Rodriguez L, Lopez-Bucio J. The role of microbial signals in plant growth and development. Plant Signal Behav. 2009;4(8):701–12.

45. Abbamondi GR, De Rosa S, Iodice C, Tommonaro G. Cyclic dipeptides produced by marine sponge-associated bacteria as quorum sensing signals. Nat Prod Commun. 2014;9(2):229–32.

6

The effects of early leaf removal and cluster thinning treatments on berry growth and grape composition in cultivars Vranac and Cabernet Sauvignon

Marina Bogicevic[1*], Vesna Maras[2], Milena Mugoša[2], Vesna Kodžulović[2], Jovana Raičević[2], Sanja Šućur[2] and Osvaldo Failla[1]

Abstract

Background: The aim of the present work was to investigate the effect of early leaf removal and cluster thinning treatments in the Mediterranean climate on berry growth and how these two techniques affect phenolic profile (especially proathocyanidins) and color characteristics for later wine production. The study was conducted in 2011 in Podgorica, Montenegro. Two grapevine cultivars were selected to compare different ability in flavonoid accumulation: Vranac, with moderate accumulation and Cabernet Sauvignon, usually showing very good accumulation of polyphenols. Four treatments were compared: only leaf removal, only cluster thinning, leaf removal combined with cluster thinning, and no treatment that was used for control (control set).

Results: Early defoliation reduced the yield in both varieties. In Cabernet Sauvignon, defoliation initially delayed berry growth, but at the end, defoliation slightly affected almost all yield parameters (cluster weight, berry weight, and number of berries per cluster), while in cultivar Vranac, defoliation did not modify the berry growth and berry weight. In both varieties, cluster thinning did not affect the berry weight. In the treatments where both defoliation and cluster thinning was applied, a reduction of the cluster weight, berry weight, and berry numbers per cluster was observed. Cabernet Sauvignon showed a greater reactivity to the applied techniques, while Vranac was less reactive. At harvest, no damaged bunches (caused by sunburn) were found in defoliated treatment.

Conclusions: It can be concluded that for both varieties, early defoliation and cluster thinning lead to better soluble solids accumulation than in the control set. The treatments lead to raised concentration of anthocyanins and proanthocyanidins in both varieties. It is confirmed that the highest content of anthocyanins and proanthocyanidins was in the skin extracts of the grapes where both treatments were applied. This is followed by the treatment where only defoliation was applied. The enhanced contents of these compounds per berry in grape variety Vranac are the result of increased synthesis, while in Cabernet sauvignon variety, increased content was due to the less berry weight. The best wine characteristics (alcohol, color intensity, color hue, total anthocyanins, total polyphenols) were found in products, where defoliation was applied.

Keywords: Early defoliation; Grape yield; Berry composition; Proanthocyanidins; Anthocyanins; Polyphenols

* Correspondence: marina.bogicevic@unimi.it
[1]Department of Agricultural and Environmental Sciences, University of Milan, Via Celoria 2, 20133 Milan, Italy
Full list of author information is available at the end of the article

Background

In Montenegro, viticulture is mainly based on autochthonous grapevine varieties among which Vranac variety occupies about 70% of total autochthonous production. Besides autochthonous varieties, big attention is given to international grapevine varieties, in the first place to Cabernet Sauvignon, i.e., to its agrobiological technological and economics characteristics in agro-ecological conditions of sub-region Podgorica.

In modern viticulture, canopy management plays a key role, and it is widely recognized as an important factor in the composition of the resulting wines. Leaf removal in the fruiting zone, both manual and mechanical, is a common practice in high vigor, vertically trained vine canopies [1,2] and it could be applied from flowering to full veraison, thus improving fruit composition and reducing herbaceous wine character [3]. Defoliation is a consolidated practice for improving source-sink relations, photosynthetic capacity, and quality of crop plants; in Mediterranean environment, leaf removal is usually carried out in July during break out of color [4]. Early leaf removal performed in early stage of fruit development is an innovative viticultural practice for regulating yield components and improving grape quality [5-9]. Palliotti et al. [10] reported that vineyard efficiency was influenced by this practice, i.e., yield per vine and cluster weight were limited. Removing the leaves from cluster zone increases the evaporative potential within the fruit zone, lowering the humidity and making the cluster microclimate less conducive for the development of fungal diseases [11]. In the Sangiovese variety, characterized by highly compact clusters, early defoliation significantly reduced the fruit set, yield per shoot, cluster weight, number of berries per cluster, and cluster compactness [6,7,12]. Autochtonous variety, Vranac is also characterized by highly compacted clusters; we decided to apply these agro-techniques in order to reduce the yield parameters hoping that it would have the same beneficial effects as in the case of Sangiovese variety. Some studies reported that the influence of leaf removal on yield components depended on the variety. Leaf removal decreased the yield per vine and cluster weight in Merlot and Sangiovese, while the berry size was unaffected in both varieties. Only in grape variety Merlot, the number of berries per cluster and cluster compactness decreased, while in Cabernet Sauvignon variety, the effect of leaf removal was restrained by berry size [13].

Cluster thinning is described as the suppression of flowers or clusters before full maturation [14]. Therefore, cluster thinning has a direct effect on the source/sink ratio; having less sinks (fruits) photosynthetic assimilation might be improved, increasing grape quality [15]. It inducts physiological adjustments in the plant, improving the maturation's kinetics. Plus, this operation

improves canopy sanitary conditions as thinning allows more enlightenment and fresh air penetration in the vegetation and clusters [16]. Cluster thinning's most evident effect is apparently crop load reduction, but its decrease is not equivalent to the thinning's intensity. Martins [17] found that, for the same intensity of thinning, in two consecutive years, the decline in production has been uneven. In fact, the vine compensates the stack lost, increasing the berry's volume and weight [18]. For this reason, Climaco et al. [19] suggests that cluster thinning must only be executed in the years when vineyards have such a fertility that may possibly undermine production quality. Cluster thinning advances grape maturity, improves grape quality, and also influences the chromatic characteristic of the wine [20].

The aim of the present work was to investigate the effect of early leaf removal and cluster thinning treatments on the berry growth and how these two techniques affect phenolic profile (especially proathocyanidins) and color characteristics on wine production in the Mediterranean climate. Two grapevine cultivars with different abilities in flavonoid accumulation were compared: Vranac with moderate accumulation of flavonoid and Cabernet Sauvignon with very good accumulation of flavonoid.

Methods

The trial was carried out during the 2011 growing season in the commercial vineyard of the Plantaze company in the Cemovsko field in Podgorica (Montenegro), planted with both local Vranac variety as well as Cabernet Sauvignon. The study was conducted in vineyards with uniform growing conditions. The treatments of variety Vranac were established in 10-year-old vineyard, grafted onto Kober 5BB rootstock, trained to a modified double Guyot training system, rows spaced 2.8 m apart and with 0.9 m between plants in the row. The grapevines of Cabernet Sauvignon were planted in 2005 (clone R5), grafted onto 1103P rootstock, and trained to a Guyot training system. The vine had a between-row and within-row spacing of 2.60 m × 0.70 m.

Winter pruning, for both varieties, was carried out leaving 14 buds per vine. In the first week of May, when the shoots reached 20 cm to each vine, shoot thinning was applied and ten shoots per plant were retained.

Yield components

At harvest, cluster number as well as the total yield was recorded on 15 tagged vines per treatment. In the laboratory, the following variables of 25 randomly selected clusters for each treatment were estimated: cluster weight, cluster and berry length and width, and berry number and weight.

Ratio skin/berry is expressed in percents where both weights were individually measured.

Grape juice analysis

The soluble solids of grape juice were determined by refractrometry and pH values achieved by pH meter and titratable acidity (TA) with 0.1 N NaOH and bromothymol blue as indicator (expressed as g/L of tartaric acid equivalents).

Grape skin analysis

For the two grape varieties studied, on days 161, 172, 185, 200, 213, and 222 and harvest time 231, 20 berries were sampled and weighed for each treatment in triplicate. Berry skins were removed manually from the pulp using a laboratory scalpel, weighed, and quickly placed in 50 ml hydro-alcoholic buffer at pH 3.2, containing 2 g/L $Na_2S_2O_5$ and 12% of ethanol. The samples were stored at $-20°C$ until analysis for phenolic compounds was carried out. The total phenolic content of skin extracts of grapes was determined using the Folin-Ciocalteu method [21]. Determination of total anthocyanins was performed using the method described by Di Stefano et al. [22] and total proanthocyanidins described by Di Stefano and Cravero [23]. Total anthocyanins, phenolics, and proanthocyanidins were expressed in mg per kg grapes and mg per berry. All analyses were repeated three times.

Wine analysis

Eight microvinifications were carried out to study the influence of the agricultural treatments to wine composition and quality. All treatments were individually harvested manually at the moment of optimal phenological and technological maturation. For each treatment, approximately 100 kg of grapes were stored overnight in a cool chamber (4°C) and the following day warmed to room temperature before being slightly crushed. All musts were immediately inoculated with selected *Saccharomyces cerevisiae* yeast (Lalvin BDX, Lallemand Inc., Montreal, Canada) and Go-ferm protect (30 g hL^{-1}), yeast nutrient, Fermaid E (25 g hL^{-1}) was added during fermentation of both varieties. Fermentations were conducted in tanks at 25°C for 10 days. The total polyphenols [21], total anthocyanins [22], and total proanthocyanidins [23] of the wine were determined by UV-vis spectrophotometry. Spectrophotometric measurements of absorbance at 420, 520, and 620 nm were made using a 1 mm quartz cuvette. Color intensity was calculated by adding absorbance values at 420, 520, and 620 nm [24]. The tonality of the wine is defined as the ratio of absorbance at 420 and 520 nm [24].

Statistical analysis

Within each variety analysis of variance ANOVA was used to test the main effect using SPSS software (IBM SPSS version19). Comparison of means was performed using Duncan test at $p < 0.05$.

Results and discussion

Berry growth

In the first phase of berry growth, with the Cabernet Sauvignon variety, early defoliation affected the berry growth, causing delay in development compared to those of the control. However, before veraison, defoliated treatment showed a higher average berry weight with no significant differences (Table 1A). During the ripening stage, the NLR-CT did not modify the growth of the berry in confront to the control. The treatment previously defoliated (LR-NCT) showed a delay in growth after veraison; even if at harvest; it did not show a difference by control and thinned treatment. The treatment defoliated and then thinned (LR-CT) stopped in advance the growth and showed an early transition in over ripening. As a result at harvest, the treatment LR-CT showed the lowest an average weight of berry (Table 1B).

In cultivar Vranac during the first part of the berry growth, until veraison, in response to defoliation, there were no significant differences in the berry weight (Table 2A). However, the treatment defoliated presented before veraison a higher berry weight. In the course of ripening, the cluster thinning did not affect the berry growth. The defoliated treatment (LR-NCT), similar to Cabernet Sauvignon, had a developmental delay but reached similar values at harvest to the other treatments.

Overall, all types of treatments showed the maximum weight of the berries in the DOY 213 and decrease in the last 20 days (Table 2B).

Yield components

Table 3 shows the yield production per vine of four experimental conditions for two varieties. The defoliation reduced the ¡yield in both varieties. The impact of this practice on the yield is higher in Cabernet Sauvignon: defoliation reduced the yield by 36% and defoliation with subsequent cluster thinning reduced it by 63%. Reduction was less pronounced in the Vranac variety:

Table 1 Berry weight (g) in Cabernet Sauvignon

DOY		161	172	185	200
A	NLR	6.01 ± 0.84^a	12.01 ± 1.42^a	15.98 ± 1.17^a	15.28 ± 0.62^a
	LR	6.62 ± 0.63^a	11.64 ± 0.75^a	12.02 ± 1.64^b	16.24 ± 1.48^a
DOY		213	222	231	
B	NLR-NCT	22.30 ± 0.60^a	21.70 ± 0.60^a	23.75 ± 1.27^a	
	LR-NCT	20.16 ± 0.34^a	18.49 ± 0.89^b	24.14 ± 0.98^a	
	NLR-CT	20.70 ± 1.48^a	22.01 ± 0.74^a	21.52 ± 1.47^{ab}	
	LR-CT	21.61 ± 2.89^a	18.47 ± 0.91^b	19.41 ± 2.15^b	

To before veraison (A) and after veraison to harvest (B). Treatments: leaf removal (LR), not leaf removal (NLR), leaf removal-not cluster thinning (LR-NCT), not leaf removal-cluster thinning (NLR-CT), leaf removal-cluster thinning (LR-CT), and control (NLR-NCT).

Table 2 Berry weight (g) in Vranac

DOY		161	172	185	200
A	NLR	6.03±1.30[a]	21.13±1.51[a]	24.16±1.24[a]	27.79±3.85[a]
	LR	7.09±1.16[a]	19.98±2.07[a]	24.56±1.93[a]	29.90±3.37[a]
DOY		213	222	231	
B	NLR-NCT	45.30±1.78[a]	40.73±3.97[a]	39.48±1.97[a]	
	LR-NCT	40.81±2.66[a]	38.05±2.25[a]	40.43±2.18[a]	
	NLR-CT	42.54±4.94[a]	45.53±5.76[a]	38.36±1.34[a]	
	LR-CT	44.43±1.71[a]	42.24±5.61[a]	39.19±1.74[a]	

To before veraison (A) and after veraison to harvest (B). Treatments: leaf removal (LR), not leaf removal (NLR), leaf removal-not cluster thinning (LR-NCT), not leaf removal-cluster thinning (NLR-CT), leaf removal-cluster thinning (LR-CT), and control (NLR-NCT).

defoliation reduced the yield by 23%, while defoliation followed by cluster thinning reduced it by 46% compared to the controlling set.

In Cabernet Sauvignon, the defoliation reduced the average weight of the bunch and total berry numbers per cluster. On the other hand, the cluster thinning had no influence on the berry weight. The treatment 'defoliated-cluster thinning' had lower bunch weight and berry weight. The defoliation treatment resulted in the lowest number of berries per cluster, while the 'cluster thinning' resulted in the highest number of berries per bunch. As a consequence, the density of the cluster is higher in the treatment NLR-CT. There are no significant differences in the ratio skin/berry (Table 3).

In cultivar Vranac, early leaf removal slightly reduced the cluster weight. This treatment also resulted in a slightly increased ratio of skin vs. berry weight. There were no recorded differences in the berry weight and in the number of berries per cluster. However, early leaf removal followed by cluster thinning resulted in a lower berry weight and number of berries which determine the lowest average cluster weight.

Grape juice analysis

In both varieties and after all tested treatments, the result was a good accumulation of sugars. Consequently, the control set showed the lowest value in the content of sugar. The best accumulation of soluble solids in Vranac was achieved by the early defoliation followed by the cluster thinning (LR-CT), while only the cluster thinning treatment (NLR-CT) resulted in a greater accumulation of sugars in the variety Cabernet Sauvignon (Table 4). Regarding the total acids, it is noticed that early defoliation induced higher content of total acids in Cabernet Sauvignon variety, when compared to Vranac variety where these treatments had no influence on this parameter.

Total polyphenols in berry skins

Early defoliation in Cabernet Sauvignon in the early stages of berry growth lead to increase in total berry polyphenols (Table 5A). Before veraison, there were no significant differences in the total polyphenols content between the treatments. Increased content in mg/kg grapes was due to the less berry weight and not due to the increase of synthesis (Table 5B).

The total phenols in the variety Vranac, up to veraison, increased due to the effect of early leaf removal (Table 6A). Polyphenols content in mg per berry was also increased in defoliated treatment, due to increased synthesis in the berries (Table 6B).

Total anthocyanins in berry skins

Table 7A shows the total anthocyanin content (mg/kg grapes) in Cabernet Sauvignon from veraison to harvest. No significant differences between the treatments were found, except at harvest time. Results for the control set (NLR-NCT) and early defoliation treatment only (LR-NCT) resulted in lower total anthocyanins when compared to results after cluster thinning only (NLR-CT) and early defoliation

Table 3 Yield components and cluster and berry characteristics at harvest recorded in Cabernet Sauvignon and Vranac

		Clusters/ Vine	Yield/vine (kg)	Cluster wt (g)	Berry wt (g)	No of berries/ cluster	Compactness index	Skins/ berry %	
Cabernet Sauvignon	NLR-NCT	18	1.48±0.48[c]	134±3.8[b]	1.13±0.02[b]	113±3.3[b]	0.18±0.01[a]	17.17±2.1[a]	
	LR-NCT	17	0.96±0.33[b]	117±3.9[b]	1.13±0.02[b]	97±3.2[ab]	0.18±0.01[a]	20.48±2.3[a]	
	NLR-CT	10	0.91±0.32[b]	163±4.2[c]	1.08±0.02[b]	152±3.9[c]	0.28±0.02[b]	18.59±2.4[a]	
	LR-CT	9	0.56±0.17[a]	86±4.3[a]	0.94±0.02[a]	89±4.4[a]	0.22±0.02[a]	17.18±3.5[a]	
Vranac	NLR-NCT	13	2.35±0.14[b]	176±8.0[a]	1.96±0.03[a]	89±4.2[a]	0.45±0.01[a]	16.10±2.8[a]	
	LR-NCT	11	1.82±0.43[a]	161±8.1[a]	1.96±0.03[a]	80±3.6[a]	0.42±0.01[a]	19.86±3.0[a]	
	NLR-CT	9	1.64±0.28[a]	170±7.4[a]	1.94±0.03[a]	89±4.1[a]	0.47±0.03[a]	14.73±3.4[a]	
	LR-CT	9	1.27±0.15[a]	147±8.2[a]	1.87±0.03[a]	75±4.5[a]	0.32±0.01[b]	16.73±3.2[a]	

Vines subjected to early defoliation (LR-NCT), cluster thinning (NLR-CT), early defoliation and cluster thinning (LR-CT), or control (NLR-NCT).

Table 4 Must composition at harvest recorded in Cabernet Sauvignon and Vranac

		TSS (Brix)	Titratable acidity (g tartaric acid/L)	pH
Cabernet Sauvignon	NLR-NCT	22.40	7.40	3.59
	LR-NCT	24.40	8.90	3.53
	NLR-CT	25.60	7.57	3.59
	LR-CT	24.80	7.95	3.58
Vranac	NLR-NCT	22.00	6.77	3.57
	LR-NCT	22.80	6.60	3.62
	NLR-CT	22.80	6.75	3.61
	LR-CT	24.60	6.53	3.59

Vines subjected to early defoliation (LR-NCT), cluster thinning (NLR-CT), early defoliation, and cluster thinning (LR-CT) or control (NLR-NCT). The reported values are from must just crushed for microvinification.

Table 6 Total polyphenols and polyphenols per berry in Vranac from berry set to before veraison

A - total polyphenols (mg/kg)

DOY	172	185	200
NLR	3006±401^a	2358±148^a	1894±326^a
LR	3505±147^a	2922±295^b	2248±357^a

B - polyphenols per berry (mg)

DOY	172	185	200
NLR	3.18±0.51^a	2.84±0.15^a	2.62±0.47^a
LR	3.49±0.26^a	3.59±0.51^b	3.32±0.26^b

Treatments: leaf removal (LR) and not leaf removal (NLR).

followed by the cluster thinning treatments (LR-CT) which resulted in higher total athocyanins concentration. This effect seems to be related to berry growth and not to the increase of the synthesis, as can be seen in Table 7B.

No significant differences between the treatments during maturation were found in the total anthocyanin content for the cultivar Vranac, except that at DOY 222, lowest content in the thinned treatment (Table 8A) is observed. However, the highest concentration was in the treatment defoliated-thinned. The concentration of anthocyanins per berry is not associated to berry growth albeit to the increase in their synthesis (Table 8B). Besides, agricultural practices defoliation and cluster thinning have had an impact to the content of anthocyanins, increasing it compared to the control.

Total proanthocyanidins in berry skins

The content of proanthocyanidins in Cabernet Sauvignon, from berry set to before veraison, reacts in the same way as polyphenols: defoliation causing retardation of the berry growth and increased content of proanthocyanidins (Table 9A). Even in this case, the effect was not due to the increased synthesis per berry (Table 9B).

After veraison, for all treatments, a further increase in the content of proanthocyanidins was observed, which could be due to the persistence in the synthesis or different extraction of proanthocyanidins from the skins. All viticultural practices led to higher content of proanthocyanidins at harvest, without significant differences (Table 10A,B).

In Vranac, before veraison, greater accumulation of total proanthocyanidins and proanthocyanidins per berry in the treatment defoliated was observed (Table 11A,B). At harvest, the highest contents per berry and in mg/kg grapes was observed in the treatment defoliated-cluster thinned. However, treatments defoliation and cluster thinning enhanced proanthocyanidins concentration compared to the controlling set (Table 12A,B).

Table 5 Total polyphenols and polyphenols per berry in Cabernet Sauvignon from berry set to before veraison

A - total polyphenols (mg/kg)

DOY	172	185	200
NLR	5377±827^a	4372±327^a	3384±303^a
LR	5772±479^a	4782±529^a	3375±450^a

B - polyphenols per berry (mg)

DOY	172	185	200
NLR	3.19±0.36^a	3.49±0.38^a	2.58±0.26^a
LR	3.35±0.31^a	2.85±0.21^b	2.73±0.19^a

Treatments: leaf removal (LR) and not leaf removal (NLR).

Table 7 Total anthocyanins and anthocyanins per berry in Cabernet Sauvignon in 2011 from veraison to harvest

A - total anthocyanins (mg/kg)

DOY	200	213	222	231
NLR-NCT	153±75^a	1229±50^b	1846±114^a	2165±107^{ab}
LR-NCT	89±13^a	1360±127^a	1845±559^a	2104±195^a
NLR-CT	86±53^a	1356±75^a	1954±167^a	2439±129^{bc}
LR-CT	96±26^a	1499±173^a	2043±65^a	2522±230^c

B - anthocyanins per berry (mg)

DOY	200	213	222	231
NLR-NCT	0.11±0.05^a	1.62±0.10^a	2.00±0.10^{ab}	2.57±0.15^a
LR-NCT	0.07±0.01^a	1.37±0.11^a	1.69±0.46^a	2.53±0.13^a
NLR-CT	0.07±0.04^a	1.40±0.08^a	2.15±0.11^b	2.62±0.19^a
LR-CT	0.08±0.03^a	1.64±0.40^a	1.89±0.13^{ab}	2.43±0.06^a

Treatments: leaf removal-not cluster thinning (LR-NCT), not leaf removal-cluster thinning (NLR-CT), leaf removal-cluster thinning (LR-CT), and control (NLR-NCT).

Table 8 Total anthocyanins and anthocyanins per berry in Vranac in 2011 from veraison to harvest

A - total anthocyanins (mg/kg)

DOY	200	213	222	231
NLR-NCT	225±88^a	950±133^a	1922±190^{ab}	2059±238^a
LR-NCT	205±16^a	1061±176^a	2173±133^a	2195±69^a
NLR-CT	217±134^a	1124±133^a	1755±175^b	2228±144^a
LR-CT	192±84^a	1169±90^a	2168±156^a	2351±173^a

B - anthocyanins per berry (mg)

DOY	200	213	222	231
NLR-NCT	0.34±0.10^a	2.14±0.21^a	3.86±0.65^a	4.06±0.51^a
LR-NCT	0.29±0.03^a	2.16±0.29^a	4.11±0.42^{ab}	4.44±0.38^a
NLR-CT	0.27±0.16^a	2.37±0.03^a	4.26±0.72^{ab}	4.25±0.70^a
LR-CT	0.29±0.11^a	2.59±0.11^a	4.55±0.30^b	4.60±0.17^a

Treatments: leaf removal-not cluster thinning (LR-NCT), not leaf removal-cluster thinning (NLR-CT), leaf removal-cluster thinning (LR-CT), and control (NLR-NCT).

Table 10 Total proanthocyanidins and proanthocyanidins per berry in Cabernet Sauvignon from veraison to harvest

A - total proanthocyanidins (mg/kg)

DOY	213	222	231
NLR-NCT	3819±511^a	4520±87^a	5119±1209^a
LR-NCT	4928±624^a	6072±586^{ab}	5414±1162^a
NLR-CT	4257±219^a	5245±1006^a	6074±802^a
LR-CT	4613±150^a	7283±1428^b	7157±999^a

B - proanthocyanidins per berry (mg)

DOY	213	222	231
NLR-NCT	5.01±0.57^a	4.90±0.20^a	6.04±1.19^a
LR-NCT	4.96±0.56^a	5.61±0.58^{ab}	6.55±1.53^a
NLR-CT	4.40±0.24^a	5.78±1.21^{ab}	6.55±1.12^a
LR-CT	4.97±0.52^a	6.71±1.24^b	6.95±1.25^a

Treatments: leaf removal-not cluster thinning (LR-NCT), not leaf removal-cluster thinning (NLR-CT), leaf removal-cluster thinning (LR-CT), and control (NLR-NCT).

Wine analysis

Table 13 shows descriptors of wines made from grapes of four experiments for two varieties. In the wines of cultivar Cabernet Sauvignon, higher alcohol content was found in the cluster thinning treatment (as a result of the best accumulation of sugar) and lowest in the control set. All the wines exhibit similar color hue. The value of total anthocyanins, polyphenols, proanthocyanidins, and color intensity was highest in the treatment defoliated-cluster thinned followed by the treatment defoliated. The lowest values of all parameters were found in the control set (no treatments). The ethanol content in Vranac wines, in accordance with sugar accumulation, was the same for the treatment 'defoliated' and treatment 'cluster thinning', and the highest in the treatment 'defoliated-cluster thinned' and the lowest in the control set. The values of color intensity and color hue were very similar for all the treatments, even though some slightly higher value was found in the treatment 'defoliated-cluster thinned'. The content of total anthocyanins, polyphenols, and proanthocyanidins is highest in the treatment 'defoliated-cluster thinned' followed by the treatment 'defoliated'. The best wine characteristics were found in products from the plots where defoliation was applied. These results could be due to better extraction of polyphenolic compounds in wine. Internal tasting of all experimental wines made of both varieties of grapes showed that wines made of grapes where leaf removal and cluster thinning were applied were characterized by fuller body, higher fruitiness aromas, and more intense color. All of this resulted in enhanced complexity of aromas.

Experimental

In the experimental design, four treatments were compared: a) not defoliated - not thinned (NLR-NCT), leaf removal - not cluster thinning (LR-NCT), not leaf removal - cluster thinning (NLR-CT), and leaf removal - cluster thinning (LR-CT).

Table 9 Total proanthocyanidins and proanthocyanidins per berry in Cabernet Sauvignon from berry set to before veraison

A - total proanthocyanidins (mg/kg)

DOY	172	185	200
NLR	9037±1205^a	7448±1000^a	5682±436^a
LR	10438±633^b	9087±1174^b	5992±1370^a

B - proanthocyanidins per berry (mg)

DOY	172	185	200
NLR	5.36±0.27^a	5.91±0.58^a	4.34±0.34^a
LR	6.08±0.58^b	5.41±0.63^a	4.80±0.48^a

Treatments: leaf removal(LR) and not leaf removal (NLR).

Table 11 Total proanthocyanidins and proanthocyanidins per berry in Vranac from berry set to before veraison

A - total proanthocyanidins (mg/kg)

DOY	172	185	200
NLR	5470±624^a	4837±582^a	3518±784^a
LR	6215±596^a	5771±730^a	3788±535^a

B - proanthocyanidins per berry (mg)

DOY	172	185	200
NLR	5.83±1.85^a	5.83±0.63^a	4.82±0.85^a
LR	6.20±0.78^a	7.14±1.41^b	5.61±0.55^a

Treatments: leaf removal (LR) and not leaf removal (NLR).

Table 12 Total proanthocyanidins and proanthocyanidins per berry in Vranac from veraison to harves

A - total proanthocyanidins (mg/kg)

DOY	213	222	231
NLR-NCT	3155 ± 235^a	3455 ± 845^a	2494 ± 455^a
LR-NCT	3250 ± 445^a	3674 ± 640^a	2780 ± 268^{ab}
NLR-CT	2933 ± 489^a	2985 ± 397^a	3010 ± 434^{ab}
LR-CT	3439 ± 606^a	2777 ± 736^a	3731 ± 756^b

B - proanthocyanidins per berry (mg)

DOY	213	222	231
NLR-NCT	7.16 ± 0.78^a	6.99 ± 1.58^a	4.94 ± 1.08^a
LR-NCT	6.67 ± 1.31^a	6.98 ± 1.24^a	5.64 ± 2.45^{ab}
NLR-CT	5.84 ± 0.69^a	7.32 ± 0.11^a	5.76 ± 0.73^{ab}
LR-CT	7.61 ± 1.12^a	6.00 ± 2.22^a	7.28 ± 1.26^b

Treatments: leaf removal-not cluster thinning (LR-NCT), not leaf removal-cluster thinning (NLR-CT), leaf removal-cluster thinning (LR-CT), and control (NLR-NCT).

Defoliated treatment was applied in full bloom on day of year (DOY) 152 corresponding to the phenophase 23 according to the grapevine growth stage classification proposed by Coombe [25], which consisted of manual removal of the first eight basal leaves of each shoot. All lateral shoots were retained. Cluster thinning was conducted on DOY 200, at mid veraison, at stage 35 [25], where the distal cluster was removed leaving one cluster per shoot. The elementary experimental plot was composed of 20 consecutive vines; each treatment was replicated in three elementary plots, randomly positioned in the vineyard.

Conclusions

Objectives of the research program were to study the effect of some environmental and physiological aspects on the intensity of flavonoid synthesis. The study was conducted in 2011 in Podgorica, Montenegro. Two grapevine cultivars were selected to compare their ability in flavonoid accumulation: autochtonous variety Vranac,

with moderate accumulation of flavonoids and Cabernet Sauvignon with good accumulation of polyphenols. The following experimental treatments were compared: early leaf removal (flowering time), cluster thinning (veraison time) and combination of both treatments. The early defoliation reduced the yield per vine in Cabernet Sauvignon and Vranac. In Cabernet Sauvignon, defoliation initially delayed berry growth, but at harvest, only the treatment 'defoliation-cluster thinning' had significantly lower berry weight. In cultivar Vranac, defoliation did not modify the berry growth and berry weight. In both varieties, cluster thinning had no effect on the berry weight. In the treatment 'defoliated-thinned', reduction of the cluster weight, berry weight, and berry number per cluster is observed. This is probably the consequence of a lower fruit set, where the defoliation had a greater impact on the first cluster. Cabernet Sauvignon showed a greater reactivity to the applied techniques, compared to Vranac. At harvest, no damaged bunches (caused by sunburn) were found in defoliated treatment. Early defoliation and cluster thinning in both varieties raised the concentration of anthocyanins and proanthocyanidins. The enhanced contents of these compounds per berry in grape variety Vranac are the result of increased synthesis, while in Cabernet sauvignon variety, increased content was due to the less berry weight. Defoliation and cluster thinning led to better soluble solid accumulation than in the control sets (no treatments applied). The skin extracts contained the highest content of anthocyanins and proanthocyanidins in the treatment defoliated-thinned followed by the treatment thinned, while these contents were higher in wines from the vineyards where defoliation was applied. It could be due to better extraction of these compounds during winemaking. Additional work is in progress to verify that early leaf removal and cluster thinning do indeed result in better quality of Vranac wine.

Table 13 Chemical composition in Cabernet Sauvignon and Vranac wines

		Alcohol content (% vol)	Colour intensity	Colour hue	Total anthocyanins (mg/L)	Total polyphenols (mg/L)	Total proanthocyanidins (mg/L)
Cabernet Sauvignon	NLR-NCT	13.43	1.42	0.63	295	1897	872
	LR-NCT	14.59	1.76	0.62	333	2311	1112
	NLR-CT	15.36	1.69	0.64	318	2028	910
	LR-CT	14.94	1.89	0.64	353	2749	1224
Vranac	NLR-NCT	13.31	1.86	0.57	389	1532	308
	LR-NCT	13.67	1.86	0.59	392	1711	314
	NLR-CT	13.68	1.85	0.59	344	1548	292
	LR-CT	14.68	2.06	0.61	467	1842	555

Treatments: leaf removal-not cluster thinning (LR-NCT), not leaf removal-cluster thinning (NLR-CT), leaf removal-cluster thinning (LR-CT), and control (NLR-NCT).

Competing interests

The authors declare that they have no competing interests.

Authors' contribution

VM was the project coordinator and together with OF and MB contributed to project conception and overall experimental design. MB participated in experimental designing, performed spectrophotometric analysis and statistical data analysis. MM, VK, JR, SŠ participated in experimental designing, performed experimental treatments, vinification and analysis of quality control of grapes and wines. VM and OF supervised final data analysis and made the concept of results and conclusions. All co-authors participated in critical reading of the manuscript and approved the final manuscript.

Acknowledgment

This research was supported by the following project, Ecophysiological and molecular aspects of the synthesis and polymerization of flavanols in the grapes of Vranac and Cabernet sauvignon cultivars and their impact on the quality of wine", co-funded by the Ministry of Science of Montenegro.

Author details

[1]Department of Agricultural and Environmental Sciences, University of Milan, Via Celoria 2, 20133 Milan, Italy. [2]"13. Jul Plantaze" a.d., Put Radomira Ivanovica 2, 81000 Podgorica, Montenegro.

References

1. Smart R (1985) Principles of grapevine canopy microclimate manipulation with implications for yield and quality: a review. Am J Enol Vitic 35:230–239
2. Bledsoe AM, Kliewer WM, Marois JJ (1988) Effects of timing and severity of leaf removal on yield and fruit composition of Sauvignon Blanc grapevines. Am J Enol Vitic 1:49–54
3. Kliewer WM, Bledsoe A (1987) Influence of hedging and leaf removal on canopy microclimate, grape composition, and wine quality under California conditions. Acta Horticult 206:157–168
4. Nicolosi E., Continella A., Gentile A., Cicala A., Ferlito F., (2012). Influence of early leaf removal on autochthonous and international grapevines in Sicily. Scientia Horticulturae, pp. 1–6
5. Poni S, Casalini L, Bernizzoni F, Civardi S, Intrieri C (2006) Effects of early defoliation on shoot photosynthesis, yield components, and grape quality. Am J Enol Vitic 57:397–407
6. Intrieri C, Filippetti I, Allegro G, Centinari M, Poni S (2008) Early defoliation (hand vs. mechanical) for improved crop control and grape composition in Sangiovese (Vitis vinifera L.). Aust J Grape Wine Res 14:25–32
7. Poni S, Bernizzoni F, Civardi S (2008) The effect of early leaf removal on whole-canopy gas exchange and vine performance of Vitis vinifera L. 'Sangiovese'. Vitis 47(1):1–6
8. Dokoozlian N, Wolpert J (2009) Recent advances in grapevine canopy management. In: Proceedings of the International Symposium. Davis, CA, p 62
9. Diago MP, Vilanova M, Tardaguila J (2010) Effects of timing of manual and mechanical early defoliation on the aroma of Vitis vinifera L. Tempranillo wine. Am J Enol Vitic 61(3):382–391
10. Palliotti A, Gatti M, Poni S (2011) Early leaf removal to improve vineyard efficiency gas exchange, source-to-sink balance, and reserve storage responses. Am J Enol Vitic 62(2):219–228
11. English JT, Bledsoe AM, Marois JJ, Kliewer WM (1990) Influence of grapevine canopy management on evaporative potential in the fruit zone. Am J Enol Vitic 41:137–141
12. Poni S, Bernizzoni F, Civardi S, Libelli N (2009) Effects of pre-bloom leaf removal on growth of berry tissues and must composition in two red Vitis vinifera L. cultivars. Aust J Grape Wine Res 15:185–193
13. Yorgos K, Georgiadou A, Tikos P, Kallithraka S, Koundouras S (2012) Effects of severity of post-flowering leaf removal on berry growth and composition of three red Vitis vinifera L. Cultivars grown under semiarid conditions. J Agric Food Chem 60:6000–6010
14. Palliotti A, Cartechini A (2000) Cluster thinning effects on yield and grape composition in different grapevine cultivars. Acta Hort (ISHS) 512:111–120
15. Reynolds A, Price S, Wardle D, Watson B (1994) Fruit environment and crop level effects on Pinot noir. I. Vine performance and fruit composition in the British Columbia. Am J Enol Vitic 45:452–459
16. Smithyman RP, Howell GS, Miller DP (1998) The use of competition for carbohydrates among vegetative and reproductive sinks to reduce fruit set and botrytis bunch rot in Seyval blanc grapevines. Am J Enol Vitic 49:163–170
17. Martins S. (2007). Monda de cachos na casta 'Touriga Nacional'. Efeitos no rendimento e qualidade. Tese Mestrado em Viticultura e Enologia. Universidade Técnica de Lisboa. Universidade do Porto.
18. Rubio J. A. (2002). Riego y aclareo de racimos: efectos en la actividad fisiolofica, en el control del rendimiento y en la calidad de la uva del cv. tempranillo (Vitis vinifera L.) Universidad Politécnica de Madrid, Escuela de Agrónomos.
19. Clímaco P, Teixeira K, Ferreirinho MC (2005) Efeitos da monda de cachos no rendimento e qualidade da cv. Alicante Bouschet. Vinea. Revista Viticultura Alentejo, Abril-Junho, pp 13–16
20. Giloz RR, Vilapez JI, Martínez-Cutillas A (2009) Effects of cluster thinning on anthocyanin extractability and chromatic parameters of syrah and tempranillo grapes and wines. J Int Sci Vigne Vin 43:45–53
21. Slinkard K, Singleton VL (1977) Total phenol analysis: automation and comparison with manual methods. Am J Enol Vitic 28:49–55
22. Di Stefano R., Cravero M. C., Gentilini N. (1989). Metodi per lo studio dei polifenoli dei vini. L'Enotecnico pp. 83–89.
23. Di Stefano R, Cravero MC (1991) Metodi per lo studio dei polifenoli. Riv Vitic Enol 2:37–45
24. Glories Y (1984) The color of red wines. Part 2. Measurement, origin, and interpretation. Conn Vigne Vin 4:253–271
25. Coombe BG (1995) Adoption of a system for identifying grapevine growth stages. Aus J Grape Wine Res 1:104–110

7

Characterization of terrestrial dissolved organic matter fractionated by pH and polarity and their biological effects on plant growth

Rachel L Sleighter[1,2]*, Paolo Caricasole[1], Kristen M Richards[1], Terry Hanson[1] and Patrick G Hatcher[1,2]

Abstract

Background: Humic substances are ubiquitous in the environment, complex mixtures, and known to be beneficial to plant growth. To better understand and identify components responsible for plant growth stimulation, a terrestrial aquatic DOM sample was fractionated according to pH and polarity, obtaining acid-soluble and acid-insoluble portions, as well as acid-soluble hydrophobic and hydrophilic fractions using C18. The various fractions were characterized then evaluated for their biological effects on plant growth using bioassays with corn at two carbon rates.

Results: Approximately 43% and 57% of the carbon, and 31% and 69% of the iron, was found in the acid-insoluble and acid-soluble fractions, respectively. Upon separating the acid-soluble portion using C18 extraction, about 64% and 36% of the carbon (and 96% and 4% of the iron) was present in the hydrophilic and hydrophobic fractions, respectively. The acid-insoluble portion was more aromatic and less oxygenated than the acid-soluble fraction. The hydrophilic filtrate was oxygen-rich and contained mostly tannin-like molecules, while the hydrophobic retentate was more aromatic and lignin-like. During bioassay testing, it was found that more hydrophilic samples (those that are more oxygenated) yielded the highest response for shoot measurements. For root measurements, the lower DOC rate (0.01 mg/L C) gave better results than the higher DOC rate (0.1 mg/L C). Also, the hydrophobic, less oxygenated acid-insoluble sample performed better than the more hydrophilic acid-soluble portion. The polarity fractions at the lower carbon application showed that larger root systems occurred when there was more hydrophobic C18 retentate material present. The opposite was true for the root system at the higher carbon application, where larger roots existed when more hydrophilic C18 filtrate material was present.

Conclusions: Compositional differences were found when comparing the acid-soluble versus acid-insoluble portions and the hydrophobic versus hydrophilic C18 fractions, and activity with respect to plant stimulation was discerned. While a carbon rate affect was observed during foliar application to corn plants (with the lower carbon rate generally yielding the best biological stimulation), the various observed trends indicate that plant response is due to not only the amount of carbon present but also the type of carbon.

Keywords: Humic substances; Dissolved organic matter; C18 solid phase extraction; EEMs; FTICR-MS; Proton NMR; Plant growth; Bioassays

* Correspondence: rsleighter@fbsciences.com
[1]Research and Development, FBSciences, Inc, 4111 Monarch Way, Suite 408, Norfolk, VA 23508, USA
[2]Department of Chemistry and Biochemistry, Old Dominion University, Norfolk, VA 23529, USA

Background

Humic substances, which are ubiquitous in the environment, are present in varying amounts in all natural waters, soils, and sediments [1-3]. They are complex mixtures, containing thousands of individual molecules, all varying in their individual structure, function, reactivity, and polarity. Humic substances have long been regarded as beneficial to soil fertility and plant growth, and there are multiple excellent reviews on the subject that encompass decades of studies [4-6]. The positive outcomes are likely a combination of both direct and indirect effects, and humic substances have been shown to suppress the effect of pathogens and fungi, enhance lateral root development and root length/density, increase the cation exchange capacity and pH buffering capacity of soils, increase the availability and uptake of micronutrients, and promote growth and development via a hormone-like mechanism [7-13]. While there are many studies that have demonstrated improved plant health by treatment with humic substances, several questions remain, such as the best way to treat plants (seed vs. soil vs. foliar application), the rate/amount for treatment, and when to perform application(s), among numerous others. Because there are so many variables associated with the treatments on an agricultural level, there are sometimes conflicting and controversial results. It is likely that the optimal method and quantity for treatment varies depending on the crop, soil properties, and environmental conditions, as the mechanism for enhanced plant growth is uncertain.

Extensive chemical characterization of humic substances, and dissolved organic matter (DOM) in general, by use of advanced analytical techniques [14] has increased our understanding of the sources and composition of these complex mixtures. While numerous studies have highlighted the ability of humic substances (in general) to enhance plant response and growth parameters, the new challenge is to ascertain what portion or specific component(s) of the humic substances is most responsible for generating the positive response [15]. Nardi et al. [16] found that the lowest molecular weight portion of a humic acid fractionated by size exclusion chromatography was most active in stimulating plant metabolism, and this fraction was the most hydrophilic, containing more carbohydrates and less lignin-derived material, when compared to the larger size fractions. Other studies also found a link between hydrophobicity and plant response, but the more hydrophobic humic acids were those most active in plant response, with some humics acids increasing root length while others increased root density [17,18]. It is not uncommon for studies to present conflicting findings, especially when the various studies perform applications of the humic substances in different ways to different plants under different growing conditions, and also varying in concentration. It is clear that further research is needed to better understand the effects of humic substances on plants, especially as the use of humic substances for agricultural purposes is becoming more widely accepted and promoted.

Here, a terrestrial aquatic DOM sample was fractionated according to pH solubility and then polarity. The whole, original sample is available commercially from FBSciences, Inc. and has been shown to yield positive plant growth responses when utilized in agricultural applications. However, there is a need to identify what fractions of this DOM are most responsible for plant responses, so that we may employ further refinements to isolate and make available a more effective product. Thus, the whole sample was acidified to obtain the acid-insoluble and acid-soluble portions (similar to the method used by the International Humic Substances Society, [19]), and then, the acid-soluble portion was fractionated further into polar and non-polar fractions using C18 solid phase extraction (simulating that which occurs during reversed phase liquid chromatography, [20-22]). Material that passes through the C18 is polar, hydrophilic, and is referred to as the filtrate, while the material that is retained by the C18 (i.e., the retentate) is non-polar and hydrophobic [23]. The various fractions were chemically characterized using an array of advanced analytical techniques. Ultraviolet visible (UV/Vis) absorbance spectroscopy and excitation emission matrix spectroscopy (EEMs) were utilized to understand the optical properties of the chromophoric and fluorescent DOM [24,25]. To understand bulk chemical functionalities, liquid-state proton nuclear magnetic resonance (^{1}H NMR) spectroscopy was employed [26]. Finally, to obtain molecular-level information, electrospray ionization Fourier transform ion cyclotron resonance mass spectrometry (ESI-FTICR-MS) was used [27-29]. After characterization, the original whole and fractionated samples were subjected to plant bioassays at two carbon concentrations. Bioassays (performed *in vivo*) are utilized as a tool for measuring the effects of the different sample types on plant activity, in an effort to correlate molecular level details with plant response (i.e., root and shoot elongation, fresh and dry weights).

The two main goals of this study were to 1) chemically characterize the composition of the DOM fractionated according to pH solubility and polarity using a variety of advanced analytical techniques, and 2) demonstrate statistically significant responses upon treating plants with the fractionated samples in comparison to a control, in order to examine the influence that fractionation of the NOM has on plant growth and biological response.

Methods

Sample Fractionation

In this study, a terrestrial aquatic DOM sample was utilized for fractionation into various components. The source of the DOM is from fresh water in Northern Europe, and the final DOM product has been concentrated by a proprietary method and is commercially available from FBSciences, Inc. The "whole" sample was acidified drop-wise with 12 M HCl (trace metal grade, Fisher Scientific) to pH < 2. Precipitation was allowed to occur at room temperature for 40 hours. The sample was centrifuged to isolate the solid "acid-insoluble" portion from the liquid "acid-soluble" portion. The "acid-insoluble" solids were rinsed with acidified ultra-high quality (UHQ) water and freeze-dried to obtain a dry powder. The "acid-soluble" portion was fractionated further using C18 solid phase extraction (SPE, [30]), to obtain the hydrophobic retentate portion (i.e., what is retained by the C18 resin) and the hydrophilic filtrate portion (i.e., that which passes through the C18 resin). The C18 extraction disk (3 M Empore) was cleaned and activated using methanol (LC-MS grade, Fisher Scientific) and acidified UHQ water. The "acid-soluble" sample was passed through the C18 disk at a flow rate <25 mL/min, and the solution that passed through the resin was collected as the "filtrate". That which was adsorbed onto the resin was rinsed with acidified UHQ water and eluted off the resin with LC-MS grade methanol, to give the "retentate" portion.

Dissolved organic carbon (DOC) and Iron (Fe) Analysis

The C18 filtrate and retentate samples were rotary evaporated to remove methanol and replaced with UHQ water. The dried acid-insoluble fraction was re-dissolved in water amended with ammonium hydroxide (NH_4OH, final concentration 0.05%, pH 9). All fractions were then analyzed for their dissolved organic carbon (DOC) and iron (Fe) concentrations. Samples were analyzed for their DOC concentrations by high temperature catalytic combustion to CO_2 on a Shimadzu TOC-V$_{CSN}$ total organic carbon analyzer calibrated with potassium hydrogen phthalate (Shimadzu). Samples were analyzed for their Fe content on a Perkin Elmer AAnalyst 200 Atomic Absorption Spectrometer calibrated with an iron reference standard solution (Fisher Scientific).

UV/Vis and EEMs Analysis

UV/Vis absorbance values were measured simultaneously as EEMs spectra were acquired on a Horiba Scientific Aqualog spectrophotometer with an excitation range of 240–600 nm at 3 nm intervals and an emission range of 213–623 at 3.368 nm intervals, with an integration time of 1 sec. All samples were analyzed at a range of DOC values (approximately 0.5-8 mg/L C), and acquired spectra were corrected for their inner filter effects and Rayleigh masking using the Aqualog V3.6 software, which was also utilized for area integrations. Humic-like peaks A and C were integrated at [Ex 240–300 nm; Em 400–500 nm] and [Ex 300–360 nm; Em 400–500 nm], respectively, while peptide-like peak T was integrated at [Ex 240–300 nm; Em 250–350 nm] [12]. Specific UV absorption at 254 nm (SUVA254) was calculated based on the absorbance at 254 nm and the DOC concentration [31], and the fluorescence index (FI) was calculated from the ratio of emission intensity at 470:520 nm at excitation 370 nm [24,32].

Cation Exchange Resin Procedure

Because these samples contained a significant amount of Fe, cation exchange resin was utilized to prepare the samples for NMR and FTICR-MS, as these analyses are known to be hindered by the presence of paramagnetic species and ionic salts, respectively. The whole, acid-soluble fraction, and C18 filtrate were prepared using cation exchange resin (AG MP-50 resin, hydrogen form, 100–200 dry mesh size, Bio-Rad), similar to the guidelines established by the International Humic Substances Society [19]. The samples were prepared in batches, according to manufacturer guidelines, where the sample was added directly to the cleaned resin, stirred for 60 minutes, and then centrifuged to remove the resin from the sample. Using 3 steps of resin addition, 80-90% of the Fe was removed. It was not necessary to use the cation exchange resin on the acid-insoluble or C18 retentate samples, as these samples contained significantly less Fe.

NMR Analysis

The liquid-state ^{1}H NMR spectra were obtained on a 400 MHz Bruker Biospin Avance III NMR equipped with a broadband inverse probe using a water suppression pulse program (optimized WATERGATE pulse sequence [26]). This sequence suppresses the large peak due to water protons (at 4.7 ppm), allowing for the detection of protons from the natural DOM that would otherwise be obscured. This program also eliminates pre-concentration and re-dissolution steps that potentially lose an unknown portion of the DOM. The acid-insoluble sample was re-dissolved using sodium deuteroxide in D_2O diluted to pH 9 with UHQ water. All samples had a final sample composition of 90:10 H_2O:D_2O. Spectra were acquired with a 1 msec recycle delay and a time domain of 16 k, with 1000 scans (giving an approximate 1 hour analysis time). The scans were co-added and the summed free induction decay (FID) signal was exponentially multiplied and zero-filled once. The spectra were processed with a line broadening of 10 Hz.

For both NMR and FTICR-MS, the C18 filtrate and retentate samples were analyzed together in filtrate: retentate ratios of 100:0, 75:25, 50:50, 25:75, and 100:0. These ratios were also used during bioassays (described in detail below), in order to better understand how differing ratios of these two very different fractional components can influence plant response.

FTICR-MS Analysis

The samples prepared with cation exchange resin, described in the NMR section above, were also utilized for FTICR-MS analysis. The acid-insoluble sample was redissolved using UHQ water amended with NH_4OH (final concentration 0.05%, pH 9). All samples were diluted to give a final sample composition of 1:1 H_2O: MeOH and were continuously infused into an Apollo II ESI ion source (operating in negative ion mode) of a Bruker Daltonics 12 Tesla Apex Qe FTICR-MS using a syringe pump operating at 120 μL/hr. ESI voltages were optimized for each analysis using a spray shield voltage of 3.3-3.5 kV and a capillary voltage of 4.0-4.2 kV, yielding consistent and stable ESI spray shield and capillary currents of 180–210 nA and 20–30 nA, respectively. Ions were accumulated for 2–5 sec in a hexapole before being transferred to the ICR cell, where 300 transients, collected with a 4 MWord time domain, were added, giving about a 30–40 min total analysis time. The summed FID signal was zero-filled once and Sine-Bell apodized prior to fast Fourier transformation and magnitude calculation using the Bruker Daltonics Data Analysissoftware.

Prior to data analysis, all samples were externally calibrated with a polyethylene glycol standard and internally calibrated with fatty acids, dicarboxylic acids, and other naturally present compounds that are part of various CH_2 homologous series [33]. All m/z lists, created using a signal to noise threshold of 3, were exported for further analysis. Mass spectra were found to be reproducible, and the various samples were significantly different from one another when compared to instrumental duplicates of the same sample [34]. A molecular formula calculator generated empirical formula matches for all samples using carbon, hydrogen, oxygen, nitrogen, and sulfur with atomic ranges of $C_{5-50}H_{5-100}O_{1-30}N_{0-1}S_{0-1}$. Molecular formulae were assigned based on previously described rules [35], and the calculated theoretical m/z values of the assigned formulae agreed with measured m/z values with an error value of ≤ 0.5 ppm. For all samples, 82-92% of all peaks were assigned a unique molecular formula (excluding contributions from ^{13}C isotopes), and these formulae accounted for 91-97% of the summed total spectral peak magnitude. From the molecular formula assignments,

average (by number and magnitude-weighted) mass spectral characteristics were calculated [23].

Statistical Analysis

In order to reveal the subtle differences among the NMR spectra and, separately, among the formulae assigned to FTICR-MS data, principal component analysis (PCA) was utilized. PCA, which assumes a linear relationship, reduces a multidimensional space into fewer dimensions, where the first dimension (i.e., the first principal component, PC1) explains the most variance. The second dimension (PC2), which is orthogonal to the first, explains most of the residual variance. The PCA output gives scores and loadings, which represent the projections of the samples and the variables, respectively, onto each PC. The loadings indicate that variable's contribution to the data variability along each PC. From these results, the samples and variables can be plotted on a two-dimensional PCA projection (a biplot), not only to group the samples according to their differences, but also to determine relationships between samples and variables. Data was compiled to create NMR and FTICR-MS matrices for each PCA.

The intensity values for the NMR spectra at 0.5-11 ppm were exported for each sample from the instrument at the resolution that the NMR measures, giving intensity values approximately every 0.0008 ppm, yielding 13,225 data points per sample. Because this level of resolution unnecessarily generates an excessively large dataset, an in-house written MatLab code was utilized to average across every 5 data points, giving a new resolution of 0.004 ppm and 2645 data points per sample. Data were normalized for each sample so that the summed total intensity equaled 1 for all samples.

For the FTICR-MS PCA, a data matrix was created for the samples and the complete set of unique CHO-only formulae using the relative magnitudes of the peaks. Relative magnitude is simply calculated by dividing the peak magnitude by the summed total peak magnitude for each sample. If a formula was not detected in that sample, a 0 was given in the matrix. Data were normalized for each sample so that the summed total magnitude equaled 1 for all samples, and PCA was conducted according to a previously published method [36].

Bioassays

Whole and fractionated samples were tested for plant response using a foliar application on corn (*Zea mays* L.) seeds treated with a pesticide package (to give healthy, pathogen-free seeds). Planting pots (18 cm diameter) filled with a vermiculite (grade 2, ULINE) substrate media were used, and 8 seeds were planted per pot. There were 3 replicate pots per treatment (including the control). Emergence was allowed to progress for 6 days, and then seedlings were thinned (to allow more space for root growth), from 24 to 10 seedlings per treatment. The seedlings continued to grow for another 8 days, and then the foliar treatment was applied. The DOC concentrations of all 8 samples (whole, acid-insoluble, acid-soluble, and the 5 filtrate : retentate ratios of 100:0, 75:25, 50:50, 25:75, and 0:100) were normalized, and then 2 DOC rates were used for the foliar application: 0.01 and 0.1 mg/L C (UHQ water was used for the control). By using the same number of sprays from a hand sprayer containing the specified DOC value of each sample, 7 mL of each sample at each concentration were sprayed onto the plant foliage of each pot, ensuring adequate and uniform coverage of the plant leaves without inducing dripping or runoff of the product from the leaf margins. The plants continued to grow for another 7 days, and then the plants were harvested to evaluate their response to the treatment. Previous experiments (data not shown) have indicated that 7 days is the optimal amount of time post foliar application for this pot size. This time frame is long enough for the root system to grow and respond without the occurrence of root knotting but not so long that the plants become stressed due to the lack of macronutrients.

The entire growing period occurred in a greenhouse during the summer of 2014, where watering occurred on a daily basis. The following day/night program was used: day from 7 AM at 21–25°C with 50-60% humidity; night from 7 PM at 16–18°C with 50-60% humidity. Plants were not watered the day prior to harvesting, to allow the substrate media to partially dry. The plants were

extracted from the substrate, the roots were carefully washed to remove the vermiculite, and then fresh weight biomasses (for all seedlings of all replicates for each treatment taken together) were recorded. Then, the longest root and shoot of each individual seedling (for each replicate and each treatment) were measured to evaluate root and shoot elongation. The seedlings were then allowed to dry overnight in an oven at 60°C on aluminum foil, prior to taking dry masses of the full root and the full shoot systems for each seedling individually.

The assessment parameters were statistically evaluated by ANOVA single factor and Duncan's multiple range tests, to determine if each treatment is statistically different than the control and if the treatments are statistically different from one another, respectively. A value of $P = 0.05$ was chosen to determine the significance of the data (if $P > 0.05$ the data are not statistically different; if $P \leq 0.05$ the data are statistically different).

Results and Discussion
Carbon and Iron Concentrations
The fractionation scheme, along with the percentage that each fraction contributes to the whole, is represented in Figure 1. In terms of carbon, the acid-insoluble fraction was 43% of the whole, and the acid-soluble was 57%. The acid-soluble portion was then fractionated further, and the C18 hydrophilic filtrate accounted for 64% of the carbon in the acid-soluble portion, and the C18 hydrophobic retentate contained 36%. Based on iron, the acid-insoluble fraction was 31% of the whole, and

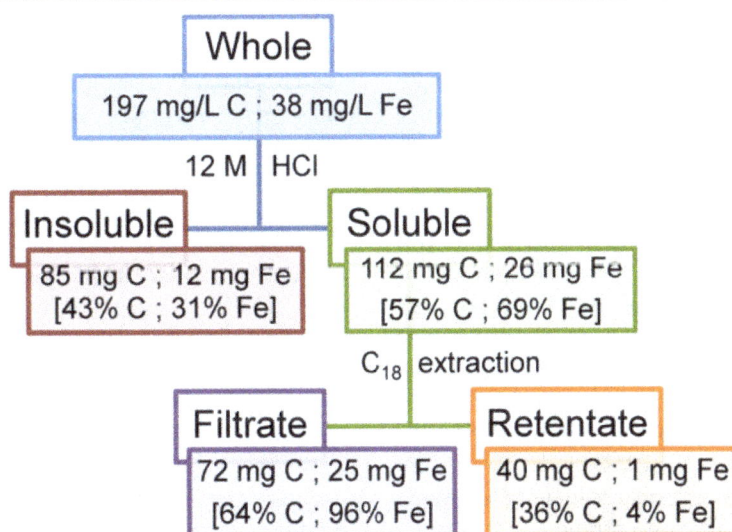

Figure 1 The analytical scheme utilized to fractionate the whole sample. After acidification, the acid-insoluble and acid-soluble components were isolated and then the acid-soluble portion was fractionated into its hydrophilic C18 filtrate and hydrophobic C18 retentate. The starting dissolved organic carbon and iron contents of the whole sample are given, along with the partitioning of carbon (and iron) between the fractions within each step.

the acid-soluble was 69%. The C18 hydrophilic filtrate accounted for 96% of the iron in the acid-soluble portion, and the C18 hydrophobic retentate contained 4%.

Characterization from Optical Measurements

Figure 2a shows the absorbance values across the entire wavelength range of 250–600 nm (for the samples analyzed at approximately 4–5 mg/L C). Typical of all DOM samples, absorbance values decrease with increasing wavelength, and there are no characteristic peaks [37]. Figure 2b shows the absorbance at 254 nm for each sample across the DOC range of 0.5-8 mg/L C. The whole and acid-insoluble samples have similar absorbance values, which are higher than the nearly identical absorbance values obtained for the acid-soluble and hydrophobic C18 retentate. The hydrophilic C18 filtrate has the lowest absorbance. Table 1 gives the SUVA254 values, and SUVA254 values are indicative of aromatic quality [31]. However, it should be noted that these SUVA254 values are somewhat higher that most natural waters, likely due to the high Fe content of the samples. FeIII absorbs at low wavelengths, which therefore inflates SUVA254 values [31]. The whole sample has the highest SUVA254 value, while the acid-insoluble portion has a higher SUVA254 value than the acid-soluble portion. The C18 retentate has a SUVA254 value higher than that of the filtrate.

Table 1 also gives the FI values, along with the percentages that humic-like peak A, humic-like peak C, and peptide-like peak T contribute to the total integrated fluorescence (for the samples analyzed at approximately 5 mg/L C). Higher SUVA254 and lower FI values are both indicators of aromaticity, with SUVA254 being indirectly proportional to FI. They are calculated from different parameters (SUVA254 using low wavelength adsorption and DOC concentration; FI using a ratio of higher wavelength fluorescence and is independent of DOC) and are often evaluated together, because FI is not subject to the same interferences as SUVA254 (e.g., Fe concentration). While Fe is known to quench fluorescence [25], FI is rarely affected by Fe concentrations because it is calculated by a ratio of emissions whose signals are decreased in the same manner. Corresponding with the SUVA254 values, the acid-insoluble portion has a lower FI than the acid-soluble sample, and the C18 retentate has a lower FI than the filtrate. While the whole sample had the highest SUVA254, its FI falls between the acid-insoluble and acid-soluble. The whole sample has the highest Fe concentration, inflating its SUVA254 value more severely than the other samples. However, in this case based on our results, FI seems to reveal the true trend of aromaticity amongst the fractionated samples, better than SUVA254.

Figure 2c gives the integrated Peak A areas from the EEMs analyses across the DOC range for all samples, while Figures 2d-f show the EEMs spectra for the whole, acid-insoluble, and acid-soluble samples. The whole and acid-soluble samples give quite similar EEMs spectra (which are also comparable to the hydrophobic retentate and hydrophilic filtrate samples whose spectra are not shown). The acid-insoluble sample appears to be somewhat different, with a wider (in the emission dimension) peak A (Ex 240–300 nm, Em 400–500 nm) and a less well defined peak C (Ex 300–360 nm, Em 400–500 nm). The hydrophobic retentate has the highest fluorescence, followed by the acid-soluble sample that fluoresces similarly to the whole sample, and the hydrophilic filtrate sample fluoresces slightly less than those. The acid-insoluble sample has the lowest fluorescence, but that is likely due to the fluorescence intensity of peak A being more widespread in this sample than in the others. As discussed above, Fe is known to increase absorbance values at low wavelengths as well as quench overall fluorescence [25]. This is likely why the hydrophobic retentate, that contains very little iron (Figure 1), has the highest fluorescence. The percentages that peaks A and C contribute to the total fluorescence are quite similar for all samples, with peak C being about 2/3 of that of peak A. Peak T (Ex 240–300 nm, Em 250–350 nm) is very low for all samples, as expected based on the terrestrial nature of these samples. This is consistent with the low FI values, which also indicate that there is little influence from microbial DOM.

Characterization from Proton NMR Measurements

Figure 3 shows the NMR spectra for the whole, acid-soluble, and acid-insoluble samples, as well as the various mixed ratios of the acid-soluble C18 filtrate : retentate (100:0, 75:25, 50:50, 25:75, and 0:100) samples. Peaks were integrated and their resulting areas are shown in Table 2. The peak at 0.0 ppm is tetramethylsilane (TMS), a reference standard added to the D_2O. It should be noted that in NMR, natural organic matter (NOM) samples give broad peaks due to high complexity. However, small molecules that are present in the NOM exist as sharp peaks (typically showing on top of the overlapping, broad NOM peaks). Acetic acid/acetate ($H_3C-COOH$, H_3C-COO^-) and formic acid/formate ($HCOOH$, $HCOO^-$), present in most dissolved NOM samples due to bacterial or photodegradation, give distinct peaks at approximately 2 and 8.3 ppm, respectively. Because these 'spikes' can alter the overall peak area integrations, giving false impressions of the total spectral intensity, they are excluded from the integration regions as marked in Table 2. Furthermore, the acetate and formate peaks appear slightly shifted from one sample to the next, because of the varying pH levels of these

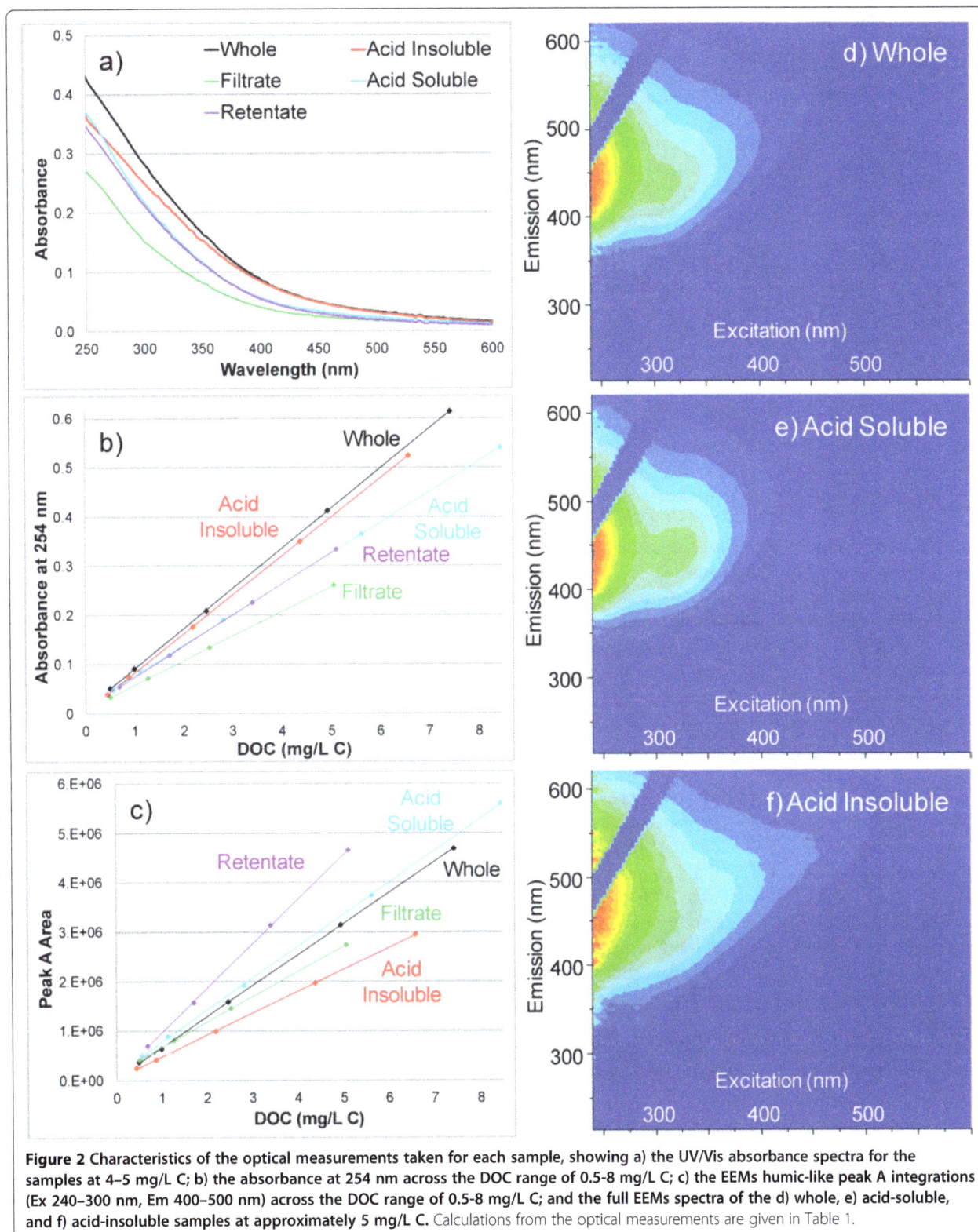

Figure 2 Characteristics of the optical measurements taken for each sample, showing a) the UV/Vis absorbance spectra for the samples at 4–5 mg/L C; b) the absorbance at 254 nm across the DOC range of 0.5-8 mg/L C; c) the EEMs humic-like peak A integrations (Ex 240–300 nm, Em 400–500 nm) across the DOC range of 0.5-8 mg/L C; and the full EEMs spectra of the d) whole, e) acid-soluble, and f) acid-insoluble samples at approximately 5 mg/L C. Calculations from the optical measurements are given in Table 1.

Table 1 Calculations from optical measurements of each sample

Sample	SUVA254 (L mg^{-1} m^{-1})	FI	EEMs Peak Areas		
			% A	% C	% T
Whole	8.35	1.34	61%	37%	2%
Acid insoluble	8.00	1.13	62%	35%	3%
Acid soluble	6.50	1.40	59%	39%	2%
C18 filtrate	5.16	1.44	56%	41%	2%
C18 retentate	6.53	1.29	61%	36%	3%

SUVA254: specific ultraviolet absorbance at 254 nm.
FI: fluorescence index, emission ratio of 420:520 nm at excitation 370 nm.
Peak A: 240–300 nm Ex; 400–500 nm Em.
Peak C: 300–360 nm Ex; 400–500 nm Em.
Peak T: 240–300 nm Ex; 250–350 nm Em.
Absorbance and EEMs fluorescence spectra are shown in Figure 2.

samples that allow these compounds to exist in both their ionic (i.e., acetate and formate) and protonated (i.e., acetic acid and formic acid) forms. For example, because the whole and acid-soluble samples are acidic (due to their preparation with the cation exchange resin), the formic acid peak is up-field of the formate peak that exists in the basic, acid-insoluble sample (Figure 3a). While methanol is also a natural biodegradation product, a small amount of residual methanol likely exists in the C18 samples, as the methanol spike at about 3.4 ppm in the C18 samples increases as more of the retentate fraction exists in the samples where ratios of filtrate : retentate are mixed (Figure 3b).

Based on the NMR spectra and area integrations, the acid-soluble sample is more oxygenated (2–5 ppm), whereas the acid-insoluble portion contains more olefins and aromatics (5–9 ppm). For the C18 samples, as the ratio of filtrate : retentate goes from 100:0 to 0:100, the aliphatic and aromatic components both increase, while the oxygenation decreases. While there is much overlap of the samples across the entire chemical shift range, the subtle differences amongst the spectra are revealed by utilizing PCA.

The original PCA executed on the NMR data revealed that statistical differences were primarily driven by the pH affect. Acidic samples were enriched in acetic acid and formic acid, while neutral/basic samples were enriched in acetate and formate. The presence or absence of methanol was also driving the PCA (particularly for the C18 samples). Because these small molecules exist as sharp peaks in the spectra, they are the first (and largest) differences found by PCA. Thus, a data matrix was created for a new PCA, ignoring certain regions of the NMR spectra. The following areas were deleted: >8 ppm, 3.2-4 ppm, and 1.8-2.4 ppm, to remove the areas of the spectrum that depend on pH, as well as the

methanol peak. The subsequent NMR PCA matrix had a dimension of 8 (samples) x 1537 (NMR intensity values as variables).

Figure 4 shows the PCA biplot of the samples' scores, along with the NMR spectra that have been reconstructed using the PC1 and PC2 variable loadings. PC1 explains 57% of the variance of the 1537 variables among the 8 samples, with PC2 explaining an additional 31% of the variance. The C18 fractionated samples all have positive PC1 scores, while the whole, acid-insoluble, and acid-soluble have negative PC1 scores. PC2 separates the acid-insoluble from the whole and acid-soluble samples, as well as the hydrophobic retentate from the hydrophilic filtrate.

The reconstructed NMR spectra elucidate the reasons for the sample groupings. From the PC1 loadings (Figure 4b), it is clear that the large aliphatic peak at about 1.5 ppm versus the oxygenated region at 2.4-3.2 ppm are the main factors driving the PC1 variance. Because the peak at 1.5 ppm has a negative PC1 loading, it is enhanced in the whole, acid-insoluble, and acid-soluble samples (as they have negative PC1 scores). The whole and acid-insoluble samples are more enriched in the peak at 1.5 ppm than in the oxygenated region (as they have larger negative PC1 scores). However, the acid-soluble sample has both of these regions at quite high intensities, giving a PC1 score that is closer to 0 (but still negative). The peak at 1.2 ppm (which is defined in all samples, Figure 3) is due to CH_x protons on aliphatic chains, while the peak at 1.5 is likely due to aliphatic protons that are attached to carbons that are next to either an olefin group or a carboxyl group ($H_2C = CH-CH_2-R$; $HOOC-CH_2- CH_2-R$). It is clear that this aliphatic peak at 1.5 ppm is more prevalent in the samples fractionated by pH and less so in the C18 samples (Figure 3). Because this peak is prevalent in the acid-soluble sample, we would also expect to detect it in the polarity-separated samples. It is likely that this region is less defined in the C18 samples, because these types of functionalities partition between the hydrophobic retentate and hydrophilic filtrate depending on the composition of the R group. The presence of a well-defined peak at 1.5 ppm in the acid-soluble sample and the lack there of in the C18 samples leads to their separation in the PC1 dimension.

The C18 samples (with positive PC1 scores) are more enriched in the oxygenated region at 2.4-3.2 ppm (as this region has positive PC1 loadings) than in the peak at 1.5 ppm. Furthermore, the oxygenation degree increases as more of the filtrate sample is present in the mixed ratios of C18 filtrate : retentate. These observations are in agreement with the expected mechanism of the extractions, where polar species (i.e., those that are more oxygenated) are not adsorbed to the resin and thus remain

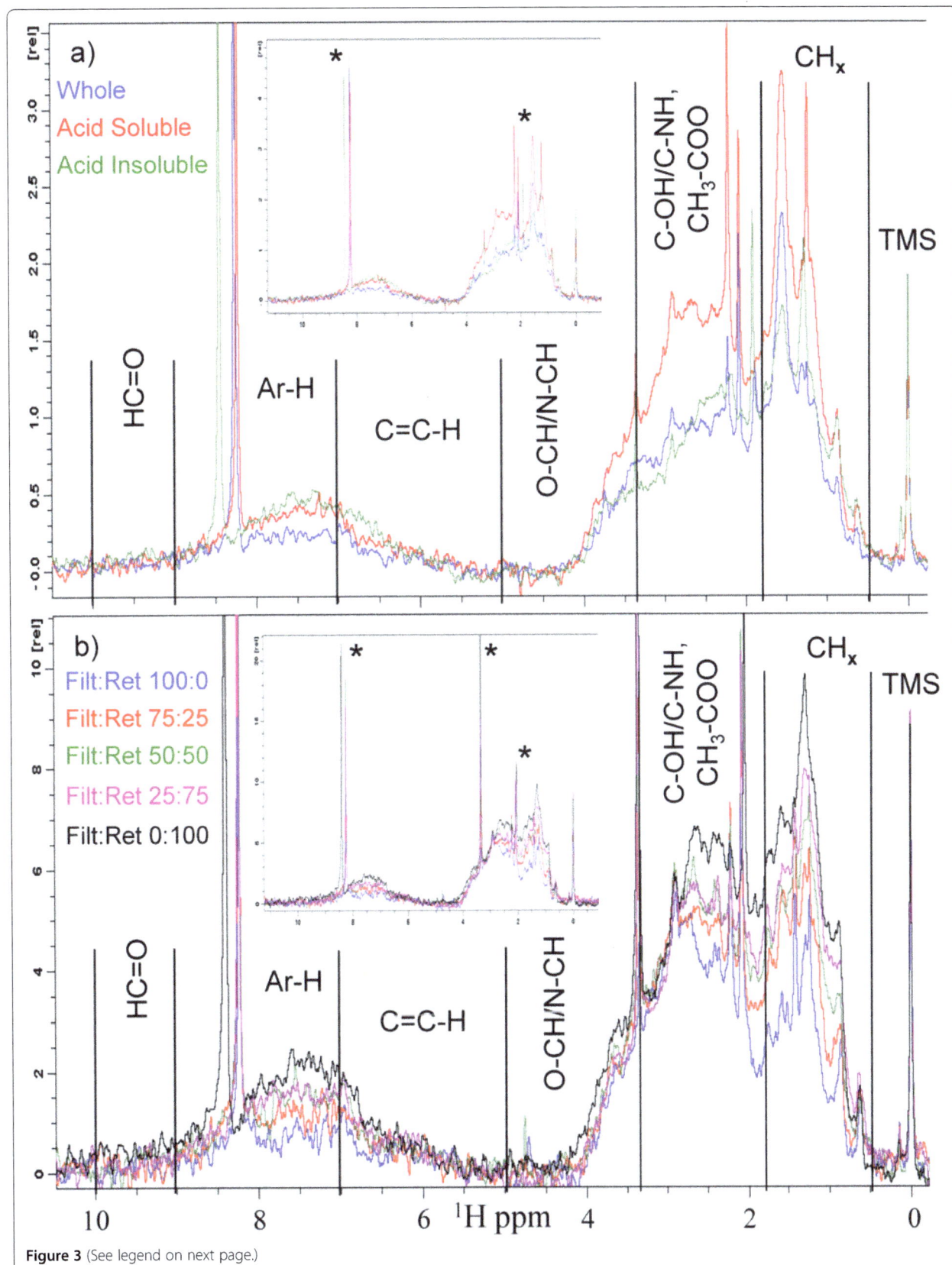

Figure 3 (See legend on next page.)

(See figure on previous page.)
Figure 3 The proton NMR spectra for all samples, showing the samples fractionated according to a) pH solubility and b) polarity using **C18 resin (the hydrophilic filtrate and hydrophobic retentate were mixed in ratios of filtrate: retentate of 100:0, 75:25, 50:50, 25:75, and 0:100).** The sample names given in color correspond to the spectral line coloring. Bulk functionalities by chemical shift are given, and the spectra are scaled to the TMS (tetramethylsilane) peak. Asterisks (*) indicate the peaks due to acetate/acetic acid, methanol, and formate/formic acid. Peak area integrations for each sample are given in Table 2.

in the filtrate portion. However, the C18 samples do not simply increase horizontally in the PC1 dimension, but rather on a diagonal, meaning that their differences are a combination of PC1 and PC2 variance.

In the PC2 dimension, the main drivers of the variance (Figure 4c) are the peak at 1.5 ppm and the oxygenated region at 2.4-3.2 ppm versus the aromatic region at 7–8 ppm and the aliphatic region at 0.5-1.2 ppm. The acid-insoluble sample has a very low PC2 score, indicating that it is particularly enriched in the aromatic region, as well as aliphatic CH_x functionalities. The other samples are more oxygenated (Figure 3, Table 2), giving higher PC2 scores. The samples with higher C18 filtrate portions having higher PC2 scores are more enriched in the oxygenated area, while the C18 retentate-dominated samples with lower PC2 scores are enriched in the aromatic and CH_x aliphatic (0.5-1.2 ppm) areas. In this PC2 dimension, the acid-soluble sample score falls between the 50:50 and 75:25 filtrate : retentate ratio sample scores, as expected based on the relative proportions of the filtrate and retentate in the original acid-soluble sample (Figure 1).

Characterization from FTICR-MS Measurements
All of the acquired FTICR mass spectra were similar to DOM samples analyzed previously [14,23,33-36,38], with most peaks existing at 300–600 m/z. The majority of peaks appear at odd m/z values, and, based upon the nitrogen rule, indicate a predominance of zero or an even number of nitrogens in the molecules. Peaks at even nominal masses are those that contain either an odd

number of nitrogen atoms or are ^{13}C isotopes of the ^{12}C peaks at the previous nominal mass [39]. The presence of the ^{13}C isotopes 1.0034 m/z units away from their corresponding ^{12}C peaks indicate that the ions are singly charged [39,40]. Numerous peaks (up to 25) were detected at each nominal mass, indicating the complexity of each sample. The formula assignments showed that the vast majority of compounds contain only C, H, and O (64-84% by number and 84-91% by magnitude), with less contribution from N and S (Table 3). There are very few formulae containing both N and S together (<1% of the total number and summed total spectral magnitude), and formulae that do contain both N and S correspond to low magnitude peaks with S/N ratios near the threshold value of 3.

The whole and acid-soluble samples are very similar with regard to their averaged elemental ratios and heteroatom composition, with the acid-insoluble material containing the highest amount of CHO-only formulae, indicating that the heteroatoms are less likely to precipitate, especially sulfur-containing components (Table 3). The C18 filtrate contains more heteroatoms than the C18 retentate (which is consistent with previous studies [23,30,36], as these polar species are not adsorbed to the C18 resin, especially the sulfur species (which are typically also associated with high oxygenation, such as sulfates, and are very polar). Also in Table 3 are the number-averaged and magnitude-weighted calculations. Again, the whole and acid-soluble samples are quite similar. The acid-insoluble sample has a much lower O/C average and an H/C average that is slightly higher than

Table 2 Proton NMR peak area integrations at the given ppm ranges, for each sample

Sample	CH$_x$	Total aliphatic	#C-OH/ C-NH	O-CH/ N-CH	Total oxygenated	C = C-H	*Ar-H	*Total aromatic	O = C-H
	0.5-1.8	0.5-1.8	2.0-3.3	3.4-5	2-5	5-7	7-9	5-9	9-10
Whole	43.0	**43.0**	30.4	7.0	**37.4**	5.5	12.3	**17.8**	1.8
Acid insoluble	38.9	**38.9**	28.0	6.8	**34.8**	7.1	16.7	**23.8**	2.5
Acid soluble	38.7	**38.7**	36.5	7.6	**44.1**	4.9	11.2	**16.1**	1.1
C18 Fil:Ret 100:0	30.4	**30.4**	42.3	8.5	**50.8**	6.3	11.3	**17.6**	1.2
C18 Fil:Ret 75:25	33.5	**33.5**	38.8	7.3	**46.1**	6.0	13.2	**19.2**	1.2
C18 Fil:Ret 50:50	36.2	**36.2**	34.5	7.1	**41.6**	6.4	14.6	**21.0**	1.2
C18 Fil:Ret 25:75	40.2	**40.2**	32.3	5.2	**37.5**	6.2	15.0	**21.2**	1.1
C18 Fil:Ret 0:100	41.2	**41.2**	30.3	5.3	**35.6**	6.3	15.9	**22.2**	1.0

#integration region excludes the peak area for acetate/acetic acid.
*integration region excludes the peak area for formate/formic acid.
Values given in bold are for the summed total peak areas in the aliphatic, oxygenated, and aromatic regions.
The ^{1}H NMR spectra are shown in Figure 3.

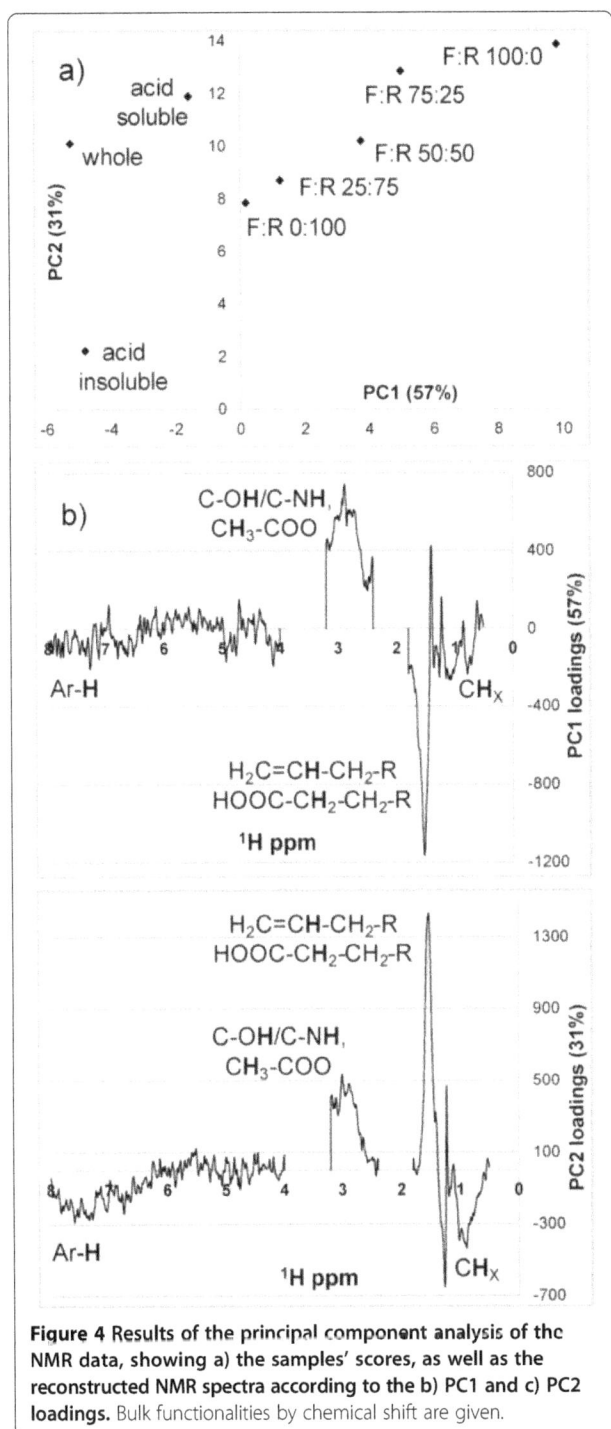

Figure 4 Results of the principal component analysis of the NMR data, showing a) the samples' scores, as well as the reconstructed NMR spectra according to the b) PC1 and c) PC2 loadings. Bulk functionalities by chemical shift are given.

From the molecular formula assignments, van Krevelen diagrams [41] were created, where molar O/C ratios are plotted against molar H/C ratios (Figure 5). From these diagrams, one can parse the formulae by regions of the van Krevelen diagram that denote molecular similarities with groups of compounds such as lipid, lignin, tannin, or condensed aromatics [42-44]. The whole and acid-soluble samples encompass all of these regions, giving van Krevelen diagrams that are very similar to one another (Figure 5a and b). The formulae in the acid-insoluble portion exist mostly at low O/C in the condensed aromatic and lipid regions, with some points aligning in the low O/C lignin-like region (Figure 5c). This van Krevelen diagram of the acid-insoluble fraction is similar to those of other types of humic acids [43,45]. There are many formulae aligning in the lipid-like region of the acid-insoluble sample, and because this was precipitated material from the whole, we would have expected to observe this cluster of formulae there as well. Highly oxygenated functional groups, like those present in the whole sample, ionize more easily in negative ion mode, and it is likely that these low O/C and high H/C compounds that are also present in the whole sample simply could not compete for a charge. The polar fractionation of the acid-soluble portion clearly separates the sample into two distinct fractions. It is obvious that the tannin-like formulae are dominant in the C18 filtrate (Figure 5d), while the lignin-like component dominates the C18 retentate (Figure 5e), with formulae aligning at O/C 0.4-0.7 and H/C 0.5-1.5 being present in both portions. The overlap of formulae in both the filtrate and retentate may indicate that certain components partition into both fractions. However, because FTICR-MS is not distinguishing between structural isomers, it is also probable that the same formulae being detected in both samples are actually different structures that have different polarities that allow them to fractionate into separate portions [23]. The variations in molecular formulae found in hydrophilic versus hydrophobic fractions is consistent with what has been observed previously [23,36]. While significant differences are apparent in these van Krevelen diagrams, PCA is utilized to reveal the more subtle differences that exist based on the relative magnitudes corresponding to the assigned formulae.

From the 8 samples, 17,388 CHO-only formulae were assigned. When removing the duplicate formulae (i.e., those that were common to more than 1 sample), a unique list of 3751 formulae existed. Figure 6a shows a frequency plot demonstrating the number (and percentage) of formulae detected either uniquely to 1 sample or commonly amongst a number of samples. These 3751 formulae account for 84-91% of the total magnitude of all formulae assigned per sample. About 15% of these

the whole and acid-soluble samples. These values indicate that the precipitated acid-insoluble material contains components with very little oxygenation. As the sample progresses from filtrate to retentate, the average O/C and DBE/C ratios decrease. This indicates that the filtrate is more oxygenated and less aromatic than the retentate portion. These averaged mass spectral characteristics agree quite well with the optical and NMR results.

Table 3 Average (by number and magnitude-weighted) mass spectral characteristics for each sample

Sample	O/C		H/C		DBE		DBE/C		%CHO		%CHON		%CHOS	
	num	mag	num	mag	num	mag	num	mag	num	mag	num	mag	num	mag
Whole	0.58	0.61	0.91	0.87	12.9	12.9	0.60	0.62	69%	87%	17%	8%	13%	5%
Acid insoluble	0.37	0.37	0.95	0.95	13.2	12.9	0.57	0.57	82%	87%	16%	11%	2%	2%
Acid soluble	0.57	0.60	0.94	0.91	12.5	12.5	0.58	0.60	70%	87%	19%	9%	11%	4%
C18 Fil:Ret 100:0	0.64	0.67	0.91	0.89	11.7	11.6	0.61	0.61	64%	84%	21%	9%	15%	7%
C18 Fil:Ret 75:25	0.57	0.60	0.94	0.90	12.3	12.3	0.58	0.60	73%	88%	18%	8%	9%	3%
C18 Fil:Ret 50:50	0.53	0.55	0.96	0.94	12.4	12.4	0.57	0.58	76%	90%	17%	7%	7%	2%
C18 Fil:Ret 25:75	0.51	0.51	0.97	0.96	12.8	12.7	0.56	0.57	75%	91%	17%	7%	8%	2%
C18 Fil:Ret 0:100	0.45	0.46	0.96	0.97	14.1	13.6	0.56	0.56	75%	91%	18%	7%	6%	2%

DBE (double bond equivalents) = $(2c + 2 + n + p - h)/2$ for any formula $C_cH_hO_oN_nS_sP_p$
DBE/C: double bond equivalents normalized to the number of carbon atoms in the formula
num indicates that the value given in number-averaged
mag indicates that the value given in magnitude-weighted
The van Krevelen diagrams are shown in Figure 5.

formulae are detected in all 8 samples, while 35% of the formulae are detected in 6–7 samples, indicating that there is a high percentage of commonality amongst some samples. Conversely, about 22% of formulae are detected in only 1 sample, as highlighted by the van Krevelen diagram in Figure 6b. Most of these unique formulae have low O/C ratios and are detected only in the acid-insoluble sample. As discussed above, we would have expected to detect these formulae in the whole sample as well, but these low O/C compounds do not ionize efficiently enough in negative ion mode amongst the lignin- and tannin-like compounds present in the whole sample to enable their detection. The formulae in all 8 samples (Figure 6c) cluster together in the low H/C (mostly <1) region where the lignin- and tannin-like groups meet at O/C 0.4-0.7. The overlap of formulae in numerous samples that have been fractionated (either by pH or polarity) is not necessarily unexpected. As mentioned above, FTICR-MS is not distinguishing between structural isomers, and formulae being detected in multiple samples may have different structures that allow them to fractionate into separate portions, but this structural differentiation is not recognized by FTICR-MS.

The relative magnitudes of these 3751 formulae are used as variables in the FTICR-MS PCA, and Figures 6d and e show the biplots of the samples' scores and the variables' loadings, respectively. PC1 explains 46% of the variance, while PC2 explains an additional 36%. In general, PC1 separates the C18 samples according to polarity, with the filtrate-dominated samples having high positive PC1 scores and the retentate-dominated samples having high negative PC1 scores. The whole and acid-soluble samples fall very closely to the filtrate: retentate 75:25 sample, which approximates the relative proportions of the filtrate and retentate in the acid-soluble sample (Figure 1). The acid-insoluble sample

separates from the other samples with high negative PC1 and PC2 scores. The corresponding van Krevelen diagram, colored according to the boxes drawn in Figure 6e, is shown in Figure 6f. Formulae in purple (with high positive PC1 loadings) are enriched in the whole, acid-soluble, and filtrate-dominated samples (that have high positive PC1 scores), and these formulae fall mostly in the tannin-like region with high O/C values (Table 4). The green formulae (with both high negative PC1 loadings and high positive PC2 loadings) are enriched in the retentate-dominated samples (that have scores in the same quadrant) and fall mostly in the lignin-like region. These formulae also have notably higher m/z values (Table 4). The red formulae (with both high negative PC1 and PC2 loadings) are enriched in the acid-insoluble sample and have low O/C values and span the H/C scale. These formulae have lower m/z values, indicating that the smaller molecules are first to precipitate out of solution under acidic conditions.

Bioassays and Plant Response

Corn plants (*Zea mays* L.) treated with foliar applications of the 8 samples at 0.01 and 0.1 mg/L C were compared to the control (that was treated with UHQ water). For all assessments (root and shoot elongation, fresh and dry masses), the whole and acid-soluble samples all gave plant growth responses that were statistically better (or at least observably higher) than the control for both carbon rates (Table 5, Figure 7). The whole sample at both concentrations was statistically better than the control for shoot elongation, and the acid-soluble sample at 0.01 mg/L C gave the highest shoot elongation of any of the samples analyzed. The acid-insoluble sample was not statistically different from the control for shoot elongation, but did give a value higher (but not statistically better) than the control for the application at 0.1 mg/L C. For the C18 fractionated samples, the filtrate: retentate

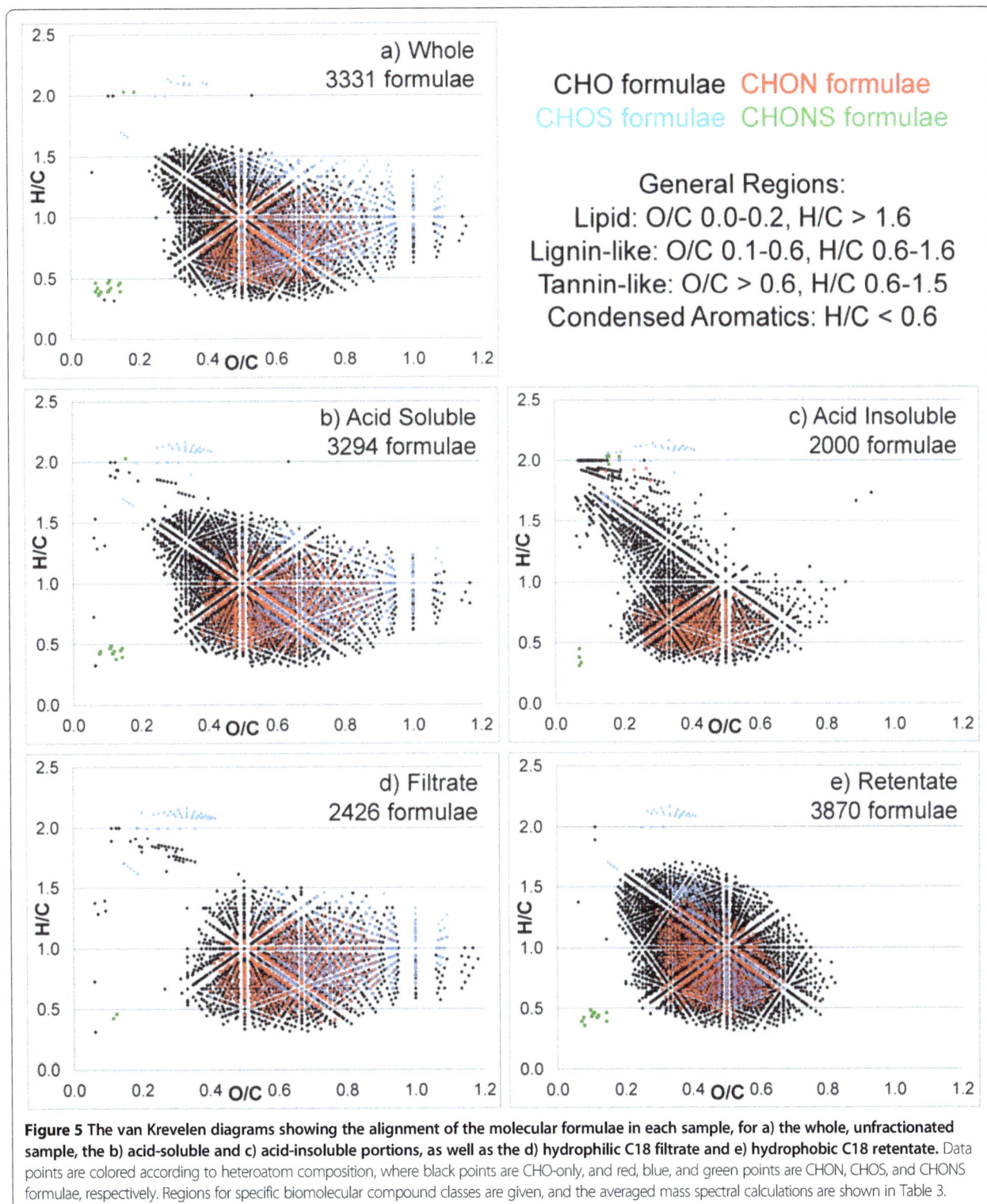

Figure 5 The van Krevelen diagrams showing the alignment of the molecular formulae in each sample, for a) the whole, unfractionated sample, the b) acid-soluble and c) acid-insoluble portions, as well as the d) hydrophilic C18 filtrate and e) hydrophobic C18 retentate. Data points are colored according to heteroatom composition, where black points are CHO-only, and red, blue, and green points are CHON, CHOS, and CHONS formulae, respectively. Regions for specific biomolecular compound classes are given, and the averaged mass spectral calculations are shown in Table 3.

mixture of 100:0 gave the highest shoot elongations and dry shoot masses at both DOC values, while the other mixtures (75:25, 50:50, 25:75, and 0:100) were not statistically different from the control for shoot elongation or dry shoot mass. Based on these results, it seems that

hydrophilic samples are most influential on shoot development, as the acid-soluble portion performed better than the acid-insoluble portion (which is less oxygenated and more hydrophobic), and the 100:0 hydrophilic filtrate : hydrophobic retentate gave higher shoot measurements

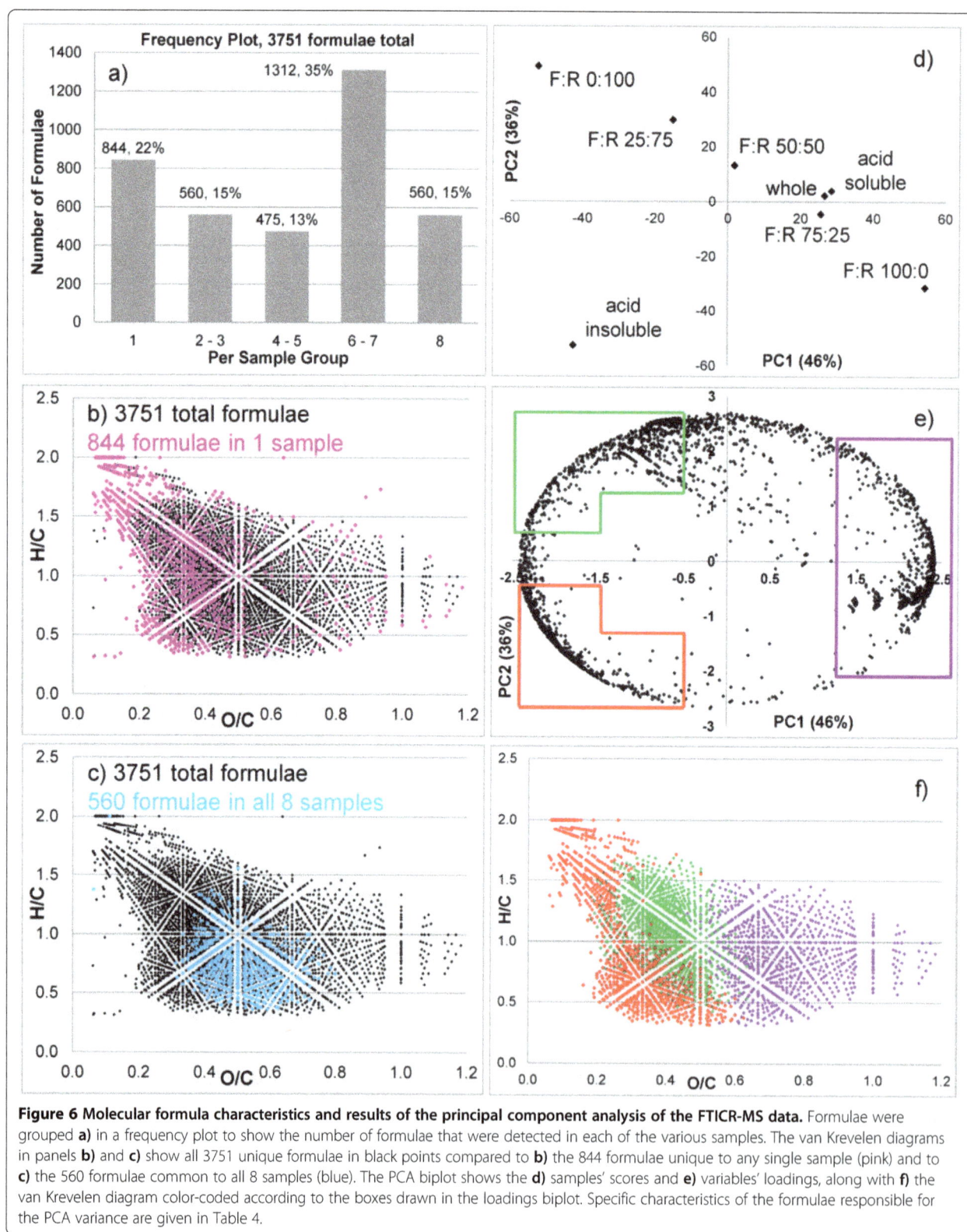

Figure 6 Molecular formula characteristics and results of the principal component analysis of the FTICR-MS data. Formulae were grouped **a)** in a frequency plot to show the number of formulae that were detected in each of the various samples. The van Krevelen diagrams in panels **b)** and **c)** show all 3751 unique formulae in black points compared to **b)** the 844 formulae unique to any single sample (pink) and to **c)** the 560 formulae common to all 8 samples (blue). The PCA biplot shows the **d)** samples' scores and **e)** variables' loadings, along with **f)** the van Krevelen diagram color-coded according to the boxes drawn in the loadings biplot. Specific characteristics of the formulae responsible for the PCA variance are given in Table 4.

Table 4 Mass spectral characteristics of the formulae responsible for the variance in the FTICR-MS PCA dataset

Parameter	General characteristics of molecular formulae from PCA					
	Purple (enriched in filtrate-dominated samples)		Green (enriched in retentate-dominated samples)		Red (enriched in acid insoluble)	
	range	average	range	average	range	average
O/C	0.40-1.18	0.74	0.18-0.63	0.40	0.06-0.63	0.29
H/C	0.32-1.50	0.90	0.43-1.70	1.10	0.32-2.00	0.99
DBE	4-20	11.2	4-26	14.1	1-26	13.1
DBE/C	0.31-0.89	0.61	0.20-0.82	0.49	0.03-0.89	0.55
m/z	233-683	451	277-699	559	227-665	429

The PCA scores and loadings are given in Figure 6d and 6e, respectively, with the corresponding color-coded van Krevelen plot shown in Figure 6 f.

than any of the samples containing the retentate. Although only 3 of the shoot measurements were statistically better than the control, 2 of these values were at the lower carbon rate of 0.01 mg/L C. While stimulation can be observed for the shoot measurements, the root system is the principal target of the fertilizer effect and these measurements demonstrate the enhanced plant growth best (as observed in previous studies as well [4-6]).

In general, the lower DOC rate (0.01 mg/L C) gave higher root elongation values, in comparison to the higher DOC rate of 0.1 mg/L C (Table 5, Figure 7). The whole, acid-soluble, and acid-insoluble samples all gave root elongation values higher than the control for both DOC rates, but none were statistically better. For the C18 fractionation samples, the 50:50 and 25:75 mixtures of the filtrate : retentate gave the highest root elongations. In regard to dry root mass, the whole sample was better, but not statistically, than the control for both DOC treatments, but the acid-soluble and acid-insoluble gave root masses that were statistically better (more so for the lower DOC rate). In this case, the acid-insoluble portion that is more hydrophobic gave the best results for the samples fractionated by pH. However, the samples fractionated by polarity display the most interesting trends for the dry root mass measurements of the bioassay. At the lower DOC rate of 0.01 mg/L C, the root masses increased as the C18 mixed samples became dominated by the hydrophobic retentate. The 100.0 filtrate : retentate sample had higher dry root mass than the control (but not statistically better), while the other 4 samples (75:25, 50:50, 25:75, 0:100 filtrate : retentate) were all statistically better and increased in that order. The opposite is true for the higher DOC rate of 0.1 mg/L C, where root mass increased as the samples became dominated by the hydrophilic filtrate. In this case, all 5 mixed samples gave root masses statistically better than the control, but the order here was reversed, where 100:0 filtrate : retentate yielded the highest root mass. It should also be noted that while all of the C18 fractionation samples gave positive results for dry root mass,

they largely gave lower dry shoot masses than the control (but not statistically worse), except for the 100:0 filtrate : retentate sample that gave higher dry shoot masses than the control. It is evident that the application was taken up easily by the foliage and that it impacted the root system most.

The results presented here are a good example highlighting that plant response is due to not only the amount of carbon applied but also the type (i.e., fraction) of carbon. Certain parts of the plant (root vs. shoot) appear to respond differently to various types of carbon (hydrophilic vs. hydrophobic). These types of contradictory effects have also been reported in other studies [6,15-18]. While the focus for this foliar application was on the amount and type of carbon, it is likely that the trace metals and other micronutrients naturally present in the DOM also play an active role in enhancing plant growth [46-48]. In some cases, the hydrophilic samples (i.e., the acid-soluble or filtrate-dominated samples) gave the best plant response (i.e., shoot elongation and dry shoot mass at both DOC rates, dry root mass for the filtrate-dominated samples at the higher DOC rate of 0.1 mg/L C). These hydrophilic samples also have the highest Fe content. Other times the hydrophobic samples (i.e., the acid-insoluble or retentate-dominated samples) were better (i.e., dry root mass, especially for the retentate dominated samples at the lower DOC rate of 0.01 mg/L C). The acid-insoluble portion contains less Fe than the acid-soluble sample, but still has a significant Fe concentration, whereas the hydrophobic C18 retentate is nearly void of Fe. In general, the hydrophobic samples have lower Fe contents. It has yet to be determined whether these plant responses are due to the type of carbon components in each of these fractions (i. e., hydrophilic polar vs. hydrophobic non-polar) or due to the varying concentrations of Fe and other micronutrients that are also present. While more research is necessary to discern the mechanism(s) responsible for these observations, this study is an example of enhanced plant growth during controlled application of humic substances that vary in their chemical components and

Table 5 Results of the bioassays employing foliar applications of the 8 samples at 2 carbon rates

Sample	Length (cm)		Fresh weight (g)	Dry Weight (mg)	
	Root	Shoot		Root	Shoot
Control	32.7 ± 3.5	28.2 ± 3.9	27.4	155 ± 19	95 ± 36
Whole, 0.01 mg/L C	**33.0 ± 3.6**	***32.6 ± 4.3***	**30.4**	165 ± 31	118 ± 23
	(0.881)	*(0.028)*		(0.437)	(0.111)
Acid soluble, 0.01 mg/L C	**36.2 ± 5.5**	***34.1 ± 3.4***	**32.8**	*189 ± 25*	125 ± 35
	(0.115)	*(0.002)*		*(0.004)*	(0.085)
Acid insoluble, 0.01 mg/L C	**36.4 ± 8.0**	27.8 ± 3.2	**28.0**	*221 ± 37*	108 ± 25
	(0.197)	(0.800)		*(0.0001)*	(0.383)
100:0 Filt : Ret, 0.01 mg/L C	***37.0 ± 3.3***	**30.6 ± 3.9**	**29.5**	184 ± 20	111 ± 32
	(0.011)	(0.190)		(0.108)	(0.332)
75:25 Filt : Ret, 0.01 mg/L C	**35.5 ± 5.6**	27.9 ± 6.4	25.7	*192 ± 53*	89 ± 36
	(0.203)	(0.894)		*(0.045)*	(0.685)
50:50 Filt : Ret, 0.01 ppm	***38.3 ± 4.4***	27.5 ± 3.6	27.4	*219 ± 21*	92 ± 21
	(0.006)	(0.668)		*(0.0001)*	(0.777)
25:75 Filt : Ret, 0.01 mg/L C	***38.8 ± 5.5***	26.3 ± 3.3	25.5	*221 ± 39*	84 ± 20
	(0.008)	(0.243)		*(0.0001)*	(0.399)
0:100 Filt : Ret, 0.01 mg/L C	**36.4 ± 3.1**	27.0 ± 3.5	**28.4**	*231 ± 24*	98 ± 25
	(0.062)	(0.457)		*(0.00004)*	(0.882)
Whole, 0.1 mg/L C	**33.0 ± 1.9**	***32.1 ± 4.1***	**29.5**	160 ± 14	112 ± 28
	(0.809)	*(0.044)*		(0.537)	(0.260)
Acid soluble, 0.1 mg/L C	**35.5 ± 5.5**	**31.1 ± 2.5**	**29.3**	173 ± 25	112 ± 23
	(0.202)	(0.063)		(0.102)	(0.247)
Acid insoluble, 0.1 mg/L C	**36.6 ± 5.6**	**29.1 ± 4.4**	**28.6**	*204 ± 30*	113 ± 39
	(0.078)	(0.634)		*(0.0005)*	(0.313)
100:0 Filt : Ret, 0.1 mg/L C	32.0 ± 3.5	**29.2 ± 5.0**	**32.3**	*229 ± 40*	118 ± 36
	(0.638)	(0.645)		*(0.0001)*	(0.185)
75:25 Filt : Ret, 0.1 mg/L C	31.9 ± 3.7	26.5 ± 4.1	26.8	*214 ± 31*	98 ± 29
	(0.607)	(0.342)		*(0.0001)*	(0.872)
50:50 Filt : Ret, 0.1 mg/L C	***37.2 ± 5.5***	26.8 ± 4.0	25.7	*214 ± 37*	90 ± 22
	(0.041)	(0.430)		*(0.0003)*	(0.697)
25:75 Filt : Ret, 0.1 mg/L C	**33.2 ± 3.8**	27.9 ± 4.4	**27.5**	*195 ± 61*	105 ± 33
	(0.764)	(0.862)		*(0.006)*	(0.561)
0:100 Filt : Ret, 0.1 mg/L C	31.7 ± 5.6	26.3 ± 3.5	26.1	*202 ± 45*	90 ± 23
	(0.621)	(0.253)		*(0.007)*	(0.691)

Average values (with the standard deviations) of the 10 individual measurements for each treatment are given for the root and shoot measurements. The fresh weight of all seedlings for a given treatment were taken together to give one measurement. Values in bold indicate a response observably higher than the control, and values in bold-italic indicate a response statistically better than the control. P values are given in parentheses for each measurement. Visual depictions of the root and shoot values are shown in Figure 7.

concentrations. Future studies that alter the DOC and micronutrient rates in a systematic manner will provide an enhanced understanding of this likely synergistic plant stimulation. Moreover, the application method (foliar in this case) may play a key role in determining the plant responses. Thus, other types of applications (such as seed treatment and soil application) are also being investigated.

Conclusions

The acid-insoluble component accounted for 43% of the carbon and 31% of the iron of the whole sample. After extraction of the acid-soluble portion, the C18 retentate portion contained 36% and 4% of the carbon and iron, respectively, while the remaining (and majority of) carbon and iron was found in the filtrate portion. The whole and acid-soluble samples were very similar, as

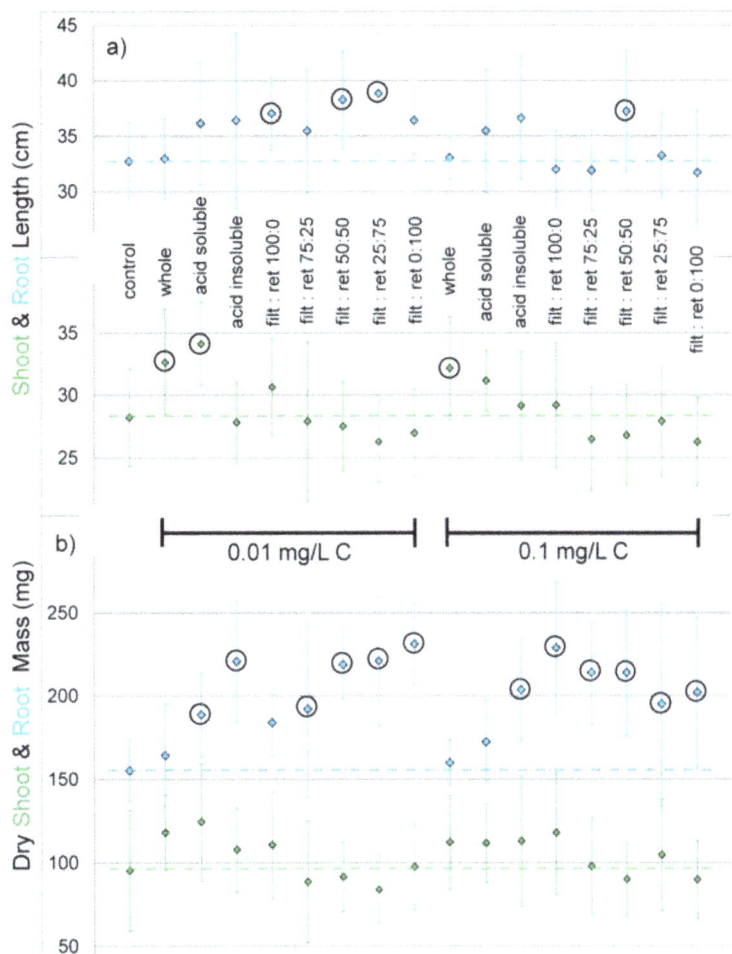

Figure 7 Results of the bioassays employing foliar applications of the 8 samples at 2 carbon rates. The DOC rates used were 0.01 and 0.1 mg/L C, and the average **a)** shoot and root lengths (cm) and **b)** dry shoot and root masses (mg) for all seedlings are given. Shoot measurements are given in green, while root measurements are given in blue. Error bars are the standard deviations of the 10 individual measurements for each treatment. Circles around data points indicate that the data point is statistically better (p ≤ 0.05) than the control. The dotted lines across the plot indicate the control average for each type of measurement. Actual measurements and p values are given in Table 5.

characterized by NMR and FTICR-MS. The acid-insoluble sample was very low in oxygen and was mostly aromatic, but still contained a significant portion of aliphatics. A clear shift in the chemical composition was observed for the C18 samples, where the filtrate and retentate samples were mixed at ratios of filtrate : retentate 100:0, 75:25, 50:50, 25:75, and 0:100. As more retentate was added, the O/C ratios decreased while the DBE increased (with H/C not changing significantly). The tannin-like component dominates the filtrate portion, while the lignin-like component dominates the retentate portion, due to the differences in their polarity and hydrophobicity. PCA assisted in revealing the more subtle differences between the whole sample fractionated according to pH, as well as the acid-soluble portion fractionated using C18 resin to obtain the filtrate and retentate samples.

Bioassays using foliar applications of the 8 samples at 0.01 and 0.1 mg/L C demonstrated that plant responses are not only due to the amount of carbon but also due to the type of carbon present. Samples of a more hydrophilic nature (those that are most oxygenated as revealed by the NMR and FTICR-MS analysis) yielded the highest response for the shoot measurements. However, the root measurements were more responsive than the shoot. The lower DOC rate of 0.01 mg/L C in general gave higher root elongation values than the higher DOC rate of 0.1 mg/L C. The acid-insoluble and acid-soluble portions individually gave higher root elongation and larger root masses than the whole sample and the control, with the hydrophobic, less oxygenated acid-insoluble sample performing better. Based on dry root masses, the C18 polarity fractions showed that larger root systems occurred when there was more hydrophobic material present at the lower carbon application (0.01 mg/L C). The opposite was true for the root system at the higher carbon application (0.1 mg/L C), where larger roots existed when more hydrophilic material was present. This study clearly demonstrates that plant growth can be improved by the controlled application of humic substances that vary in their chemical components and concentrations. However, mechanisms for these enhancements remain unclear. Further work is necessary to discern whether the DOC type and concentration are responsible for the improved plant growth or whether micronutrients are more significant, but it is likely a combined effect in which both variables are important.

Competing interests

The authors declare that they have no competing interests.

Authors' contributions

RLS designed the chemical experiments and acquired characterization data, with help from KMR. PC designed and executed the bioassays. All authors contributed to data interpretation and integration, as well as writing the manuscript. All authors read and approved the final manuscript.

Acknowledgments

We thank the College of Sciences Major Instrumentation Cluster at Old Dominion University for their assistance with NMR and FTICR-MS data acquisition. We also thank the two anonymous reviewers and editor Dr. Marios Drosos whose comments significantly improved the quality of this manuscript.

References

1. Aiken GR, MacCarthy P, Malcolm RL, Swift RS (1985) Humic Substances in Soil, Sediment, and Water. Wiley, New York
2. Stevenson JF (1994) Humus Chemistry: Genesis, Composition, Reactions. Wiley, New York
3. Perdue EM, Ritchie JD (2003) Dissolved organic matter in freshwaters. In: Drever JI (ed) Treatise on Geochemistry: Surface and Ground Water, Weather, and Soils, volume 5. Elsevier, Oxford
4. Chen Y, Aviad T (1990) Effects of humic substances on plant growth. In: MacCarthy P, Clapp CE, Malcolm RL, Bloom RR (ed) Humic Substances in Soil and Crop Sciences: Selected Readings. American Society of Agronomy, Madison
5. Nardi S, Pizzeghello D, Muscolo A, Vianello A (2002) Physiological effects of humic substances on higher plants. Soil Biol Biochem 34:1527–1536
6. Rose MT, Patti AF, Little KR, Brown AL, Jackson WR, Cavagnaro TR (2014) A meta-analysis and review of plant growth response to humic substances: Practical implications for agriculture. In: Sparks DL (ed) Advances in Agronomy, volume 124. Elsevier, San Diego
7. Canellas LP, Olivares FL, Okorokova-Facanha AL, Racanha AR (2002) Humic acids isolated from earthworm compost enhance root elongation, lateral root emergence, and plasma membrane H⁺-ATPase activity in maize roots. Plant Physiol 130:1951–1957
8. Pascual JA, Garcia C, Hernandez T, Lerma S, Lynch JM (2002) Effectiveness of municipal waste compost and its humic fraction in suppressing *Pythium ultimum*. Microb Ecol 44:59–68
9. Chen Y, De Nobili M, Aviad T (2004) Stimulatory effects of humic substances on plant growth. In: Magdoff F, Weil RR (ed) Soil organic matter in sustainable agriculture. CRC Press, Boca Raton
10. Zandonadi DB, Canellas LP, Facanha AR (2007) Indolacetic and humic acids induce lateral root development through a concerted plasmalemma and tonoplast H⁺ pumps activation. Planta 225:1583–1595
11. Loffredo E, Berloco M, Senesi N (2008) The role of humic fractions from soil and compost in controlling the growth in vitro of phytopathogenic and antagonistic soil-borne fungi. Ecotox Environ Safe 69:350–357
12. Ferrara G, Brunetti G (2010) Effects of the times of application of a soil humic acid on berry quality of table grape (*Vitis vinifera* L.) cv Italia. Span J Agric Res 8:817–822
13. Silva-Matos RRS, Cavalcante IHL, Junior GBS, Albano FG, Cunha MS, Beckmann-Cavalcante (2012) Foliar spray of humic substances on seedling production of watermelon cv. Crimson Sweet. J Agron 11: 60–64
14. Hertkorn N, Ruecker C, Meringer M, Gugisch R, Frommberger M, Perdue EM, Witt M, Schmitt-Kopplin P (2007) High-precision frequency measurements: Indispensable tools at the core of the molecular-level analysis of complex systems. Anal Bioanal Chem 389:1311–1327
15. Muscolo A, Sidari M, Nardi S (2013) Humic substance: Relationship between structure and activity. Deeper information suggests univocal findings. J Geochem Explor 129:57–63
16. Nardi S, Muscolo A, Vaccaro S, Baiano S, Spaccini R, Piccolo A (2007) Relationship between molecular characteristics of soil humic fractions and glycolytic pathway and krebs cycle in maize seedlings. Soil Biol Biochem 39:3138–3146
17. Canellas LP, Spaccini R, Piccolo A, Dobbs LB, Okorokova-Facanha AL, de Araujo SG, Olivares FL, Facanha AR (2009) Relationships between chemical characteristics and root growth promotion of humic acids isolated from Brazilian Oxisols. Soil Sci 174:611–620
18. Dobbs LB, Canellas LP, Olivares FL, Aguiar NO, Peres LEP, Axevedo M, Spaccini R, Piccolo A, Facanha AR (2010) Bioactivity of chemically transformed humic matter from vermicompost on plant root growth. J Agric Food Chem 58:3681–3688
19. Thurman EM, Malcolm RL (1981) Preparative isolation of aquatic humic substances. Environ Sci Technol 15:463–466
20. Koch BP, Ludwichowski K-U, Kattner G, Dittmar T, Witt MR (2008) Advanced characterization of marine dissolved organic matter by combining reversed-phase liquid chromatography and FT-ICR-MS. Mar Chem 111:233–241
21. Stenson AC (2008) Reversed-phase chromatography fractionation tailored to mass spectral characterization of humic substances. Environ Sci Technol 42:2060–2065
22. Liu Z, Sleighter RL, Zhong J, Hatcher PG (2011) The chemical changes of DOM from black waters to coastal marine waters by HPLC combined with ultrahigh resolution mass spectrometry. Estuar Coast Shelf Sci 92:205–216
23. Sleighter RL, Hatcher PG (2008) Molecular characterization of dissolved organic matter (DOM) along a river to ocean transect of the lower Chesapeake Bay by ultrahigh resolution electrospray ionization Fourier transform ion cyclotron resonance mass spectrometry. Mar Chem 110:140–152
24. McKnight DM, Boyer EW, Westerhoff PK, Doran PT, Kulbe T, Andersen DT (2001) Spectrofluorometric characterization of dissolved organic matter for indication of precursor organic material and aromaticity. Limnol Oceanogr 46:38–48
25. Fellman JB, Hood E, Spencer RGM (2010) Fluorescence spectroscopy opens new windows into dissolved organic matter dynamics in freshwater ecosystems: A review. Limnol Oceanogr 55:2452–2462
26. Lam B, Simpson AJ (2008) Direct ¹H NMR spectroscopy of dissolved organic matter in natural waters. Analyst 133:263–269
27. Gaskell SJ (1997) Electrospray: principles and practice. J Mass Spectrom 32:677–688
28. Cech NB, Enke CG (2001) Practical implications of some recent studies in electrospray ionization fundamentals. Mass Spectrom Rev 20:362–387
29. Marshall AG, Hendrickson CL, Jackson GS (1998) Fourier transform ion cyclotron resonance mass spectrometry: A primer. Mass Spectrom Rev 17:1–35
30. Dittmar T, Koch B, Hertkorn N, Kattner G (2008) A simple and efficient method for the solid phase extraction of dissolved organic matter (SPE-DOM) from seawater. Limnol Oceanogr Meth 6:230–235
31. Weishaar JL, Aiken GR, Bergamaschi BA, Fram MS, Fujii R, Mopper K (2003) Evaluation of specific ultraviolet absorbance as an indicator of the chemical composition and reactivity of dissolved organic carbon. Environ Sci Technol 37:4702–4708
32. Cory RM, Miller MP, McKnight DM, Guerard JJ, Miller PL (2010) Effect of instrument-specific response on the analysis of fulvic acid fluorescence spectra. Limnol Oceanogr Meth 8:67–78
33. Sleighter RL, McKee GA, Liu Z, Hatcher PG (2008) Naturally present fatty acids as internal calibrants for Fourier transform mass spectra of dissolved organic matter. Limnol Oceanogr Meth 6:246–253

34. Sleighter RL, Chen H, Wozniak AS, Willoughby AS, Caricasole P, Hatcher PG (2012) Establishing a measure of reproducibility of ultrahigh-resolution mass spectra for complex mixtures of natural organic matter. Anal Chem 84:9184–9191

35. Stubbins A, Spencer RGM, Chen H, Hatcher PG, Mopper K, Hernes PJ, Mwamba VL, Mangangu AM, Wabakanghanzi JN, Six J (2010) Illuminated darkness: Molecular signatures of Congo River dissolved organic matter and its photochemical alteration as revealed by ultrahigh precision mass spectrometry. Limnol Oceanogr 55:1467–1477

36. Sleighter RL, Liu Z, Xue J, Hatcher PG (2010) Multivariate statistical approaches for the characterization of dissolved organic matter analyzed by ultrahigh resolution mass spectrometry. Environ Sci Technol 44:7576–7582

37. Helms JR, Stubbins A, Ritchie JD, Minor EC, Kieber DJ, Mopper K (2008) Absorption spectral slopes and slope ratios as indicators of molecular weight, source, and photobleaching of chromophoric dissolved organic matter. Limnol Oceanogr 53:955–969

38. Reemtsma T (2009) Determination of molecular formulas of natural organic matter molecules by (ultra-) high-resolution mass spectrometry: Status and needs. J Chromatogr A 1216:3687–3701

39. Sleighter RL, Hatcher PG (2011) Fourier transform mass spectrometry for the molecular level characterization of natural organic matter: Instrument capabilities, applications, and limitations. In: Nikolic G (ed) Fourier Transforms- Approach to Scientific Principles. InTech, Vienna. available from: http://www.intechopen.com/articles/show/title/fourier-transform-mass-spectrometry-for-the-molecular-level-characterization-of-natural-organic-matt

40. Stenson AC, Landing WM, Marshall AG, Cooper WT (2002) Ionization and fragmentation of humic substances in electrospray ionization Fourier transform-ion cyclotron resonance mass spectrometry. Anal Chem 74:4397–4409

41. Kim S, Kramer RW, Hatcher PG (2003) Graphical method for analysis of ultrahigh-resolution broadband mass spectra of natural organic matter, the van Krevelen diagram. Anal Chem 75:5336–5344

42. Hockaday WC, Purcell JM, Marshall AG, Baldock JA, Hatcher PG (2009) Electrospray and photoionization mass spectrometry for the characterization of organic matter in natural waters: a qualitative assessment. Limnol Oceanogr Meth 7:81–95

43. Ohno T, He Z, Sleighter RL, Honeycutt CW, Hatcher PG (2010) Ultrahigh resolution mass spectrometry and indicator species analysis to identify marker components of soil- and plant biomass- derived organic matter fractions. Environ Sci Technol 44:8594–8600

44. Sleighter RL, Cory RM, Kaplan LA, Abdulla HAN, Hatcher PG (2014) A coupled geochemical and biogeochemical approach to characterize the bioreactivity of dissolved organic matter from a headwater stream. J Geophys Res Biogeosci 119:1520–1537

45. Ikeya K, Sleighter RL, Hatcher PG, Watanabe A (2012) Compositional features of Japanese Humic Substances Society standard soil humic and fulvic acids by Fourier transform ion cyclotron resonance mass spectrometry and X-ray diffraction profile analysis. Humic Substances Res 9:25–33

46. Pinton R, Cesco S, De Nobili M, Santi S, Varanini Z (1998) Water- and pyrophosphate-extractable humic substances fractions as a source of iron for Fe-deficient cucumber plants. Biol Fert Soils 26:23–27

47. Chen Y, Clapp CE, Magen H (2004) Mechanisms of plant growth stimulation by humic substances: The role of organo-iron complexes. Soil Sci Plant Nutr 50:1089–1095

48. Santiago A, Delgado A (2007) Effects of humic substances on iron nutrition of lupin. Biol Fert Soils 43:829–836

8

Effect of arbuscular mycorrhizal fungi (AMF) and water stress on growth, phenolic compounds, glandular hairs, and yield of essential oil in basil (*Ocimum gratissimum* L)

Zakaria Hazzoumi[1*], Youssef Moustakime[1], El hassan Elharchli[2] and Khalid Amrani Joutei[1]

Abstract

Background: Water stress is one of the most adverse conditions that may affect growth, and synthesis of essential oils in aromatic and medicinal plants. To overcome these climatic conditions, mycorrhiza is an adaptation strategy developed by plants to help them cope with these adverse conditions. For this purpose, we studied the influence of mycorrhizal fungi (*Glomus intraradices*) and water stress on the growth of basil plants (*Ocimum gratissimum* L), the yield of essential oils, and the abundance of glandular hairs.

Results: The analyses show that AMF increases the yield of oils with a maximum recorded in stressed mycorrhizal plants (0.33%) and the lowest in non-stressed non-mycorrhizal plants (0.22%). The contents of total phenolic compounds increase in non-mycorrhizal plants under stress (104% in leaves and 97% in the roots) unlike the mycorrhiza which did not stimulate the synthesis of these compounds, Moreover, the contents of chlorophyll pigments decrease with the application of stress in non-mycorrhizal plants (53%) and increase in mycorrhizal plants. The proline contents increased significantly with the application of water stress; this increase is more pronounced in non-mycorrhizal plants than mycorrhizal plants.

Conclusions: Water stress limits the growth and leads to a decrease in morphological parameters, this reduction is accompanied by a synthesis of several molecules in particular proline and phenolic compounds, However, the AMF stimulates growth, and drives the water status in plants at an optimal level, thus confirming the role of mycorrhizal symbiosis in plant defense against biotic and abiotic stress.

Keywords: Chlorophyll; Essential oils; Glandular hairs; *Glomus intraradices*; Phenolic compounds; Proline; *O. gratissimum*; Water stress

Background

Permanent or temporary water deficit plays an important role in the distribution of natural vegetation and cultivated in performance more than any other environmental factors plants [1]. It limits the growth and leads to changes in the metabolism of plants. Water stress causes the accumulation of phenols in *Bermuda grass* (grass) and in *Gomphocarpus fruticosus* (cotton) [2].

Wahid and Ghazanfar [3], Wahid and Close [4] confirmed these data in other plant species, with increased synthesis of phenolic compounds, flavonoids, and phenylpropanoids. Stress-induced increase in the activity of phenylalanine ammonia-lyase (PAL) may be regarded as the beginning of the acclimation of cells facing water stress.

Other studies have shown that severe water stress has a negative effect on the synthesis of these compounds in tobacco; it will stop the synthesis of anthocyanins in flowers and polyphenols in the leaves [5]. A similar phenomenon is observed for the flowers and leaves of *Begonia gracilis* [6].

* Correspondence: Zakaria.hazzoumi@yahoo.fr
[1]Laboratory of Bioactive Molecules, Structure and Function, Faculty of Science and Technology Fez, B.P. 2202-Road of Imouzzer, Fez 30000, Morocco
Full list of author information is available at the end of the article

Studies on osmoregulation indicate that proline plays an important role in the fight against water deficit. Ain-Lhout et al. [7,8] reported that the levels of these molecules increased in *Pistacia lentiscus* L and *Halimium halimifolium* L when they are exposed to water stress. These results are confirmed by the work of AJ Delauney and DPS Delauney and Verma [9], Handa et al. [10], and Heuer [11].

Furthermore, Cornic and Fresneau [12] and Kim et al. [13] suggested that the water stress had a negative effect on the chlorophyll contents, which indicates a decrease in photosynthesis due to a disturbance in the metabolism of carbon and of certain enzymes involved in the regulation of these photosynthetic reactions.

To address these climate hazards and struggle against drought, plants develop several defense strategies; mycorrhizal association with soil fungi is an example [14]. These associations improve plant nutrition, mainly nitrogen and phosphate by increasing water intake in plants by increasing the exchange surface root-soil [15]. Many experiments on the effects of mycorrhizae on plant growth showed that the rate of photosynthesis is higher in mycorrhizal plants compared to non-mycorrhizal plants Auge [16], Kucey and Paul [17], and Levy and Krikun [18]. Most studies suggest that AM fungi contribute to increased rate of photosynthesis in improving (P) nutrition in plants, [5,19,20].

For the establishment of this association, it is necessary that both partners agree. This step is initiated via a cross-talk molecule and a change in gene expression [21]. Furthermore, the AMF infection induces a change in the metabolism of host plants resulting in the induction of chemical defense [22,23]. This change affects the synthesis of several families of molecules, terpenoids [24,25], EOs [26-28], and glucosinolates [29], phytoalexins [30,31], and phenolic compounds [32,33]. Several studies have been carried on the synthesis of these molecules on medicinal and aromatic plants: basil (*Ocimum basilicum*) [34,35,26,27], oregano (*Origanum onites*) [28], mint (*Mentha requienii* [36,37] and *Mentha arvensis* [36,38]), dill (*Anethum graveolens*) [39], fennel (*Foeniculum vulgare*) [40], coriander (*Coriandrum sativum*) [40,41], lavender (*Lavandula angustifolia*) [42], pelargonium (*Pelargonium peltatum*) [43], and sage (*Salvia officinalis*) [44].

Ocimum gratissimum L is one of the most used plants in traditional medicine in Morocco. Many studies have been done on this plant [45-47]. This work has focused on improving the yield of EOs. However, the influence of water stress and mycorrhiza on growth and chemical characteristics of this plant is very little studied.

The aim of this work was to evaluate the influence of water stress and inoculation with *Glomus intraradices* on some physiological and biochemical parameters of *O. gratissimum* L. and synthesis of EOs as well as the abundance of secreting glands of these oils.

Material and methods
Pregermination of seeds
The seeds of basil (*O. gratissimum* L) were surface disinfected by a passage in ethanol 95° (1 min 30 s) and immersed in a mercury hypochlorite solution (1%) for 3 to 4 min. Seeds were then rinsed several times in sterile distilled water before being placed on agar medium or on filter paper moistened with sterile distilled water in Petri dishes. Then, placed in an oven at 26°C, in the dark, to allow germination of the seeds.

Culture basil seedlings
After germination, seedlings were transplanted into plastic pots (3 kg capacity) containing the growth substrate in an amount of 50 to 60 plants per pot and were grown in a greenhouse at temperature comprising between 25 and 34°C.

The basil plants (M or NM) are watered daily. After the 50th day, the plants are subjected to different water regimes:

a) Water regime unstressed (NS): the plants are not deprived of water throughout their growth.
b) Application of water stress (S): plants are deprived of water for 2 weeks.

Estimation of mycorrhizal root infection
The identification of mycorrhizal root infection is in optical microscopy through a technique not vital staining with trypan blue (TB), described by Hayman [48], revealing the set of fungal biomass.

Root samples, taken at random, are thoroughly rinsed to remove adhering substrate. Then, the root fragments are digested in a solution of potassium hydroxide (KOH) at 10% for 45 min at 90°C in an oven, in order to empty the cell of their cytoplasmic contents which facilitates their coloring. Then, the roots are thoroughly rinsed with distilled water and placed in a solution lactophenol trypan blue (0.5%) at 90°C in the oven for 15 min.

Estimation of mycorrhization
We used the technique described by Trouvelot et al. [49]. This method allows to judge the state of mycorrhization and reflects the potential of the symbiotic system. Colored roots were cut into fragments of approximately 1 cm in length. Thirty random fragments are assembled and crushed between slide and cover slip in lactoglycerol, with 15 fragments per slide. Estimating the endomycorrhizal infection is by observing under light microscope. Several parameters are evaluated as follows:

$$F\% = (\text{Number of mycorrhizal fragments}/N) \times 100$$

- (F%): The frequency of mycorrhization reflects the importance and the percentage of fragments of infected roots.

with: N = Total number of root fragments observed

- (M%): The colonization of the cortex intensity expresses the portion of the cortex colonized with respect to the entire root system.

with: n5, n4,..., n1 = number of fragments respectively denoted as 5, 4,..., 1.

$$\%m = M \times (\text{total number})/(\text{number of mycorrhizal fragments})$$
$$= M \times 100/F.$$

Mycorrhizal intensity:

$$\%A = a \times (M/100).$$

- (A%): Frequency of arbuscular in the root system.

$$\% a = (100\,mA3 + 50\,mA2 + 10\,mA1)/100$$

Arbuscular intensity of the mycorrhizal part:

where mA3, mA2, and mA1% m are respectively assigned as A3 notes, A2, and A1.

with: mA3 = ((95n5A3 70n4A3 + + + 30n3A3 5n2A3 n1A3 +)/mycorrhizal number) × 100/m.

Similarly for A2 and A1.

$$\%M = (95\,n5 + 70\,n4 + 30\,n3 + 10\,n2 + n1)/N.$$

Relative water contents

The relative water content (RWC) is measured on the seventh or eighth leaf fully developed using the following formula according to Bandurska [50]:

$$RWC\% = 100 \times [(FW-DW)/(WT-DW)]$$

With

FW

the weight of fresh leaf material.

WT

the weight of fresh material from the turgid leaf was submerged in distilled water for 4 h.

DW

dry weight of the sheet material placed in an oven at 70°C for 24 h.

Extraction and determination of total phenols

Oxidation of phenols reduces this reactant in a mixture of the blue oxides of tungsten and molybdenum. The color intensity is proportional to the rate of oxidized phenolic compounds.

- Extraction of TPC.

Fragments of leaves and roots (0.5 g) were ground in a mortar containing a specific volume usually 5 ml, ethanol 50% (water-alcohol solution). Then, we collect the extracts in tubes with lids and well numbered, then leave the tubes in the refrigerator overnight to allow time for ethanol to extract the maximum amount of phenol present in the extract.

In the tubes containing the leaf extracts, there was a risk of the existence of chlorophylls; we tried to eliminate it by adding in 3 ml of extract 0.5 ml of chloroforms the tubes are vortexed and centrifuged 5 min at 5 × 1,000 mtp; two phases were separated, a phase supernatant and pellet.

-Determination of TPC.

The assay of total phenols using the method based on the Folin-Ciocalteu reagent, described by Ribereau-Gayon and Stonestreet [51].

• Prepare in test tubes the following mixture: 0.5 ml of extract, 3 ml of water, 0.5 ml of Na_2CO_3 (20%); mix, wait 3 min, and then add 0.5 ml of Folin-Ciocalteu reagent. Mix and place the tubes for 30 min at 40°C, reading absorbance at 760 nm.

The amount of phenolic compounds was calculated using gallic acid for the standard curve and expressed in milligrams per gram of fresh leaf matter.

Dosage of proline

Proline content was determined according to Bates et al. [52] by measuring the quantity of the colored reaction product of proline with ninhydric acid. The absorbance was read at 520 nm. The amount of proline was calculated using L-proline (Panreac) for the standard curve and expressed in micrograms per gram of fresh leaf matter.

Determination of chlorophylls

Fragments of leaves (1 g) were ground in a mortar previously placed in ice with a pinch of magnesium carbonate and 5 g of anhydrous sodium sulfate. Then, 10 ml of acetone 80% are poured into the ground material, which is filtered on a Buchner; the residue is recovered in tubes essai; and further extractions are carried out with acetone to obtain a filtrate colorless (devoid of all traces of chlorophyll pigments) which the final volume is specified.

OD measurements were made with a spectrophotometer at wavelengths of around 663 to 645 nm for chlorophyll a and chlorophyll b.

McKinney [53] has established systems of equations that calculate the concentrations (g/l) of chlorophyll from absorbance at 663 and 645 nm of an extract of acetone:

$$\text{Chlorophyll a} = (0.0127\ D.O663) - (0.00269\ OD\ 645)$$

$$\text{Chlorophyll b} = (0.0229\ OD\ 645) - (663\ 0.00468\ OD)$$

$$\text{Total chlorophyll} = (0.0202\ OD\ 645) + (0.00802\ OD\ 663)$$

Extraction of EOs

One hundred grams of dried aerial parts of *O. gratissimum* were submitted to hydrodistillation with a Clevenger-type apparatus [54] and extracted with 2 l of water for 180 min (until no more EO was obtained). The EO was collected, dried under anhydrous sodium sulfate, and stored at 4°C until analyzed. The EO yield is given by the following formula:

$$YEO \ (ml/100 \ g \ Dm) \ = \ (V/Dm \ \times \ 100) \ \pm \ (\Delta V/\ Dm \ \times \ 100)$$

YEO
  essential oil yield of dry matter.
V
  the volume of essential oils collected (ml).
ΔV
  reading error.
Dm
  dry plant mass (g).

Description of environmental scanning electron microscopy

The observations were performed using a Scanning Electron Microscope Environmental Quanta 200 (FEI Company, Hillsboro, OR, USA) category. The microscope is equipped with electron gun tungsten. The analyses are carried out under a partial pressure of water vapor.

Statistical analysis

One-way analysis of variance was carried out for each parameter studied. Tukey's post hoc multiple mean comparison test was used to test for significant differences between treatments (at 5% level). Univariate analysis was used to test significant differences in treatments, accessions, and their interaction for an individual parameter. All statistical analyses were performed with IBM.SPSS statistics, Version 19. The results of each experiment (biochemical essays) were repeated three times (20 times for morphological essays).

Results

Mycorrhizal colonization

Table 1 shows mycorrhizal colonization of basil plants subjected to continuous irrigation (NS) and water stress (S) after disclosure by the trypan blue. This mycorrhizal colonization, estimated by the mycorrhizal frequency (F%), mycorrhizal intensity (M%), and arbuscular richness (A%), showed no differences between the stressed plant and non-stressed plant differences.

As against, control plants which were grown in a sterilized soil showed no mycorrhizal colonization.

Effect of mycorrhizae and water stress on plant growth and water contents

Mycorrhization and stress influenced significantly on the growth of aerial and root part; this elongation of basil plants are largely driven by mycorrhiza with maximum growth recorded in MNS plants. By cons, we find that water stress inhibits the growth of these two parts mainly in NM plants (Figure 1). In stressed mycorrhizal plants, water supply deficit is offset by the AMF which generates growth at the root and aerial part.

The same observations can be made on the water contents in leaves of basil plant; we recorded a significant variation between the different treatments (Figure 2). In MNS plants, mycorrhiza increases the levels of water to a value which can reach 93%. Non-mycorrhizal plants showed the lowest levels with a dramatic decrease recorded in NMS plants (81%). In MS plants, there have been relatively high water contents (91%) reflecting the role of mycorrhizal fungi to withstand water deficit in stressed plants.

Effect of mycorrhizae and water stress on the contents of chlorophyll pigments

The estimated levels of chlorophyll pigments show that in the absence of water stress, the content of total chlorophyll (Cha + Chb) is slightly higher (10%) in mycorrhizal plants (2.59 mg/g MF) than in NM plants (2.36 mg/g MF). This content is reduced by water stress significantly, especially in NM plants. Furthermore, when comparing the levels of these pigments between NM.S and MS plants, we see that mycorrhizal plants accumulate 72.72% more of chlorophyll (Figure 3).

The same observations can be made about the Chla/Chlb ratio. In the absence of water stress, this ratio is higher in mycorrhizal plants than in NM plants and decreases in times of stress (Figure 3).

Effect of mycorrhizae and water stress on proline contents

During growth, the proline contents in the aerial part significantly change although in case of irrigation or not, in one hand, and mycorrhization or not in the other (Figure 4).

However, in case of water stress, proline contents increased and this increase is more pronounced in NM plants (55 μg/g FM) than in M plants (41 μg/g FM).

At root parts, proline shows the same trend of accumulation for the aerial parts. Thus, proline contents were comparable in unstressed plants inoculated whether or not, while the lack of water leads to an increase in the

Table 1 Mycorrhizal colonization of basil plants (*O.gratissimum*) roots subjected to continuous irrigation (NS) and water stress (S)

Mycorrhizal colonization	NS	S
Mycorrhizal frequency (F%)	78.5	77.3
Mycorrhizal intensity (M%)	35	32
Arbuscular richness (A%)	20	17

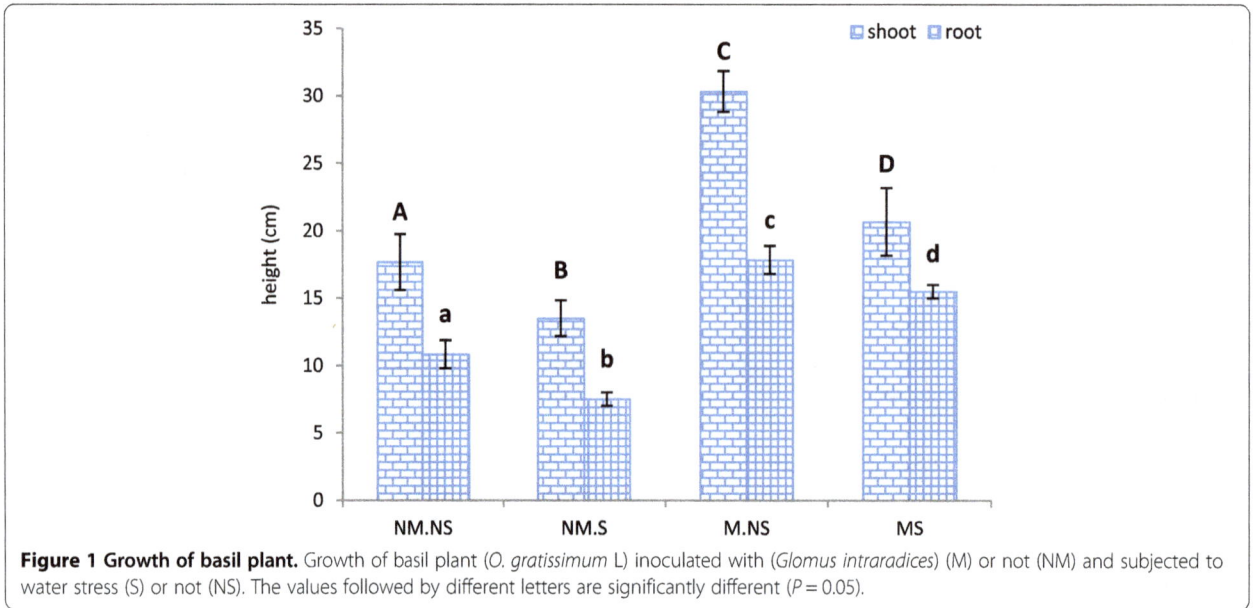

Figure 1 Growth of basil plant. Growth of basil plant (*O. gratissimum* L) inoculated with (*Glomus intraradices*) (M) or not (NM) and subjected to water stress (S) or not (NS). The values followed by different letters are significantly different (*P* = 0.05).

synthesis of proline mainly in non-mycorrhizal plants (31 μg/g FM) (Figure 4).

Effect of mycorrhizae and water stress on phenolic compounds contents

The contents of TPC in aerial parts do not change significantly when the plants are mycorrhized and in the presence or absence of water stress; however, in non-mycorrhizal plants, these levels increase significantly in stressed plants (4.3 mg/g FM) (Figure 5).

In roots, the levels of TPC increase slightly in the NM.S plants. However, in the other treatment, a difference in the level of this compound was not observed.

EO contents

The yield of EOs is very influenced by the mycorrhization since we noted a higher synthesis in both cases stress and irrigation. This increase reaches values up to 37% more than those found in non-mycorrhizal plants; however, water stress did not show a significant

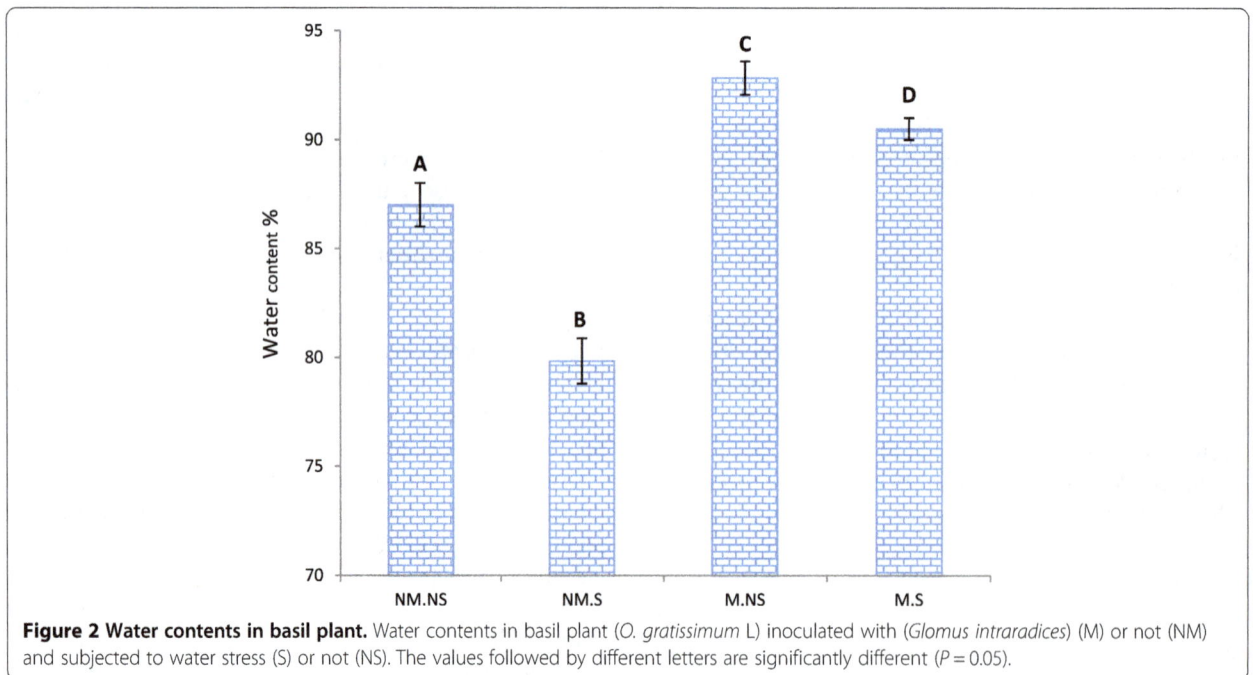

Figure 2 Water contents in basil plant. Water contents in basil plant (*O. gratissimum* L) inoculated with (*Glomus intraradices*) (M) or not (NM) and subjected to water stress (S) or not (NS). The values followed by different letters are significantly different (*P* = 0.05).

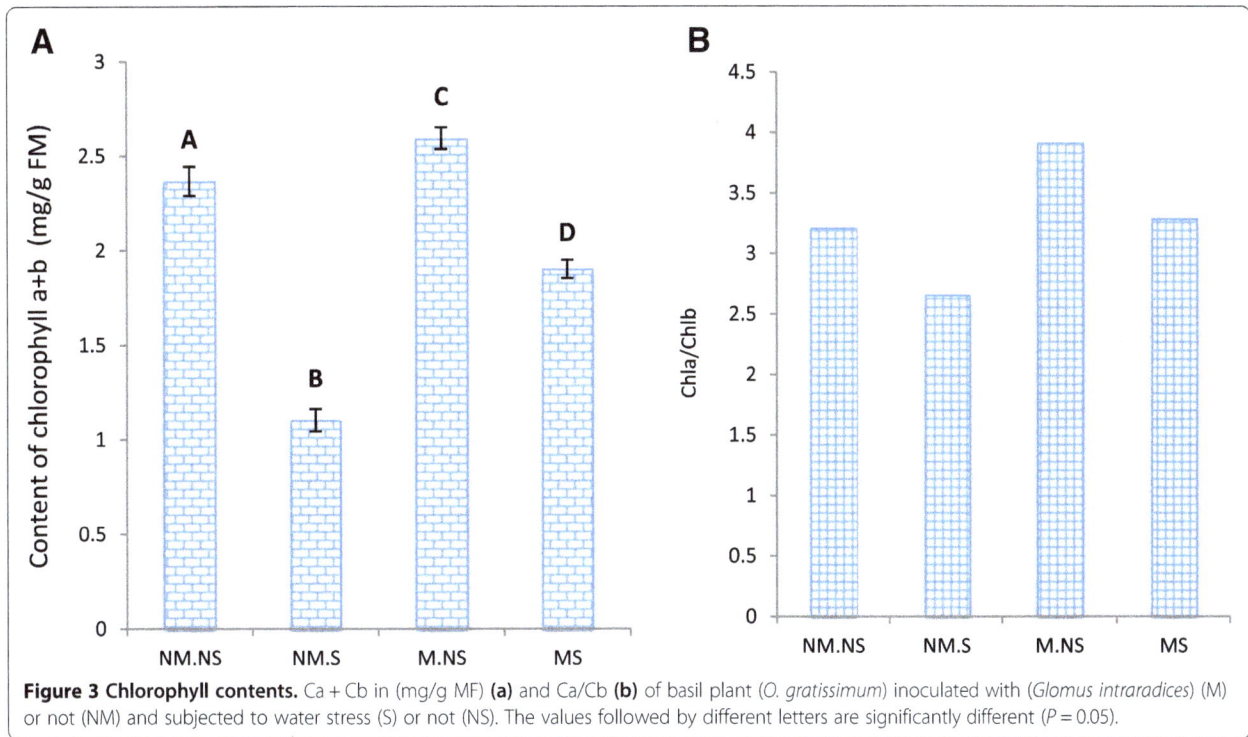

Figure 3 Chlorophyll contents. Ca + Cb in (mg/g MF) **(a)** and Ca/Cb **(b)** of basil plant (*O. gratissimum*) inoculated with (*Glomus intraradices*) (M) or not (NM) and subjected to water stress (S) or not (NS). The values followed by different letters are significantly different ($P = 0.05$).

influence on the synthesis of EO because we noticed a slight increase (9%) between plants NM and (10%) between M plants. Moreover, the combination of mycorrhization and stress (MS) increases this value to 50%, greater than those found in NM.NS plants (Table 2).

These results can be confirmed by the increase in the number of glandular hairs in basil leaves (Figure 6). The observations in environmental scanning electron microscopy show an increase in the number of glandular hairs in the basal part of mycorrhizal plants compared to basal

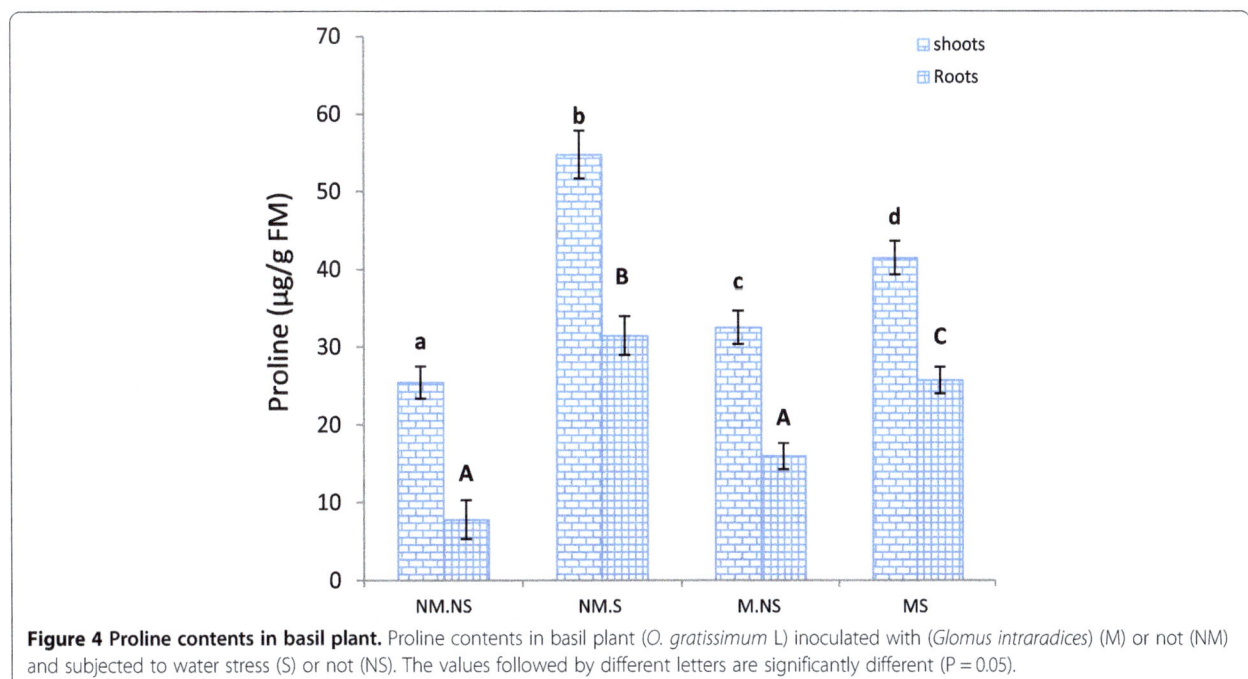

Figure 4 Proline contents in basil plant. Proline contents in basil plant (*O. gratissimum* L) inoculated with (*Glomus intraradices*) (M) or not (NM) and subjected to water stress (S) or not (NS). The values followed by different letters are significantly different (P = 0.05).

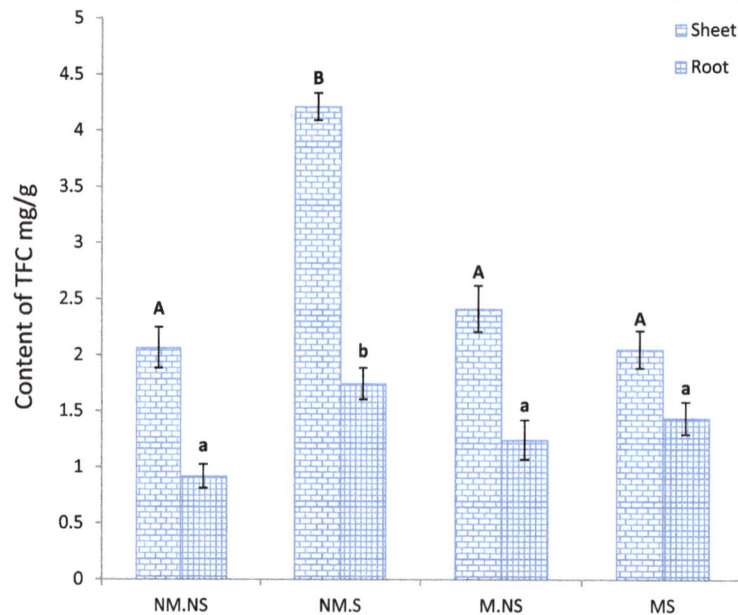

Figure 5 Total phenolic compound contents. Total phenolic compound (TPC) contents in basil plant (*O. gratissimum* L) inoculated with (*Glomus intraradices*) (M) or not (NM) and subjected to water stress (S) or not (NS). The values followed by different letters are significantly different (P = 0.05).

areas of non-mycorizel plant. The mycorrhization can double the number of the glandular hairs.

Discussion

The effect of mycorrhizae on basil plant tolerance to water stress is studied using the fungus *G. intraradices* as mycorrhizal inoculum. Plants inoculated with *G. intraradices* show better growth than non-inoculated plants, both in terms of irrigation, in conditions of water stress. In irrigated conditions, inoculation with *G. intraradices* causes an increase in the length of the aerial part and root of the irrigated plants, respectively, 71% and 75%, while in conditions of water stress, the growth of NM.S plants is limited because stress alters the growth of root parts which influence the water and nutritional status of plants. Under these conditions, mycorrhiza increases the growth of aerial parts of 51% and root parts of nearly 130%. Similar results are observed by many authors, although the magnitude of adverse effects caused by the water stress and the positive effects of mycorrhizal is very

Table 2 variation of essential oils contents under conditions of stress and mycorrhizal in *O. gtatissimum* plants

	NM.NS	NM.S	M.NS	MS
Oils content (%)	0.22 a	0.24 a	0.3 b	0.33 c

Three replications were taken for the extraction of essential oils, the values followed by different letters are significantly different (P = 0.05).

variable: Ferahani et al. [41] showed that mycorrhizal stressed coriander present a normal growth. An-Dong et al. [20], Baslam et al. [2]), Rasouli-Sadaghianil et al. [55], Toussaint et al. [35], and Zolfaghari et al. [56] showed that mycorrhizal fungi *Glomus fasciculatum*, *Glomus etuonicatum*, and *G. intraradices* stimulate the growth and nutritional status of several plants *Lactuca sativa* L., *Lonicera confusa*, *O. basilicum* L, and *Vicia faba* L. Water stress induced a reduction of water contents (WC) of non-mycorrhizal plants. In mycorrhizal plants, there has also been a reduction, but this reduction is very small. It maintains water contents at higher levels under conditions of water deficit through the mycorrhizal association is due to a better exploitation of available water in the soil by fungal hyphae [57].

In the case of irrigation, the plants do not seem affected by mycorrhiza as non-mycorrhizal plants have chlorophyll levels comparable to those of mycorrhizal plants. However, in plants subjected to water stress, mycorrhizal colonization by *G. intraradices* reduced the magnitude of the reduction in chlorophyll and maintains its contents at much higher than those of non-mycorrhizal plants since the improvement rate can reach 72%. These results confirm those found by Kucey and Paul [17], Levy and Krikun [18], Snellgrove et al. [58], and Allen et al. [59]; they have shown that mycorrhizal combination stimulates photosynthesis of many aromatic and medicinal plants. These authors reported stimulation of photosynthesis in improving

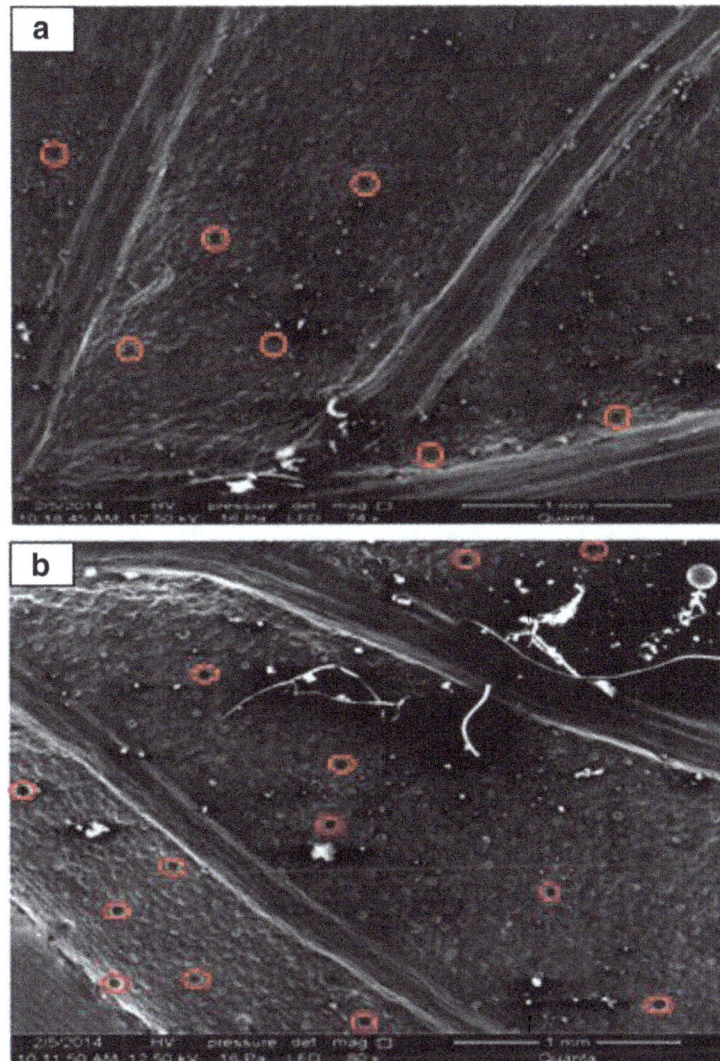

Figure 6 Observation by environmental scanning electron microscopy. Observation by environmental scanning electron microscopy shows the abundance of glandular hairs at the basal part of an unstressed plant and subjected to M **(a)** or not NM **(b)**.

P nutrition necessary for reactions of CO2 assimilation in plants. This stimulation can also be linked to an increase in leaf area [60] or associated with increased hydration leaves [61].

Our results show also, with water stress, proline contents as well as in the aerial parts in root parts increased compared to non-stressed plants. This increase is higher in non-mycorrhizal than in mycorrhizal plants. This latter accumulates proline 39% less than non-inoculated plants. Mycorrhiza thus mitigates the increase in proline contents in water stress conditions. Similar results are shown by many authors [62,63] who showed that water stress induced proline accumulation in many plants leaves. Proline is a compound that is considered an osmotic regulator [7,8] to promote tolerance to water stress in the stressed plants by maintaining turgor through osmotic adjustment.

We noted also that in case of stress, the proline contents increased slightly in the aerial and root parts when plants are mycorrhizal. This slight increase was due to the defense mechanism set up by the plant response to mycorrhizal infection. According to Giovannetti and Avio [64], the mycorrhizal colonization would initially be perceived by the plant as a stress or an attack at the very location of colonization by endomycorrhizal, where the synthesis of secondary compounds in the early stages of colonization.

We also showed that the levels of TPC increased mainly in the leaves of NMS plants, in response to water stress. These results are in agreement with those of Iker et al. [65] who showed that the levels of flavanols, which are the monomers of condensed tannins (flavanols epicatechin (EC), epicatechin gallate (ECG), and epigallocatechin gallate

(EGCG), increase in the leaves of *Cistus clusii* in response to hydric deficit. Most phenols plants are also considered stress metabolites [66,67] and their accumulation in plants is affected by several factors.

We also found in this work that the yield of EO increases significantly when the plants are mycorrhizal. This increase may be related to the number of glandular hairs in the leaves. ESM observations of leaves show an increase in glandular hairs in mycorrhizal plants. These results are confirmed by the work of Andrea Copetta et al. [26,27] who showed that Mycorrhizal colonization of *O. basilicum* L plants by *Glomus mosseae*, *Gigaspora margarita*, and *Gigaspora rosea* increases the number of glandular hairs in the leaves compared to non-mycorrhizal plants. Zolfaghari et al. [56], Gupta et al. [38], Freitas et al. [36], and Mucciarelli [68] reported an increase in yield of EOs in *Mentha arvensis* and *Mentha piperita* L inoculated with mycorrhizal fungi, and this increase related to a change in plant secondary metabolism and the number of glands per leaf; this increase in the levels of EO may also be related to fungal colonization. According to Copetta et al. [26,27], the production of terpenoids and constituents of EOs possess fungicidal properties which are considered a defensive response to colonization by the fungus and causing increased yield and increased production of these metabolites in mycorrhizal plants.

Conclusion

This study helps us to understand the reaction of *O. gratissimum* vis-a-vis mycorrhizae and water stress. In the case of non-mycorrhizae, water stress limits the growth and photosynthetic capacity and leads to a reduction of water contents in the plant. This decrease in morphological parameters is accompanied by a synthesis of several molecules in particular proline and phenolic compounds, known for their role in plant tolerance to adverse conditions. However, the AMF stimulates growth and photosynthesis and drives the water status in plant at an optimal level. The results showed a decrease in levels of proline and phenolic compounds, thus confirming the role of mycorrhizal symbiosis in plant defense against biotic and abiotic stress. Note also that this symbiosis leads to an increase in the yield of EOs, regardless of the water status of the plant (stressed or not), and this increase is correlated with the increase of the glandular hairs abundance.

Abbreviation

AMF: arbuscular mycorrhizal fungi; Cha: chlorophyll (a); Chb: chlorophyll (b); ChT: chlorophyll (total); EO: essential oil; M.NS: mycorrhizel unstressed; MS: mycorrhizel stressed; NM.NS: non-mycorrhizel unstressed; NM.S: non-mycorrhizel stressed; TPC: total phenolic compounds.

Competing interests

The authors declare that they have no competing interests.

Authors' contributions

YM, EhE, and KAJ participated in the data processing and in designing the study and performed the statistical analysis and drafting of the manuscript. All authors read and approved the final manuscript.

Acknowledgements

We would like to thank Pr Naima EL GHACHTOULI from Laboratory of microbial Biotechnology who provided for us the mycorrhizal inoculum, also a special thank for Pr Zain El Abidine Fatemi, from National Institute for Agricultural Research (INRA), Plant Breeding Unit, Meknes-Morocco for statistical help (SPSS software and technical help).

Author details

[1]Laboratory of Bioactive Molecules, Structure and Function, Faculty of Science and Technology Fez, B.P. 2202-Road of Imouzzer, Fez 30000, Morocco. [2]Laboratory of microbial Biotechnology, Faculty of Science and Technology Fez, B.P. 2202-Road of Imouzzer, Fez 30000, Morocco.

References

1. Hong-Bo S, Li-Ye Chu D, Cheruth Abdul J, Chang-Xing Z (2008) Water-deficit stress-induced anatomical changes in higher plants C. R. Biogeosciences 331:215–225
2. Baslam M, Garmendia I, Goicoechea N (2011) Arbuscular mycorrhizal fungi (AMF) improved growth and nutritional quality of greenhouse grown lettuce. J Agric Food Chem 59:5504–5515
3. Wahid A, Ghazanfar A (2006) Possible involvement of some secondary metabolites in salt tolerance of sugarcane. J. Plant Physiol. 163:723–730
4. Wahid A, Close TJ (2007) Expression of dehydrins under heat stress and their relationship with water relations of sugarcane leaves. Biol. Plant. 51:104–109
5. Sivak MN, Walker DA (1986) Photosynthesis in vivo can be limited by phosphate supply. New Phytol 102:499–512
6. Zubek S, Mielcarek S, Turnau K (2012) Hypericin and pseudo hypericin concentrations of a valuable medicinal plant Hypericum perforatum L. are enhanced by arbuscular mycorrhizal fungi. Mycorrhiza 22:149–156
7. Ain-Lhout F, Zunzunegui M, Diaz Barradas MC, Tirado R, Clavijo A, Garcia Novo F (2001) Comparison of proline accumulation in two mediterranean shrubs subjected to natural and experimental water deficit. Plant Soil 230:175–183
8. Ain-Lhout F, Zunzunegui M, Diaz Barradas MC, Tirado R, Clavijo A, Garcia F (2001) Novo Comparison of proline accumulation in two mediterranean shrubs subjected to natural and experimental water deficit. Plant Soil 230:175–183
9. Delauney AJ, Verma DPS (1993) Proline biosynthesis and osmoregulation in plants. Plant J 4:215–223
10. Handa S, Handa AK, Hasegawa PM, Bressan RA (1986) Proline accumulation and the adaptation of cultured plant cells to salinity stress. Plant Physiol 80:938–945
11. Heuer B (1994) Osmoregulatory role of proline in water and salt-stressed plants. In: Pessarakli M (ed) Handbook of Plant and Crop Stress. Marcel Dekker, Inc., New York, pp 363–381
12. Cornic G, Fresneau C (2002) Photosynthetic carbon reduction and carbon oxidation cycles are the main electron sinks for photosystem II activity during a mild drought. Ann Bot 89:887–894
13. Kim JY, Mahe A, Brangeon J, Prioul JL (2000) A maize vacuolar invertase, IVR2, is induced by water stress. Organ/tissue specificity and diurnal modulation of expression. Plant Physiol 124:71–84
14. Wang B, Qiu YL (2006) Phylogenetic distWANG, B., and QIU, Y. L. 2006. Phylogenetic distribution and evolution of mycorrhizas in land plants. Mycorrhiza 16:299–363. ribution and. (s.d.)
15. Smith SE, Read DJ (1997) Mycorrhizal Symbiosis. Academic, London
16. Auge R (2001) Water relations, drought and VA mycorrhizal symbiosis. Mycorrhiza 11:3–42
17. Kucey RMN, Paul EA (1982) Carbon flow, photosynthesis and N2 fixation in mycorrhizal and nodulated fababeans (*Vicia faba* L). Soil Bioi Biochem 14:407–412
18. Levy Y, Krikun J (1980) Effect of vesicular-arbuscular mycorrhiza on Citrus jambhiri water relations. New Phytol 85:25–31

19. Marschner H (1995) Mineral Nutrition of Higher Plants. Academic, London

20. An-Dong SH, Qian L, Jian-Guo H, Ling Y (2013) Influence of arbuscular mycorrhizal fungi on growth, mineral nutrition and chlorogenic acid contents of lonicera confusa seedlings under field conditions. Pedosphere 23(3):333–339

21. Harrison M, van Buuren M (1995) A phosphate transporter from the mycorrhizal fungus Glomus versiforme. Nature 378:626–629

22. Pozo M, Azco'n-Aguilar C (2007) Unraveling mycorrhiza-induced resistance. Curr Opin Plant Biol 10:393–398

23. Slezak S, Dumas-Gaudot E, Paynot M, Gianinazzi S (2000) Is a fully established arbuscular mycorrhizal symbiosis required for bioprotection of Pisum sativum root against Aphanomyces euteiches? Mol Plant Microbe Interact 13:238–241

24. Akiyama K, Hayashi H (2002) Arbuscular mycorrhizal funguspromoted accumulation of two new triterpenoids in cucumber roots. Biosci Biotechnol Biochem 66:762–769

25. Rapparini F, Llusia J, Penuelas J (2008) Effect of arbuscular mycorrhizal (AM) colonization on terpene emission and contents of Artemisia annua L. Plant Biol 10:108–122

26. Copetta A, Lingua G, Berta G (2006) Effects of three AM fungi on growth, distribu-tion of glandular hairs, and essential oil production in Ocimum basilicum L. Var Genovese Mycorrhiza 16:485–494

27. Copetta A, Lingua G, Berta G, MASOERO G (2006) Three arbuscular mycorrhizal fungi differently affect growth, distribution of glandular trichomes and essential oil composition in Ocimum basilicum var. Genovese. Proceedings of the 1st International Symposium on the Labiatae: Advances in Production. Biotechnol Utilisation 723:151–156

28. Khaosaad T, Vierheilig H, Nell M, Zitterl-Eglseer K, Novak J (2006) Arbuscular mycorrhiza alter the concentration of essential oils in oregano (Origanum sp., Lamiaceae). Mycorrhiza 16:443–446

29. Vierheilig H, Gagnon H, Strack D, Maier W (2000) Accumulation of cyclohexenone derivatives in barley, wheat and maize roots in response to inoculation with different arbuscular mycorrhizal fungi. Mycorrhiza 9:291–293

30. Sundaresan P, Raja NU, Gunasekaran P (1993) Induction and accumulation of phytoalexins in cowpea roots infected with the mycorrhizal fungus Glomus fasciculatum and their resistance to Fusarium wilt disease. J Biosci 18:291–301

31. Yao MK, Desilets H, Charles MT, Boulanger R, Tweddell RJ (2003) Effect of mycorrhization on the accumulation of rishitin and solavetivone in potato plantlets challenged with Rhizoctonia solani. Mycorrhiza 13:333–336

32. Morandi D (1996) Occurrence of phytoalexins and phenolic compounds on endomycorrhizal interactions, and their potential role in biological control. Plant Soil 185:241–251

33. Rojas-Andrade R, Cerda-Garcia-Rojas CM, Frias-Hernandez JT, Dendooven L, Olalde-Portugal V, Ramos-Valdivia AC (2003) Changes in the concentration of trigonelline in a semi-arid leguminous plant (Prosopis laevigata) induced by an arbuscular mycorrhizal fungus during the presymbiotic phase. Mycorrhiza 13:49–52

34. Pascual-Villalobos MJ, Ballesta-Acosta MC (2003) Chemical variation in an Ocimum basilicum germplasm collection and activity of the essential oil on Callosobruchus maculates. Biochem Syst Ecol 31:673–679

35. Toussaint JP, Smith FA, Smith SE (2007) Arbuscular mycorrhizal fungi can induce the production of photochemicals in sweet basil irrespective of phosphorus nutrition. Mycorrhiza 17.291–297

36. Freitas MSM, Martins MA, Curcino Vieira IJ (2004) Yield and quality of essential oils of Mentha arvensis in response to inoculation with arbuscular mycorrhizal fungi. Pesqui Agropecu Bras 39:887–894

37. Cabello M, Irrazabal G, Bucsinszky AM, Saparrat M, Schalamuk S (2005) Effect of an arbuscular mycorrhizal fungus, Glomus mosseae, and a rock-phosphate-solubilizing fungus, Penicillium thomii, on Mentha piperita growth in a soilless medium. J Basic Microbiol 45:182–189

38. Gupta ML, Prasad A, Ram M, Kumar S (2002) Effect of the vesicular-arbuscular mycorrhizal (VAM) fungus Glomus fasciculatum on the essential oil yield related characters and nutrient acquisition in the crops of different cultivars of menthol mint (Mentha arvensis) under field conditions. Bioresour Technol 81:77–79

39. Kapoor R, Chaudhary V, Bhatnagar AK (2007) Effects of arbuscular mycorrhiza and phosphorus application on artemisinin concentration in Artemisia annua L. Mycorrhiza 17:581–587

40. Kapoor R, Giri B, Mukerji KG (2004) Improved growth and essential oil yield and quality in Foeniculum vulgare mill on mycorrhizal inoculation supplemented with P-fertilizer. Bioresour Technol 93:307–311

41. Ferahani HA, Lekaschi MH, Hamidi A (2008) Effects of arbuscular mycorrhizal fungi, phosphorus and water stress on quantity and quality characteristics of coriander. Adv Nat Appl Sci 2:55–59

42. Tsuro M, Inoue M, Kameoka H (2001) Variation in essential oil components in regenerated lavender (Lavandula vera DC) plants. Sci Hortic 88:309–317

43. Perner H, Schwarz P, Bruns C, Maeder P, George E (2007) Effect of arbuscular mycorrhizal colonization and two levels of compost supply on nutrient uptake and flowering of pelargonium plants. Mycorrhiza 17:469–474

44. Nell M, Vötsch M, Vierheilig H, Steinkellner S, Zitterl-Eglseer K, Franz C, Novak J (2009) Effect of phosphorus uptake on growth and secondary metabolites of garden sage (Salvia officinalis L.). J Sci Food Agric 89:1090–1096

45. Hazzoumi Z, Moustakime Y, Khalid A (2014) Effect of gibberellic acid (GA), indole acetic acid (IAA) and benzylaminopurine (BAP) on the synthesis of essential oils and the isomerization of methyl chavicol and trans-anethole in Ocimum gratissimum L. Springer Plus 3:321

46. Madeira SVF, Rabelo M, Soares PMG, Souzaa EP, Meireles AVP, Montenegro C, Limaa RF, Assreuya AMS, Criddle DN (2005) Temporal variation of chemical composition and relaxant action of the essential oil of Ocimum gratissimum L. (Labiatae) on guinea-pig ileum. Phytomedicine 12:506–509

47. Yayi E, Gbenou JD, Léon Akanni A, Mansour M, Jean Claude Chalchat O (2004) gratissimum L., siège de variations chimiques complexes au cours du développement C. R. Chimie 7:1013–1018

48. Hayman P e (1970) Improved procedures for clearing and staining parasite and vesiculaire-arbuscular mycorrhizal fungi for rapid assassment of infection trans. Brit Mycolsoc 55:158–161

49. Trouvelot A, Kough JL, Gianinazzi-Pearson V (1986) Meesure du taux de mycorhization VA d'un système radiculaire. Recherche des méthodes d'estimation ayant une signification fonctionnelle. In: Bay G-P, Gianinazzi S (ed) Physiological aspect of mycorrizea. INRA, Paris, pp 217–221

50. Bandurska H (1991) Akumulacja wolnej proliny jako przejaw metabolicznej reakcji roœelin na dzia³anie stresu wodnego. Wiad. Bot. 35:35–46

51. Ribereau-Gayon P, Stonestreet E (1966) Les composes phénoliques des végétaux

52. Bates LS, Waldren RP, Teare JD (1973) Rapid determination of proline for water stress studies. Plant Soil 39:205–207

53. McKinney (1941) Absorption of light by chlorophyll solutions. J. Biol. Chem. 140:315–332

54. Clevenger JF (1928) Determination of volatile oil. J Ann Pharm Assoc 17 (4):346–351

55. Rasouli-Sadaghianil MH, Hassani A, Barin M, Danesh YR, Sefidkon F (2010) Effects of arbuscular mycorrhizal (AM) fungi on growth, essential oil production and nutrients uptake in basil. J Med Plants Res 4(21):2222–2228

56. Zolfaghari M, Nazeri V, Sefidkon F, Rejali F (2013) Effects Effect of arbuscular mycorrhizal fungi on plant growth and essential oil contents and composition of Ocimum basilicum L. Iran J Plant Physiol 3(2):643–650

57. Davies J, Potter JR, Linderman RG (1993) Drought resistance of mycorrhizal pepper plants independent of leaf P concentration–response in gas exchange and water relations. Physiol Planta 87:45–53

58. Snellgrove RC, Stribley DP, Tinker PB, Lawlor DW (1986) The effect of vesicular-arbuscular mycorrhizal infection on photosynthesis and carbon distribution in leek plants. See Ref 13:421–424

59. Allen MF, Smith WK, Moore TS, Christensen M (1981) Comparative water relations and photosynthesis of mycorrhizal and non-mycorrhizal Bouteloua gracilis (HBK) Lag ex Steud. New Phytol 88:683–693

60. Fredeen AL, Terry N (1988) Influence of vesicular-arbuscular mycorrhizal infection and soil phosphorus level on growth and carbon metabolism of soybean. Can J Bot 66:2311–2316

61. Snellgrove RC, Splittstoesser WE, Stribley DR, Tinker RB (1982) The distribution of carbon and the demand of the fungal symbiont in leek plants with vesicular-arbuscular mycorrhizas. New Phytol 92:75–S7

62. Ruiz-Lozano JM (2003) Arbuscular mycorrhizal symbiosis and alleviation of osmotic stress. New Perspect Mol Stud Mycorrhiza 13:309–317

63. Rosa LP, dos Santos MA, Matvienko B, dos Santos EO, Sikar E (2004) Greenhouse gases emissions by hydroelectric reservoirs in tropical regions. Climatic Change 66(1–2):9–21

64. Giovannetti M, Avio L (2002) Biotechnology of arbuscular mycorrhizas. Mycorrhizas. In: Khachatourians GG, Arora DK (ed) Applied Mycology and Biotechnology, Vol. 2. Agriculture and Food Production. Elsevier, Amsterdam, pp 275–310

65. Iker H, Leonor A, Sergi M-b (2004) Drought-induced changes in flavonoids and other low molecular weight antioxidants in Cistus clusii grown under Mediterranean field conditions. Tree Physiol 24:1303–1311
66. Strack D, Fester T, Hause B, Schliemann W, Walter MH (2003) Arbuscular mycorrhiza: biological, chemical and molecular aspects. J Chem Ecol 29:1955–1979
67. Sheppard JW, Peterson JF (1976) Chlorogenic acid and Verticillium wilt of tobacco. Can J Plant Sci 56:157–160
68. Mucciarelli M, Scannerini S, Bertea C, Maffei M (2003) In vitro and in vivo peppermint (Mentha piperita) growth promotion by nonmycorrhizal fungal colonization. New Phytologist 158(3):579–591

9

Alterations in mineral nutrients in soybean grain induced by organo-mineral foliar fertilizers

Vesna Dragičević[1*], Bogdan Nikolić[2], Hadi Waisi[3], Milovan Stojiljković[4], Sanja Đurović[2], Igor Spasojević[1] and Vesna Perić[1]

Abstract

Background: Chemical composition of soybean grain may be modified by application of foliar fertilizers. The aim of this study was to test the effect of different organo-mineral foliar fertilizers: Zlatno inje, Bioplant Flora, Algaren BZn, Zircon, as well as plant growth regulator Epin Extra, on potential availability of mineral elements (Mg, Fe, Mn and Zn) from grain of three commercial soybean varieties: ZP-015, Nena and Laura (variety lacking in Kunitz trypsin inhibitor). In addition, phytate (Phy) and β-carotene contents were determined.

Results: ZP-015 achieved the highest P, Mg, Fe, Mn and β-carotene contents. Laura had the highest Phy level, which might reflect the diminished availability of nutrients from grain. Compared to control, most of the applied fertilizers increased β-carotene and decreased Mn content in all three soybean varieties. Increase in β-carotene content was followed by increase in Fe content, mainly in grains with larger weight, as a part of improved yielding potential.

Conclusions: Positive effect of Zircon application was evident on increased grain weight, and β-carotene and Fe content. These parameters together with the lowest values found for Phy/β-carotene and Phy/Mg ratios may explain the enhanced Mg and Fe bioavailability. On the other hand, positive effects of Epin Extra were mostly reflected by a decrease of Phy and an increase in Fe and Mn, thus becoming more bio-available. Accordingly, the organo-mineral foliar fertilizers based mainly on phenolic acids (Zircon) and bioregulator (Epin Extra) are to be recommended for soybean fortification.

Keywords: Organo-mineral foliar fertilizer; Grain composition; Mineral elements; Phytic phosphorus; *Glycine max* (L.) Merr

Background

Nutrition is crucial factor in reduction of hunger, malnutrition and obesity [1]. Human body requires more than 22 mineral elements, which can be provided by adequate diet. On the other hand, nutritional deficiencies (e.g. in iron, zinc, vitamin A) account for almost two-thirds of the childhood deaths worldwide [2]. These deficiencies can be surpassed by increase of mineral nutrients in food through supplementation, food fortification or plant breeding [3,4].

Iron and zinc are considered to be the most important mineral elements in vegetarian diets. Elimination of meat from diet, along with increased intake of whole grain cereals and legumes rich in anti-nutrients, like phytate, significantly decrease Fe and Zn absorption [5]. The most prevalent among mineral elements deficiencies is Fe deficiency (anemia), affecting approximately 30% of the world's

population. Zn is essential element, involved in the immune system, activation of many enzymes and the growth. Zn deficiency has been detected in cases of inadequate dietary supply, abnormal blood losses or high physiological requirements for growth, as well as during puberty, pregnancy and lactation [4,5]. As a part of the antioxidant system of defense in mitochondria, manganese is also essential element for humans and is involved in metabolism, bone development and wound healing. It has been shown that Mg has protective role against various diseases. However, numerous studies indicated that Mg concentration in human body is usually insufficient [6]. According to Nielsen [7], low level of Mg has been associated with pathological conditions characterized as a chronic inflammatory stress, being widely associated with obesity, atherosclerosis, hypertension, osteoporosis, diabetes mellitus, and cancer.

According to present knowledge, it is necessary to increase content of mineral nutrients in edible parts of plants. Accumulation of mineral elements in seeds and grains is controlled by a number of processes including

* Correspondence: vdragicevic@mrizp.rs
[1]Maize Research Institute, Slobodana Bajića 1, 11185 Zemun Polje, Serbia
Full list of author information is available at the end of the article

root-cell uptake, root-shoot transfer, and the ability to deliver these nutrients to developing seeds and grains [8]. Designing of cultivation systems, in order to improve nutrition and health, should become an integral part of goals in modern agriculture. It is mainly concerned to cultivation on poor soils, where micronutrient element enhancement can contribute to increased crop yield. According to Graham *et al.* [9], probably half of all soils are deficient in micronutrients and even though plant production is not limited, humans and animals whose diets are mainly based on crops can be potentially deficient in essential micronutrients. Incorporation of important mineral elements into soil by fertilizers could be problematic due to their pathway in soil. For instance, Fe from fertilizers could be quickly oxidized and became insoluble in soil, so Fe deficiency is mainly a consequence of Fe deficient soils [9]. Welch [8] reported significant impact of fertilizers containing N, P, K, S and Zn on accumulation of nutrients in edible plant products, including grains. Other micronutrient fertilizers were shown to have very small effect on the amount of micronutrients accumulated in edible seeds and grains when applied to soils.

Increased content of mineral elements in crops presents only the first step in making them improved sources of nutrients for humans [10], since not all mineral elements in plant foods are bio-available to humans and animals. Plant food can contain anti-nutrients, which interfere with the absorption of mineral nutrients in humans and animals. The question of bio-availability must be taken into consideration when enrichment of plant food with mineral elements was employed. This also takes into account enhancing substances - promoters (e.g. ascorbic acid, S-containing amino acids, etc.) that promote micronutrient bioavailability and/or suppress anti-nutrient substances (e.g. phytate, polyphenolics, etc.) that inhibit micronutrient bioavailability [2,11]. Thus, it is essential to decrease content of various antinutrients in foods and to increase content of promoters [9].

Phytic acid - Phy (myo-inositol 1,2,3,4,5,6-hexakisphosphate) is the major phosphorus storage compound in grains (accounting for up to 80% of total P) and it can acts as anti-nutritional factor that chelate essential elements including Ca, Zn and Fe [12]. As content of phytic acid in diet increases, the intestinal absorption of Zn, Fe and other mineral nutrients decreases [12], while the reduction in phytic acid content in food is likely to result in improved Fe, Zn and Mn content [3,13]. β-carotene is considered to be a promoter due to positive effect on mineral nutrients absorption. Lönnerdal [3] stated that β-carotene can enhance Fe absorption in humans. Luo and Xie [14] found that addition of food rich in β-carotene or pure β-carotene, can significantly enhance Fe and Zn bioavailability from the grains. Moreover, Noh and Koo [15] reported that low β-carotene absorption is associated with low Zn intake or slight Zn deficiency. Different cultivation

practices, including macronutrient treatments (N, P and Mg), can result in increased concentration of β-carotene (by 42%) and micronutrients in carrots [8].

Soybean is important dietary source of proteins, lipids, minerals, vitamins, fiber and bioactive compounds. However, commonly high levels of phytate in soybean grain could negatively affect its nutritive value. Variability of mineral elements in soybean grain is significant and it also depends on applied cultivation systems [16,17]. Since Zn bioavailability from some soya products is low, application of an adequate cultivation system becomes important. However, compared to other plant foods with lower phytate contents, the Fe availability from soya flour and soya isolates is higher.

The aim of this experiment was to investigate the effect of applied foliar fertilizers on mineral nutrients content (i.e. Mg, Fe, Mn and Zn), along with contents of phytate as anti-nutritive factor and β-carotene as promoter, in chosen soybean varieties differing in chemical composition of grain.

Experimental
Plant material
Two commercial soybean varieties with standard grain composition - ZP-015 and Nena, and the variety lacking in Kunitz trypsin inhibitor - Laura, were the objectives of the present study.

Soil
The field trial was carried out in Zemun Polje (44°52'N 20° 20'E), vicinity of Belgrade, Serbia (in rain-fed conditions). Soil was a slightly calcareous chernozem with 0.0 % coarse, 53.0 % sand, 30.0 % silt, 17.0 % clay, 3.3 % organic matter, 7.0 pH KCl and 7.17 pH H_2O. The texture was silty clay loam, containing: 37.45 mg kg^{-1} N, 10.70 37.45 mg kg^{-1} P, 107.40 37.45 mg kg^{-1} K, 327.95 37.45 mg kg^{-1} Mg, 0.65 37.45 mg kg^{-1} Fe and < 0.02 37.45 mg kg^{-1} Zn in 0–30 cm layer, before fertilizer application. A split-plot experimental design in four replications was used in the experiment. Size of elementary plot was 5 m x 5 m.

Foliar fertilizers
Experimental trial included application of different foliar fertilizers in recommended doses, at the beginning of flowering (first half of June): 1. Zlatno inje (liquid extract of cow's manure, with 0.8% of organic matter, 0.004% N and 0.0004% P), in amount of 4 L ha^{-1}; 2. Bioplant Flora (organic fertilizer with 8% humic acids, isolated from vermicompost, with 1.0% N, 1.5% P, 48.35 mg L^{-1} Mg, 2.41 mg L^{-1} B, 13.14 mg L^{-1} Cu, 212.8 mg L^{-1} Zn, 1.64 mg L^{-1} Co, 462 mg L^{-1} Mn, 775.6 mg L^{-1} Mo and 500 mg L^{-1} Fe), in the amount of 1 L ha^{-1}; 3. AlgarenBZn (organic fertilizer based on *Ecklonia maxima* algae extract with 2% of B and 3% of Zn), in an amount of

0.834 L ha^{-1}; 4. Zircon (extract of medicinal plant *Echinacea purpurea* L., that contains a mixture of 0.1 g L^{-1} of phenolic acids: 3,4-dihydroxycinnamic (caffeic) acid (IUPAC: 3-(3, 4-dihydroxyphenyl)-2-propenoic acid; CAS No 331-95-5), chlorogenic acid (IUPAC: (1*S*,3*R*,4*R*,5*R*)-3-{[(2*Z*)-3-(3,4-dihydroxyphenyl)prop-2-enoyl]oxy}-1,4,5-trihydroxycyclohexanecarboxylic acid; CAS No 327-97-9), cichoric acid (IUPAC: (2*R*,3*R*)-2,3-bis{[(*E*)-3-(3,4-dihydroxyphenyl)prop-2-enoyl]oxy}butanedioic acid; CAS No 327-97-9), as active ingredients identical to *Echinacea purpurea* L. plant extract), in the amount of 0.12 L ha^{-1}; 5. plant growth regulator Epin Extra (based on 0.025 g L^{-1} of 24-epibrassinolide (IUPAC: (22R 23R 24S)-2α, 3α, 22, 23 tetra hydroxy-24-metyl 5α-holestan-6-on; CAS No 72962-43-7), in the amount of 0.136 L ha^{-1}. All these organo-mineral fertilizers were applied with a dose of 400 L ha^{-1} of water.

Methods

Chemical analyses

After harvesting, 1,000 grain weight was measured and contents of different metabolites in soybean grain were determined. Contents of inorganic phosphorus (P_i) and phytic phosphorus (P_{phy}) were determined colorimetrically after extraction with 5% trichloroacetic acid, by method of Dragicevic *et al.* [18]: P_{phy} was determined with Wade

reagent and P_i with vanado-molybdate reagent. β-carotene content was also determined colorimetrically, after extraction with saturated butanol [19]. Content of total phosphorus (P_{tot}) was analysed with vanado-molybdate colorimetric method after wet digestion with $HClO_4$ + HNO_3, by method of Pollman [20]. The same digested samples were used for determination of mineral elements (i.e. Fe, Mn, Zn, and Mg) by Inductively Coupled Plasma - Optical Emission Spectrometry.

Statistical analysis

All analyses were performed in four replicates (n = 4) and the results were presented as mean ± standard deviation (SD). The differences among soybean varieties and applied treatments, based on mean values of observed parameters, were evaluated by using Principle Component Analysis (PCA). Statistical analysis was performed by SPSS 15.0 for Windows Evaluation version. Correlation analyses were performed using Pearson's correlation coefficient.

Results and Discussion

Grain weight and chemical composition of the grain

Results presented in Table 1 indicated that the greatest average 1,000 grain weight was achieved by the Laura

Table 1 The effect of different foliar fertilizers on chemical composition of grain in three soybean varieties

	Treatment	1,000 grain weight (g)	P_{tot}** (g kg^{-1})	P_i (g kg^{-1})	P_{phy} (g kg^{-1})	β-carotene (mg kg^{-1})
ZP-015	Control	178.1 ± 12.1*	16.12 ± 0.08	0.30 ± 0.01	12.88 ± 0.23	13.53 ± 0.04
	Zlatno inje	198.2 ± 6.7	16.87 ± 0.00	0.38 ± 0.03	12.71 ± 0.02	11.95 ± 0.19
	Epin Extra	180.9 ± 9.4	16.53 ± 0.03	0.30 ± 0.03	12.14 ± 0.16	13.92 ± 0.17
	Zircon	174.2 ± 6.9	16.25 ± 0.03	0.34 ± 0.03	12.06 ± 0.05	18.73 ± 0.09
	Bioplant Flora	182.7 ± 10.5	16.96 ± 0.11	0.38 ± 0.01	12.59 ± 0.35	15.96 ± 0.06
	AlgarenBZn	176.4 ± 11.9	17.12 ± 0.03	0.46 ± 0.00	12.61 ± 0.20	14.87 ± 0.09
	Average	181.7 ± 9.6	16.64 ± 0.04	0.36 ± 0.02	12.50 ± 0.17	14.83 ± 0.11
Nena	Control	171.9 ± 13.5	13.69 ± 0.08	0.49 ± 0.02	13.49 ± 0.03	14.84 ± 0.11
	Zlatno inje	170.4 ± 6.0	14.27 ± 0.11	0.47 ± 0.00	12.50 ± 0.36	10.94 ± 0.17
	Epin Extra	172.9 ± 8.7	15.42 ± 0.11	0.47 ± 0.00	12.33 ± 0.40	14.08 ± 0.11
	Zircon	191.0 ± 7.0	14.40 ± 0.00	0.29 ± 0.01	12.70 ± 0.16	12.40 ± 0.13
	Bioplant Flora	172.6 ± 0.8	14.38 ± 0.08	0.31 ± 0.02	12.43 ± 0.14	15.78 ± 0.19
	AlgarenBZn	171.1 ± 8.2	14.97 ± 0.03	0.32 ± 0.02	12.95 ± 0.25	12.61 ± 0.06
	Average	175.0 ± 8.7	14.52 ± 0.07	0.39 ± 0.01	12.73 ± 0.22	13.44 ± 0.13
Laura	Control	207.0 ± 11.4	14.68 ± 0.13	0.46 ± 0.00	12.46 ± 0.00	10.99 ± 0.11
	Zlatno inje	194.9 ± 6.7	14.29 ± 0.08	0.41 ± 0.03	12.43 ± 0.20	12.22 ± 0.11
	Epin Extra	197.5 ± 8.7	14.79 ± 0.00	0.47 ± 0.00	12.18 ± 0.00	11.44 ± 0.02
	Zircon	206.4 ± 5.5	14.13 ± 0.24	0.46 ± 0.00	11.93 ± 0.01	12.93 ± 0.06
	Bioplant Flora	208.1 ± 10.2	13.67 ± 0.19	0.49 ± 0.00	12.36 ± 0.01	10.33 ± 0.09
	AlgarenBZn	207.9 ± 8.8	14.08 ± 0.03	0.48 ± 0.00	12.42 ± 0.06	11.99 ± 0.21
	Average	203.6 ± 8.5	14.27 ± 0.11	0.46 ± 0.01	12.29 ± 0.05	11.65 ± 0.10

*The results are represented as mean ± SD (standard deviation) in four replicates.
**P_{tot}, total P; P_i, inorganic P; P_{phy}, phytic P.

variety (25.25 g larger than ZP-015 and Nena). Sudarić *et al.* [21] indicated that 1,000 grain weight is a very important yielding parameter, so that its increase should be considered as the main target for any applied cultivation measure. Compared to control, an average 1,000 grain weight increase of 4.49 g was achieved in Zircon treatment, for all three genotypes. Concerning the individual effect of each applied fertilizer, Zlatno inje increased the 1,000 grain weight mainly in ZP-015, while Epin Extra did so in Nena, and Bioplant Flora in the Laura variety.

The largest average P_{tot} was recorded in ZP-015 (2.02 g kg^{-1} greater than Nena and Laura), as presented in Table 1. A significant effect of fertilizers on increased P_{tot} content was obtained by Algaren BZn, mainly exhibited in grains of ZP-015 and Nena, while the P_{tot} content increase in Laura grain was achieved only by applying Epin Extra. It is known that the majority of P pool in soybean grain is present as P_{phy}, while the minor part is due to inorganic P_i. It is interesting to underline that in grain of ZP-015, only 75% of average P_{tot} consisted of P_{phy}, while it was 86-88% in other two varieties. Nevertheless, such decreased P_{phy} in ZP-015 did not determine an increase of P_i, that is the main source of available P from grains. Hence this genotype may be considered for further P_{phy} decrease by breeding [3,22].

The largest average content of P_{phy} was found in Nena grain, with slight variations of P_{phy} and P_i content in response to the applied foliar fertilizers. Bioplant Flora and Algaren BZn induced slight increase in P_i content in grain of ZP-015 and Laura, while Epin Extra and Zircon decreased P_{phy} in grains of all three varieties. It is also noticed that the largest average P_{phy} was found in control. In relation to P_{phy}, β-carotene variation was to a larger extent, with the largest average values observed in ZP-015 grain. Among the applied fertilizers, the greatest increase in β-carotene content was achieved by the Zircon application on ZP-015 and Laura, and by Bioplant Flora on Nena grain.

Investigated soybean varieties differed to a larger extent in mineral composition of their grain. The largest average Mg, Fe and Mn content was observed in ZP-015 grains, while the relatively greatest average Zn content was found in the Laura grains (Table 2). In parallel with P_{phy} decrease (Table 1), foliar fertilizers mainly exhibited a positive impact on the increase of mineral nutrients content. This could be positively related to a larger Fe, Zn and Mn status [3,13]. Compared to control and for all three genotypes, average increase for Mg and Zn contents (by 3% and 20%, respectively) was found under Bioplant Flora, for Fe content (7%) under Zircon, and

Table 2 The effect of different foliar fertilizers on mineral element contents in soybeans grain

	Treatment	Mg (mg kg^{-1})			Fe (mg kg^{-1})			Mn (mg kg^{-1})			Zn (mg kg^{-1})		
ZP-015	Control	2284.4	±	48.6*	65.66	±	0.49	29.66	±	1.37	34.41	±	2.78
	Zlatno inje	2331.3	±	0.0	70.13	±	0.18	29.97	±	1.46	45.41	±	3.23
	Epin Extra	2459.4	±	4.4	71.47	±	0.84	31.53	±	0.31	36.00	±	4.15
	Zircon	2215.6	±	39.8	78.41	±	0.35	25.78	±	0.57	48.13	±	2.08
	Bioplant Flora	2371.9	±	30.9	68.69	±	1.99	26.75	±	0.35	44.91	±	1.15
	AlgarenBZn	2356.3	±	35.4	67.91	±	0.66	28.56	±	1.33	38.13	±	2.12
	Average	2336.5	±	26.5	70.38	±	0.75	28.71	±	0.90	41.16	±	2.59
Nena	Control	2221.9	±	13.3	57.34	±	0.40	26.00	±	0.00	37.44	±	2.92
	Zlatno inje	2106.3	±	35.4	60.75	±	2.25	23.56	±	1.02	35.63	±	0.09
	Epin Extra	2162.5	±	17.7	64.34	±	0.80	26.28	±	0.53	32.09	±	0.13
	Zircon	2321.9	±	39.8	62.31	±	0.62	23.78	±	0.49	35.09	±	2.34
	Bioplant Flora	2320.9	±	39.8	63.16	±	0.75	22.97	±	1.64	40.38	±	2.65
	AlgarenBZn	2181.3	±	0.0	58.81	±	0.62	21.94	±	0.09	30.56	±	3.36
	Average	2219.1	±	24.3	61.12	±	0.91	24.09	±	0.63	35.20	±	1.92
Laura	Control	2172.5	±	25.6	60.09	±	1.37	26.59	±	1.02	42.84	±	2.52
	Zlatno inje	2215.9	±	13.3	66.13	±	0.22	27.31	±	0.75	59.72	±	8.62
	Epin Extra	2234.7	±	45.1	59.34	±	1.06	25.81	±	0.71	56.59	±	11.62
	Zircon	2138.1	±	25.2	57.13	±	0.62	24.78	±	0.49	54.81	±	4.95
	Bioplant Flora	2177.8	±	85.3	56.72	±	3.58	25.38	±	1.15	57.44	±	1.94
	AlgarenBZn	2090.6	±	19.0	58.00	±	0.88	20.47	±	0.09	45.03	±	3.31
	Average	2187.8	±	35.6	59.57	±	1.29	25.06	±	0.70	52.74	±	5.49

*The results are represented as mean ± SD (standard deviation) in four replicates.

for Mn content (only 2% of its increase) under Epin Extra treatment. Zlatno inje increased Fe, Mn and Zn content mainly in Laura grain, while Epin Extra showed the greatest impact on Mg increase in ZP-015 and Laura grains, and on Mn increase in ZP-015 and Nena grain, as well as on Fe increase in Nena grain. Zircon was the most efficient for Fe and Zn increase in ZP-015 and for Mg increase in Nena grains. Among the applied fertilizers, only Bioplant Flora was responsible for Zn content increase in Nena grains.

Availability of mineral nutrients

Variations in ratio between phytate (Phy) and β-carotene may indicate possible availability of nutrients [12,23]. This trait is an important parameter for the characterisation of the investigated genotypes. From this point, ZP-015 could be considered as a favourable variety for breeding towards an improved nutritive quality, having the lowest ratios for Phy/Mg, Phy/Fe and Phy/Mn in the control (Table 3). This indicates an additional quality of this genotype for a possible increase of mineral nutrients availability, under application of organo-mineral foliar fertilizers. This is important, since Luo and Xie [14] and Hess et al. [24] found that food rich in β-carotene can significantly enhance Fe and Zn bioavailability from

grain, while either low phytate level or its degradation may enhance Mn availability [16]. Besides ZP-015, Laura variety was characterised by the lowest P_{phy}/P_i and Phy/Zn ratios. This may become a possible indicator for further P_{phy} decrease during breeding [3,22], although further research is required. When the individual impact of each foliar fertilizer on all three genotypes was considered, Zlatno inje decreased P_{phy}/P_i and Phy/Mn ratios to the largest extent, while Zircon was the most prominent for Phy/β-carotene, Phy/Mg and Phy/Zn ratios. Epin Extra was mostly efficient in Phy/Fe decrease. Since Fe and Zn deficiencies are common worldwide [5], it is very important that Zircon and Epin Extra, that are organo-mineral foliar fertilizers which are primarily dedicated to fortification, increased average Fe and Zn content in soybean grains. Furthermore, each fertilizer had a specific site of action: Zircon and Epin Extra decreased $P_{phy}/β$-carotene, Phy/Mg, Phy/Fe and Phy/Zn ratios in grain of ZP-015, while Bioplant Flora and Epin Extra decreased $P_{phy}/β$-carotene, Phy/Mg and Phy/Fe ratios in Nena grain. In Laura variety, application of Zlatno inje induced a decrease in $P_{phy}/β$-carotene, Phy/Fe, Phy/Mn and Phy/Zn ratios.

Different responses of the examined soybean varieties to the applied treatments can be better visualised by PC

Table 3 The effect of different foliar fertilizers on investigated ratios in grain of three soybean varieties

	Treatment	P_{phy}/P_i*	Phy/β-carot	Phy/Mg	Phy/Fe	Phy/Mn	Phy/Zn
ZP-015	Control	43.17	774.3	0.48	16.60	36.74	31.67
	Zlatno inje	33.15	865.5	0.46	15.34	35.90	23.70
	Epin Extra	40.63	709.2	0.42	14.37	32.57	28.53
	Zircon	35.44	523.6	0.46	13.01	39.57	21.20
	Bioplant Flora	33.22	641.6	0.45	15.51	39.83	23.72
	AlgarenBZn	27.61	689.9	0.45	15.71	37.36	27.99
	Average	34.77	685.7	0.45	15.03	36.84	25.69
Nena	Control	27.24	739.4	0.51	19.90	43.89	30.48
	Zlatno inje	26.87	929.5	0.50	17.42	44.91	29.70
	Epin Extra	26.41	712.2	0.48	16.22	39.70	32.51
	Zircon	44.42	833.1	0.46	16.88	45.19	29.98
	Bioplant Flora	40.19	640.8	0.45	17.02	46.79	26.62
	AlgarenBZn	40.14	835.0	0.50	18.63	49.94	35.85
	Average	32.58	770.5	0.49	17.63	44.73	30.61
Laura	Control	27.25	922.2	0.49	17.54	39.64	24.60
	Zlatno inje	30.27	827.1	0.50	15.90	38.50	17.61
	Epin Extra	25.71	865.8	0.48	17.36	39.92	18.21
	Zircon	25.78	750.5	0.45	17.67	40.73	18.41
	Bioplant Flora	25.31	973.9	0.48	18.45	41.23	18.22
	AlgarenBZn	26.14	842.1	0.47	18.12	51.34	23.34
	Average	26.66	858.5	0.48	17.47	41.52	19.73

*Ratios: P_{phy}/P_i, phytic and inorganic P; Phy/β-carot, phytate and β-carotene; Phy/Mg, Phy/Fe, Phy/Mn, Phy/Zn, phytate and mineral nutrients.

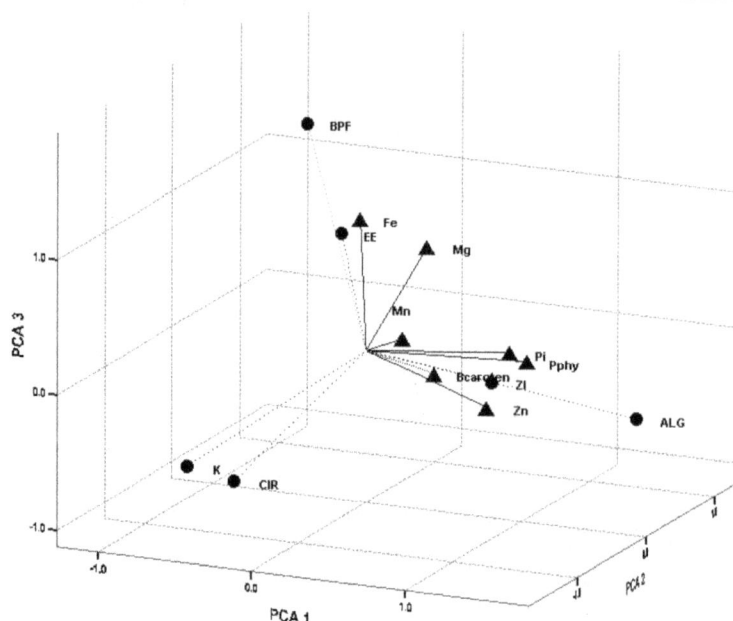

Figure 1 Principal component analysis (PCA) for chemical composition of grain in ZP-015 variety (P_{phy}, phytic phosphorus; P_i, inorganic phosphorus, K, control; ZI, Zlatno inje; EE, Epin Extra; CIR, Zircon; BPF, Bioplant Flora; ALG, Algaren BZn).

analysis. In grains of ZP-015, results indicated the Epin Extra treatment as the most efficient for Fe and Mg content increase (Figure 1). Positive effects of Zlatno Inje and Algaren BZn were shown mostly on the increase of P_i and P_{phy} content, and, partially, on β-carotene and Zn content. In Nena's grain, a most pronounced increase was found in β-carotene, Mg and Fe contents under the

application of the Zircon treatment, thus positively affecting the bioavailability of these mineral elements (Figure 2). However, application of investigated foliar fertilizers did not demonstrate the expected positive effect on an improved content and bioavailability of Zn. In Laura's grain, Epin Extra was the most efficient foliar fertilizer, leading to an increase in P_{phy}, Fe and Zn

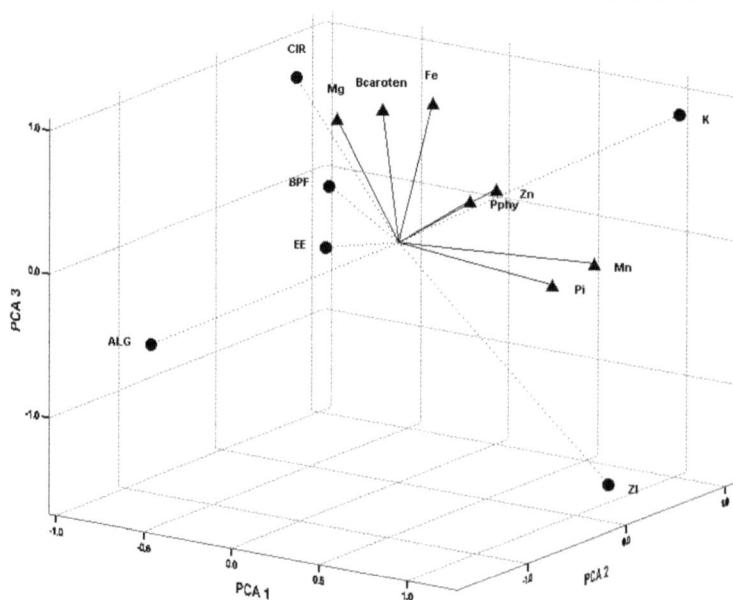

Figure 2 Principal component analysis (PCA) for chemical composition of grain in Nena variety (P_{phy}, phytic phosphorus; P_i, inorganic phosphorus, K, control; ZI, Zlatno inje; EE, Epin Extra; CIR, Zircon; BPF, Bioplant Flora; ALG, Algaren BZn).

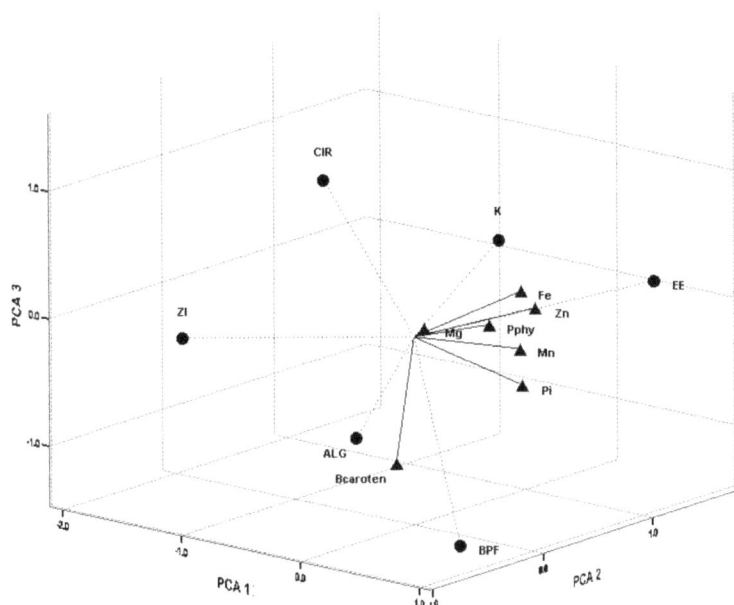

Figure 3 Principal component analysis (PCA) for chemical composition of grain in Laura variety (P_{phy}, phytic phosphorus; P_i, inorganic phosphorus, K, control; ZI, Zlatno inje; EE, Epin Extra; CIR, Zircon; BPF, Bioplant Flora; ALG, Algaren BZn).

content, and, by a lower extent, to that in P_i and Mn content, whereas treatment with Algaren BZn showed a β-carotene content increase (Figure 3). Such results suggest that measures to increase mineral nutrients in soybean grain (such as fortification) were related to alterations of P_{phy} and β-carotene. Moreover, results obtained on maize and beans did not support a lower Fe availability from grains linked to slightly larger phytate content, while a significant effect on Fe availability is commonly atributed to promoting factors [25].

Interactions between examined parameters may be important for understanding potential availability of mineral elements. Results presented in Table 4 indicated that a large grain weight correlated significantly and positively with Zn content but negatively with β-carotene. Conversely, significant and negative correlation

between P_{phy} and Zn may be directly related to an improved Zn availability, as shown by the results of Lönnerdal [3] and Underwood and Suttle [13], who ascertained that any reduction in phytic acid content in food is likely to result in improved Fe, Zn and Mn status. Improved Fe availability could be supported by its significant and positive correlation with β-carotene. Significant and positive correlation between Mg, Fe and Mn, as well as with P_{tot} indicates that an improved mineral nutrition under application of organo-mineral foliar fertilizers, may increase the concentration of individual elements in grain, althogh their availability may be still questionable. Opposite to Mg, Fe and Mn, the Zn increase was primarily related to a yield parameter, such as the 1,000 grains weight, that was mainly increased by Bioplant Flora (Table 1).

Table 4 Correlations between investigated parameters in grain of three soybean varieties

	1,000 grain weight	P_{tot}**	P_i	P_{phy}	β-carot.	Mg	Fe	Mn
P_{tot}	−0.310*							
P_i	0.367	−0.394						
P_{phy}	−0.370	−0.063	−0.052					
β-carot.	−0.596*	0.475*	−0.377	−0.016				
Mg	−0.264	0.674*	−0.601*	0.091	0.389			
Fe	−0.359	0.801*	−0.513*	−0.233	0.654*	0.581*		
Mn	−0.066	0.660*	−0.179	−0.015	0.135	0.651*	0.571*	
Zn	0.683*	−0.256	0.389	−0.494*	−0.203	−0.204	−0.082	0.006

*Correlation is significant at 0.05 level.
**P_i, inorganic P; P_{phy}, phytic P; P_{tot}, total P.

Conclusions

Our findings showed that ZP-015 can be generally considered as a favourable variety for increased bioavailability of Mg, Fe and Mn (due to the lowest ratios of Phy/Mg, Phy/Fe and Phy/Mn found in control samples). Moreover, in all varieties, an improvement in grain yielding potential and grain quality was achieved through foliar application of organo-mineral fertilizers. Positive effect of Zircon application was evident on the increased grain weight, and β-carotene and Fe content. The latter which may, along with the lowest values obtained for Phy/β-carotene and Phy/Mg ratios, have determined an increase in Mg and Fe bio-availability, mainly for Nena grain. On the other hand, positive effect of Epin Extra was observed mostly because of both Phy decrease and Fe and Mn increase, thus contributing to their increased bio-availability. Accordingly, organo-mineral foliar fertilizers based mainly on phenolic acids (Zircon) and bioregulators (Epin Extra) should be recommended to be used for soybean fortification.

Competing interests
The authors declare that they have no competing interests.

Authors' contributions
Bogdan Nikolić and Vesna Dragičević designed the research. Bogdan Nikolić, Vesna Dragičević, Hadi Waisi, Igor Spasojević and Vesna Perić performed field trials. Milovan Stojiljković, Vesna Dragičević and Sanja Đurović performed chemical analyses. Vesna Dragičević and Igor Spasojević analyzed the data. Vesna Dragičević and Bogdan Nikolić wrote the paper. All authors read and approved the final manuscript.

Acknowledgement
This work was supported by Project TR31037 from the Ministry of Education, Science and Technological Development, Republic of Serbia.

Author details
[1]Maize Research Institute, Slobodana Bajića 1, 11185 Zemun Polje, Serbia. [2]Institute for Plant Protection and Environment, Teodora Drajzera 9, 11000 Belgrade, Serbia. [3]Institute for the Development of Water Resources, "Jaroslav Černi", Jaroslava Černog 80, 11226 Belgrade, Serbia. [4]Vinca Institute of Nuclear Sciences, 52211001 Belgrade, Serbia.

References
1. Campanini B (Ed) (2002) The World Health report (2002) Reducing Risks. Promoting Healthily Life. World Health organization, Geneva, Switzerland, pp 1–168
2. Welch RM, Graham RD (2004) Breeding for micronutrients in staple food crops from a human nutrition perspective. J Exp Bot 55(396):353–364
3. Lönnerdal B (2003) Genetically modified plants for improved trace element nutrition. J Nutr 133(5):1490S–1493S
4. White PJ, Broadley MR (2005) Biofortifying crops with essential mineral elements. Trends Plant Sci 10(12):586–593
5. Hunt JR (2003) Bioavailability of iron, zinc, and other trace minerals from vegetarian diets. Am J Clin Nutr 78(suppl):633S–639S
6. Vormann J (2003) Magnesium: nutrition and metabolism. Molecul Aspects Med 24(1–3):27–37
7. Nielsen FH (2010) Magnesium, inflammation, and obesity in chronic disease. Nutrition Rev 68(6):333–340
8. Welch RM (2003) Farming for Nutritious Foods: Agricultural Technologies for improved Human Health. IFA-FAO Agriculture Conference "Global Food Security and the Role of Sustainable Fertilization" Rome, Italy, pp.1-24
9. Graham RD, Welch RM, Saunders DA, Ortiz-Monasterio I, Bouis HE, Bonierbale M, et al. (2007) Nutritious Subsistence Food Systems. Advances Agron 92:1–74
10. Welch RM, Graham RD (1999) A new paradigm for world agriculture: meeting human needs - Productive, sustainable, nutritious. Field Crops Res 60:1–10
11. Hurrell RF (2003) Influence of vegetable protein sources on trace element and mineral bioavailability. J Nutr 133:2973S–2977S
12. Walter Lopez H, Leenhardt F, Coudray C, Remesy C (2002) Minerals and phytic acid interactions: is it a real problem for human nutrition? Internat J Food Sci Technol 37:727–739
13. Underwood EJ, Suttle NF (1999) Manganese. The Mineral Nutrition of Livestock, CABI Publishing, USA, In, pp 397–420
14. Luo YW, Xie WH (2012) Effects of vegetables on iron and zinc availability in cereals and legumes. Internat Food Res J 19(2):455–459
15. Noh SK, Koo SI (2003) Low zinc intake decreases the lymphatic output of retinol in rats infused intraduodenally with beta-carotene. J Nutr Biochem 14(3):147–53
16. Dragicevic V, Oljaca S, Dolijanovic Z, Stojiljkovic M, Spasojevic I, Nisavic M (2013) Effect of intercropping systems and fertilizers on maize and soybean grain composition. Cost Action FA0905 Mineral Improved Crop Production for Healthy Food and Feed, 4TH Annual Conference. Norwegian University of Life Sciences, Aas, Book of Abstracts, p 31
17. Jaffe G (1981) Phytic acid in soybeans. J Am Oil Chem Soc 58(3):493–495
18. Dragičević V, Sredojević S, Perić V, Nišavić A, Srebrić M (2011) Validation study of a rapid colorimetric method for the determination of phytic acid and norganic phosphorus from grains. Acta Period Technol 42:11–21
19. American Association of Cereal Chemists Method (1995) Approved Methods of the AACC. The association: St. Paul, Minnesota, USA, AACC Method, pp 14–50
20. Pollman RM (1991) Atomic absorption spectrophotometric determination of calcium and magnesium and colorimetric determination of phosphorius in cheese. Collaborative study. J Assoc Official Analytic Chem 74(1):27–30
21. Sudarić A, Vratarić M, Duvnjak T (2002) Quantitative genetic analysis of yield components and grain yield for soybean cultivars. Poljoprivreda (Osijek) 8 (2):11–16
22. Dragicevic V, Kovacevic D, Sredojevic S, Dumanovic Z, Drinic Mladenovic S (2010) The variation of phytic and inorganic phosphorus in leaves and grain in maize populations. Genetika 42:555–563
23. Dragičević V, Mladenović Drinić S, Stojiljković M, Filipović M, Dumanović Z (2013) Variability of factors that affect availability of iron, manganese and zinc in maize lines. Genetika 45:907–920
24. Hess SY, Thurnham DI, Hurrell RF (2005) Influence of provitamin A carotenoids on iron, zinc, and vitamin A status. HarvestPlus Technical Monography 6. International Food Policy Research Institute (IFPRI) and International Center for Tropical Agriculture (CIAT), Washington, DC and Cali
25. Beiseigel JM, Hunt JR, Glahn RP, Welch RM, Menkir A, Maziya Dixon BB (2007) Iron bioavailability from maize and beans: a comparison of human measurements with Caco-2 cell and algorithm predictions. Am J Clin Nutr 86:388–396

10

Optimized procedure for the determination of P species in soil by liquid-state ^{31}P-NMR spectroscopy

Meng Li[1,2], Pierluigi Mazzei[1], Vincenza Cozzolino[1], Hiarhi Monda[1], Zhengyi Hu[2*] and Alessandro Piccolo[1*]

Abstract

Background: Liquid-state ^{31}P-NMR spectroscopy becomes progressively an important role for studying phosphorus (P) dynamics in soil. Soils of different origin and organic matter content were used to optimize sample preparation and re-dissolution procedures to improve characterization of P species in soil by ^{31}P-NMR spectroscopy. The efficiency of P extraction from an untreated fresh soil was compared to that from freeze-dried and air-dried soil samples.

Results: A freeze-drying pretreatment not only provided the greatest extraction yields of total and organic P from both farmland and forest soils but also enhanced the intensity of signals for inorganic and organic P species in ^{31}P-NMR spectra, except for polyphosphates. Re-dissolution of freeze-dried soil extracts in relatively dilute alkaline solution and addition of a small aliquot of concentrated HCl to the NMR tube prior to analysis improved the quality of NMR spectra. Finally, the visibility of relatively weak P signals, such as for phosphorus diesters, phosphonates, polyphosphate, phospholipids, and DNA were reproducibly enhanced when ^{31}P-NMR spectra were generated after at least 15 h of acquisition time.

Conclusion: The optimized procedure presented here ensured the greatest detectability of inorganic and organic P species by liquid-state P-NMR spectroscopy in soil extracts.

Keywords: Liquid-state ^{31}P-NMR spectroscopy; Soil pretreatment before extraction; Re-dissolution of P extracts; Inorganic and organic P species

Background

Solution-state ^{31}P-NMR spectroscopy represents a useful tool to follow the cycle of phosphorus (P) in the environment [23]. However, due to the relatively low concentration of P compounds present in environmental compartments (soils and sediments), natural organic matter (NOM), and microbial biomass, the procedures for sample preparation for the detection of P content by NMR require a careful setup of experimental conditions, including the removal of paramagnetic species [25,27,11]. In fact, an optimization of sample pretreatment and NMR experimental parameters become critical to improve the final quality of ^{31}P NMR

spectra and minimize variations in the determination of P compounds [7,18,21].

Several works were devoted to improve procedures in order to obtain meaningful and reproducible liquid-state ^{31}P-NMR spectra. In particular, the extracting solution may affect the quantitative solubilization of P species [3,20]. Cade-Menun and Preston [5] found that an aqueous solution containing both NaOH and EDTA extracted the greatest diversity of P forms and the largest percentage of total phosphorus in soil. Recently, it has been shown that a 0.25 M NaOH and 0.05 M EDTA extracting solution was most efficient in solubilizing a maximum range of organic P forms from sediments [26,28], while minimizing the co-extraction of Fe (III) and Mn (II) metals, which may significantly affect the resolution and intensity of ^{31}P NMR spectra [14, 1]. Addition of HCl or Chelex to extracts was also found to be useful in reducing interferences [13,24]. Recently, the capacity of bicarbonate dithionate (BD), EDTA, Chelex-100, and 8-hydroxyquinoline

* Correspondence: zhyhu@ucas.ac.cn; alessandro.piccolo@unina.it
[2]College of Resources and Environment, Sino-Danish College, University of Chinese Academy of Sciences, Beijing 100049, People's Republic of China
[1]Interdepartmental Research Centre on Nuclear Magnetic Resonance for the Environment, Agro-Food, and New Materials (CERMANU), University of Naples Federico II, Portici 80055, Italy

(HQ) were compared in order to prevent solubilization of Fe and Mn from environmental samples [9]. It was found that HQ was the most efficient treatment, although it was also observed that excessive HQ in the extracting solution produced an insoluble residue during subsequent freeze-drying of extracts.

Extracted matter is normally concentrated prior to solution-state ^{31}P-NMR analysis by either rotary evaporation or freeze-drying [10]. However, the latter procedure was reported to lead to considerable degradation of polyphosphate and glucose 6-phosphate and significant reduction of NMR signals of phospholipids extracted from a calcareous sediment [4]. The disappearance of phopholipids and pyrophosphate in extracts from a sediment from Taihu Lake in China was also noted after freeze-drying of samples [26]. On the other hand, while rotary evaporation of aqueous extracts is a time-consuming enrichment procedure as compared to freeze-drying, it also fails to sufficiently enrich P content in samples with very low P concentration. This becomes evident for soil-water leachates or dissolved organic matter, for which a 1,000- to 2,000-fold enrichment is usually required [4,19,8]. In these cases, sample freeze-drying followed by re-dissolution in the proper solution for NMR analysis may be the most useful method.

The aim of this work was to set up a reproducible procedure to extract and concentrate organic P species solubilized from soil samples of different organic matter content. Sample extraction and re-dissolution and spectral acquisition time were assayed to optimize detection of P species by ^{31}P-NMR spectroscopy.

Methods
Materials and chemical properties
A surface (0 to 20 cm) layer of an alluvial clayey loam agricultural soil was collected at the Experimental Farm of the University of Naples Federico II at Castel Volturno (CE). This soil was used without (sample A) and with (sample B) addition of 125 q ha^{-1} of farm compost as P source. A volcanic silty loam soil sample (0 to 20 cm) was collected from the forested area around the Royal Palace of Portici (NA) (sample C). The overlaying litter material was collected from the same forest soil (sample D). The compost used here to amend the agricultural soil and extract P (sample E) was derived from a pilot composting plant build at the Experimental Farm of the University of Naples Federico II at Castel Volturno (CE). The pH of the samples was determined by a pH meter (HANNA Instruments, Padova, Italy) in a soil:water suspension of 1:2.5 ratio. The C, H, and N content of samples was determined by an elemental analyzer (Eager 200, Fisons, Ipswich, UK), while the content of the organic matter (OM) was achieved by the Walkley-Black method. Organic P was calculated in soils

and compost by difference in P content before and after ignition at 550°C and followed by extraction with 1 M HCl [2]. Total P in both original sample materials and their extracts (TP$_E$) was obtained by first digesting samples in concentrated H$_2$SO$_4$ and HClO$_4$ for 2 h and the measuring total dissolved P in digested solutions by a UV-vis spectrometer (Lambda 25, Perkin Elmer, Waltham, MA, USA) by the molybdate colorimetry method [17].

Sample treatments and preparation for NMR acquisition
Soils, litter, and compost samples used in this study were cleaned from plant remains, passed through a 1-mm sieve, and divided into three aliquots. One aliquot (fresh) was subjected to direct extraction of P species, the second aliquot was first freeze-dried, and a third aliquot was left to air-dry (20°C to 25°C) for 2 weeks before extraction.

Extractions
About 4 g of each aliquot of soil and compost was mixed with 0.25 M NaOH and 0.05 M EDTA solution at a 1:8 soil:solution ratio and shaken at 20°C for 16 h. The mixture was centrifuged at 12,000 g (20°C) for 30 min, and the clear supernatant solution was collected into a 50-mL centrifuge tube. An amount (1.1 mL for each 10 mL extract) of a 3% aqueous 8-hydroxyquinoline (HQ) solution was added to the tube to remove Fe and Mn metals [9]. The solution pH was adjusted to 9.0 ± 0.1, and, after at least 30 min, the solution was centrifuged. Thirty millimeters of the supernatant was separated, frozen at −80°C, and freeze-dried. The resulting powder was finely ground before further analysis.

Re-dissolutions
The freeze-dried powder obtained with samples extraction was divided into two aliquots of about 1 g each. One aliquot was re-dissolved in 2 mL of a 1 M NaOH solution, while a second aliquot was re-dissolved in 2 mL of a 10 M NaOH solution. After about 2 h, the samples were centrifuged at 12,000 g for 30 min. Then, 930 μL of the supernatant were transferred into a 5-mm NMR tube, together with a deuterated solution of metylendiphosphonic acid-P,P′-disodium salt (MDP) (Epsilon Chimie, Guipavas, France) as internal standard (δ = 16.62 ppm), for a final 2.65 mM concentration. Moreover, in order to flocculate possible suspended particles, which may be present in the redissolved extracts and interfere with NMR analyses, 100 μL of concentrated HCl was further added to the NMR tube prior to acquisition of ^{31}P-NMR spectra. All samples were prepared in triplicates for NMR analyses.

^{31}P-NMR analysis
The P containing solutions were analyzed by a 400-MHz Bruker Avance spectrometer (Bruker AXS, Inc., Madison,

WI, USA), equipped with a 5-mm Bruker broadband inverse (BBI) probe, operating at ^{31}P resonating frequency of 161.81 MHz, applying 6 s initial delay and a 45° pulse length ranging between 8.5 and 9.5 µs (−2 dB power attenuation). The ^{31}P-NMR spectra of sample solutions were acquired from 5, 10, and 15 h of acquisition time. The 5 h of acquisition time comprised 3,000 transients and 5,461 time domain points, while the ^{31}P spectra for the 15-h acquisition time consisted in 9,000 transients, 16,384 time domain points, and a spectral width of 250 ppm (40,650 Hz). Except for the samples used to determine the best acquisition time, the rest of the ^{31}P-NMR spectra of this study were acquired with 15 h of acquisition time. An inverse gated pulse sequence, with 80-µs length Waltz16 decoupling scheme, with around 15.6 dB as power level, was employed to decouple phosphorous from proton nuclei [16]. All spectra were baseline-corrected and processed by MestReC software (v. 4.9.9.9). The free induction decays (FID) for solution-state ^{31}P-NMR spectra were transformed by applying a fourfold zero filling and a line broadening of 6 Hz. Signals were assigned according to literature [6,15,22,12]. The relative proportions of P species were estimated by integration of ^{31}P-NMR spectral peaks and expressed in respect to the concentration of total P in extracts (TP$_E$).

Results and discussion

Basic properties

The farmland soils and compost had pH between 8.82 and 8.97, while pH values for litter and forest soil were neutral and weakly alkaline, respectively (Table 1). The largest organic matter (OM) content was found in both compost (522 mg g^{-1}) and litter (284 mg g^{-1}) (Table 1). The content of total P (TP) in compost and litter was significantly greater than that in the forest soil, whereas organic P (OP) was largest in litter and farmland soil. The relatively low amount of both TP and OP in forest soil may be due its lighter texture and sloping topography.

Phosphorus in extracts

The ^{31}P-NMR spectra of extracts obtained from fresh, air-dried, and freeze-dried samples are shown in Figure 1. Spectra reveal that P signals were more visible and intense in extracts from freeze-dried samples than from other sample treatments, being the improvement by freeze-drying more extensive for farmland and forest soils than for forest litter. However, signal enhancement in freeze-dried samples was not as large as that reported by Xu et al. [26], who observed a 50% increase for freeze-dried sediment samples.

TP$_E$ and OP in extracts were larger for freeze-dried than those for either air-dried or fresh samples (Table 2). This difference was generally reflected in the amount of P species measured by ^{31}P-NMR spectra. Exceptions were the phosphonates (18 to 24 ppm) and polyphosphates (−19 to 21 ppm) for the forest soil and the orthophosphates (5.6 ppm) for the forest litter, which were larger in the fresh samples than in both freeze-dried and air-dried samples (Table 2). As for the phosphonates, their reduced content in freeze-dried samples may be explained by their easy degradability, possibly accelerated by samples manipulation.

Air-drying decreased considerably the signals for P monoesters, phospholipids (PL), DNA, and polyphosphates in farmland soil, either with or without compost addition, while these species were not as extensively decreased in the fresh soil (Table 2). Advanced microbial degradation of organic P may be the cause of the reduced values in air-dried samples.

Influence of re-dissolution

The powder materials resulting from freeze-drying soil extracts required to be re-dissolved prior to analysis by liquid-state ^{31}P-NMR spectroscopy. We compared re-dissolution of extracts in either a 1 M or a 10 M NaOH solution.

^{31}P-NMR spectra showed that re-dissolution in a 1 M NaOH solution provided larger values for both TP and OP species than for the 10 M NaOH solution (Figure 2). In particular, approximately 17.0 mg kg^{-1} of organic P and 63.6 mg kg^{-1} of inorganic P were lost during re-dissolution of extract from forest soil with 10 M NaOH. While all P signals showed a general relevant reduction when re-dissolved in 10 M NaOH solution, the phospholipid (PL) signal almost disappeared (Figure 2). This effect was particularly evident in spectra of the humus-rich

Table 1 Some properties of the sample materials used in this study

Materials	pH	OM (mg g^{-1})	TN (mg g^{-1})	C (mg g^{-1})	TP (mg g^{-1})	OP (mg g^{-1})
Farmland soil	8.82	56	0.56	14.7	0.57	0.17
Farmland soil added with compost	8.87	64	0.68	16.5	0.86	0.23
Forest soil	7.47	83	0.36	10.2	0.72	0.08
Forest litter	6.97	284	4.1	50.3	1.18	0.29
Compost	8.97	522	21.9	206	2.18	0.35

OM, organic matter; TN, total nitrogen; C, elemental carbon; TP, total phosphorus; OP, organic phosphorus.

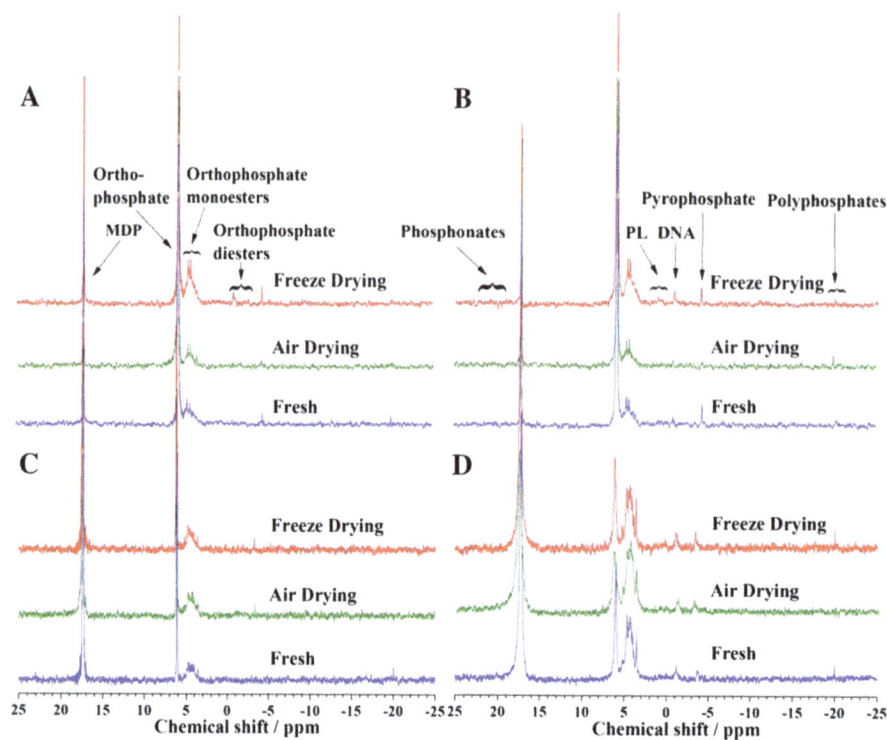

Figure 1 ^{31}P-NMR spectra of extracts from samples subjected to different treatments and 15 h acquisition time. **(A)** Farmland soil. **(B)** Farmland soil added with compost. **(C)** Forest soil. **(D)** Forest litter. MDP, metylendiphosphonic acid-P,P'-disodium salt as internal standard; PL, phospholipids.

Table 2 Total P (TP$_E$) and organic P (OP$_E$) in soil extracts (mg kg^{-1}) and relative distribution of P

P species	Phos-P	Orth-P	Monoesters	PL	DNA	Diesters[a]	Pyro-P	Poly-P	OP$_E$	TP$_E$
ppm intervals	18 to 24	5.6	5.3 to 3.3	1.6 to −0.3	0.8	−1.1 to −1.6	−4.6	−19 to −21		
Farmland soil										
Freeze-drying	0.5	116.7	14.7	0.6	1.0	1.6	0.2	0.1	16.8	133.8
Air-drying	ND	106.1	7.8	0.5	0.3	0.9	0.2	0.1	8.8	115.1
Fresh	1.4	98.7	8.4	0.9	0.3	1.1	0.2	0.1	11.0	109.9
Farmland soil added with compost										
Freeze-drying	0.5	160.3	16.0	1.7	1.4	3.2	0.4	0.2	19.9	180.7
Air-drying	0.5	153.7	8.6	0.3	0.2	0.4	0.1	0.7	10.3	164.1
Fresh	0.3	153.8	12.8	1.6	0.6	2.2	0.8	0.7	16.0	170.5
Forest soil										
Freeze-drying	4.1	172.0	74.8	3.0	2.5	5.5	0.9	0.4	84.9	257.9
Air-drying	4.5	159.1	70.7	2.0	2.1	4.1	1.0	0.2	79.5	239.7
Fresh	5.2	127.8	52.7	1.2	ND	1.2	ND	2.0	61.1	188.9
Forest litter										
Freeze-drying	8.1	38.6	80.3	2.6	6.7	9.4	2.5	1.4	97.1	140.2
Air-drying	7.8	31.0	83.4	1.9	6.0	7.9	2.5	ND	99.1	132.7
Fresh	3.3	46.2	80.6	2.1	4.9	7.0	1.2	1.0	91.9	139.4

Species in ^{31}P-NMR spectra acquired for 15 h, as referred to TP$_E$ and expressed in milligrams per kilograms. Phos-P, phosphonates; Ortho-P, orthophosphate; Mono-esters, orthophosphate monoesters; PL, phospholipids; Diesters, orthophosphate diesters; Pyro-P, pyrophosphate; Poly-P, polyphosphate; OP$_E$, organic phosphorus in extracts; TP$_E$, total phosphorus in extracts; ND, not detected. [a]Extracted orthophosphate diesters were calculated by the sum of phospholipids and DNA.

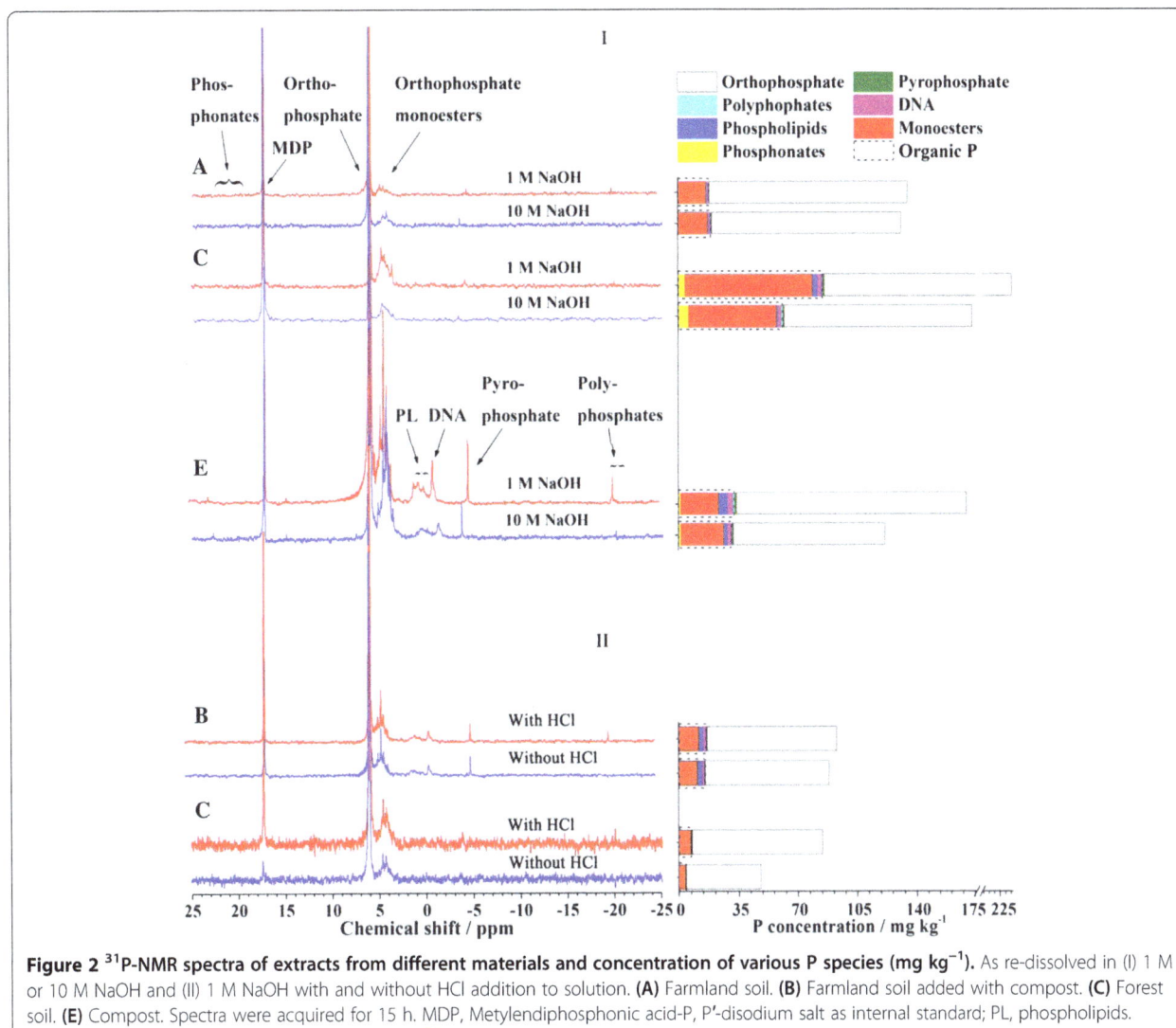

Figure 2 **³¹P-NMR spectra of extracts from different materials and concentration of various P species (mg kg⁻¹).** As re-dissolved in (I) 1 M or 10 M NaOH and (II) 1 M NaOH with and without HCl addition to solution. **(A)** Farmland soil. **(B)** Farmland soil added with compost. **(C)** Forest soil. **(E)** Compost. Spectra were acquired for 15 h. MDP, Metylendiphosphonic acid-P, P′-disodium salt as internal standard; PL, phospholipids.

compost, whereby PL, DNA, and polyphosphate signals were also significantly less intense in the 10 M NaOH solution than those in the dilute alkaline solution. Such lowering of OP signals in the strongly alkaline solution may be due to an enhanced hydrolysis of P from organic compounds.

It was also noted that a displacement of signals chemical shifts in the 10 M NaOH solution (Figure 2). The chemical shift of pyrophosphate was approximately 0.5 ppm downfield as compared to the sample re-dissolved in 1 M NaOH, while those of polyphosphates and DNA were 0.6 ppm downfield. Signals of orthophosphate monoesters were found in the 5.3 to 3.3 ppm range in 1 M NaOH, while they were reduced in a smaller 5.1 to 3.5 ppm interval in 10 M NaOH.

It was also observed that complete sample re-dissolution in alkaline solutions may form suspended particles, which may reduce spectral quality and undermine TP and OP detectability. To alleviate this problem, samples re-dissolved in 1 M NaOH were added with 100 μL HCl before undergoing NMR analyses. This procedure was found to decrease the amount of particles suspended in the alkaline solution and successful in providing more intense signals in ³¹P-NMR spectra (Figure 2).

NMR acquisition time

As indicated earlier, it is crucial to reach the adequate acquisition time during ³¹P-NMR experiments in order to generate spectra with sufficient signal intensity to observe differences in sample treatments [6]. Here, we acquired ³¹P-NMR spectra for 5, 10, and 15 h of acquisition time and compared spectral results (Figure 3). We found that the longest acquisition time (15 h) brought about the greatest response for signals of diesters, phosphonates, and polyphosphate for samples rich in organic matter, such as forest litter and compost (Figure 3). Such long acquisition time is required to increase spectral

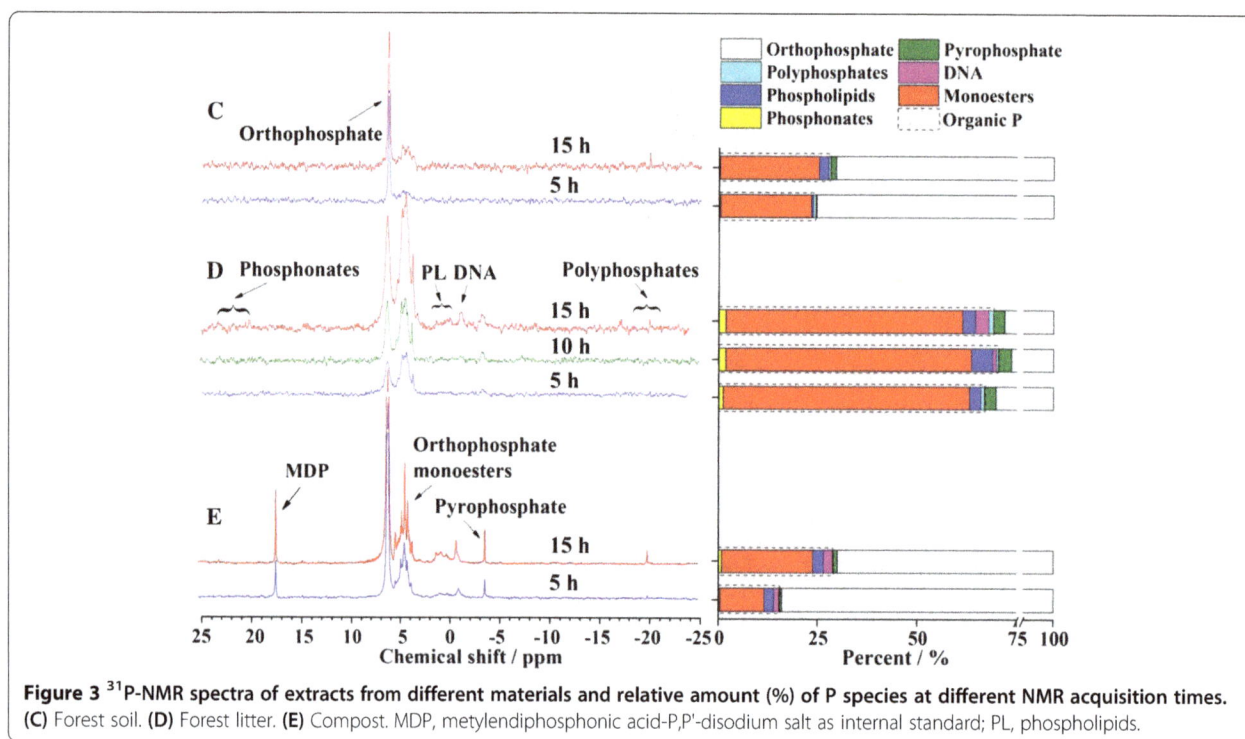

Figure 3 ^{31}P-NMR spectra of extracts from different materials and relative amount (%) of P species at different NMR acquisition times. **(C)** Forest soil. **(D)** Forest litter. **(E)** Compost. MDP, metylendiphosphonic acid-P,P'-disodium salt as internal standard; PL, phospholipids.

quality, especially for relatively weak signals. In fact, P signals for DNA and polyphosphate in extracts from forest litter and forest soil, respectively, could not be yet detected after a 5-h long acquisition time, whereas they were visible after 15 h (Figure 3).

Conclusions

Solution-state ^{31}P-NMR spectra obtained in this work revealed that soil and compost samples which were freeze-dried before NMR analysis ensured a larger detectability of total and organic P than samples undergone NMR experiments either after air-drying or directly without pretreatment. Thus, a freeze-drying pretreatment represents the method of choice for natural samples with low P concentration and whose NMR detectability of P signals is poor.

We also showed that re-dissolution of extracts from natural samples into 1 M NaOH solution makes P signals more visible by ^{31}P-NMR spectroscopy than when a stronger alkaline solution is used. Furthermore, when suspended particles may interfere with spectral quality, addition of 100 μL of concentrated HCl to the NMR tube containing the alkaline re-dissolved extracts, the intensity of NMR signals are improved. Finally, we showed that at least 15 h of NMR acquisition time were needed in order to reach the sufficient intensity of P signals and enable distinction among sample treatments. We believe that the procedure optimized here to obtain ^{31}P-NMR spectra for natural samples such as soil and humus-rich materials (litter and compost) may become useful to foster studies on total and organic P dynamics in the environment.

Competing interests

The authors declare that they have no competing interests.

Authors' contributions

ML performed soil extraction and sample manipulation before NMR analyses and wrote the first manuscript draft. PM ran the NMR experiments. VC supervised the soil and compost collection and characterization. HM characterized soil and compost properties. ZH contributed to design the experiments. AP supervised the experimental design and finalized the manuscript. All authors read and approved the final manuscript.

Acknowledgements

This work was supported by the EU-7FP project BIOFECTOR (grant agreement no. 312117). The scientific work at CERMANU of the first author was sponsored by the National Natural Science Fund Projects of China (No. U1133604), China Scholarship Council (CSC).

References

1. Ahlgren J, De Brabandere H, Reitzel K, Rydin E, Gogoll A, Waldebäck M (2007) Sediment phosphorus extractants for phosphorus-31 nuclear magnetic resonance analysis. J Environ Qual 36:892–898

2. Aspila KI, Agemian H, Chau ASY (1976) A semi-automated method for the determination of inorganic, organic and total phosphate in sediments. Analyst 101:187–197

3. Cade-Menun BJ, Liu CW, Nunlist R, McColl JG (2002) Soil and litter phosphorus-31 nuclear magnetic resonance spectroscopy: extractants, metals, and phosphorus relaxation times. J Environ Qual 31:457–465

4. Cade-Menun BJ, Navaratnam JA, Walbridge MR (2006) Characterizing dissolved and particulate phosphorus in water with ^{31}P nuclear magnetic resonance spectroscopy. Environ Sci Technol 40:7874–7880

5. Cade-Menun BJ, Preston CM (1996) A comparison of soil extraction procedures for ^{31}P-NMR spectroscopy. Soil Sci 63:770–785

6. Cade-Menun BJ (2005) Characterizing phosphorus in environmental and agricultural samples by ^{31}P nuclear magnetic resonance spectroscopy. Talanta 66:359–371

7. Cheesman AW, Turner BL, Inglett PW, Reddy KR (2010) Phosphorus transformations during decomposition of wetland macrophytes. Environ Sci Technol 44:9265–9271

8. Condron LM, Frossard E, Newman RH, Tekeley P, Morel J-L (1997) Use of 31P NMR in the study soils and the environment. In: Nanny MA, Minear RA, Leenheer JA (ed) Nuclear magnetic resonance spectroscopy in environment chemistry. Oxford University Press, New York. USA

9. Ding SM, Xu D, Li B, Fan CX, Zhang CS (2010) Improvement of ^{31}P NMR spectral resolution by 8-hydroxyquinoline precipitation of paramagnetic Fe and Mn in environmental samples. Environ Sci Technol 44:2555–2561

10. Ding SM, Bai XL, Fan CX, Zhang L (2010) Caution needed in pretreatment of sediments for refining phosphorus-31 nuclear magnetic resonance analysis: results from a comprehensive assessment of pretreatment with ethylenediaminetetraacetic acid. J Environ Qual 39:1668–1678

11. Dong L, Yang Z, Liu X, Liu G (2012) Investigation into organic phosphorus species in sediments of Baiyangdian Lake in China measured by fractionation and ^{31}P NMR. Environ Monit Assess 184:5829–5839

12. Doolette AL, Smernik RJ, Dougherty WJ (2011) Overestimation of the importance of phytate in NaOH-EDTA soil extracts as assessed by ^{31}P NMR analysis. Org Geochem 42:955–964

13. Makarov MI, Haumaier L, Zech W, Malysheva TI (2004) Organic phosphorus compounds in particle-size fractions of mountain soils in the northwestern Caucasus. Geoderma 118:101–114

14. Makarov MI, Haumaier L, Zech W, Marfenina OE, Lysak LV (2005) Can ^{31}P NMR spectroscopy be used to indicate the origins of soil organic phosphates? Soil Biol Biochem 37:15–25

15. Makarov MI, Haumaier L, Zech W (2002) Nature of soil organic phosphorus: an assessment of peak assignments in the diester region of ^{31}P NMR spectra. Soil Biol Biochem 34:1467–1477

16. Mazzei P, Piccolo A (2012) Quantitative evaluation of noncovalent interactions between glyphosate and dissolved humic substances by NMR spectroscopy. Environ Sci Technol 46(11):5939–5946

17. Murphy J, Riley JP (1962) A modified single solution method for the determination of phosphate in natural waters. Anal Chim Acta 27:31–36

18. Reitzel K, Ahlgren J, DeBrabandere H, Waldebäck M, Gogoll A, Tranvik L, Rydin E (2007) Degradation rates of organic phosphorus in lake sediment. Biogeochemistry 82:15–28

19. Toor GS, Condron LM, Di HJ, Cameron KC, Cade-Menun BJ (2003) Characterization of organic phosphorus in leachate from a grassland soil. Soil Biol Biochem 35:1317–1323

20. Turner BL, Cade-Menun BJ, Condron LM, Newman S (2005) Extraction of soil organic phosphorus. Talanta 66:294–306

21. Turner BL, Condron LM, Richardson SJ, Peltzer DA, Allison VJ (2007) Soil organic phosphorus transformations during pedogenesis. Ecosystems 10:1166–1181

22. Turner BL, Mahieu N, Condron LM (2003) Phosphorus-31 nuclear magnetic resonance spectral assignments of phosphorus compounds in soil NaOH-EDTA extracts. Soil Sci Soc Am J 67:497–510

23. Turner BL, Newman S (2005) Phosphorus cycling in wetland soils: the importance of phosphate diesters. J Environ Qual 34:727–733

24. Turner BL (2004) Optimizing phosphorus characterization in animal manures by solution phosphorus-31 nuclear magnetic resonance spectroscopy. J Environ Qual 33:757–766

25. Turner BL (2008) Soil organic phosphorus in tropical forests: an assessment of the NaOH-EDTA extraction procedure for quantitative analysis by solution ^{31}P NMR spectroscopy. Eur J Soil Sci 59:453–466

26. Xu D, Ding SM, Li B, Jia F, He X, Zhang CS (2012) Characterization and optimization of the preparation procedure for solution P-31 NMR analysis of organic phosphorus in sediments. J Soils Sediments 12:909–920

27. Zhang R, Wu F, He Z, Zheng J, Song B, Jin L (2009) Phosphorus composition in sediments from seven different trophic lakes, China: a phosphorus-31 NMR study. J Environ Qual 38:353–359

28. Zhang WQ, Shan BQ, Zhang H, Tang W (2013) Assessment of preparation methods for organic phosphorus analysis in phosphorus-polluted Fe/Al-rich Haihe river sediments using solution 31P-NMR. PLoS One 8:e76525

11

Humic matter elicits proton and calcium fluxes and signaling dependent on Ca^{2+}-dependent protein kinase (CDPK) at early stages of lateral plant root development

Alessandro C Ramos[1], Leonardo B Dobbss[2], Leandro A Santos[3], Mânlio S Fernandes[3], Fábio L Olivares[4], Natália O Aguiar[4] and Luciano P Canellas[4*]

Abstract

Background: The humic acid (HA) fraction of soil organic matter (SOM) exerts an effective plant growth promotion through a complex mechanism involving a coordinated activation of several key ion transport and signaling systems. We investigated the effects of HA on H$^+$ and Ca^{2+} cellular dynamics at the early stages of lateral plant root development.

Results: Emergence of lateral root in rice seedlings were related to specific H$^+$ and Ca^{2+} fluxes in the root elongation zone underlying an activation of the plasma membrane H$^+$-ATPase and of the Ca^{2+}-dependent protein kinase (CDPK). The latter was coupled with an increased expression of the voltage-dependent *OsTPC1* Ca^{2+}channels and two stress responsive CDPK isoforms, such as *OsCPK7* and *OsCPK17*.

Conclusions: HA act as molecular elicitors of H$^+$ and Ca^{2+} fluxes, which seem to be upstream of a complex CDPK cell-signaling cascade. These findings shed light on the first ion signaling events involved in the mechanism of action of HA on plant growth and development.

Keywords: Calcium ion; Anion channel; H$^+$-ATPase; Rice; Ion-selective vibrating probe; pH signaling; Physiological effects of humic substances

Background

Humic substances (HS) represent the major components of soil organic matter (SOM) and are complex supramolecular associations of biotically and abiotically transformed biomolecules released in soil after cell lyses [1]. They exert diverse morphological, physiological, and biochemical influence on plant growth [2]. HS effects may be either stimulating or inhibiting plant activities, depending on their origin, molecular composition, and concentration, and also varying with plant species and interactions with symbiotic microorganisms [3]. Despite prolonged efforts aimed to correlate molecular composition and structural arrangements of humic substances

with their biological activity on higher plants, the involved mechanisms of action are yet to be elucidated [4-7].

Tip-growing plant cells, as root hair and cells in elongation zone, exhibit markedly polarized internal gradients and/or external fluxes of important regulatory ions (e.g., calcium, protons, and anions). Calcium (Ca^{2+}) ions are since long recognized to represent important second messengers in many signaling pathways, while increasing evidences have accounted pH changes to several Ca^{2+}-dependent signal transduction in plant and fungal cells [8,9].

Protons (H$^+$) are major contributors to cell currents [10] and H$^+$ fluxes through membranes control the cytosolic pH regulation, the secondary active transport of organic and inorganic nutrients, turgor-regulation, and cell wall plasticity. These processes are described by the 'acid-growth theory' [11] that suggests a close relationship

---

* Correspondence: canellas@uenf.br
[4]Universidade Estadual do Norte Fluminense Darcy Ribeiro (UENF) Núcleo de Desenvolvimento de Insumos Biológicos para Agricultura (NUDIBA), Av. Alberto Lamego 2000, Campos dos Goytacazes 28013-602, Brazil
Full list of author information is available at the end of the article

between H^+ pump activity and wall acidification and cell expansion. This mechanism has been also associated with the auxin effects on seed embryo growth [12] and early growth responses to gravistimulation of root cells [13].

Auxin cell signaling elicited by HS was previously showed by Trevisan et al. [14] using the *DR5::GUS* construct in *Arabidopsis*, and evidence for their involvement in plasma membrane H^+-ATPase activation has been recently reported [7,15]. This enzyme creates the electrochemical gradient of H^+ that energizes secondary transport processes, including ion and nutrient uptake [16]. The influence of HS on H^+-ATPases and H^+ transport has been extensively reported [17,18].

Protein kinases have also paramount importance in cell signal transduction [19,20]. They are responsible for the post-translational control of target proteins and act as critical regulators of various signaling cascades [19]. Furthermore, many plant receptors are endowed with protein kinase activity, through the participation in processes like growth, hormone perception, and stress responses [21]. Cytoplasmic Ca^{2+} oscillations also play a key role in cell signaling, either through activation of ion channels or as secondary messenger [22]. Signaling occurs when the cell is stimulated to release the Ca^{2+} ions stored in intracellular compartments, or when Ca^{2+} ions enter the cell via the plasma membrane, resulting in cytosolic waves in which the Ca^{2+} concentration increases by 1 or 2 orders of magnitude over a short period of time [22,23].

The aim of this work was to verify whether humic matter is capable to regulate the elongation of lateral roots in rice seedlings by a mechanism of integrated modulation of membrane Ca^{2+} and H^+ fluxes and specific calcium-dependent protein kinase (CDPK) activations.

Methods

Isolation of humic acids from vermicompost

A vermicompost was obtained from cattle manure. The organic residues were mixed and earthworms were added at a ratio of 5 kg earthworms (*Eisenia foetida*) per m^3 of organic residue. A bed of worms and organic residues was first prepared in a container and additional layers of organic residues were periodically placed over the pile, as a function of temperature, until the pile reached 50 cm. At the end of the transformation process (3 months after addition of the last organic residues), worms were transferred into a pile of fresh organic residue (plant + cattle manure) placed in a corner of the container and the vermicompost removed and sieved through a 2-mm sieve. The vermicompost had the following properties: pH 7.8, 46.5 g kg^{-1} total organic carbon and 17.3 g kg^{-1} humic acid (HA) carbon. HA were isolated from vermicompost and purified as reported elsewhere [17]. The

HA were suspended in distilled water and titrated to pH 7.0 by automatic titrator (VIT 909 Videotitrator, Copenhagen, Denmark) with a 0.1 NaOH solution under N2. The resulting sodium-humates were then passed through a 0.45-μm Millipore filter (Millipore, Billerica, MA, USA), dialyzed against ultrapure water and freeze-dried.

HA characterization

Total organic carbon (OC), nitrogen (TN), and hydrogen (H) contents were determined by dry combustion using an automatic CHN analyzer (Perkin-Elmer 2400 Series, Perkin-Elmer, Norwalk, CT, USA). The oxygen (O) content was obtained by difference after deducting the ash content (2.1%). The C functionality distributions of the humic acid samples were determined by solid-state CP-MAS ^{13}C-NMR spectroscopy. The spectra were acquired with an Avance 500 MHz (Bruker, Karlsruhe, Germany) spectrometer equipped with a 4-mm-wide bore MAS probe and operating at ^{13}C and ^{1}H frequencies of 125 MHz and 500 MHz, respectively. The samples (100 to 200 mg) were packed in 4-mm zirconia rotors with Kel-F caps, which were spun at 13 ± 1 kHz. The spectra were acquired by the ramped CP-MAS method, with linear amplitude variation of the ^{1}H pulse. The experiments were carried out using a cross-polarization time of 1.0 ms, an acquisition time of 25 ms, a cycle delay of 2 s and a high-power two-pulse phase modulation (TPPM) proton decoupling of 70 KHz. The Bruker Topspin 1.3 software (Bruker Biospin, Karlsruhe, Germany) was used to collect and process the spectra. All the free induction decays (FIDs) were transformed by applying a 4 k zero filling and a line broadening of 75 Hz. The spectra were normalized by area and integrated in the following ^{13}C chemical shift intervals: 190 to 160 ppm (carbonyls of ketones, quinones, aldehydes, and carboxyls), 160 to 140 ppm (phenols and O-substituted aromatic C), 140 to 110 ppm (unsubstituted aromatic C and olefinic C), 110 to 95 ppm (anomeric C), 95 to 65 ppm (O-alkyl systems), 65 to 45 ppm (methoxy substituent; N-alkyl groups), and 45 to 0 ppm (alkyl C, mainly CH_2 and CH_3). The relative areas of the alkyl (45 to 0 ppm) and aromatic (160 to 110 ppm) components were summed to represent the proportion of hydrophobic C in the humic samples (degree of hydrophobicity, HB). Similarly, the summation of the relative areas in intervals related to polar groups (190 to 160 ppm and 110 to 45 ppm) indicates the degree of C hydrophilicity (HI); the HB/HI ratio was then calculated.

Germination of *Oryza sativa*

Rice (*O. sativa* var. Nipponbare) seeds provided by Plant Mineral Nutrition Laboratory of UFRRJ, were surface sterilized by soaking in 0.5% (*w/v*) NaClO for 30 min, followed by rinsing and then soaking in water for 6 h. Afterward, the seeds were sown on wet filter paper and germinated in the dark at 28°C. A preliminary assay was conducted by using 4-day-old seedlings with roots approximately 5 cm long. These were transferred into a solution containing 2 mM CaCl2 and HA at varying concentrations: 0, 0.5, 1.0, 2.0, 4.0, 6.0, and 8.0 mM C L^{-1}. A minimal medium (2 mM CaCl2) has been used to avoid any interference from nutrients that may act synergistically with humic matter on plant growth and metabolism. After regression analysis, a new experiment was carried out using the best concentration (best dose = 3.5 mM).

Root growth measurement
On the 7th day, 20 seedling roots for each treatment were collected to estimate root area using Delta-T scan software (Delta-T Devices, Cambridge, UK).

Frequency of sites of lateral root emergence
Seeds of rice were germinated for 4 days in wet filter paper and rooted in a medium containing 0 or 3.5 mM C HA L^{-1}. The whole root systems (three replicates) of both treatments were harvested every day during a period of 7 days to evaluate the number of mitotic sites after mild digestion by boiling at 75°C for 20 min in KOH (0.5%, w/v) and hematoxylin treatment according to Canellas et al. [17].

PM-enriched vesicles
Plasma membrane (PM) vesicles were isolated from rice roots grown with and without 3.5 mM C HA L^{-1} using differential centrifugation. In brief, about 15 g (fresh weight) of rice roots was homogenized using a mortar and pestle in 30 mL of ice-cold buffer containing 250 mm Suc, 10% (w/v) glycerol, 0.5% (w/v) polyvinylpyrrolidone-40 (40 kD), 2 mM ethylenediaminetetraacetic acid (EDTA), 0.5% (w/v) bovine serum albumin, and 0.1 M Tris-HCl buffer, pH 8.0. Just before use, 150 mM KCl, 2 mM dithiothreitol (DTT), and 1 mM phenylmethylsulfonyl fluoride were added to the buffer. The homogenate was strained through four layers of cheesecloth and centrifuged at 8,000×g for 10 min. The supernatant was centrifuged once more at 8,000×g for 10 min and then at 100,000×g for 40 min. The pellet was resuspended in a small volume of ice-cold buffer containing 10 mM Tris-HCl (pH 7.6), 10% (v/v) glycerol, 1 mM DTT, and 1 mM EDTA. The suspension containing membrane vesicles was layered over a 20%/30%/42% (w/w) discontinuous Suc gradient that contained, in addition to Suc, 10 mM Tris-HCl buffer (pH 7.6), 1 mM DTT, and 1 mM EDTA. After centrifugation at 100,000×g for 3 h in a swinging bucket, the vesicles that sedimented at the interface between 30% and 42% (w/w) Suc were collected, diluted with three volumes of ice-cold water, and centrifuged at 100,000×g for 40 min. The pellet was resuspended in a buffer containing 10 mM Tris-HCl

(pH 7.6), 10% (v/v) glycerol, 1 mM DTT, and 1 mM EDTA. The vesicles were either used immediately or frozen under liquid N2 and stored at –70°C until use. Protein concentrations were determined by the method of Lowry et al. [24].

ATPase activity
ATPase activity in PM vesicles was determined by measuring the release of Pi colorimetrically. Between 80% and 95%, ATPase activity in PM vesicles, as measured at pH 6.5, was inhibited by vanadate (0.1 mM), a very effective inhibitor of the PM P-type H^+-ATPase. In all experiments, the ATPase activity was measured at 30°C, with and without vanadate, and the difference between these two activities was attributed to the PM H^+-ATPase.

ATPase H^+ pumping
The electrochemical H^+ gradient generated by the H^+-ATPase was estimated from the initial rate of quenching of the fluorescent pH probe 9-amino-6-chloro-2-methoxyacridine (ACMA) (2.5 μM, 415/485 nm excitation/emission) and expressed in percentage of quenching per minute. The assay medium contained 10 mM HEPES-KOH (pH 6.5), 100 mM KCl, 3 mM MgCl2, 2.5 μM ACMA, and 0.05 mg L^{-1} PM vesicle proteins. The reaction was triggered by addition of 1 mM ATP.

Measurements of H^+ and Ca^{2+} anion fluxes
Measurements of ion fluxes in rice plants treated with HA were conducted by using the ion-selective vibrating probe system. For H^+ and Ca^{2+} measurements, rice roots were grown in hydroponic conditions and placed in plastic Petri dishes (140 × 140 mm) filled-up with 30 mL of modified Clark solution at ¼ strength supplemented with 0 mM or 3.5 mM C HA, excepted for Ca^{2+} measurements when 100 μM Ca^{2+} was used. Three to four days after germination, ion fluxes were analyzed by the vibrating probe system. H^+-specific vibrating microelectrodes were produced as described by Feijó et al. [8]. Micropipettes were pulled from 1.5-mm borosilicate glass capillaries with a Sutter P-98 Flaming Brown electrode puller (www.sutter.com). These were then baked in covered dishes at 250°C for 8 to 12 h and vaporized with dimethyl dichlorosilane (Sigma, St. Louis, MO, USA) for 30 min, and the covers were removed before further baking at 250°C for 1 h. After silanization, the capillaries were backfilled with a 15- to 20-mm column of electrolyte (15 mM KCl and 40 mM KH_2PO_4, pH 6.0 for H^+; 100 mM KCl for anions; 100 mM $CaCl_2$ for Ca^{2+}) then front-loaded with a 20- to 25-μm column of a pH selective liquid exchange cocktail from Fluka (Milwaukee, WI, USA). This contained protons (hydrogen ionophore II-cocktail B, no. 95223), chloride (chloride ionophore cocktail A, no 24902), and calcium (calcium ionophore I

cocktail A, no 21048). This solution was drawn into the tip of the microelectrode by application of suction to the basal end of the pipette. We used Cl^- electrodes to measure the anion fluxes, although this electrode has poor selectivity for Cl^- under our experimental conditions. According to Messerli et al. (2004), 1 mM NO_3^- is enough to diminish the Cl^- selectivity. Indeed, under low Cl^- concentration, Cl^- detection was influenced by NO_3^-, SO_4^{2-} and PO_4^{2-}, suggesting a low selectivity of this ionophore under our experimental conditions. The Cl^- electrode calibration with different anions also showed that this electrode could detect NO_3^- but not SO_4^{2-} and PO_4^{2-}. Therefore, our measurements reflected changes in the 'anionic' concentration rather than in Cl^- fluctuations. An Ag/AgCl wire electrode holder (World Precision Instruments, Sarasota, FL, USA) was inserted into the back of the microelectrode in order to establish electrical contact with the bathing solution. The ground electrode was a dry reference (DRIREF-2, World Precision Instruments) that was inserted into the sample bath. In order to obtain a calibration line, microelectrodes were calibrated at the beginning and end of each experiment using standard solutions covering the experimental range of each ion. Both the slope and intercept of the calibration line were used to calculate the respective ion concentration from the mV values measured during the experiment near and distant from the root surface. A detailed description of the ion-selective vibrating probe system was reported by Feijó et al. [8] and Ramos et al [25,26]. The vibrating-electrode system was attached to a Nikon Eclipse TE-300 inverted microscope (Nikon, Melville, NY, USA) and housed inside a copper-sheet Faraday cage over a vibration-free platform. For routine experiments, an X20 Plan Apo objective under differential interference contrast (DIC) was used.

Kinase activity assays

Protein kinase activity was measured by incorporating radioactive phosphate from ATP into syntide 2 (Sigma) [27]. Protein kinase activity was assayed in 20 μl of 30 mM Hepes-Tris buffer (pH 7.0), 2.5 mM free Mg^{2+}, 1.5 mM EGTA and 1.5 mM N-hydroxyethylethylenediaminetriacetic acid (HEDTA), 10 μM free Ca^{2+}, 50 μM syntide 2, and 0.2 mM [γ-32P]ATP (3.77 TBq mol^{-1} from Amersham Bioscience (Piscataway, NJ, USA), unless stated otherwise). Reactions were started by adding 1 μl of a kinase-containing solution and incubated for 30 min at room temperature. At the end of incubation, the reaction mixture was spotted on 2 cm × 2 cm Whatman P81 phosphocellulose paper pieces. These were then washed three times with 75 mM H3PO4 (for 10 min), rinsed for 5 min in ethanol, air dried, placed in vials with scintillation liquid and levels of radioactivity determined.

Rice growth conditions for gene expression experiment

Rice plants was held in a growth chamber, with a light cycle of 12/12 h (light/dark), photosynthetic photons flux of 400 μmol m^{-2} s^{-1}, relative humidity of 70%, and temperatures of 28°C/24°C (day/night). Rice seeds (*O. sativa* L. cv. Nipponbare) were previously disinfected in 2% sodium hypochlorite solution for 15 min and then washed several times with distilled water. These seeds were germinated only in distilled water. Five days after germination, the seedlings were transferred to 0.3 liter pots, where five plants per pot received Hoagland solution modified to contain 1/4 of the total ionic strength and 1.5 mM of N (1.25 mM NO_3^--N and 0.25 mM NH_4^+-N) pH 5.8. This nutrient solution was replaced every 2 days. Thirteen days after germination, the plants received the same solution with or without humic acid (3.5 mM of C). The experiment was conducted in a completely random design with three replications. Root samples were collected at 24 h for total RNA extraction and real-time PCR.

Total RNA extraction and cDNA synthesis

Total RNA was extracted according to Gao et al. [28] with modifications using NTES buffer (0.2 M Tris-HCl pH 8.0; 25 mM EDTA; 0.3 M NaCl; 2% SDS). Three root samples were ground in N_2 and homogenized in a mixture containing 1 mL NTES buffer and 0.7 mL of phenol: chloroform (1:1). Homogenized samples were centrifuged at 18,000×g for 20 min at 4°C and each supernatant transferred to a new tube. Total RNA was precipitated by adding 1/10 volume of 3 M sodium acetate pH 5.2 ($NaOAc_{DEPC}$) and one volume of cold isopropanol. Samples were placed at −80°C for 1 h and centrifuged at 18,000×g for 20 min after that time. Pellets were re-suspended in 0.5 mL of H_2O_{DEPC} and precipitated overnight at 4°C by the addition of 0.5 mL of 6 M lithium chloride ($LiCl_{DEPC}$). After centrifugation at 18,000×g for 20 min, pellets were re-suspended with 0.5 mL H_2O_{DEPC} and precipitated again with addition of 0.5 mL LiCl 6 M for 1 h on ice. After centrifugation at 18,000×g for 20 min, pellets were re-suspended in 0.5 mL of H_2O_{DEPC}, precipitated with two volumes of ethanol for 1 h at −80°C and washed with 70% ethanol. The pellets were dried on ice for 10 min and dissolved in 30 μL of H_2O_{DEPC}.

The quality of total RNA extracted was verified spectrophotometrically through A_{260}/A_{230} and A_{280}/A_{260} ratios and visualization in agarose gel (1%) with ethidium bromide. Total RNA samples used for cDNA synthesis were treated with DNAse I (Invitrogen, Inc., Waltham, MA, USA) by following the manufacturer's instructions. Single strand cDNA was synthesized by using the 'TaqMan Reverse Transcription Reagents' (Applied Biosystems, Inc., Loughborough, UK) and oligo dT primer according to the manufacturer's instructions.

Real-time PCR

Real-time PCR reactions were performed in duplicate, using 'SYBR® Green PCR Master Mix' kit (Applied Biosystems, Inc.) according to the manufacturer's instructions. PCR reactions were as follows: 10 min at 95°C, 45 amplification cycles at 95°C for 15 s and 60°C for 1 min (annealing, extension, and fluorescence detection), followed by the 'melting curve' in order to verify the specificity of the reaction. Actin gene (NM 001057621.1) was used as an endogenous control. The primer sequences for *OsCPK7*, *OsCPK17*, and *OsTPC1* were designed with Primer Express software (Applied Biosystems) (Additional file 1: Table S1) and OsA7 [29]. The specificity of the primer sequences was analyzed using BLAST at TIGR (http://rice.plantbiology.msu.edu/) and in NCBI (http://www.ncbi.nlm.nih.gov) as well as experimentally at the end of the PCR reaction through the melting curve. Relative expression was performed according to Livak and Schmittgen [30], using the nutrient solution without humic acid as reference.

Results

HA characterization

The C content of the HA was 44.7%, whereas N and H content were 3.51% and 5.36%, respectively. The ^{13}C CPMAS NMR spectrum of HA is shown in supplementary data (Additional file 2). It was characterized by strong signals in the O-alkyl-C (56 to 110 ppm) and alkyl-C (45 to 0 ppm) regions, revealing a molecular composition dominated by carbohydrates and aliphatic components. The signals related to O-Alkyl-C and alkyl-C components represent in fact the majority of the total organic C, accounting respectively for the 40.9% and 21.9% of the total area of the spectra. The different resonances in the O-alkyl-C region (56 to 110 ppm) are currently assigned to monomeric units in oligo- and polysaccharidic chains of plant tissues. The signal centered at 72 ppm corresponds to the overlapping resonances of carbons 2, 3, and 5 in the pyranoside structure in cellulose and hemicellulose. These signals may represent the di-O-alkyl carbon of polysaccharidic chains other than cellulose such as the hemicellulose components contained in cell wall of the vascular tissues, like xylan, glucomannas, etc. Moreover, the shoulders around 20/23 ppm could be assigned to the methyl group in acetyl substituent in hemicellulose components.

The broad peak in the Alkyl-C region (30 ppm) of NMR spectra indicated the presence of alkyl chains (-CH$_2$- groups) derived mainly from various lipid compounds, plant waxes, and plant polyesters like cutin and suberin. The main signal at 56 ppm may be associated with either the methoxyl substituent on aromatic rings of guaiacyl and syringyl units in lignin, or with C-N bonds in amino acid moieties. The broad band around 131 ppm

may be mainly related to unsubstituted and C-substituted phenyl carbons of lignin monomers of guaiacyl and syringyl units as well as to condensed aromatic moieties, while the signals shown in the phenolic aromatic region (152 ppm) indicated the presence of O-substituted ring carbon derived from different aromatic structures. The resonances included in the 148 to 155 ppm chemical shift range are in fact usually assigned to carbons 3, 4, and 5 in the aromatic ring in lignin components, carbons 3 and 5 being coupled to methoxyl substituents (see Additional file 2: Figure S1. However, the prominent peak found at 145 and 157 ppm in the NMR spectra, suggest also the presence of tannin derivatives. The aromatic carbon represents 27.9% of spectrum. Finally, the broad signal at 174 ppm (8.3%) indicates the content of carbonyl groups of aliphatic acids and amino acid moieties, as well as that of acetyl groups in hemicellulose components.

Root growth promotion by HA

A quadratic model ($R^2 = 0.95$, $P < 0.001$) describes the effects of different concentrations of HA on number of lateral roots emerged from rice seedlings (Figure 1). The optimal concentration of HA on a carbon content basis was 3.5 mM. The quantitative results of root area and number of mitotic sites were normalized in respect to control. The elongation/differentiation zone of the root includes small, densely meristematic cells in continuous metabolic activity that is more susceptible to lateral root formation. A marked effect was observed on root surface area after exposure to HA (data not shown).

HA effect on H$^+$-ATPase activity

Treatment of PM vesicles isolated from rice roots for 7 days with 3.5 mM C HA L^{-1} produced an increase in vanadate-sensitive ATP hydrolysis (Figure 2A) as well as a steeper ATP-dependent proton gradient, as measured by quenching of ACMA fluorescence (Figure 2B). The initial rate of gradient formation and ATP hydrolysis were enhanced by two- to threefold in response to treatment with HA.

HA effects on H$^+$, Ca^{2+} and anion fluxes, and on CDPK

Analysis of the root H$^+$ flux rate using a noninvasive technique revealed a differential pattern of H$^+$ fluxes along the lateral rice roots when treated with HA (Figure 3A). In HA-treated rice roots, apical (0 to 100 μm), meristematic (100 to 300 μm), and elongation (300 to 800 μm) zones showed a considerable increase on H$^+$ efflux if compared to the apex zone. In contrast, the control roots presented strong H$^+$ influxes (Figure 3A). A sixfold stimulation on H$^+$ effluxes was observed at the elongation zone in the presence of 3.5 mM C HA ($P < 0.001$), that resulted in the lowest root surface pH value (Figure 3B). These superior H$^+$ effluxes found at the

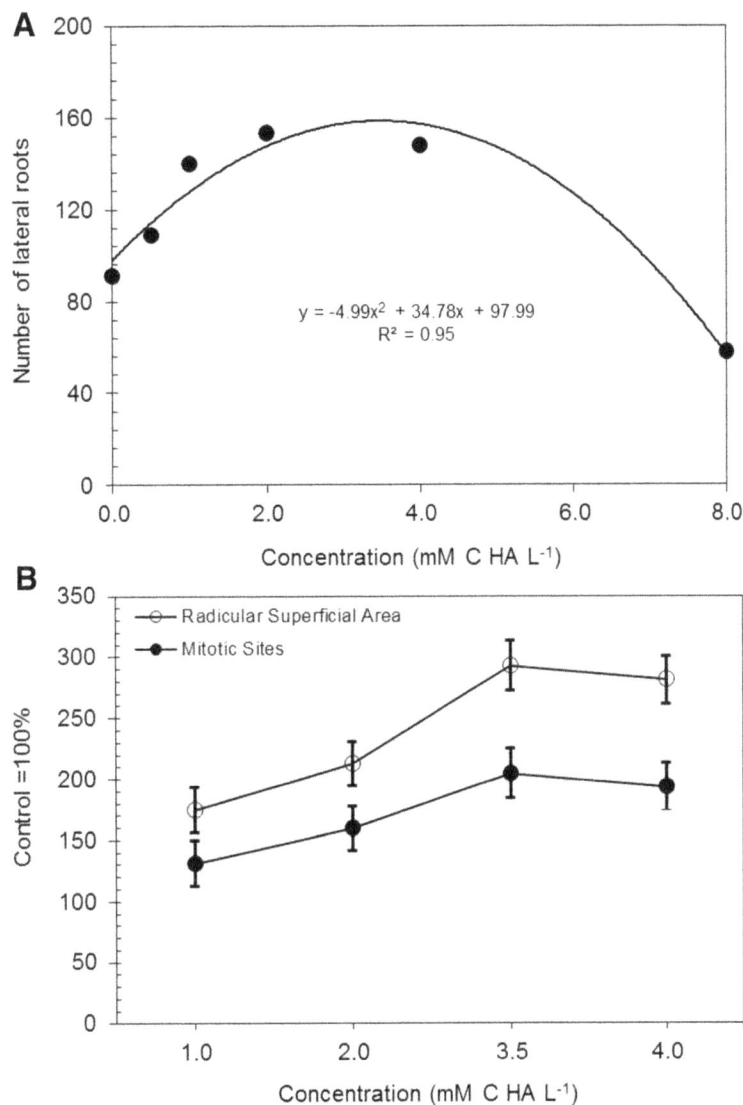

Figure 1 Effect of different HA concentrations on number of lateral roots which emerged from principal axis (A). Data are the mean of 25 rice seedlings. **(B)** radicular superficial area and number of mitotic sites.

elongation zone of HA-treated roots were vanadate-sensitive which suggests the H^+-ATPase activity triggers ion dynamics in the presence of HA (Figures 2 and 3A).

Patterns of the Ca^{2+} fluxes in the control and HA-treated roots revealed a quite different scenario. Treatment with HA also increased Ca^{2+} influxes at the elongation zone (Figure 4), thus leading to a possible increase in free cytosolic Ca^{2+} concentration. At such root zone, a negative correlation can be observed between Ca^{2+} and H^+ ion fluxes ($r = -0.8951$ $P < 0.001$). On the other hand, a significant (tenfold) increase in anion efflux was observed in all root zones of rice seedlings treated with HA (Figure 5). Furthermore, we observed a clear effect of HA on the enhancement of CDPK activity in the experiments described here (Figure 6).

Relative gene expression of Ca^{2+} channels (OsTPC1), PM H^+-ATPase (OsA7), and calcium-dependent protein kinase (OsCPK7 and OsCPK17)

To address the aforementioned increase in Ca^{2+} efflux, we examined the expression level of *OsTPC1* voltage-dependent Ca^{2+} channel in rice seedlings at the early stage of HA treatment. OsTPC1 encoded a putative voltage-gated Ca^{2+} channel from rice, ubiquitously expressed in mature leaves, shoots, and roots [31]. In a study for functional characterization of *OsTPC1*, Kurusu et al. [31] suggested that OsTPC1 has Ca^{2+} transport activity across the plasma membrane and is involved in the regulation of growth and development in rice. *OsTPC1* significantly increased upon HA treatment in respect to control plants (Figure 7D). In addition, we observed a concomitant

A

TREATMENTS	SPECIFIC ACTIVITY mM Pi / mg PTN / min	H⁺ PUMP STIMULATION %
CONTROL	1.18	100
HA (1.0 mM C L⁻¹)	1.29	109
HA (2.0 mM C L⁻¹)	1.37	116
HA (3.5 mM C L⁻¹)	2.86	242
HA (4.0 mM C L⁻¹)	2.51	213

B

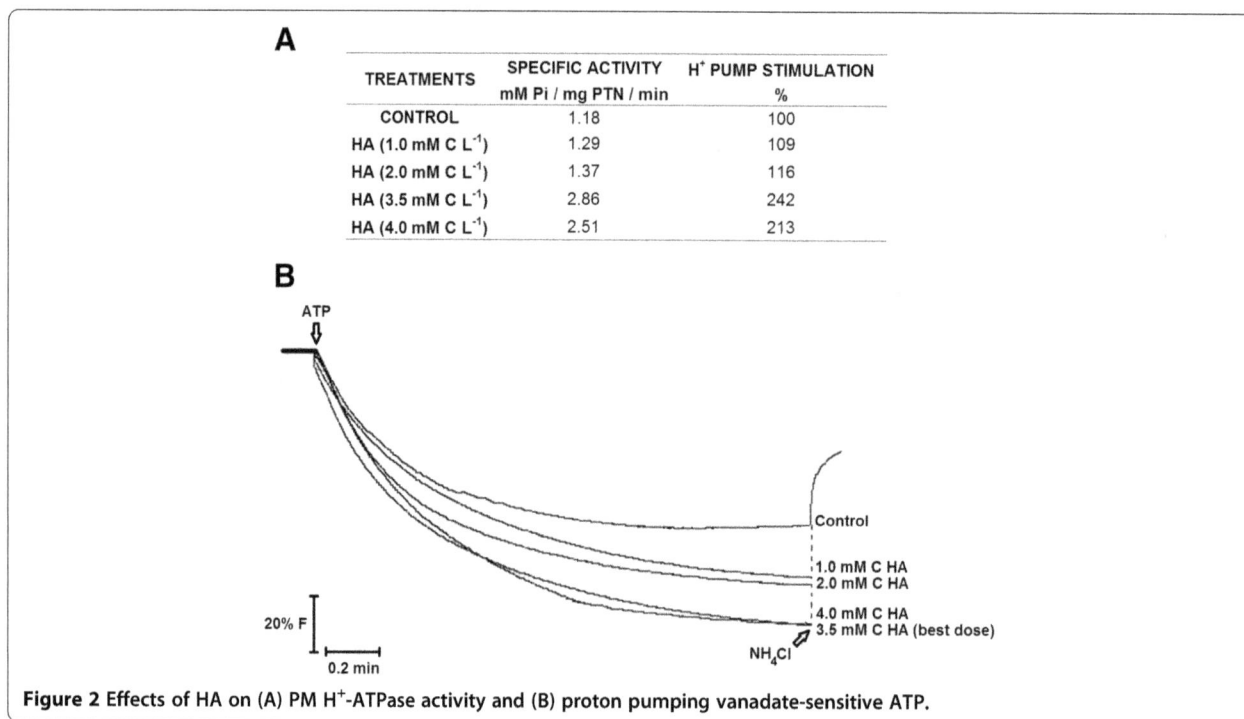

Figure 2 Effects of HA on (A) PM H⁺-ATPase activity and (B) proton pumping vanadate-sensitive ATP.

enhancement of *OsA7* PM H⁺-ATPase expression in rice seedlings treated with HA (Figure 7C). The *OsA7* isoform is a highly expressed PM H⁺-ATPase member of the subfamily II [32]. Sperandio et al. [29] showed that *OsA7* is upregulated in response to nitrogen resupply and suggested that *OsA7* could be involved in N nutrition in rice. We also evaluated the expression of two CDPK genes, *OsCPK7* and *OsCPK17*, both constitutively expressed in roots, stems, leaves, and panicles and responsive to various stress stimuli [33]. *OsCPK7* was upregulated by salt stress in roots [33], jasmonic acid [34], cold, and gibberellin [35] and suppressed by rice blast infection [33]. *OsCPK17* is downregulated by cold, drought, and salt stress in the rice seedlings [33]. Both *OsCPK7* and *OsCPK17* showed an increase of expression at 24 and 72 h after HA treatment as compared to untreated plants (Figure 7A,B).

Discussion

We used in this study a humic acid-like substance extracted from a vermicompost obtained from cattle manure with a large presence of OCH₃ groups with NMR spectral resonance at 56 ppm (Additional file 1) that is a trait for promotion of lateral root emergence, as shown previously [36]. The best dose-response concentration for lateral root induction of rice plantlets was used (Figure 1), and it was found that the HA treatment changed ion fluxes pattern over the root axis. Such changes for H⁺ efflux was well characterized by other research reports which used using molecular and biochemical approaches involving H⁺-ATPase enzymatic activity measurements

to establish a relation with HA bioactivity in plants [2,3,7,15,17]. Here, using a vibration probe device, we specifically detected and unequivocally showed live H⁺ flux over different anatomical regions of the rice root axis displayed by the presence of HA (Figure 3A).

In comparison with control, the largest rates of H⁺ efflux coupled with acidic surface pH following the HA treatments were located at the elongation/differentiation zones of rice roots, where plant cells with increased growth rates (reflected by cell volumetric expansion) and large cell differentiation are found (Figure 3). As quoted in literature, these effluxes are dependent on PM H⁺-ATPase (PMA) activity (Figure 2). In fact, it has been shown that this zone presents significantly larger enzyme immunolocalization and activity levels of PMA than root cap and meristematic zones [37]. The H⁺ efflux mediated by PMA is important for the regulation of cytoplasmic pH [10] and the activation of cell wall-loosening enzymes and proteins, both phenomena occurring through acidification of the apoplastic compartment [38]. In agreement with this, we observed a high epidermal cells H⁺ efflux in HA-treated roots at elongation/differentiation zone (Figure 3A), leading to an increased capacity of these roots to acidify the local medium (Figure 3B). These effluxes proved to be dependent on PMA, since they were vanadate-sensitive and activated by fusicoccin (data not shown).

In addition, parallel to H⁺ efflux, we described for the first time a simultaneous Ca⁺² influx and anion efflux modulated by HA (Figures 4 and 5). Again, such ionic

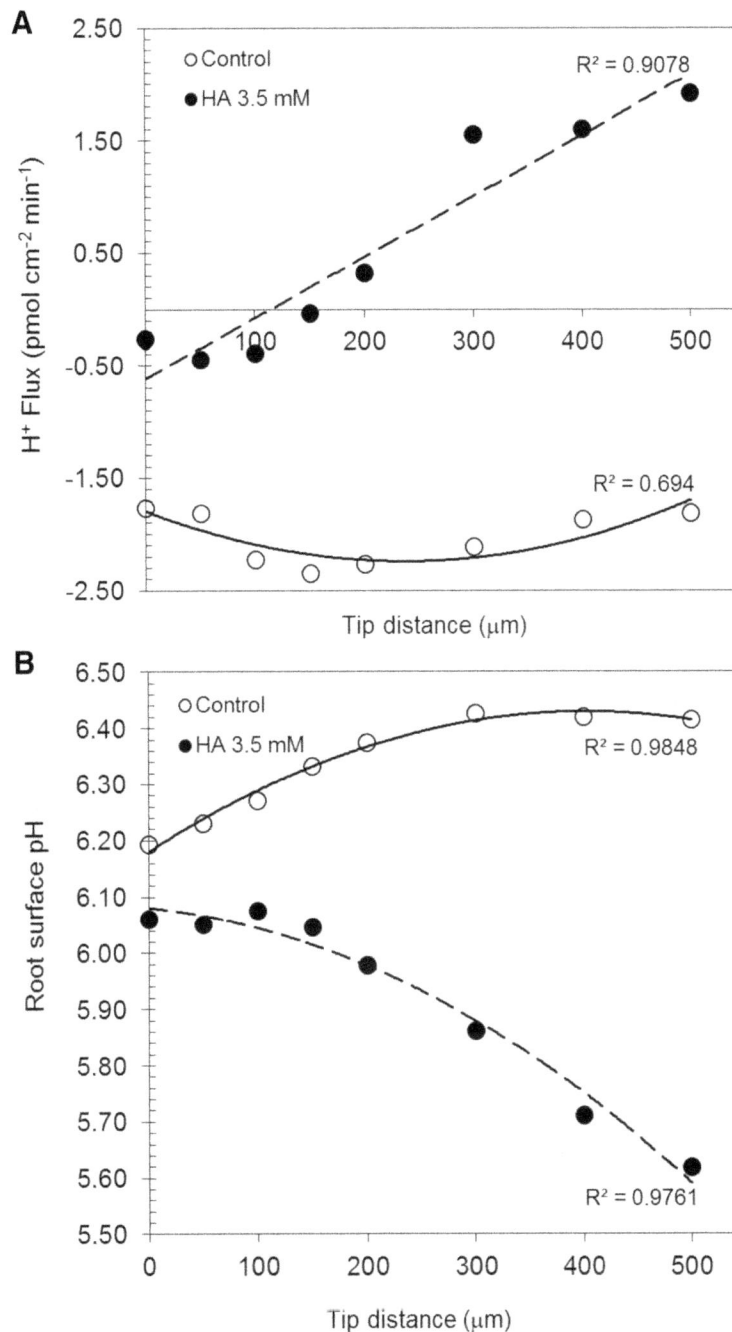

Figure 3 Proton fluxes and surface pH measured along the apical and elongation zones of rice roots. Proton (H⁺) fluxes **(A)** and surface pH **(B)** measured along the apical (0 to 100 μm) and elongation (200 to 500 μm) zones of rice roots before (o) and after (●) incubation with 3.5 mM C L⁻¹ humic acid (HA).

flux behavior was greater at elongation/differentiation root zone. The relevance of the coordinated spatial bidirectional fluxes for H⁺, Ca^{+2}, and anions with HA bioactivity could only be partially explained by the present study, while future attempts using specific probes or inhibitors for relevant pathways and plant mutagenesis in

some of these functions may better clarify signaling process related to lateral root induction by HA.

A relationship between proton and anion efflux may be established on the basis of plant nutrition or cytoplasm ionic balance view. Indeed, it is well known that an increase in root surface concentration of H⁺ generates a

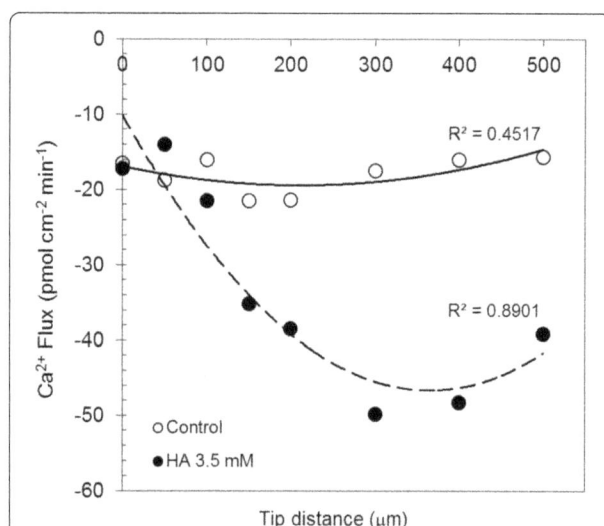

Figure 4 Calcium fluxes in the apical and elongation zones of rice roots. Calcium (Ca^{2+}) fluxes in the apical (0 to 100 µm) and elongation (200 to 500 µm) zones of rice roots, before (o) and after (•) incubation with 3.5 mM C L^{-1} humic acid (HA).

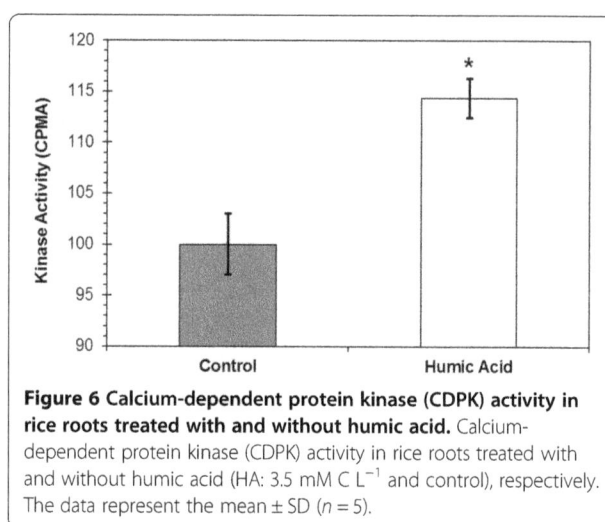

Figure 6 Calcium-dependent protein kinase (CDPK) activity in rice roots treated with and without humic acid. Calcium-dependent protein kinase (CDPK) activity in rice roots treated with and without humic acid (HA: 3.5 mM C L^{-1} and control), respectively. The data represent the mean ± SD ($n = 5$).

proton-motive force that is required to drive the secondary transport of NO_3^-, SO_4^{2-}, Cl^-, Ca^{2+}, and K^+ [10]. It has been reported that the PMA enzyme is stimulated by anions in plant cells [39-41] and host cell-parasite interactions [42]. Therefore, changes in root surface pH are most likely to produce rapid changes in root growth rate [43], thus resulting in rapid volume expansion of nutrient-demand cells followed by symplastic charge balance control.

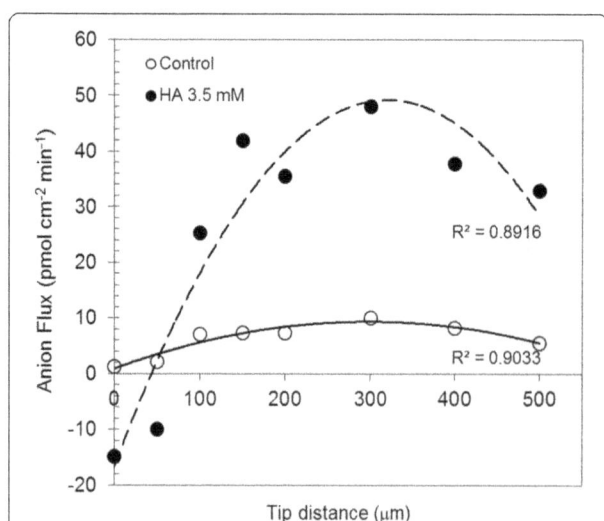

Figure 5 Anion fluxes measurements in the apical and elongation zones of rice roots. Anion fluxes measurements in the apical (0 to 100 µm) and elongation (200 to 500 µm) zones of rice roots, before (o) and after (•) incubation with 3.5 mM C L^{-1} humic acid (HA).

Other candidates that may contribute to the control of extracellular H^+ flux in HA are identified with anions in the growth medium, which are reported to act as stimulators of the PM H^+-ATPase [44]. This effect and the H^+ effluxes are closely related to auxin-induced cell growth, as proposed by the 'acid-growth theory' by Rayle and Cleland [11]. Previous reports described the auxin-like effect of HA using different approaches [7,17,18]. However, changes in organic acid exudation from maize roots induced by HA solutions were previously observed [45]. The relationship between organic acids and proton exudation, earlier described by Ohno et al. [46], was recently observed also by Tomasi et al. [47]. In line with these findings, Puglisi et al. [48] reported an enhancement of organic acid exudation in maize seedlings following a humic matter treatment.

Observed HA-modulated calcium influx (Figure 4) was supposed to be accompanied by an increased Ca-cytosolic concentration, as revealed in this study by both enhanced Ca-transporter and CDPK activity under transcriptional level and biochemical CDPK activity (Figures 6 and 7).

Transient elevations in cytosolic Ca^{2+} concentration are believed to be involved in a multitude of physiological processes, including responses to abiotic stresses, hormones, and pathogens [21]. In the case of HA-treated seedlings, we found a significant increase of Ca^{2+} influx in roots (Figure 4), as well as an enhanced expression of *OsTPC1* voltage-dependent Ca^{2+} channels (Figure 7) that was concomitant to an increased CDPK activity (Figure 6). In fact, it is known that Ca^{2+} channels are known to contribute to short transient Ca^{2+} influx in response to various stimuli, including chilling and microbial interaction [49]. Among the possible effects attributed to the cytosolic increase of Ca^{2+} concentration and down-streaming signaling events, we may include a regulatory H^+ efflux activity in the elongation/differentiation root zone. Indeed,

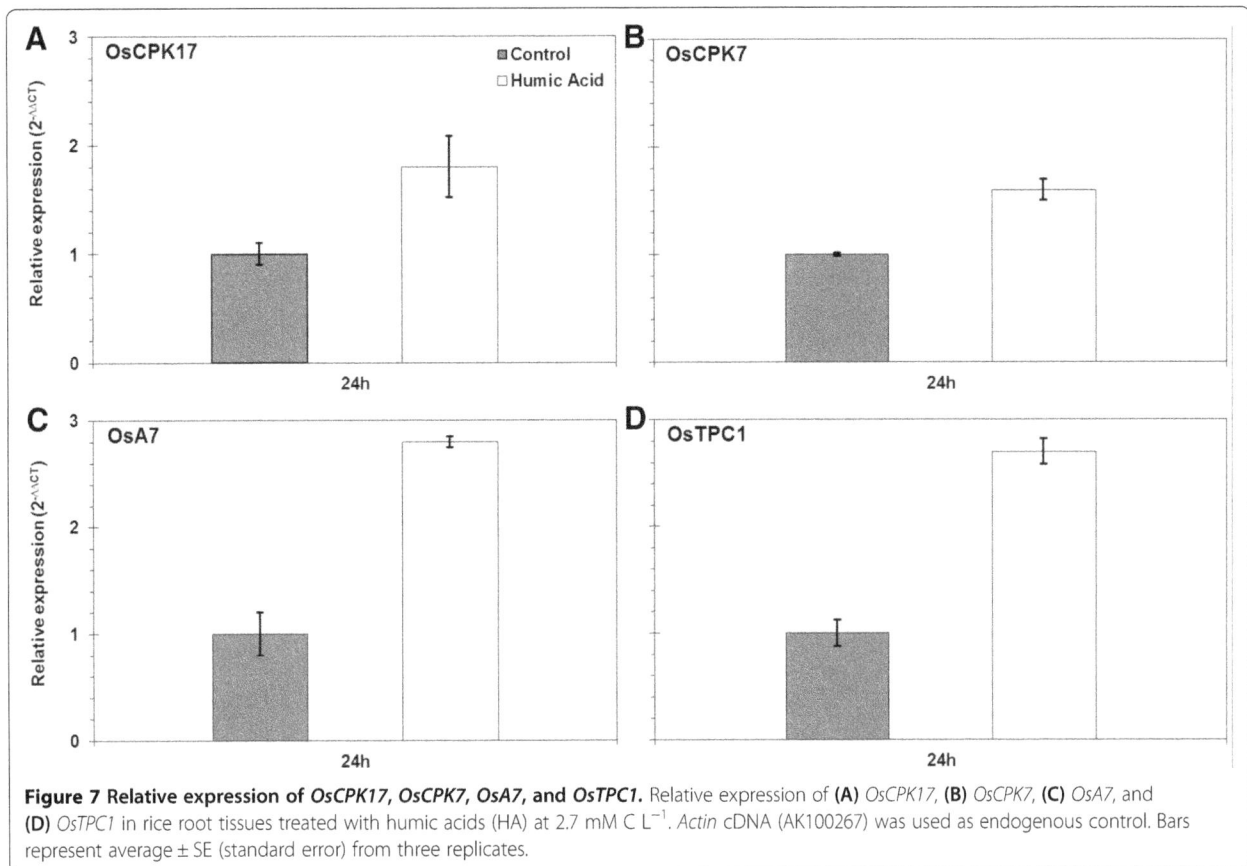

Figure 7 Relative expression of *OsCPK17, OsCPK7, OsA7,* and *OsTPC1*. Relative expression of **(A)** *OsCPK17,* **(B)** *OsCPK7,* **(C)** *OsA7,* and **(D)** *OsTPC1* in rice root tissues treated with humic acids (HA) at 2.7 mM C L^{-1}. *Actin* cDNA (AK100267) was used as endogenous control. Bars represent average ± SE (standard error) from three replicates.

H^+-ATPase enzyme is stimulated by external anion concentrations [39,40] and inhibited by Ca^{2+} [27]. Hypothetically, during the plant cell cycle, transient cell expansion (growth phase) related to apoplastic proton efflux may be followed by its inhibition by the calcium influx. A cascade signalization of protein-kinase-mediated protein phosphorylation may result in cell and tissue morphogenesis (differentiation phase).

The H^+ gradient generated by PMAs provides the energy necessary to drive the secondary transport [16]. Accordingly, we found that the changes in H^+ efflux attributable to HA were related to Ca^{2+} influx and anion efflux. However, roots treated with HA displayed strong anion and H^+ effluxes primarily at the elongation zone (Figure 3). An explanation may reside in the large net negative charge provided by the HA adsorbed on the root surface, that may be temporarily balanced by PMA activity, thus inducing an anion efflux. In fact, it has been reported that PMA is stimulated by anions in plants [41] and that consequent root surface acidifications are necessary for the mechanism of nutrient uptake [50], as this occurs via PM co-transporters [25]. An increase of PMA activity and extracellular acidification were observed (Figures 2 and 3). The activity of the PM is known to be regulated by reversible protein phosphorylation in a complex manner, including CDPKs [51].

All these observations point toward a role of CDPK during plants response to HA treatments as regulators of PM H^+-ATPase. The role of CDPK is remarkable in the context of plant signal transduction. This protein is responsible for post-translational control of target proteins acting as critical regulation of many signaling cascades. Many plant protein kinases act as receptors and participate in processes such as disease resistance, growth development, hormone perception, and stress responses [21]. Increased expression and activity of CDPK in plants treated with HA (Figures 6 and 7) indicate that these proteins can act as elicitors of responses mediated by stress resulting in altered root geometry.

Wan et al. [33] showed a response of many *CDPK* genes under different environmental stress and indentified multiple stress-responsive cis-elements in the promoter region (1 kb) upstream of these genes. Here, we found a concomitant increased expression of *OsTPC1* and *OsCPK7/17* at 24 h after exposure of plants to HA treatment (Figure 7). Correa-Aragunde et al. [52] reported that nitric oxide (NO) mediates auxin lateral root development and the hypothesis of a NO signal downstream of auxin signaling during lateral root formation has been strongly supported [53] and induced by Ca^{2+} waves. Zandonadi et al. [15] showed that maize root architecture is similarly modified by auxin, HA, and sodium nitroprusside (SNP), thus leading to an increase of lateral root abundance, root density, and PM H^+-ATPase activation.

The role of CDPK, NO, Ca^{2+}, and calmodulin were studied in the auxin-induced adventitious roots in cucumber [54]. It was shown that CDPK was stimulated by auxin and depended on NO, Ca^{2+}, and calmodulin during formation of roots.

The present study suggested a new role for HA-stimulated plant physiology that reveals a potentially important aspect of coevolution between humified organic matter and the plant growth and development. Changing the ion fluxes across the root plasma membrane implies modifications in transmembrane electrical potential. The latter is generated by the electrogenic H^+ pumps involved in the ion transport systems as well as by the signal transduction related to root morphogenesis.

Conclusions

Humic matter acts a plant elicitor by triggering a series of cell signaling events. Our results suggest a model for cell signaling in roots that is directly linked to nutrient uptake and lateral root emergence. We have shown that HA induce a positive modulation of PM H^+-ATPase activity and expression that controls the H^+ efflux, root surface pH, and consequently triggers modifications in the anion fluxes. This hyperpolarization, in turn, induces a pH signal that modulates Ca^{2+} transport by increasing cytosolic free Ca^{2+} concentration, which then acts as a second messenger though the mediation of a variety of cellular responses like CDPK activities and anion channel activation. A large anion efflux is activated by the HA negative charges which are built up on the root surface and re-induce H^+-ATPase activity. Within the effort to elucidate the complex HA-plant response mechanism, our results show that HA have an effect on various entry points of the cell signaling machinery, such as fluxes of H^+ and Ca^{2+}, H^+-ATPase and CDPK activity, as well as the expression of Ca^{2+} transporters. Overall, our findings of this work indicate that plant nutrient uptake and growth promotion mediated by humic matter may consist in a pH-dependent phenomenon.

Competing interests
The authors declare that they have no competing interests.

Authors' contributions
ACR carried out the ion flux measurement and drafted the manuscript. LBD carried out the CDPK activity and drafted the manuscript. LAS and MSF carried out the molecular genetic studies and drafted the manuscript.

NOA did the humic acid characterization. FLO and LPC conceived of the study and participated in its design and coordination and helped to draft the manuscript. All authors read and approved the final manuscript.

Acknowledgements
The authors wish to thank Fundação de Amparo à Pesquisa do Estado do Rio de Janeiro (FAPERJ) and Conselho Nacional de Pesquisa (CNPq), INCT for Biological Nitrogen Fixation, and International Foudation of Science (IFS) for financial support.

Author details
[1]Laboratório de Bioquímica e Fisiologia de Microrganismos, Universidade Estadual do Norte Fluminense Darcy Ribeiro (UENF), Av. Alberto Lamego 2000, Campos dos Goytacazes 28013-602, Brazil. [2]Laboratório de Microbiologia Ambiental e Biotecnologia, Universidade de Vila Velha (UVV), Rua Comissário José Dantas de Melo 21, Boa Vista, Vila Velha, Espírito Santo, Brazil. [3]Departamento de Solos da Universidade Federal Rural do Rio de Janeiro (UFRRJ), Seropédica, km 7 BR 465, Seropédica, Rio de Janeiro CEP 23851-970, Brazil. [4]Universidade Estadual do Norte Fluminense Darcy Ribeiro (UENF) Núcleo de Desenvolvimento de Insumos Biológicos para Agricultura (NUDIBA), Av. Alberto Lamego 2000, Campos dos Goytacazes 28013-602, Brazil.

References
1. Orsi M (2014) Molecular dynamics simulation of humic substances. Chem Biol Technol Agr 1:10
2. Nardi S, Carletti P, Pizzeghello D, Muscolo A (2009) Biological activities of humic substances. In Senesi N,Xing B Huang PM (ed) Biophysico-Chemical Processes Involving Natural Non Living Organic Matter in Environmental Systems. Vol 2, part 1: fundamentals and impact of mineral-organic biota interactions on the formation, transformation. Turnover and storage of natural nonliving organic matter (NOM). Wiley, Hoboken, pp305-340
3. Canellas LP, Olivares FL (2014) Physiological responses to humic substances as plant growth promoter. Chem Biol Technol Agr 1:3
4. Mora V, Bacaicoa E, Zamarreño AM, Agguirre E, Garnica M, Fuentes M, García-Mina JM (2010) Action of humic acid on promotion of cucumber shoot growth involves nitrate-related changes associated with the root-to-shoot distribution of cytokinins, polyamines and mineral nutrients. J Plant Physiol 167:633–642
5. Canellas LP, Dobbss LB, Oliveira AL, Chagas JG, Aguiar NO, Rumjanek VM, Novotny EH, Olivares FL, Spaccini R, Piccolo A (2012) Chemical properties of humic matter as related to induction of plant lateral roots. Eur J Soil Sci 63:315–324
6. Mora V, Bacaicoa E, Baigorri R, Zamarreño AM, García-Mina JM (2014) NO and IAA key regulators in the shoot growth promoting action of humic acid in Cucumis sativus L. J Plant Growth Regul 33:430–439
7. Dobbss LB, Canellas LP, Olivares FL, Aguiar NO, Peres LEP, Azevedo M, Spaccini R, Piccolo A, Façanha AR (2010) Bioactivity of chemically transformed humic matter from vermicompost on plant root growth. J Agr Food Chem 58:3681–3688
8. Feijó JA, Sainhas J, Hackett GR, Kunkel JG, Hepler PK (1999) Growing pollen tubes posses a constitutive alkaline band in the clear zone and a growth-dependent acidic tip. J Cell Biol 144:483–496
9. Konrad KR, Wudick MM, Feijó JA (2012) Calcium regulation of tip growth: new genes for old mechanisms. Curr Opin Plant Biol 14:721–730
10. Felle HH (2001) pH: signal and messenger in plant cells. Plant Biol 3:577–591
11. Rayle DL, Cleland RE (1992) The acid growth theory of auxin-induced cell elongation is alive and well. Plant Physiol 99:1271–1274
12. Rober-Kleber N, Albrechtová JTP, Fleig S, Huck N, Michalke W, Wagner E, Speth V, Neuhaus G, Fischer-Iglesias C (2003) Plasma membrane H^+-ATPase is involved in auxin-mediated cell elongation during wheat embryo development. Plant Physiol 131:1302–1312
13. Fasano JM, Swanson SJ, Blancaflor EB, Dowd PE, Kao T, Gilroy S (2001) Changes in root cap pH are required for the gravity response of the Arabidopsis root. Plant Cell 13:907–92
14. Trevisan S, Pizzeghello D, Ruperti B, Francioso O, Sassi A, Palme K, Quaggiotti S, Nardi S (2010) Humic substances induce lateral root formation

and expression of the early auxin-responsive IAA19 gene and DR5 synthetic element in *Arabidopsis*. Plant Biol 12:604–614

15. Zandonadi DB, Santos MP, Dobbss LB, Olivares FL, Canellas LP, Binzel ML, Okorokova-Façanha AL, Façanha AR (2010) Nitric oxide mediates humic acids-induced root development and plasma membrane H⁺-ATPase activation. Planta 231:1025–1036

16. Palmgren MG (2001) Plant plasma membrane H⁺-ATPases: powerhouses for nutrient uptake. Annl Rev Plant PhysiolPlant Mol Biol 52:817–845

17. Canellas LP, Façanha AO, Olivares FL, Façanha AR (2002) Humic acids isolated from earthworm compost enhance root elongation, lateral root emergence, and plasma membrane H⁺-ATPase activity in maize roots. Plant Physiol 130:1951–1957

18. Quaggiotti S, Rupert B, Pizzeghello D, Francioso O, Tugnoli V, Nardi S (2004) Effect of low molecular size humic substances on nitrate uptake and expression of genes involved in nitrate transport in maize (*Zea mays* L.). J Exp Bot 55:803–813

19. Salomon D, Bonshtien A, Sessa G (2009) A chemical-genetic approach for functional analysis of plant protein kinases. Plant Signal Behav 4:645–647

20. Zhang Y, Gao P, Yuan JS (2010) Plant protein-protein interaction network and 58. Interactome Curr Genomics 11:40–46

21. Kudla J, Batistic O, Hashimoto K (2010) Calcium signals: the lead currency of plant information processing. Plant Cell 22:541–563

22. Hamada H, Kurusu T, Okuma E, Nokajima H, Kiyoduka M, Koyano T, Sugiyama Y, Okada K, Koga J, Saji H, Miyao A, Hirochika H, Yamane H, Murata Y, Kuchitsu K (2012) Regulation of a proteinaceous elicitor-induced Ca²⁺ influx and production of phytoalexins by a putative voltage-gated cation channel, OsTPC1, in cultured rice cells. J Biol Chem 23:9931–0039

23. Pei ZM, Murata Y, Benning G, Thomine S, Klusener B, Allen GJ, Grill E, Schroeder JI (2000) Calcium channels activated by hydrogen peroxide mediate abscisic acid signalling in guard cells. Nature 406:731–734

24. Lowry OH, Rosebrough NJ, Farr AL, Randall RJ (1951) Protein measurement with the Folin phenol reagent. J Biol Chem 193:265–275

25. Ramos AC, Façanha AR, Lima PT, Feijó JA (2008) pH signature for the responses of arbuscular mycorrhizal fungi to external stimuli. Plant SignalBehav 3:850–852

26. Ramos AC, Lima PT, Dias PN, Kasuya MCM, Feijó JA (2009) A pH signaling mechanism involved in the spatial distribution of calcium and anion fluxes in ectomycorrhizal roots. New Phytol 181:448–462

27. Lino B, Baizabal-Aguirre VM, González de la Vara LE (1998) The plasma-membrane H⁺-ATPase from beet root is inhibited by a calcium-dependent phosphorylation. Planta 204:352–359

28. Gao J, Liu J, Li B, Li Z (2001) Isolation and purification of functional total RNA from blue-grained wheat endosperm tissues containing high levels of starches and flavonoids. Plant Mol Biol Rep 19:185–1185

29. Sperandio MVL, Santos LA, Bucher CA, Fernandes MS, Souza SR (2011) Isoforms of plasma membrane H + -ATPase in rice root and shoot are differentially induced by starvation and resupply of NO₃⁻ or NH₄⁺. Plant Sci 180:251–258

30. Livak KJ, Schmittgen TD (2001) Analysis of relative gene expression data using real-time quantitative PCR and the 2 − ΔΔCt method. Methods 25:402–408

31. Kurusu T, Sakurai Y, Miyao A, Hirochika H, Kuchitsu K (2004) Identification of a putative voltage-gated Ca²⁺-permeable channel (OsTPC1) involved in Ca²⁺ influx and regulation of growth and development in rice. Plant Cell Physiol 45:693–702

32. Arango M, Gevaudant F, Oufattole M, Boutry M (2003) The plasma membrane proton pump ATPase: the significance of gene subfamilies. Planta 216:355–365

33. Wan B, Lin Y, Mou T (2007) Expression of rice Ca²⁺-dependent protein kinases (CDPKs) genes under different environmental stresses. FEBS Lett 581:1179–1189

34. Akimoto-Tomiyama C, Sakata K, Yazaki J, Nakamura K, Fujii F (2003) Rice gene expression in response to Nacetylchitooligosaccharide elicitor: comprehensive analysis by DNA microarray with randomly selected ESTs. Plant Mol Biol 52:537–551

35. Abbasi F, Onodera H, Toki S, Tanaka H, Komatsu S (2004) OsCDPK13, a calcium-dependent protein kinase gene from rice, is induced by cold and gibberellin in rice leaf sheath. Plant Mol Biol 55:541–552

36. Aguiar NO, Novotny EH, Oliveira AL, Rumjanek VM, Olivares FL, Canellas LP (2013) Prediction of humic acids bioactivity using spectroscopy and multivariate analysis. J Geochem Explor 129:95–102

37. Enriquez-Arredondo C, Sanchez-Nieto S, Rendon-Huerta E, Gonzalez-Halphen D, Gavilanes-Ruiz M, Diaz-Pontones D (2005) The plasma membrane H⁺-ATPase of maize embryos localizes in regions that are critical during the onset of germination. Plant Sci 169:11–19

38. Hager A (2003) Role of the plasma membrane H⁺-ATPase in auxin induced elongation growth. Historical and new aspects. J Plant Res 116:483–505

39. Churchill KA, Sze H (1984) Anion-sensitive, H⁺ pumping ATPase of oat roots: direct effects of Cl⁻, NO₃⁻ and a disulfonic stilbene. Plant Physiol 76:490–497

40. Ullrich CI, Novacky AJ (1990) Extra- and intracellular pH and membrane potential changes induced by K+, Cl⁻, H2PO4-, NO3-, uptake and fusicoccin in root hairs of Limnobium storoniferum. Plant Physiol 94:1561–1567

41. Zimmermann S, Thomine S, Guern J, Barbier-Brygoo H (1994) An anion current at the plasma membrane of tobacco protoplasts shows ATP-dependent voltage regulation and is modulated by auxin. Plant J 6:707–716

42. Vieira L, Slotki I, Cabantchik ZI (1995) Chloride conductive pathways which support electrogenic H⁺ pumping by *Leishmania major* Promastigotes. J Biol Chem 270:5299–5304

43. Peters WS, Felle HH (1999) The correlation of profiles of surface pH and elongation growth in maize roots. Plant Physiol 121:905–912

44. Garnett TP, Shabala SN, Smethurst PJ, Newman IA (2001) Kinetics of ammonium and nitrate uptake by eucalypt roots and associated proton fluxes measured using ion selective microelectrodes. Funct Plant Biol 30:1165–1176

45. Canellas LP, Teixeira Junior LRL, Dobbss LB, Silva CA, Médici LO, Zandonadi DB, Façanha AR (2008) Humic acids cross interactions with root and organic acids. Ann Appl Biol 153:157–166

46. Ohno T, Nakahira S, Suzuki Y, Kani T, Hara T, Koyama H (2004) Molecular characterization of plasma membrane H⁺ATPase in a carrot mutant cell line with enhanced citrate excretion. Plant Physiol 122:265–274

47. Tomasi N, Kretzschmar T, Espen L, Weisskopf L, Thoe Fuglsang A, Palmgren G, Neumann G, Varanini Z, Pinton R, Martinoia E, Cesco E (2009) Plasma membrane H⁺ATPase-dependent citrate exudation from cluster roots of phosphate-deficient white lupin. Plant Cell Environ 32:465–475

48. Puglisi E, Fragoulis G, Del Re AM, Spaccini R, Gigliotti G, Said-Pullicino D, Trevisan M (2008) Carbon deposition in soil rhizosphere following amendments with soluble fractions, as evaluated by combined soil-plant rhizobox and reporter gene systems. Chemosphere 73:1292–1299

49. Thion L, Mazars C, Nacry P, Bouchez D, Moreau M, Ranjeva R, Thuleau P (1998) Plasma membrane depolarizationactivated depolarizationactivated calcium channels, stimulated by microtubule-depolymerizing drugs in wild-type Arabidopsis thaliana protoplasts, display constitutively large activities and a longer half-life in ton 2 mutant cells affected in the organization of cortical microtubules. Plant J 13:603–610

50. Forde BG (2000) Nitrate transporters in plants: structure, function and regulation. v. Biochim Biophys Acta 1465:219–235

51. Vera-Estrella R, Barkla BJ, Higgins VJ, Blumwald E (1994) Plant defense response to fungal pathogens. Activation of host-plasma membrane H⁺-ATPase by elicitor-induced enzyme dephosphorylation. Plant Physiol 104:209–215

52. Correa-Aragunde N, Graziano M, Lamattina L (2004) Nitric oxide plays a central role in determining lateral root development in tomato. Planta 218:900–905

53. Kolbert Z, Bartha B, Erdei L (2008) Exogenous auxin-induced NO synthesis is nitrate reductase-associated in *Arabidopsis thaliana* root primordia. J Plant Physiol 165:967–975

54. Lanteri ML, Pagnussat GC, Lamattina L (2006) Calcium and calcium-dependent protein kinases are involved in nitric oxide- and auxin-induced adventitious root formation in cucumber. J Exp Bot 57:1341–1351

12

Variations in level of oil, protein, and some antioxidants in chickpea and peanut seeds

Vesna Dragičević[1*], Suzana Kratovalieva[2], Zoran Dumanović[1], Zoran Dimov[3] and Natalija Kravić[1]

Abstract

Background: Chickpea and peanut are two legume species not frequently used in human diets. Chickpea is rich in starch and proteins, while peanut is mainly a source of oils and proteins and they could be successfully used as protein sources in vegetarian diets.

Seeds of 19 chickpea and 13 peanut landraces were colorimetrically analyzed in respect to antioxidant content (i.e., free soluble phenolics, total glutathione, and phytate). Oil and protein contents in grain were also determined.

Results: Free soluble phenolics content varied in range from 520 to 1,050 mg kg^{-1} in peanut and from 720 to 1,370 mg kg^{-1} in chickpea. Total glutathione content ranged from 1,495 to 2,365 mmol kg^{-1} in peanut and from 955 to 1,232 mmol kg^{-1} in chickpea. Relatively low content of phytic phosphorus was found in grain of both species, ranging from 2.5 to 4.5 g kg^{-1} in peanut and from 1.4 to 3.0 g kg^{-1} in chickpea, respectively. Considering the lack of data for phytate variability in Macedonian chickpea and peanut local landraces up to date, the observed high variation in phytic phosphorus content could represent the great basis for further breeding programs for phytate decrease in seeds of those genotypes. This is significant, since phytate is an important antinutrient which affects availability of mineral elements. Regression analysis revealed positive and highly significant interdependence between oil content and total glutathione in chickpea seeds, as well as between oil content and phytic phosphorus in peanut seeds. In chickpea, significant and negative correlation between oil and phytic phosphorus content was also observed.

Conclusions: Results obtained indicated that chickpea genotypes with higher oil content could have increased nutritional value due to higher glutathione and lower phytate content observed. However, lower level of phytate content, along with higher level of soluble phenolics and total glutathione found in peanut seeds with lower oil content, indicated higher digestibility and increased antioxidant activity of those genotypes.

Keywords: Antioxidants; Nutritive value; Phenolics; Phytic phosphorus; Total glutathione

Background

Chickpea is a valued legume in Afro-Asian countries due to its nutritive seed composition high in protein content and of better protein quality compared to other legumes, thus increasingly used as a substitute for animal protein. Except of sulfur-containing amino acids, chickpea is rich in all the essential amino acids, being with a balanced content [1]. It has been shown that *in vitro* protein digestibility from protein isolates ranged between 95.6% and 96.1% [2]. Besides, it is also important to emphasize the high antioxidant activity of protein hydrolysates in chickpea [3]. Chickpea seeds contain less than 7% of oil, with linoleic and oleic acid as predominant

[4]. According to Chitra et al. [5], chickpea contains relatively low phytic acid content, compared to other legumes. Significant and negative correlation between phytic acid and *in vitro* digestibility made chickpea seeds necessary in human diet. Relatively high genotypic variability is present in chickpea seed composition, including phytic acid and other antioxidants. Rincón et al. [6] found that Desi biotypes revealed lower fat and phytic acid contents, whereas Kabuli biotypes showed lower total dietary fiber, insoluble dietary fiber, and tannin content. Leading role in protection of chickpea seeds against fungal attack was given to phenolic substances [7], which could also contribute to increased nutritional value of chickpea seeds. Beside of particular nutritive parameter, larger seed with light color are considered as desirable traits for chickpea breeding programs [8].

* Correspondence: vdragicevic@mrizp.rs
[1]Maize Research Institute, Slobodana Bajića 1, 11185 Zemun Polje, Serbia
Full list of author information is available at the end of the article

Since peanut flour is rich in oil and proteins, it is widely used in different foods: as a replacement for animal source proteins, in breakfast snack foods and cereals, as an improver of cereal flours, and it can be used to produce textured vegetable proteins or can be used directly in ground meats to provide adequate moisture and fat binding characteristics [9]. As an oilseed crop, peanut is characterized with proteins of high quantity and quality, as well as with high caloric value. It is also high in phytic acid and contains fibers and perhaps other binding agents which reduce mineral bioavailability from the seeds [10]. Chemical properties of grain are under high genotypic and environmental impact, reflected in induced variations in oil content, individual fatty acid contents, and derived oil quality parameters [11]. Dwivedi et al. [12] reported significant and negative correlation between oil and protein contents, as well as significant linear increase in oil content followed by seed mass increase. However, no such relationship was observed for protein content. Similarly to chickpea, phenolics from peanut seeds and particularly peanut skin have high antifungal and antioxidative activity [13-15]. Positive effect of peanut on human health can be confirmed by the studies of Emekli-Alturfan et al. [16], who ascertained that addition of peanut to the diet did not significantly change blood lipids, protrombin time, activated partial thromboplastin time, or fibrinogen levels, both in control and in hyperlipidemic groups. Peanut consumption improved glutathione (GSH) and high-density lipoprotein (HDL-C) levels and decreased thiobarbituric acid reactive substances (TBARS), without increasing other blood lipids in experimental hyperlipidemia.

As previously mentioned, peanut seeds are rich in phytic acid which is a strong chelating agent that can bind mono- and divalent metal ions, inducing poor bioavailability of minerals such as zinc, calcium, magnesium, iron, and phosphorus [10]. On the other hand, Chung and Champagne [17] found that phytic acid formed insoluble complexes with the major peanut allergens, resulting in peanut extract with reduced allergenic potency and suggested that phytic acid may find its use in the development of hypoallergenic peanut-based products.

According to findings reported, some of the antinutrients may play important beneficial roles in human diets by acting as anticancerogens or by promoting health in other ways such as in decreasing the risk of heart disease or diabetes. Thus, plant breeders and molecular biologists should be aware of the possible negative consequences of changing antinutrients in major plant foods [18]. In parallel, polyphenols, as well-known classes of phytochemicals, are considered to be important components in human diet. Several studies on cancer cell lines and animal models of carcinogenesis have shown that a wide range of polyphenols possess anticancerogenous

and apoptosis-inducing properties [19]. GSH, as protein antioxidant, has important role in free radical scavenging, prevention from stress [20] and could reduce activity of trypsin inhibitors (e.g., Kunitz trypsin inhibitor) [21].

Caloric value of grain, rich in proteins and phytochemicals, enables chickpea and peanut to be broadly and successfully used in vegetarian diets. Hence, to achieve the lower antinutrient content and adequate level of nutritive factors in grain is of great importance.

Those findings prompted us to evaluate a set of 19 chickpea and 13 peanut local landraces in order to determine the content of the main seed constituents (i.e., oil and proteins), as well as seed antioxidant content (i.e., phytate, free soluble phenolics, and GSH). The aim of this investigation was to select the most promising genotypes as sources for further breeding programs for grain quality increase.

Methods

Average sample of each landrace was presented with 100 uniform seeds. Samples were milled on Perten 120 - Sweden (particle size <500 μm). Oil content was determined as subtraction after extraction with petroleum ether. For protein determination, samples (4 × of 0.20 g) were digested with 5 ml of mixture $H_2SO_4 + H_3PO_4$ (50:1) with addition of 2.5 ml H_2O_2 on 420°C. After that, micro-Kjeldahl procedure [22] was applied for protein determination. Phytic P (P_{phy}) and total GSH content were determined after extraction: four replicates of each sample (0.25 g) were treated with 10 mL of 5% trichloroacetic acid for 1 h at room temperature in a rotary shaker. The extract was centrifuged on 14,000 rpm for 15 min, and the supernatant was decanted and diluted. Phytic P was determined colorimetrically by the method of Dragičević et al. [23], based on the pink colour of the Wade reagent ($FeCl_3$ + 5-sulfosalicylic acid), formed upon the reaction of ferric ion and sulfosalicylic acid. The absorbance of reaction product was determined at 500 nm. GSH was determined from the same extract as P_{phy}, by adding 0.2 M potassium phosphate buffer (pH = 8.0) and 10 mM DTNB (5.5′-dithio(2-nitrobenzoic acid)) and measuring the absorbance at 415 nm [24]. Free soluble phenolics were determined after 1 h extraction with bi-distilled water by method of Simić et al. [25], based on a slightly modified Prussian blue method where 0.05M $FeCl_3$ in 0.1 M HCl and 0.008 M $K_3Fe(CN)_6$ were added to sample solution. After 25 min, the absorbance of the reaction product was determined at 722 nm.

Statistical analysis

All analyses were performed in four measurements ($n = 4$), and the results were presented as mean ± standard deviation (SD). The differences among chickpea and peanut local landraces, based on mean values of observed parameters, were evaluated using regression analysis and principle component analysis (PCA). Statistical analysis was performed by SPSS 15.0 for Windows Evaluation version.

Results and discussion

Results presented in Table 1 showed that 1,000 seed weight varied in wide range for both species: from 211.7 to 363.6 g for chickpea landraces and from 420.5 to 661.4 g for peanut genotypes. According to Toker and Cagirgan [26], seed weight of chickpea is negatively correlated with yield, thus hindering the breeding for high yielding plants, particularly those with larger seeds. However, weight of peanut seeds is a trait, highly influenced by genotypic effect in compare to other factors, including stress [27].

Variations in oil content were insignificant for chickpea seed, ranging from 4.44% to 5.16%, while for peanut seeds, those variations were higher (i.e., from 43.5% to 52.4%). Our findings were in line with Zia-Ul-Haq et al. [4], who also reported less than 7% of oil content in chickpea seeds. Besides the differences in oil content between chickpea and peanut, there were also the differences in protein content: in peanut seed, protein content was higher, varying in wider range (i.e., from 18.4% to 29.1%) compared to chickpea seeds, where it varied from 11.3% to 17.6%. Obtained results indicated that seeds of examined peanut landraces could be considered as high oil and protein food, consisting of about 66.6% to 74.2% oil + protein content. Dwivedi et al. [12] ascertained that oil and protein content negatively correlate in peanut seeds, which was confirmed with our results, where high oil genotypes, like P1, P7, P8, P9, and P10 have also the lowest protein content. Compared to genotypic variations, environmental factors have shown to have more pronounced effect on variations in oil content of peanut seeds [11].

Group of biomolecules such as antioxidants could additionally improve the nutritional value of produced seeds. Phytic acid, as important antioxidant, could be also considered as antinutrient. According to the results presented in Table 2, examined chickpea landraces have P_{phy} in wide range from 2.39 to 4.46 mg g^{-1}, while in peanut seeds, P_{phy} content ranged from 1.44 to 2.96 mg g^{-1}, being in average by 27% lower compared to chickpea. Since Duhan et al. [28] also determined that chickpea seeds are rich in phytic acid, they recommended soaking, cooking, autoclaving, or sprouting as methods for successful phytate degradation. However, Chung and Champagne [17] found that phytic acid formed complexes with main allergenic proteins from peanut seeds and even suggested that addition of phytic acid to meals that contain peanut could reduce its allergenic properties. Nevertheless, relatively lower P_{phy} implied potentially

Table 1 Weight of 1000 seed, oil and protein content in investigated chickpea and peanut local landraces

Local landraces	1,000 seed weight (g)	Oil (%)	Protein (%)
Chickpea			
C1	330.1 ± 29.9*	4.69 ± 0.43	13.68 ± 0.32
C2	270.3 ± 28.7	4.85 ± 0.49	12.36 ± 0.46
C3	286.6 ± 28.0	4.99 ± 0.43	12.60 ± 0.23
C4	325.7 ± 28.2	4.89 ± 0.46	12.39 ± 0.12
C5	293.2 ± 30.2	4.72 ± 0.46	13.70 ± 0.39
C6	339.5 ± 35.9	4.66 ± 0.66	15.19 ± 0.30
C7	227.5 ± 30.7	4.44 ± 0.74	13.22 ± 0.33
C8	237.8 ± 28.8	5.16 ± 0.48	13.02 ± 0.36
C9	302.1 ± 32.1	4.95 ± 0.43	14.61 ± 0.29
C10	278.3 ± 26.5	4.67 ± 0.43	12.18 ± 0.82
C11	211.7 ± 29.1	4.78 ± 0.64	12.81 ± 0.44
C12	310.1 ± 25.8	4.81 ± 0.53	11.26 ± 0.15
C13	280.1 ± 31.4	5.14 ± 0.76	12.82 ± 0.26
C14	296.0 ± 33.8	4.44 ± 0.82	14.60 ± 0.31
C15	286.4 ± 35.5	4.91 ± 0.56	15.17 ± 0.38
C16	341.3 ± 33.9	4.74 ± 0.45	14.80 ± 0.29
C17	311.1 ± 37.2	4.65 ± 0.70	16.43 ± 0.23
C18	363.6 ± 39.6	4.64 ± 0.50	17.63 ± 0.12
C19	320.8 ± 32.6	4.71 ± 0.46	14.88 ± 0.31
Peanut			
P1	460.0 ± 40.6	52.39 ± 0.40	18.56 ± 0.14
P2	505.7 ± 53.2	45.29 ± 0.29	24.51 ± 0.20
P3	487.8 ± 51.0	47.44 ± 0.30	23.96 ± 0.35
P4	426.1 ± 58.0	43.50 ± 0.31	26.77 ± 0.13
P5	529.4 ± 59.3	45.22 ± 0.51	26.84 ± 0.15
P6	420.5 ± 61.8	45.03 ± 0.41	28.67 ± 0.17
P7	532.3 ± 44.0	48.89 ± 0.51	19.71 ± 0.08
P8	463.3 ± 46.6	48.52 ± 0.48	21.21 ± 0.15
P9	504.1 ± 17.9	48.18 ± 0.44	18.37 ± 7.19
P10	498.1 ± 46.2	48.54 ± 0.43	20.97 ± 0.04
P11	482.8 ± 47.8	46.83 ± 0.43	21.94 ± 0.14
P12	661.4 ± 63.3	44.32 ± 0.35	29.10 ± 0.17
P13	523.2 ± 58.9	47.33 ± 0.42	26.91 ± 0.14

*The results are represented as mean ± SD (standard deviation) in four measurements.

higher quality of peanut seeds. According to results of our investigation, chickpea genotypes with $P_{phy} < 2.5$ mg g^{-1}, such as C2, C3, C9, and C10 landraces, could be used for breeding program for further phytate decrease, i.e., increased bioavailability of mineral elements.

GSH is protein, but it is not obligatory that seeds rich in proteins have high content of thiolic groups and GSH. Dragičević et al. [29] underlined the importance of GSH

Table 2 Phytic P, total glutathione, and phenolics content in seeds of chickpea and peanut local landraces

Local landraces	P_{phy} (mg g^{-1})	GSH (nmol g^{-1})	Phenolics (µg g^{-1})
Chickpea			
C1	2.57 ± 0.09*	1,496.8 ± 21.47	522.7 ± 12.8
C2	2.43 ± 0.08	1,687.3 ± 8.27	571.2 ± 25.9
C3	2.45 ± 0.08	2,197.5 ± 14.34	670.8 ± 3.2
C4	2.77 ± 0.04	1,652.1 ± 13.21	670.6 ± 16.8
C5	2.86 ± 0.02	1,614.2 ± 757.64	678.5 ± 26.9
C6	3.85 ± 0.10	2,008.6 ± 8.25	686.7 ± 7.4
C7	4.04 ± 0.08	1,820.2 ± 10.61	648.5 ± 7.4
C8	2.72 ± 0.07	2,354.8 ± 54.99	962.6 ± 34.8
C9	2.39 ± 0.03	2,364.3 ± 12.60	1,026.6 ± 42.0
C10	2.45 ± 0.02	2,302.2 ± 22.79	1,075.1 ± 25.6
C11	3.48 ± 0.05	2,035.3 ± 4.87	951.0 ± 14.1
C12	3.02 ± 0.12	1,936.2 ± 18.18	851.0 ± 10.2
C13	4.12 ± 0.10	2,099.9 ± 14.58	789.2 ± 15.7
C14	4.46 ± 0.05	1,976.5 ± 762.23	932.0 ± 27.7
C15	3.68 ± 0.05	2,314.0 ± 7.69	1,005.8 ± 52.3
C16	3.01 ± 0.05	2,066.6 ± 20.62	1,051.1 ± 28.7
C17	4.16 ± 0.04	2,210.5 ± 15.15	909.6 ± 30.5
C18	3.39 ± 0.09	1,953.9 ± 6.97	984.4 ± 39.8
C19	3.08 ± 0.02	1,891.4 ± 11.68	992.3 ± 26.4
Peanut			
P1	2.75 ± 0.11	996.5 ± 4.58	801.1 ± 16.8
P2	1.47 ± 0.11	1,002.0 ± 19.88	812.7 ± 553.8
P3	1.44 ± 0.20	1,034.4 ± 8.43	727.3 ± 26.0
P4	1.62 ± 0.23	1,078.4 ± 5.99	993.2 ± 15.7
P5	2.45 ± 0.14	968.3 ± 8.16	916.2 ± 14.8
P6	2.68 ± 0.11	988.6 ± 3.72	751.2 ± 28.6
P7	2.32 ± 0.15	907.5 ± 5.82	1,036.1 ± 26.8
P8	2.95 ± 0.10	1,042.8 ± 5.75	1,369.6 ± 6.2
P9	2.81 ± 0.19	956.5 ± 5.91	1,286.9 ± 27.1
P10	2.68 ± 0.09	1,008.7 ± 3.31	1,020.8 ± 41.5
P11	2.73 ± 0.13	1,016.4 ± 8.10	983.9 ± 17.9
P12	1.75 ± 0.13	1,006.5 ± 2.94	1,214.2 ± 7.7
P13	2.96 ± 0.10	1,232.5 ± 4.12	941.4 ± 20.2

*The results are represented as mean ± SD (standard deviation) in four measurements. P_{phy}, phytic P; GSH, total glutathione.

and other thiolic proteins in soybean grain. In this study, GSH from peanut seeds ranged from 727.3 to 1,369.6 nmol g^{-1} (Table 2), while in chickpea seeds, it was in range from 1,496.8 to 2,364.3 nmol g^{-1}, contributing to higher antioxidant potential of chickpea proteins. Obtained results could be partly supported by the finding of Li et al. [3], who ascertained high antioxidant activity

of chickpea protein hydrolysates. Also, relatively high GSH level in examined chickpea landraces (particularly in C8, C9, C10, and C15) could be considered as a good source of thiolic amino acids, contrary to Jukanti et al. [1], who established poor sourcing of chickpea grains with thiolics.

Soluble phenolics content varied in high range in seeds of both species, with slightly higher values observed in peanut landraces. In chickpea seeds, phenolics varied in range from 522.7 to 1,075.1 µg g^{-1} and in peanut seeds in range from 727.3 to 1,369.6 µg g^{-1}. Chérif et al. [7] underlined the importance of phenolics as antifungal factor for chickpea, as well as Yu et al. [14] and Nepote et al. [15] for peanut. The same authors found that peanut skin rich in phenolics content also have high antioxidative activity.

Interactions between main seed constituents, such as oil, proteins, and antioxidants in chickpea seeds, revealed in significant and negative correlation between P_{phy} and oil ($R^2 = 0.159$; Figure 1), as well as in significant and positive correlation between P_{phy} and proteins ($R^2 = 0.165$; Figure 2). Chitra et al. [5] also observed positive correlation between phytate and proteins, but with lower significance compared to other legume seed. This could mean that irrespectively to relatively low oil and protein content in chickpea seed (Table 1), its nutritional quality could be increased by slight oil increase in parallel with phytate decrease. It is also important to underline that this trend was supported by significant and positive interdependence between oil and GSH content. Oppositely, in peanut seeds, highly positive interdependence was found between phytate and oil ($R^2 = 0.273$; Figure 1), as well as between proteins and GSH ($R^2 = 0.149$; Figure 2). In addition, negative interdependence was observed between phytate and protein content ($R^2 = 0.145$). This could indicate that in genotypes with slightly reduced oil and increased protein content, nutritive value of seeds could be higher, due to lower phytate and increased GSH content. This was supported by the findings of Dwivedi et al. [12], who also observed negative correlation between oil and protein content in peanut seeds.

Projection of variables in PCA revealed that in chickpea seeds, GSH and phenolics contributed to PCA1, which explained 39.40% of the total variability (Table 3). The second axis (PCA2), which explained 34.60% of the variation, was defined only with oil content. This means that traits, such as GSH and phenolics, vary simultaneously. In peanut seeds, oil and proteins contributed to PCA1, which explained 46.60% of the total variability while P_{phy} and GSH contributed to PCA2, which explained 23.10% of the variation. According to the results presented, oil and proteins in peanut seeds vary simultaneously, but in opposite directions. Independent to

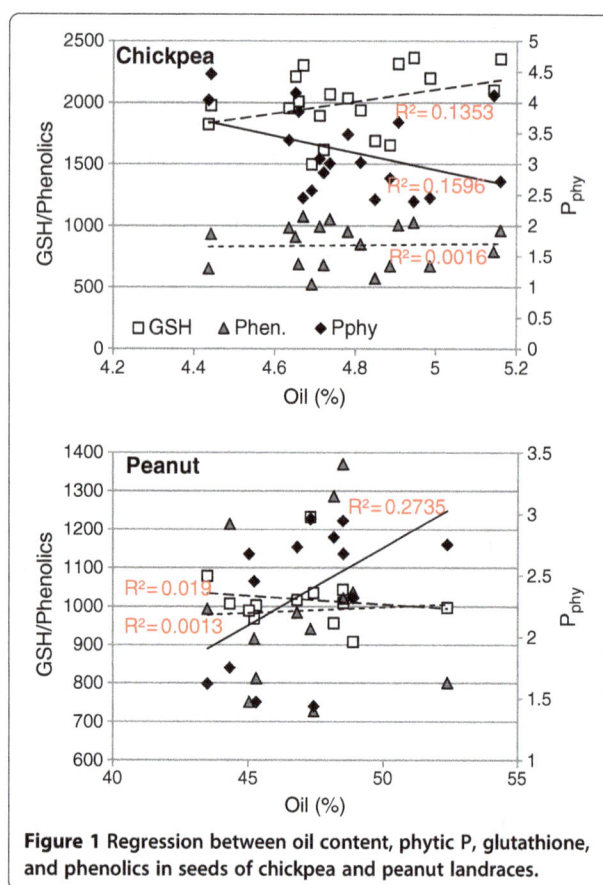

Figure 1 Regression between oil content, phytic P, glutathione, and phenolics in seeds of chickpea and peanut landraces.

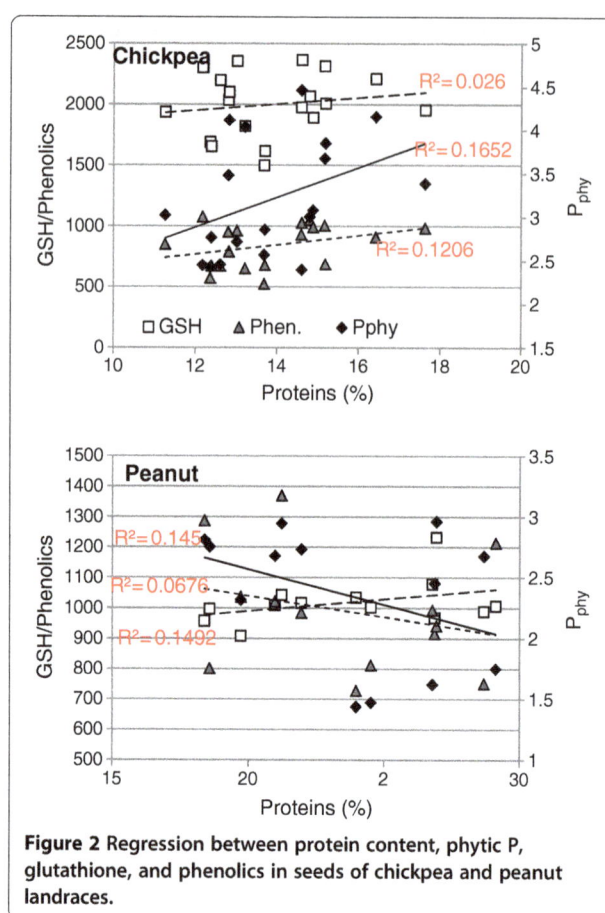

Figure 2 Regression between protein content, phytic P, glutathione, and phenolics in seeds of chickpea and peanut landraces.

main constituents of peanut seed, antioxidants such as phytate and GSH vary simultaneously with higher significance of GSH. This could be important in further breeding programs considering investigated genotypes, since increase in GSH implies P_{phy} increase to some extent.

Experimental

Plant material

Nineteen chickpea and 13 peanut local landraces from Macedonian gene bank (harvested in 2011) were the objectives of the present study.

Conclusions

Based on the results obtained, it could be concluded that investigated chickpea and peanut genotypes could be considered as highly valuable foods, particularly in vegetarian diets. This could underline the necessity for further breeding, especially if increase in the level of antioxidants is taken into account.

Chickpea landraces, with relatively low oil and protein content could be additionally improved by breeding with slight oil increase, which is related to phytate decrease. It is important to observe that protein composition could be enhanced by GSH increase, along with increase of phenolics, which could reflect in increased antioxidative

capacity. Obtained results indicate that C8, CP9, C10, and C15 local landraces could be used as potential source of increased thiolic content in further breeding programs.

Guidelines for further peanut breeding could be associated with oil and/or protein content increase. Parallel increase in proteins and GSH is connected with phytate reduction, which could have positive impact on the availability of mineral elements. However, if positive impact of

Table 3 Principal component analysis for seed chemical composition of chickpea and peanut local landraces

Variable	Chickpea		Peanut	
	PCA1*	PCA2	PCA1	PCA2
Oil	0.097	0.647	0.579	−0.024
Proteins	−0.485	−0.336	−0.600	0.220
P_{phy}**	−0.356	−0.450	0.443	0.522
GSH	−0.522	0.449	−0.213	0.763
Phenolics	−0.597	0.254	0.250	0.313
Explained variance	1.971	1.732	2.329	1.157
Proportion of total variance (%)	39.40	34.60	46.60	23.10

*Synthetic variables: PCA1, principal component axis 1 and PCA2, principal component axis 2. PCA, principal component analysis; **P_{phy}, phytic P; GSH, glutathione.

phytate on lowering of allergenic properties of peanut seeds was taken into consideration, it's increase, along with GSH and phenolic increase, could additionally raise antioxidative properties of peanut seeds.

Competing interests

The authors declare that they have no competing interests.

Authors' contributions

VD and SK designed the research. SK and ZD (Dimov) provided seeds of chosen chickpea and peanut local landraces for the analyses. VD performed chemical analyses. ZD (Dumanović) analyzed the data. VD and NK wrote the paper. All authors read and approved the final manuscript.

Acknowledgements

This study was supported by Project TR31068 'Improving the quality of maize and soybean by conventional and molecular breeding' from the Ministry of Education, Science and Technological Development, Republic of Serbia and by COST Action FA 0905 'Mineral Improved Crop Production for Healthy Food and Feed'.

Author details

[1]Maize Research Institute, Slobodana Bajića 1, 11185 Zemun Polje, Serbia. [2]State Phytosanitary Laboratory, Ministry of Agriculture, Forestry and Water Economy, Aminta Treti 2, 1000 Skopje, Republic of Macedonia. [3]Faculty of Agricultural Sciences and Food, University Ss Cyril and Methodius, Aleksandar Makedonski bb, 1000 Skopje, Republic of Macedonia.

References

1. Jukanti AK, Gaur PM, Gowda CL, Chibbar RN (2012) Nutritional quality and health benefits of chickpea (Cicer arietinum L.): a review. Br J Nutr 108(Suppl 1):S11–S26
2. Sánchez-Vioque R, Clemente A, Vioque J, Bautista J, Millán F (1999) Protein isolates from chickpea (Cicer arietinum L.): chemical composition, functional properties and protein characterization. Food Chem 64(2):237–243
3. Li Y, Jiang B, Zhang T, Mu W, Liu J (2008) Antioxidant and free radical-scavenging activities of chickpea protein hydrolysate (CPH). Food Chem 106(2):444–450
4. Zia-Ul-Haq M, Ahmad M, Iqbal S, Ahmad S, Ali H (2007) Characterization and compositional studies of oil from seeds of desi chickpea (Cicer arietinum L.) cultivars grown in Pakistan. J Am Oil Chem Soc 84(12):1143–1148
5. Chitra U, Vimala V, Singh U, Geervani P (1995) Variability in phytic acid content and protein digestibility of seed legumes. Plant Food Human Nutr 47(2):163–172
6. Rincón F, Martínez B, Ibáñez MV (1998) Proximate composition and antinutritive substances in chickpea (Cicer arietinum L.) as affected by the biotype factor. J Sci Food Agric 78(3):382–388
7. Chérif M, Arfaoui A, Rhaiem A (2007) Phenolic compounds and their role in bio-control and resistance of chickpea to fungal pathogenic attacks. Tunisian J Plant Protect 2:7–21
8. Graham J, Matassa V, Panozzo J, Starick N (2001) Genotype and environment interaction for wholeseed colour in chickpea. In: Proceedings of the 4th European Conference on Seed Legumes, 8-12 July 2001. Cracow, Poland, pp 372-373
9. Yu J, Ahmedna M, Goktepe I (2007) Peanut protein concentrate: production and functional properties as affected by processing. Food Chem 103(1):121–129
10. Erdman JW (1979) Oilseed phytates: nutritional implications. J Am Oil Chem Soc 56(8):736–741
11. Dwivedi SL, Nigam SN, Jambunathan R, Sahrawat KL, Nagabhushanam GVS, Raghunath K (1993) Effect of genotypes and environments on oil content and oil quality parameters and their correlation in peanut (Arachis hypogaea L.). Peanut Sci 20(2):84–89
12. Dwivedi SL, Jambunathan R, Nigam SN, Raghunath K, Ravi Shankar K, Nagabhushanam GVS (1990) Relationship of seed mass to oil and protein contents in peanut (Arachis hypogaea L.). Peanut Sci 17(2):48–52
13. Sanders TH (1977) Changes in tannin-like compounds of peanut fruit parts during maturation. Peanut Sci 4(2):51–53
14. Yu J, Ahmedna M, Goktepe I (2005) Effects of processing methods and extraction solvents on concentration and antioxidant activity of peanut skin phenolics. Food Chem 90(1–2):199–206
15. Nepote V, Grosso NR, Guzmán CA (2005) Optimization of extraction of phenolic antioxidants from peanut skins. J Sci Food Agric 85(1):33–38
16. Emekli-Alturfan E, Kasikci E, Yarat A (2007) Peanuts improve blood glutathione, HDL-cholesterol level and change tissue factor activity in rats fed a high-cholesterol diet. Eur J Nutr 46(8):476–482
17. Chung SY, Champagne E (2007) Effects of phytic acid on peanut allergens and allergenic properties of extracts. J Agric Food Chem 55(22):9054–9058
18. Graham RD, Welch RM, Bouis HE (2001) Addressing micronutrient malnutrition through enhancing the nutritional quality of staple foods: principles, perspectives and knowledge gaps. Adv Agron 70:77–142
19. Khan HY, Zubair H, Ullah MF, Ahmad A, Hadi SM (2012) A prooxidant mechanism for the anticancer and chemopreventive properties of plant polyphenols. Curr Drug Targets 13:1738–1749
20. Santos CVD, Rey P (2006) Plant thioredoxins are key actors in the oxidative stress response. Trends Plant Sci 11:329–334
21. Kobrehel KS, Yee BC, D'Buchanans B (1991) Role of the NADP/thioredoxin system in the reduction of a-amylase and trypsin inhibitor proteins. JBC 266:16135–16140
22. AOAC. (1984) Official Methods of Analysis of the Association of Official Analytical Chemists. S. Williams (Ed.). Association of Official Analytical Chemists, Arlington, Virginia, USA
23. Dragićević V, Sredojević S, Perić V, Nišavić A, Srebrić M (2011) Validation study of a rapid colorimetric method for the determination of phytic acid and inorganic phosphorus from grains. Acta Period Technol 42:11–21
24. Sari Gorla M, Ferrario S, Rossini L, Frova C, Villa M (1993) Developmental expression of glutathione-S-transferase in maize and its possible connection with herbicide tolerance. Euphytica 67:221–230
25. Simić A, Sredojević S, Todorović M, Ðukanović L, Radenović Č (2004) Studies on the relationship between content of total phenolics in exudates and germination ability of maize seed during accelerated aging. Seed Sci Technol 32:213–218
26. Toker C, Cagirgan MI (2004) The use of phenotypic correlations and factor analysis in determining characters for seed yield selection in chickpea (Cicer arietinum L.). Hereditas 140(3):226–228
27. Pandey RK, Herrera WAT, Pendleton JW (1984) Drought response of grain legumes under irrigation gradient: I. Yield and yield components. Agron J 76(4):549–553
28. Duhan A, Chauhan BM, Punia D, Kapoor AC (1989) Phytic acid content of chickpea (Cicer arietinum) and black gram (Vigna mungo): varietal differences and effect of domestic processing and cooking methods. J Sci Food Agric 49(4):449–455
29. Dragićević V, Perić V, Srebrić M, Žilić S, Mladenović-Drinić S (2010) Some nutritional and anti-nutritional factors of ZP soya bean varieties. J Agric Sci 55(2):141–146

13

Preharvest foliar applications of glycine-betaine protects banana fruits from chilling injury during the postharvest stage

Luis C Rodríguez-Zapata[1*], Francisco L Espadas y Gil[1], Susana Cruz-Martínez[1], Carlos R Talavera-May[1], Fernando Contreras-Marin[1], Gabriela Fuentes[1], Enrique Sauri-Duch[2] and Jorge M Santamaría[1]

Abstract

Background: Banana plantations are affected by environmental factors such as chilling injury, which reduces the quality of fruits and causes losses of up to 50% in the yield of banana and it will be more important in terms of global climate change. Chilling injury of the fruits can also occur during transport and storage at low temperatures, particularly in tropical fruits. In banana, losses of up to 20% can occur during postharvest handling. Given this situation, it is necessary to explore alternatives that might reduce chilling injury, such as the use of compatible solutes including glycine-betaine (GB).

Results: In the present work, experiments were performed to analyze the possible role of preharvest foliar applications of GB, to prevent the subsequent development of chilling injury of banana fruits during the postharvest storage at low temperatures. After 3 days of the preharvest application of 100 mM GB over banana leaves (250 ml/plant), the fruits were harvested and first stored at 10°C for 6 h and then transferred to 23°C ± 1°C until they reached commercial ripening. A second group of plants were not treated with GB during the preharvest stage, but their fruits were exposed to 10°C for 6 h before transferring them to 23°C ± 1°C until they reached commercial ripening. A control group was untreated with GB during the preharvest stage, and fruits were not exposed to low temperatures but they were kept at 23°C until they reached commercial ripening.

Conclusions: The results showed that the preharvest foliar application of GB (100 mM) to banana plants reduced the biochemical and physiological alterations caused by chilling injury on harvested fruits.

Keywords: Glycine-betaine; Chilling injury; Banana; Browning

Background

One of the factors that limits the establishment of banana plantations at certain latitudes, and has an impact on the development of plants and fruit production, is chilling damage. This loss in fruit production intensifies at the postharvest stage, because losses between 15% and 25% of production can occur at that stage [1]. Bananas are sensitive to chilling and suffer physiological damage when exposed to environment temperatures below the critical point (11°C) regardless of the type of cultivar [2,3]. It is considered that the primary cause of injuries caused by chilling are the changes in the properties of the cell membrane [4] that causes a cascade of secondary reactions,

which include the increase in the production of ethylene, increase in anaerobic respiration, and reduced photosynthesis [4-6]. Chilling-tolerant plants use evasion and tolerance mechanisms such as the synthesis of low-molecular-weight compounds called compatible solutes [4,7,8]. Compatible solutes include sugars such as trehalose and mannitol, amino acids such as proline, and amino acid derivatives such as glycine-betaine (GB) [9]. Plants capable of accumulating these osmolytes are able to survive under conditions of stress by drought, salinity, and chilling. GB is a quaternary ammonium compound that is a very effective compatible solute [10-12] and is found in a wide range of foods [13]. In plants that synthesize GB, which are known as GB accumulators (spinach, maize, and barley), this compatible solute accumulates in leaves in response to drought and salinity, as well as during

* Correspondence: lcrz@cicy.mx
[1]Unidad de Biotecnología, Centro de Investigación Científica de Yucatán, Mérida 97200, México
Full list of author information is available at the end of the article

acclimation to cold [10,14,15]. Moreover, GB has been shown *in vitro* to stabilize membranes of the oxygen-evolving photosystem II complex [16,17]. GB also stabilizes the activity of ribulose 1,5-bisphosphate carboxylase/oxygenase in a transgenic cyanobacterium *in vivo* [18]. In higher plants, GB is synthesized from choline (Cho) via betaine aldehyde (BA). The first and second steps in the biosynthesis of GB are catalyzed by the enzymes choline monooxygenase (CMO) and betaine aldehyde dehydrogenase (BADH), respectively [10]. There is evidence suggesting that GB plays a role in stress tolerance in some species of plants [19]. However, to date, there are no studies on the effect of GB application on climacteric tropical fruits. It has been reported that the browning (low-temperature injury symptom) in banana peel is caused by the oxidation of polyphenols caused primarily by polyphenol oxidase (PPO) (Yang et al. [20]). Nguyen et al. [21] report that banana fruits exposed to low temperatures experience an increased activity of both phenylalanine ammonia lyase (PAL) and PPO. In chilling-sensitive plants, oxidative stress is a major component of chilling stress and active oxygen species (AOS) such as hydrogen peroxide, superoxide radicals, and hydroxyl radicals that can react very rapidly with DNA, causing severe damage to cellular proteins and lipids [22]. Plants have mechanisms to tolerate different types of stress such as chilling that also causes cellular dehydration. Commonly, plants synthesize compatible soluble compounds to retain water in the cell or to protect cellular components from chilling-induced dehydration [23]. Exogenous applications of GB have been successful in increasing the tolerance to abiotic stresses in various species including *Arabidopsis thaliana* [24-27]. Moreover, the overexpression of genes codifying for one of the enzymes involved in GB biosynthesis resulted in increased tolerance to various abiotic stresses including chilling [28-31].

Therefore, in the present study, it was intended to evaluate a possible protective role of the preharvest foliar applications of GB on the subsequent physiological and biochemical disorders caused by postharvest low-temperature storage of banana fruits (cv Giant Dwarf).

Methods

Plant material

This work was conducted in banana *Musa acuminata* cv 'Giant-Dwarf' plants, grown at the nursery of CICY at Mérida Yucatán, Mexico. The fruits were harvested at the first stage of ripening, according to the Customers Services Department Chiquita Brands, Inc., Cincinnati, OH. [32].

Experimental

Three different groups of five plants each were formed. In the first group (control), distilled water (250 ml/plant) was sprayed over the first two leaves near the inflorescence for three consecutive days prior to harvesting the fruits that were kept at 25°C until they reach commercial ripening. A second group of plants (chilled fruits) was also treated with distilled water for three consecutive days; the fruits were then harvested, but they were exposed to 10°C for 6 h before they were kept at 23°C until they reached commercial ripening. The third group of plants (GB-treated chilled fruits) was treated with 250 ml per plant of a solution of 100 mM GB, sprayed over the first two leaves near the inflorescence, for three consecutive days. The fruits were then harvested and exposed to 10°C for 6 h before they were kept at 23°C until they reached commercial ripening. In all temperature treatments, an incubator (Forma Scientific, Diurnal Growth Chamber, Marietta, Ohio, USA) under dark conditions was used. Samples of five fruits from each treatment were taken every 24 h for 8 or 9 days, to measure the various biochemical and physiological parameters.

Enzymatic activity of polyphenol oxidase (PPO)

Five grams (fresh weight (FW)) of fruit peels from the three different groups were homogenized using a polytron (Poly science model X-520, Niles, Illinois, USA) with 10 ml of 0.1 M sodium phosphate buffer pH 7 (0.1 M phosphate monobasic sodium monohydrate and 0.1 M phosphate dibasic hepta hydrated sodium) with protease inhibitors (5 μg ml^{-1} of aprotinin, 5 μg ml^{-1} leupeptine, and 10 mM of phenyl-methyl sulphonyl fluoride (PMSF)), 10 mM ascorbic acid, and 3% of polyvinyl-polypirrolidone (PVPP). Subsequently, the homogenate was centrifuged to 13,200 rpm for 30 min at 4°C. The specific enzymatic activity of PPO was determined according to Tamayo [33] and Arzápalo [34] and expressed as μmol min^{-1} mg^{-1} protein.

Quantification of soluble phenols

One gram (FW) of fruit peels were crushed in a mortar in the presence of liquid nitrogen until obtaining a fine powder. The macerate was transferred to a test tube and 5 ml of a mixture of chilled ethanol (1:1 *v/v*) was added. The homogenate was agitated for 1 min, and it was placed in ice for 6 h. Subsequently, it was centrifuged to 10,000 rpm (Centrifuge 5415 D Eppendorff) for 10 min at 4°C; the supernatant was then separated, and this fraction was extracted for the quantification of soluble phenols. Phenol determination was made by the reaction of Folin-Ciocalteu reagent as described by Singlenton and Rossi [35] with slight modifications [21].

Soluble protein quantifications

One gram (FW) of fruit peels was macerated in a mortar by adding liquid nitrogen and 1.5% of PVPP (0.015 g ml^{-1}) to obtain a fine powder. The macerate was transferred to an Eppendorf tube containing 2 ml of extraction buffer (50 mM NaCl, 1 mM of EGTA, 50 mM Tris pH 7.4, 250 mM sucrose, 10% glycerol) and protease inhibitors

(5 μg ml^{-1} of aprotinin, 5 μg ml^{-1} leupeptine, and 10 mM PMSF). The homogenate was centrifuged to 10,400 rpm at 4°C for 30 min. The supernatant was used to determine protein content. Spectrophotometric soluble protein quantification was made as described in Bradford [36].

Measurement of chlorophylls and total carotenoids
One gram (FW) of fruit peels was crushed in a mortar with liquid nitrogen. Chilled acetone (80%) was added (10 ml per gram of tissue). The homogenate was vacuum filtered using a funnel with filter paper. The obtained filtrate was left to stand for 30 min under chilling temperatures and darkness, and it was centrifuged for 10 min at 12,000 rpm. Spectrophotometric (Beckman Coulter DU 650, Beckman Coulter, Inc., Fullerton, CA, USA) measurements of the acetonic extract were made at three different wavelengths (470, 645, and 663 nm), and chlorophyll and carotenoids concentrations were calculated according to Wellburn [37].

Ethylene and respiration measurement
Two fruits were placed in a 1-l glass container hermetically closed and remained in the dark at ambient temperature (23°C ± 1°C) for 1 h. Then, 3 ml of the mixture of gases was taken from the glass container and analyzed in a gas chromatograph (Varian Star 3400 cx, Walnut Creek, CA, USA) fitted with two detectors placed in series; the first detector was a thermal conductivity detector (TCD) for CO_2, and the second was a flame ionization detector (FID) for ethylene, using hydrogen and air as flame and helium as carrier gas. Standard mixtures of carbon dioxide (CO_2) 5,000 μmol mol^{-1}, ethylene 50 μmol mol^{-1}, and nitrogen were used for calibration. Production of ethylene and respiratory rate assessments were carried out according to McCollum and Mc Donald

[38] and Liu et al. [39] with some modifications from Chillet and Lapeyre de Bebellaire [40].

Cell membrane stability (CMS); electrolyte leakage
Cell membrane stability (CMS) was measured as electrolyte leakage according to McCollum and McDonald [38] and Prohens et al. [41]. Ten disks (10-cm diameter) of fruit peels from all treatments were placed on 25 ml of mannitol 0.4 M and incubated under agitation for 24 h at room temperature. Samples were sterilized at 121°C to break all tissues and free the electrolytes. Subsequent measurements of electrical conductivity were made with a conductimeter (Orion model 162), using 0.4 M mannitol and de-ionized water solutions as blanks.

Chlorophyll fluorescence
Fruits were exposed to darkness for 30 min prior to taking the fluorescence measurements. Photochemical efficiency of photosystem II (PSII) measured as the ratio of variable fluorescence over maximal fluorescence (Fv/Fm) was determined with a fluorometer (PEA, Hansatech, Norfolk UK) as reported in Maxwell and Johnson [42].

Statistical analysis
Each data point is the mean and SD of at least five fruits. The significant difference between treatments were detected by ANOVA with a 95% confidence level ($P = 0.05$) using StatGraphics Plus 4.1 program. For each parameter measured, ANOVA was performed on the data obtained from the last observation of the experiment.

Results and discussion
Symptoms of chilling damage
Figure 1A shows that in control fruits no change occurred in the coloring of the skin (nor on the subepidermal

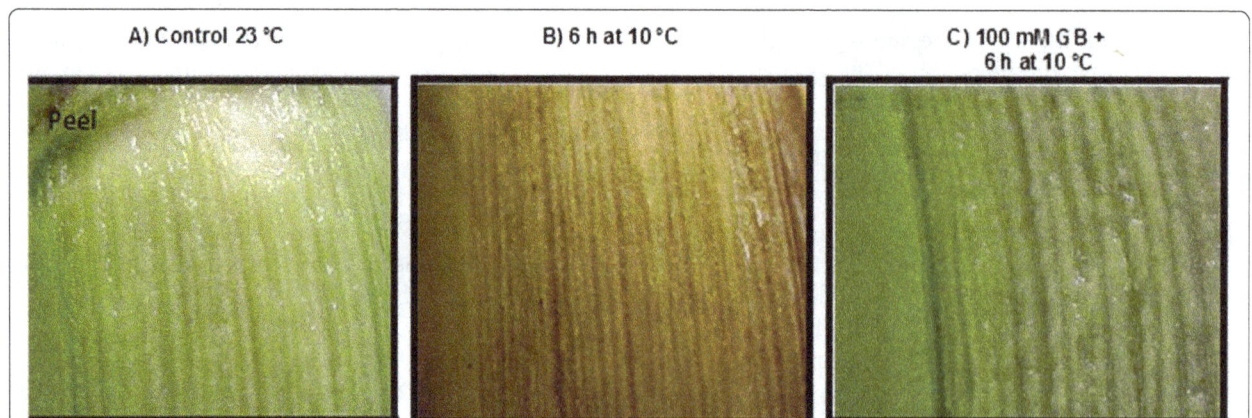

| A) Control 23 °C | B) 6 h at 10 °C | C) 100 mM G B + 6 h at 10 °C |

Figure 1 Sub-epidermal tissue from fruits from the 3 treatments. Photographs taken after 8 days of being harvested. The sub-epidermal tissue of banana fruits from the three different treatments: **A)** control fruits from plants treated with distilled water during the preharvest stage and kept at 23°C, **B)** fruits from plants treated with distilled water during the preharvest stage but exposed to 10°C for 6 h and then kept at 23°C, and **C)** fruits from plants treated during the preharvest stage with 100 mM GB and exposed to 10°C for 6 h and then kept at 23°C.

tissue) within 8 days of being stored at 23°C. When fruits were exposed to 10°C for 6 h (Figure 1B), they showed obvious signs of chilling damage (browning of the subepidermal tissues). However, when GB was applied during the preharvest stage, the postharvest chilling damage effects (browning) on the fruit was prevented (Figure 1C). These results suggest that banana (cv Giant Dwarf) suffers damage when exposed to chilling (10°C) even for a short exposure of 6 h, showing uneven ripening and epidermal browning acceleration; similar results were obtained by Morrelli et al. [43], who worked with five different varieties of banana exposed to temperatures of 5°C, 7°C, and 10°C, but for as long as 7 days.

Our results also suggest that exogenous GB (when applied to leaves prior to harvesting the fruits) may have been translocated to fruits and had prevented chilling-induced subepidemal browning of fruits. No data on endogenous content of GB are available in our experiment; however, in tomato plants, it was shown that GB was translocated to meristematic tissues including flowers and fruits of tomato when applied exogenously to the leaves and that resulted in increased tolerance to chilling [44].

Total phenols and activity of PPO

In control fruits, the concentration of total phenols increased four times as the process of ripening progressed (Figure 2A). However, when fruits were exposed for 6 h to temperatures of 10°C, the contents of phenols only increased two times irrespective of whether they were treated or not with GB during the preharvest stage.

In terms of PPO-specific activity in control fruits, the PPO-specific activity remained almost constant after 8 days of being harvested (Figure 2B). In chilled fruits, the activity of PPO showed a large peak at Day 3 to later return to basal levels by Day 4 and remained low at Day 8 of being harvested (Figure 2B).

On the contrary, for chilled fruits from plants treated with GB during the preharvest stage, although they also showed an increased PPO activity in response to chilling, this was very much attenuated (i.e., 2,500 μmol min^{-1} mg^{-1} protein in GB-treated chilled fruits vs. 6,000 μmol min^{-1} mg^{-1} protein in non-GB-treated chilled fruits).

Our results are in line with findings of other authors in apples and bananas. In apple, some cultivars showed a positive correlation between the degree of browning and enzyme activity. In some cultivars (Classic Delicious, RI Greening, McIntosh, and Cortland), PPO activity was directly related to degree of browning while in others (Empire, Rome, and Golden Delicious), the degree of browning was related more to the final phenolic concentration. High-performance liquid chromatography analysis of the phenolics in apple showed that the types of phenolic compounds in all cultivars were similar and that no one particular compound could account for the

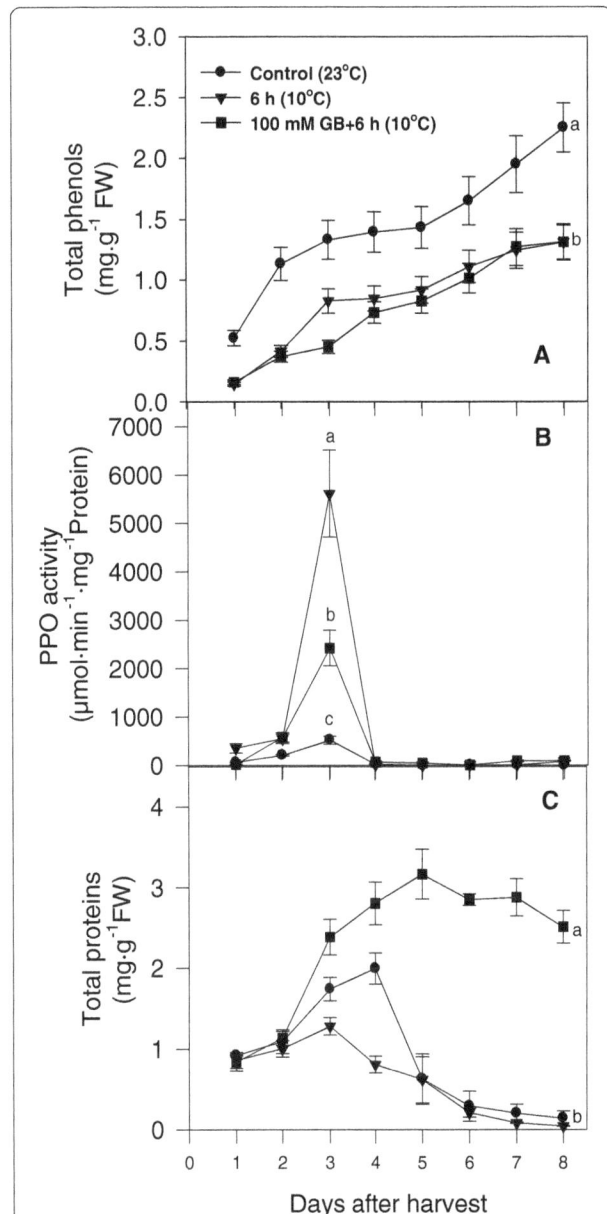

Figure 2 Soluble phenols, specific activity of PPO, and total soluble protens. Soluble phenols **(A)**, specific activity of PPO **(B)**, and total soluble proteins **(C)** from banana fruit peels during 8 days after being harvested from fruits of the three different treatments. control fruits from plants treated with distilled water during the preharvest stage and kept at 23°C (circles), fruits from plants treated with distilled water during the preharvest stage but exposed to 10°C for 6 h and then kept at 23°C (inverted triangles), and fruits from plants treated during the preharvest stage with 100 mM GB and exposed to 10°C for 6 h and then kept at 23°C (squares). Each point is the mean and SD of at least five fruits. Points with different letters at the last date of observation are statistically different.

differences observed in browning [45]. They concluded that the increase of the phenolic compounds and the activity of enzymes could be the factor that causes the increase in the degree of browning of fruits when exposed to low temperatures [45]; although, this was not the case in all varieties tested.

In banana, on the other hand, Nguyen et al. [21] noted that the development of chilling damage, which causes the fruit skin browning, was due to increased activities of the enzymes PAL and PPO. They found that the banana varieties Kluai Khai and Kluai Hom Tom, exposed to temperatures of 6°C and 10°C, had increased activity of PPO and a greater browning in the fruits. In contrast, Nguyen et al. [46] found that, in another variety of banana, PPO activity did not correlate with phenol content but that it was more related to PAL.

Protein content

In control fruits, protein content increased in the first 4 days after removal from the plant but declined afterwards being almost 0, 8 days after being harvested (Figure 2C). The decline in protein content was not associated to the climacteric peak as it occurred at Day 9 in those fruits (see Figure 3). Chilling caused a reduction in the protein content found in fruits 3 and 4 days after being harvested, when compared to control fruits, followed by a similar decline from Days 5 to 8 after harvest. In contrast, the protein content of chilled fruits from plants preharvestly treated with GB not only prevented the protein degradation associated with the exposure to low temperatures but also perhaps increased the rate of protein synthesis or prevented the ripening-related protein degradation, as protein content in those fruits increased about three times (from 1 to 3 mg of protein g^{-1} FW) at Day 5 but remained high 8 days after being harvested. These results strongly point towards a protecting effect of GB on the rate of proteolysis caused by both the ripening process in control fruits and the chilling-induced proteolysis observed in chilled fruits. GB perhaps promoted the synthesis of new proteins related to the protection of the membrane or reduced the rate of protein degradation or both. There are few reports in banana documenting the changes in the pattern of proteins associated with fruit ripening, but certainly, the observed variation in the protein concentration should be related to the synthesis and hydrolysis of proteins involved in the ripening process. Brady and O'Connell [47] reported that much of the increase in early climacteric-phase protein synthesis is the result of an increase in the turnover of preexisting proteins. On the other hand, Hubbard et al. [48] mentioned that during the climacteric phase of banana fruits there is an increased synthesis of sucrose phosphate synthase (SPS).

Pigments content
Total chlorophylls
In control fruits, total chlorophyll content of peels was reduced from 55 to 20 µg g^{-1} FW, only 1 day after being harvested, and decreases further from 20 to 0 µg g^{-1} FW

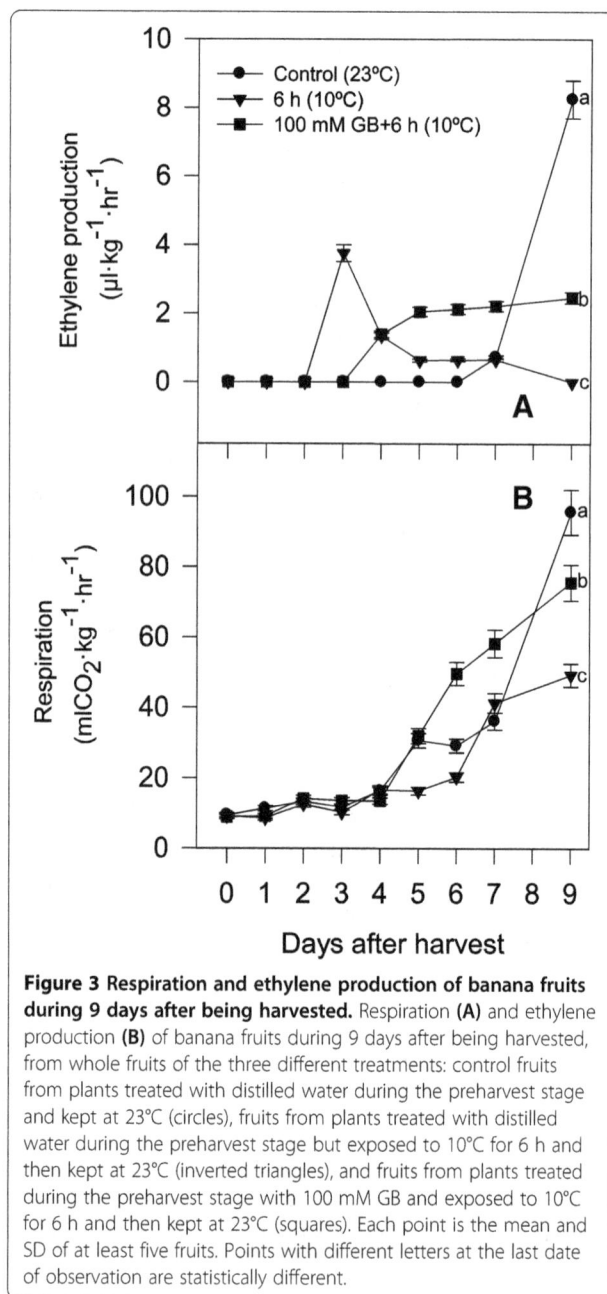

Figure 3 Respiration and ethylene production of banana fruits during 9 days after being harvested. Respiration **(A)** and ethylene production **(B)** of banana fruits during 9 days after being harvested, from whole fruits of the three different treatments: control fruits from plants treated with distilled water during the preharvest stage and kept at 23°C (circles), fruits from plants treated with distilled water during the preharvest stage but exposed to 10°C for 6 h and then kept at 23°C (inverted triangles), and fruits from plants treated during the preharvest stage with 100 mM GB and exposed to 10°C for 6 h and then kept at 23°C (squares). Each point is the mean and SD of at least five fruits. Points with different letters at the last date of observation are statistically different.

in the following 7 days (Figure 4A). In chilled fruits, the exposure to temperatures of 10°C for 6 h caused a reduction in the rate of chlorophyll degradation during the first 5 days after being harvested; however, it falls at Day 6 to values close to those of the control fruits. In contrast, in chilled fruits from plants treated with GB, the chlorophyll degradation rate was the lowest of the three treatments, remaining at values of 30 µg g^{-1} FW from Days 5 to 8 after being harvested (Figure 4A). A protective effect of GB on chlorophyll degradation has been reported previously. Blackbourn et al. [49] reported that GB can be

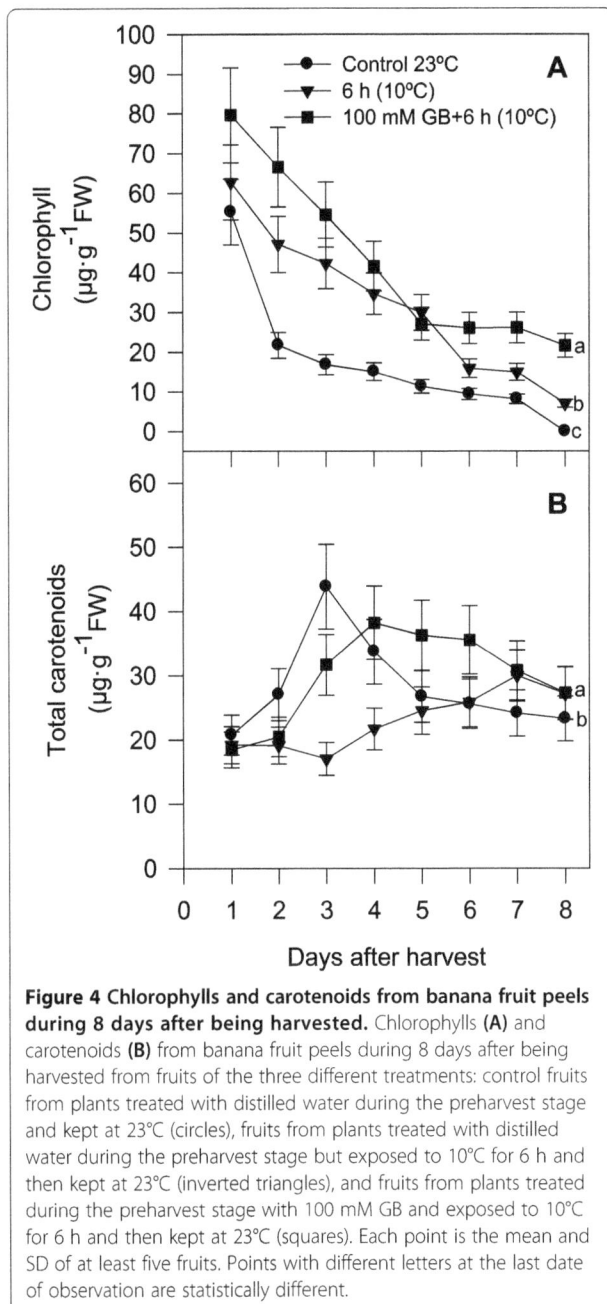

Figure 4 Chlorophylls and carotenoids from banana fruit peels during 8 days after being harvested. Chlorophylls **(A)** and carotenoids **(B)** from banana fruit peels during 8 days after being harvested from fruits of the three different treatments: control fruits from plants treated with distilled water during the preharvest stage and kept at 23°C (circles), fruits from plants treated with distilled water during the preharvest stage but exposed to 10°C for 6 h and then kept at 23°C (inverted triangles), and fruits from plants treated during the preharvest stage with 100 mM GB and exposed to 10°C for 6 h and then kept at 23°C (squares). Each point is the mean and SD of at least five fruits. Points with different letters at the last date of observation are statistically different.

found in chloroplasts and prevents membrane damage caused by peroxidation associated to ROS.

Total carotenoids

In control fruits, the carotenoid content of peels increased from 20 to 45 µg g^{-1} FW during the first 3 days but declined later reaching again values of 20 µg g^{-1} FW, 8 days after being harvested. In chilled fruits, exposure to temperatures of 10°C for 6 h caused no damage on the carotenoid content during the first 3 days after being harvested. However, it increased from Days 5 to 8 to values close to

30 µg g^{-1} FW. In contrast, in chilled fruits from plants treated with GB, during the first 4 days, the carotenoid content increased to values close to those of control fruits, remaining at values around 40 µg g^{-1} FW, 6 days after being harvested and only decreased to values of 30 µg g^{-1} FW at Day 8 (Figure 4B).

Production of ethylene and respiration

In control fruits, no ethylene was detected during the first 6 days after being harvested; ethylene increased slightly at Day 7 after harvest but showed a significant increase (eightfold) 8 days after being harvested. In chilled fruits, the exposure to temperatures of 10°C for 6 h caused an ethylene peak reaching values of 4 µl C_2H_4 Kg^{-1} hr^{-1} as early as Day 3 after being harvested, falling again at Day 4 reaching values near 0 µl C_2H_4 Kg^{-1} hr^{-1} at Day 9 after being harvested.

In contrast, in chilled fruits from plants treated with GB, ethylene also increased by Days 4 and 5 but remained as high as 2 µl C_2H_4 Kg^{-1} hr^{-1} even after 9 days of being harvested (Figure 3A). This result indicates that the chilling caused a premature ethylene peak that resulted in an uneven ripening of the fruit. The preharvest application of GB mitigated this effect, and fruits, despite being subjected to chilling, did not show the premature ethylene peak observed in non-GB-treated chilled fruits. However, chilled fruits from GB-treated plants showed intermediate values of ethylene content between the non-chilled controls and those of the non-GB-treated chilled fruits.

In the case of respiration, in control fruits, low-CO_2 rates (10 ml CO_2 Kg^{-1} hr^{-1}) were detected during the first 4 days after being harvested; they increased slightly during Days 5 to 7, and by Day 9, they showed a significant rise to values near 100 ml CO_2 Kg^{-1} hr^{-1}. Chilled fruits (exposed to temperatures of 10°C for 6 h) again showed low CO_2 during the first 4 days; they did not show an early CO_2 peak as the control fruits did. Respiration rates in chilled fruits only increased to values of 30 ml CO_2 Kg^{-1} hr^{-1} at Day 7 and to values of 50 ml CO_2 Kg^{-1} hr^{-1} by Day 9. In contrast, in chilled fruits from plants treated with GB, CO_2 also remained low during the first 4 days, but from Day 5 onwards, it showed a continuous increase reaching values of 70 ml CO_2 Kg^{-1} hr^{-1} by Day 9 after being harvested (Figure 3B). Those values were again intermediate between the high values observed in control plants and the low values observed in non-GB-treated chilled fruits. There were statistically significant differences between treatments at Day 9. GB perhaps protected proteins involved in the respiration process from chilling injury.

Under conditions of chilling stress, the rate of ethylene and CO_2 production usually increases [50-52]. Our results show that in fruit exposed to chilling, respiration declined 48% with respect to control. While the production of ethylene showed a premature peak causing early and uneven ripening of the fruit. These disturbances in the metabolism of the fruit agree with previous reports [4,53]. According to Jiang et al. [54], in fruits stored at low temperatures, chilling injuries are due to the reduction in the ability to respond to the ethylene signal, and it has an impact on an abnormal maturation of the fruit. Banana fruit exposed to chilling are perhaps less sensitive to ethylene causing abnormal maturation. Climacteric fruit ripening is a process which is coordinated by the presence of ethylene perception by receptors.

Cell membrane stability

Exposure of banana fruits to chilling temperatures resulted in changes in the CMS measured as electrolyte leakage. Chilled fruits had electrolyte leakage values of 51% compared to values of 38% found in control fruits 10 days after being harvested. Chilling did not affect fruits that were previously treated with GB, as they showed a lower electrolyte leakage (29.45%) value that was even lower than that observed in control fruits (Figure 5A). This suggests that GB was protecting the integrity of the membrane perhaps by GB taking action as an osmoregulator that allowed membrane stabilization.

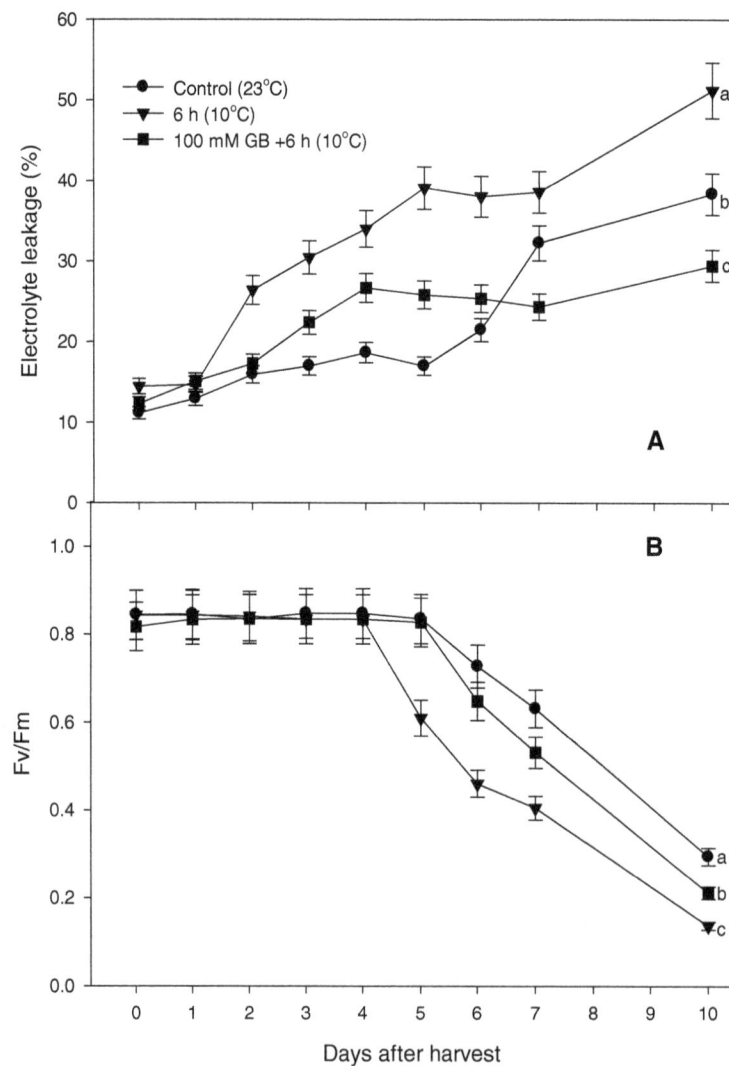

Figure 5 Electrolyte leakage and efficiency of photosystem II from banana fruit peels. Electrolyte leakage **(A)** and efficiency of photosystem II **(B)** from banana fruit peels during 10 days after being harvested from fruits of the three different treatments: control fruits from plants treated with distilled water during the preharvest stage and kept at 23°C (circles), fruits from plants treated with distilled water during the preharvest stage but exposed to 10°C for 6 h and then kept at 23°C (inverted triangles), and fruits from plants treated during the preharvest stage with 100 mM GB and exposed to 10°C for 6 h and then kept at 23°C (squares). Each point is the mean and SD of at least five fruits. Points with different letters at the last date of observation are statistically different.

Other authors also found that the exogenous application of GB allowed chilled tomato plants to maintain lower electrolyte leakage than controls [55]. More recently, Hu et al. [56] also found that exogenous GB reduced electrolyte leakage in ryegrass and contributed to ameliorate the adverse effects of salt stress.

Activity of photosystem II (Fv/Fm)

During the first 4 days after harvest, fruits from all three treatments showed no damage to the efficiency of PSII as measured by the ratio-variable chlorophyll fluorescence to maximal fluorescence (Fv/Fm). Fv/Fm values remained above 0.8 that indicate no damage to the efficiency of PSII by Day 4. In fruits exposed to chilling temperatures (10°C for 6 h), Fv/Fm did fall abruptly by Day 5 reaching values as low as 0.14 by Day 10 of being harvested (Figure 5B). Preharvest application of GB lowered the decline of Fv/Fm in chilled fruits, as by Day 5, Fv/Fm remained as high as in control fruits. By Day 6, Fv/Fm began to fall, but this group always showed intermediate Fv/Fm values between chilled and control groups from Day 6 to Day 10 after being harvested. This might indicate that the application of GB reduced the damage to membranes caused by chilling, perhaps by showing a protective effect on the structure and function of the PSII complex. Previous reports exist of GB's ability to protect the photosynthetic oxygen-evolving complex [16,17].

Although it is argued that a decreased Fv/Fm in fruits is more associated to the process of fruit senescence [57], resulting from chlorophyll degradation during ripening [58,59], in banana fruits, it has been argued that Fv/Fm declines when chlorophyll content decreases as reported by Sanxter et al. [60] or by membrane degradation during ripening as suggested by Marangoni et al. [61]. Fv/Fm has been mentioned as a good indicator for membrane damage caused by chilling [62,63], and other authors also found that the exogenous application of GB allowed chilled tomato plants to maintain higher Fv/Fm values than non-GB-treated chilled fruits [55].

In conclusion, exposure of banana fruits to chilling stress caused fruit browning, an increase in the activity of the PPO when compared to control fruits and consequently a reduction in the concentration of phenols. Thus, the browning of the chilled fruits was not related to the content of phenols. Chilling also caused uneven ripening, an increased degradation of proteins, a reduction in the carotenoid content, and an increase in the content of chlorophylls, membrane damage (increased electrolyte leakage), and reduced efficiency of photosystem II (as shown by low Fv/Fm values). Chilled fruits also showed an acceleration of ethylene production and a reduction in fruit respiration.

On the contrary, the preharvest foliar application of GB (100 mM) to banana plants reduced the biochemical and physiological alterations caused by chilling injury on their harvested fruits. When fruits from GB-treated plants were exposed to 10°C for 6 h, the sub-epidermal browning was attenuated and the accumulation of total phenols was delayed. Similarly, the electron efficiency of photosystem II (Fv/Fm) as well as the CMS remained higher (24% and 27%, respectively) than in chilled fruits from non-GB-treated plants. It might be possible that the reduction of chilled-induced fruit browning by GB could be associated to a protection of the membrane integrity and protection of photosystem II. GB also seemed to protect proteins from chilling damage, as GB-treated fruits not only did not decrease protein content but even had increased protein content than both the chilled fruits or even the control fruits. GB also increased respiration and prevented the accelerated ripening associated with a premature appearance of an ethylene peak in non-GB-treated chilled fruits. However, it did not have effect on the chilling-related changes on the concentration of phenols or chlorophyll content, but it promoted the synthesis of carotenoids.

Further research is needed to elucidate the cellular and molecular mechanisms that may explain how the preharvest foliar application of GB induced a subsequent protection of fruits to chilling injury. The protection was related to processes of membrane stabilization, electrolyte leakage prevention, and protection to photosystem II, maintaining the photosynthetic efficiency of this protein complex.

In other species such as tomato, it has been suggested that, in addition to protecting macromolecules and membranes directly, GB may enhance chilling tolerance by inducing H_2O_2-mediated antioxidant mechanisms, e.g., enhanced catalase expression and catalase activity [55]. These results suggest that GB not only can protect temperate plants from chilling injury including A. thaliana (see review by Ashraf and Foolad [27]), but also has a protective effect in tropical plants such as banana. The results of the present work suggest that foliar application of GB might be used in banana plantations to reduce the fruit damage caused by their postharvest storage at low temperature. Our results also might serve as the basis to further evaluate whether the foliar application of GB could also reduce the damage caused by frost-associated chilling injury on banana fruits in the field.

Finally, this study also might give support to the possibility to improve chilling tolerance in this monocot tropical fruit by engineering its capacity to over accumulate GB, in a similar way as it has been achieved in maize [64], rice [28,31], and more recently in sweet potato [65].

Competing interests

The authors declare that they have no competing interests.

Author's contribution

Luis Carlos Rodríguez-Zapata provided financial support of the experiments from a research grant. Francisco Espadas y Gil assisted in the conduction of field experiments. Susana Cruz Martínez conducted field experiments and most of lab determinations. Carlos Talavera May and Fernando Contreras-Marin assisted in the conduction of field experiments. Gabriela Fuentes contributed to the experimental design and in the manuscript writing. Enrique Sauri-Duch offered expert advice during the conduction of the enzymatic activity assays. Jorge M. Santamaría provided general conception and coordination of the experiments, interpretation of results and manuscript writing. All authors read and approved the final manuscript.

Author details

[1]Unidad de Biotecnología, Centro de Investigación Científica de Yucatán, Mérida 97200, México. [2]Instituto Tecnológico de Mérida, Mérida 97200, México.

References

1. Pólit C., P. (2005). http://www.sica.gob.ec/agronegocios/sistema%20valor/postcosecha_hortifuticolas.htm. (Accessed: October 2, 2005).
2. Salunkhe DK, Desai BB (1984) Postharvest biotechnology of fruits. Lib Congress Florida, United States I:43–58
3. Jones DR (2000) Ed. Diseases of banana, abaca and emset. Cab Publishing. 560pp. ISBN 0-85199-355-9
4. Levitt J (1980) Responses of plants to environmental stresses. I Chilling, partly and high temperature stresses, vol 1, Secondth edn. Academic, New York, p 497
5. Fitter AH, Hay RKM (1987) Environmental physiology of plants, Secondth edn. Academic Press, San Diego California, pp 197–221
6. Taiz L, Zeiger E (2010) Plant physiology, 5th edn. Sinauer Associates, Inc, USA, pp 755–782
7. Hällgren J, Orquist G (1990) Adaptations to low temperatures. In: Stress responses in plants: adaptation and acclimation mechanism. Wiley-Liss, Inc, Umeå, Sweden, pp 265–293
8. Iba K (2002) Acclimative response to temperature stress in higher plants: approaches of gene engineering for temperature tolerance. Annu Rev Plant Physiol Plant Mol Biol 53:225–245
9. Sakamoto A, Murata N (2000) Genetic engineering of glycinebetaine synthesis in plants: current status and implications for enhancement of stress tolerance. J Exp Bot 51(342):81–88
10. Rhodes D, Hanson AD (1993) Quaternary ammonium and tertiary sulfonium compounds in higher plants. Annu Rev Plant Physiol Plant Mol Biol 44:357–384
11. Rathinasabapathi B (2000) Metabolic engineering for stress tolerance: installing osmoprotectant synthesis pathways. Ann Bot 86:709–716
12. Chen TH, Murata N (2002) Enhancement of tolerance of abiotic stress by metabolic engineering of betaines and other compatible solutes. Curr Opin Plant Biol 5:250–257
13. de Zwart FJ, Slow S, Payne RJ, Lever M, George PM, Gerrard JA, Chambers ST (2003) Glycine betaine and glycine betaine analogues in common foods. Food Chem 83:197–204
14. McCue KF, Hanson AD (1990) Drought and salt tolerance: towards understanding and application. Trends Biotechnol 8:358–362
15. Kishitani S, Watanabe K, Yasuda S, Arakawa K, Takabe T (1994) Accumulation of glycinebetaine during cold acclimation and freezing tolerance in leaves of winter and spring barley plants. Plant Cell Environ 17:89–95
16. Murata N, Mohanty PS, Hayashi H, Papageorgiou GC (1992) Glycinebetaine stabilizes the association of extrinsic proteins with the photosynthetic oxygen-evolving complex. FEBS Lett 296:187–189
17. Papageorgiou GC, Murata N. 1995. The unusually strong stabilizing effects of glycine betaine on the structure and function of the oxygen-evolving photosystem II complex. Photosynthesis Res 44:243–252
18. Nomura M, Hibino T, Takabe T, Sugiura T, Yokota A, Miyake H, Takabe T (1998) Transgenically produced glycinebetaine protects ribulose 1,5-bisphosphate carboxylase/oxygenase from inactivation in Synechococcus sp. PCC7942 under salt stress. Plant and Cell Physiol 39:425–432
19. Giri J (2011) Glycinebetaine and abiotic stress tolerance in plants. Plant Signal Behav 6(11):1746–1751
20. Yang C, Nong L, Lu J-L, Lu L, Xu J-S, Han Y-Z, Li Y-J, Fujita S (2004) Banana polyphenol oxidase: ocurrence and change of polyphenol oxidase activity in some banana cultivars during development. Food Sci Tech RES 10(1):75–78
21. Nguyen TBT, Ketsa S, Van Doorn WG (2003) Relationship between browning and the activities of polyphenol oxidase and phenylalanine ammonia lyase in banana peel, during low temperature storage. Postharvest Life Technol 30:187–193
22. Van Breusegem F, Slooten L, Stassart JM, Moens T, Botterman J, Van Montagu M, Inze D (1999) Overproduction of Arabidopsis thaliana FeSOD confers oxidative stress tolerance to transgenic maize. Plant Cell Physiol 40:515–523
23. Robinson SP, Jones GP (1986) Accumulation of glycinebetaine in chloroplast provides osmotic adjustment during salt stress. Aust J Plant Physiol 13:659–668
24. Like VL, Wilen RW, Fu P (1996) Low-temperature stress tolerance: the role of abscisic acid, sugars and heat stable proteins. Hort Sci Vol 31(1):39–46
25. Chang MY, Chen SL, Lee CF, Chen YM (2000) Chilling acclimation and root protection from chilling injury in chilling-sensitive temperature mungbean (Vigna radiata L) seedling. Bot. B. Academic Sw 42:53–59
26. Sakamoto A, Murata N (2002) The role of glycinebetaine in the protection of plants from stress: clues from transgenic plants. Plant Cell and Environ 25:163–171
27. Ashraf M, Foolad MR (2007) Roles of glycine betaine and proline in improving plant abiotic stress resistance. Environ Exp Bot 59:206–216
28. Sakamoto A, Alia MN (1998) Metabolic engineering of rice leading to biosynthesis of glycinebetaine and tolerance to salt and cold. Plant Mol Biol 38:1011–1019
29. Huang J, Hirji R, Adam L, Rozwadowski KL, Hammerlindl JK, Keller WA, Selvaraj G (2000) Genetic engineering of glycine-betaine production towards enhancing stress tolerance in plants: metabolic limitations. Plant Physiol 122:747–756
30. Quan R, Shang M, Zhang H, Zhao Y, Zhang J (2004) Improved chilling tolerance by transformation with betA gene for the enhancement of glycinebetaine synthesis in maize. Plant Sci 166:141–149
31. Shirasawa K, Takabe T, Takabe T, Kishitani S (2006) Accumulation of glycinebetaine in rice plants that overexpress choline monooxygenase from spinach and evaluation of their tolerance to abiotic stress. Ann Bot 98:565–571
32. Clendennen SK, May DG (1997) Differential genes expression ripening banana fruit. Plant Physiol 115:463–469
33. Tamayo CJA (2002) Study of the poliphenol oxidase system of the fruit of the chicozapote (Achras sapota), PhD thesis. Instituto Tecnológico de Mérida, Mérida, Yucatán. México, p 81
34. Arzápalo MAM (2002) Mono and diphenolase activity from crude extracts of chicozapote (Achras sapota, Manilkara achras Mill.) Master's thesis in Biotechnology. Instituto Tecnológico de Mérida, Merida Yucatan, p 76 p
35. Singleton VL, Rossi JA (1965) Colorimetry of total phenolics with phosphomolybdic-phosphotungstic acid reagents. Am J Enol Vitic 16:144–158
36. Bradford MM (1976) A rapid and sensitive method for the quantitation of microgram protein utilizing the principle of protein dye binding. Anal Biochem 72:248–254
37. Wellburn AR (1994) The spectral determination of chlorophyll a and b, as well as total wholefood, using various solvents with spectrophotometers of different resolution. J Plant Physiol 144:307–313
38. McCollum TG, McDonald RE (1991) Electrolyte leakage, respiration, and ethylene production as index of chilling injury in grapefruit. HortSc 26(9):1191–1192
39. Liu X, Shiomi S, Nakatsuka A, Kubo Y, Nakamura R, Inaba A (1999) Characterization of ethylene biosynthesis associated with in banana fruit ripening. Plant Physiol. Vol 121. 1257–1265
40. Chillet M, Lapeyre Bebellaire L (2002) Variability in the production of ethylene in banana from wound the French West Indies. Sci-fi Hort 96:127–137
41. Prohens J, Miró R, Rodriguez-Burruezo A, Chiva S, Verdú G, Nuez F (2004) Temperature, electrolyte, ascorbic acid content and sunscald in two cultivars of cucumber, Solanum muricatum. J Hort Sci Biotechnol 79(3):375–379
42. Maxwell K, Johnson GN (2000) Chlorophyll fluorescence -a practical guide. J Exp Bot 51:659–668

43. Morrelli KL, Hess-Pierce BM, Kader AA (2003). Genotypic variation in chilling sensitivity of mature-green bananas and plantains. Hort Technology 13(2):328–332.

44. Park EJ, Jeknic Z, Sakamoto A, DeNoma J, Yuwansiri R, Murata N, Chen THH (2004) Genetic engineering of glycinebetaine synthesis in tomato protects seeds, plants, and flowers from chilling damage. Plant J 40:474–487

45. Coseteng M, Lee CY (1987) Changes in apple polyphenol-oxidase and polyphenol concentration in relation to degree of browning. J Food Sci 52:985–989

46. Nguyen TBT, Ketsa S, Van-Doorn WG (2004) Effect of modified atmosphere packaging on chilling-induced peel browning in banana. Postharvest Biol Technol 31:313–317

47. Brady CJ, O'Connell PBH (1976) On the significance of increased protein synthesis in ripening banana fruits. Aust J Plant Physiol 3:301–310

48. Hubbard NL, Mason Pharr D, Huber SC (1990) Role of sucrose phosphate synthase in sucrose biosynthesis in ripening bananas and its relationship to the respiratory climacteric. Plant Physiol 94:201–208

49. Blackbourn HD, Jeger MJ, John P, Thompson AK (1990) Inhibition of degreening in the peel of bananas ripened at tropical temperatures. Ann App Biol 117:175–186

50. Lyons JM, Breidenbach RW (1990) Relation of chilling stress to respiration. Ed. C.Y. Wang. Chilling injury of horticultural crops. CRC Press, In, pp 223–233

51. Field RJ (1990) Influence of chilling stress on ethylene production. Ed. C.Y. Wang. Chilling injury of horticultural crops. CRC Press. Mouse palate, In, pp 235–253

52. Leshem YY, Kaddish D (1992) Membrane related effects of chilling stress on ethylene production in green peppers stored at 4°C as compared to 17°C (shelf temperature). Res. Report. BARD Foundation, u.s.-Israel, pp 1–23

53. Lyons JM (1973) Chilling injury in plants. Annu Rev Plant Physiol Plant Mol Biol 24:445–466

54. Jiang Y, Joyce DC, Jiang W, Lu W (2004) Effect of chilling injury on binding bye banana fruit ethylene. Plant Growth Regul 43:109–115

55. Park EJ, Jeknic Z, Chen TH (2006) Exogenous application of glycinebetaine increases chilling tolerance in tomato plants. Plant Cell Physiol 47(6):706–714

56. Hu L, Hu T, Zhang X, Pang H, Fu J (2012) Exogenous glycine betaine ameliorates the adverse effect of salt stress on perennial ryegrass. J Amer Soc Hort Sci 137:38–46

57. DeEll JR, van Kooten O, Prange RK, Murr DP (1999) Applications of chlorophyll fluorescence techniques in postharvest physiology. Hortic Rev 23:69–107

58. Holden M (1961) The breakdown of chlorophyll by chlorophyllase. Biochem J 78:359–364

59. Tucker GA (1993) Introduction. In: Seymour GB, Taylor JE, Tucker GA (eds) Biochemistry of fruit ripening. Chapman & Hall, London, pp 1–51

60. Sanxter SS, Yamamoto HY, Fisher DG, Chan HT (1992) Development and decline of chloroplasts in exocarp of Carica papaya. Can J Bot 70:364–373

61. Marangoni AG, Palma T, Stanley DW (1996) Membrane effects in postharvest physiology. Postharvest Biol Technol 7:193–217

62. Lurie S, Ronen R, Meier S (1994). Determining chilling injury induction in green peppers using nondestructive pulse amplitude modulated (PAM) fluorometry. J Am Soc Hortic Sci 119:59–62

63. Van Kooten O, Mensink MGJ, Otma EC (1992) Chilling damage of dark stored cucumbers (Cucumis sativus L.) affects the maximum quantum yield of photosystem 2. In: MURATA N (ed) Progress in photosynthesis research. Kluwer, The Netherlands, pp 161–164

64. Quan R, Shang M, Zhang H, Zhao Y, Zhang J (2004) Engineering of enhanced glycine betaine synthesis improves drought tolerance in maize. Plant Biotechnol J 2:477–486

65. Fan W, Zhang M, Zhang H, Zhang P (2012) Improved tolerance to various abiotic stresses in transgenic sweet potato (Ipomoea batatas) expressing plant signaling. Behav 6(11):1746–1751

14

Humic substances stimulate maize nitrogen assimilation and amino acid metabolism at physiological and molecular level

Silvia Vaccaro[1], Andrea Ertani[1*], Antonio Nebbioso[2], Adele Muscolo[3], Silvia Quaggiotti[1], Alessandro Piccolo[2] and Serenella Nardi[1]

Abstract

Background: The effects of a humic substance (HS) extracted from a volcanic soil on the nitrate assimilation pathway of *Zea mays* seedlings were thoroughly examined using physiological and molecular approaches. Plant growth, the amount of soluble proteins and amino acids, as well as the activities of the enzymes involved in nitrogen metabolism and Krebs cycle, were evaluated in response to different HS concentrations (0, 1 and 5 mg C L^{-1}) supplied to maize seedlings for 48 h. To better understand the HS action, the transcript accumulation of selected genes encoding enzymes involved in nitrogen assimilation and Krebs cycle was additionally evaluated in seedlings grown for 2 weeks under nitrogen (N) sufficient condition and N deprivation.

Results: HS at low concentration (1 mg C L^{-1}) positively influenced nitrate metabolism by increasing the content of soluble protein and amino acids synthesis. Furthermore, the activity and transcription of enzymes functioning in N assimilation and Krebs were significantly stimulated.

Conclusions: HS treatment influenced the gene expression of *Zea mays* plants at transcriptional level and this regulation was closely dependent on the availability of nitrate in the growth medium.

Keywords: Humic substances; Biological activity; Gene expression; Nitrogen metabolism

Background

Humic substances (HS) are the major components of soil organic matter (SOM) and positively affect crop production [1] by playing pivotal roles on both soil-plant system [2] and plant metabolism and development [3-6]. The growing concern for sustainable agriculture, whose goals are to ensure lower environmental costs and long-term productivity of agro-ecosystems [7], highlights the importance of developing management strategies to maintain and 1protect soil resources and, in particular, SOM [8]. Despite this present awareness, the second half of the past century has seen a massive and indiscriminate use of high-energy input in agriculture and in particular of N

fertilizers during the so-called green revolution. These practices led to the selection of genotypes, which were highly productive when fed with high nitrogen inputs, though not efficient in nitrogen use [9], while they concomitantly caused a progressive impoverishment of soil physical and biological properties, as well as of SOM and nutrients' bioavailability.

Because of the importance of HS in soil fertility, a larger number of works focused on both the chemical-physical structure and the biological properties of humic matter. HS have been traditionally described as polydispersed heterogeneous organic compounds with large molecular weight [10]. However, recent experimental findings reached a new understanding of the humic conformational nature that is regarded as a supramolecular association of heterogeneous molecules with relatively low molecular weight (≤1.5 kD) held together in only apparently large molecular sizes by weak interactions, such as hydrogen and hydrophobic bondings. The metastable conformations of humic

---

* Correspondence: andrea.ertani@unipd.it
[1]Department of Agronomy, Food, Natural Resources, Animal and Environment, University of Padua, Agripolis - Viale dell'Università, 16, Legnaro, Padua 35020, Italy
Full list of author information is available at the end of the article

associations can be reversibly disrupted by interactions with small amounts of organic acids [11,12], while the same amount of acids do not alter conformation of the true macropolymers stabilized by covalent bonds [13]. The supramolecular nature of humic matter and its response to organic acids have been advocated to explain the bioactivity of humus on plants [6], its molecular dynamics [14] and the slow release of sorbed contaminants [15]. Furthermore, the recognition that NOM is composed by supramolecular associations rather than macropolymers has allowed the development of a fractionation strategy, called humeomics, that enables the analytical detection of most of the single molecules which constitute the supramolecular assembly [16-18]. These findings had profound implications on our understanding of SOM function and reactivity.

Several studies have reported the positive effect of HS on crop yield [19-21] and on root and shoot development [6,22]. In addition, leaf chlorophyll content [1], nutrient uptake [3,23-26] and the activities of enzymes involved in several physiological pathways, such as nitrogen assimilation [3,24,25,27] and energy metabolism [28-30], seem to be positively affected by HS. However, because of the complexity of HS' nature, the relationship between their molecular structure and biological activity is not yet completely clear. In the last decades, several studies correlated biological activity to humic chemical features [31-41]. The positive effects of different HS in relation to their chemical structure were shown by Muscolo et al. [35] and Nardi et al. [36] on *Pinus nigra* callus metabolism and *Zea mays* seedling growth. Moreover, Zancani et al. [42,43] successfully related humic matter molecular features to metabolism of tobacco BY-2 suspension cell cultures and to changes of cellular ATP and glucose-6-phosphate during embryogenesis of *Abies cephalonica*.

Furthermore, the complexity of the biological action exerted by HS on plant metabolism suggested the existence of hormonal mechanisms, in particular of an auxin-like effect [3,4]. The presence of low amount of IAA in different soil humic substances has been reported [33,44]. Moreover, these substances seem to have a physiological effect on both a maize isoform of H^+ ATPase *Mha2* [34,37,38], which is a specific auxin target, and a pea phospholipase A_2 [45], which is a component of an auxin-dependent signalling.

The objective of this work was to investigate the effect of humic matter on maize nitrate assimilation and amino acids metabolism when the humic concentration was varied. The response of growth, protein and free amino acid content of maize seedlings to humic supply was evaluated here. Moreover, the activity of the main enzymes involved in the Krebs cycle and nitrogen metabolism was determined, together with the expression of genes encoding the enzymes involved in these two metabolic pathways.

Methods

Soil humic matter extraction and chemical characterization

A humic acid (HA) from a Fulvudand soil of the volcanic caldera of Vico, near Rome, Italy, was isolated by standard methods as reported elsewhere [46]. The HA was titrated to pH 7.2 with a 0.5 M KOH solution in an automatic titrator (VIT 90 video titrator, Radiometer, Copenhagen, Denmark) under N_2 atmosphere and stirring. After reaching the constant pH 7.2, the solution containing potassium humate was left under titration for 2 h, filtered through a glass microfibre filter (Whatman GF/C; GE Healthcare, Buckinghamshire, UK) and freeze-dried. The elemental analyses and the molecular characterization of this HA by online thermochemolysis-GC-MS and CPMAS-NMR spectroscopy were reported earlier [36].

Plant material, growth solutions and HS treatment

Maize seeds (*Z. mays* L. var. DK 585) were soaked in distilled water for one night in running water and germinated for 60 h in the dark at 25°C on a filter paper wet with 1 mM $CaSO_4$ [47]. Then, the maize seedlings were raised in hydroponic conditions for 14 days in growth chamber with 450 mL of a complete nutrient solution (µM) containing KH_2PO_4 (40), $Ca(NO_3)_2$ (200), KNO_3 (200), $MgSO_4$ (200), FeNaEDTA (10), H_3BO_3 (4.6), $CuCl_2 \cdot 2H_2O$ (0.036), $MnCl_2 \cdot 4H_2O$ (0.9), $ZnCl_2$ (0.09) and $NaMoO_4 \cdot 2H_2O$ (0.01). The solution was renewed every 48 h. Growth chamber conditions were the following: day/night period of 14/10 h, air temperature of 27°C/21°C, relative humidity of 6%0/80% and photon flux density (PFD) of 280 µmol m^{-2} s^{-1}.

To evaluate morphometric parameters, protein and free amino acids content and enzyme activities, on the 12th day of growth, the plants were treated for 48 h with aqueous solutions in which the freeze-dried potassium humate was dissolved at different concentrations expressed as carbon weight per litre: 0 (control), 1 and 5 mg C L^{-1}.

To evaluate the transcript accumulation, the HS were applied on the 13th day at 1 mg C L^{-1} for 24 h to two different control solutions: a complete nutrient solution (see above) and a nutrient solution N-deprived, where N sources KNO_3 and $Ca(NO_3)_2$ $4H_2O$ were replaced with KCl and $CaCl_2 \cdot 2H_2O$ at the same concentrations. The expression analyses were performed by semi-quantitative reverse transcriptase-polymerase chain reaction (RT-PCR) on cDNA obtained from leaves and root tissues.

Determination of protein content

For the extraction of soluble protein, frozen foliar and root tissue (0.5 g) were ground in liquid nitrogen, stirred with the extraction buffer (100 mM Tris-HCl pH 7.5, 1 mM Na_2EDTA, 5 mM DTT) and centrifuged at 14,000 g. The supernatants were mixed with 10% (*w/v*) trichloroacetic acid (TCA) and centrifuged. The pellets obtained were then re-suspended in 0.1 N NaOH. The amount of total proteins in the leaves was determined by using the Bradford method [48] with bovine serum albumin as a standard. The results were expressed as mg protein/gramme fresh tissue.

Enzyme extraction and assay conditions

The enzymes were solubilized from leaves by manually crashing vegetal tissues in a mortar added with a 100 mM Hepes-NaOH solution at pH 7.5, a 5 mM $MgCl_2$ solution, and a 1 mM dithiothreitol (DTT) solution. The ratio of plant material to mixture solution was 1:3 w/v. The extract was filtered through two layers of muslin and clarified by centrifugation at 20,000 g for 15 min. The supernatant was used for enzymatic analysis. All steps were carefully performed at 4°C, while a Jasco V-530 UV-vis spectrophotometer (Jasco Co., Midrand, Gauteng, South Africa) was used for measuring enzyme activity.

The nitrate reductase (NR EC 1.7.1.1) activity was assayed according to Lewis and collaborators [49] and expressed as unit per milligramme protein. One unit corresponds to the production of 1 μmol of NO_2^- per minute at 25°C.

The nitrite reductase (NiR EC 1.7.1.4) activity was determined on the basis of the drop in NO_2^- concentration in the reaction medium [50]. After incubation at 30°C for 30 min, the NO_2^- content was determined colorimetrically at A_{540} and the activity was expressed as unit per milligramme protein. One unit corresponds to the consumption of 1 μmol of NO_2^- per minute at 25°C.

To evaluate the glutamine synthetase (GS EC 6.3.1.2) activity, the mixture for the assay contained 90 mM imidazole-HCl (pH 7.0), 60 mM hydroxylamine (neutralized), 20 mM Na_2KAsO_4, 3 mM $MnCl_2$, 0.4 mM ADP, 120 mM glutamine and the appropriate amount of enzyme extract. The assay was performed in a final volume of 750 μL. The enzymatic reaction was developed for 15 min at 37°C. γ-Glutamyl hydroxamate (gh) was colorimetrically determined by addition of 250 μL of a mixture (1:1:1) of 10% (w/v) $FeCl_3.6H_2O$ in 0.2 M HCl, 24% (w/v) trichloroacetic acid and 50% (w/v) HCl. The optical density was recorded at A_{540} [51]. The activity was expressed as unit per milligramme protein. One unit corresponds to the production of 1 μmol mg protein per minute at 37°C.

NADH-dependent glutamate synthase (NADH-GOGAT EC 1.4.1.14) assay contained 25 mM Hepes-NaOH (pH 7.5), 2 mM L^{-1} glutamine, 1 mM α-ketoglutaric acid, 0.1 mM β-NADH, 1 mM Na_2 EDTA and 100 μL of enzyme extract in a final volume of 1.1 mL. GOGAT was assayed spectrophotometrically by monitoring β-NADH oxidation at A_{340}. The activity was expressed as unit per milligramme protein. One unit corresponds to the oxidation of 1 μmol of β-NADH per minute at 37°C.

The activity of aspartate aminotransferase (AspAT EC 2.6.1.1) was measured spectrophotometrically by monitoring β-NADH oxidation at A_{340} at 30°C. The assay medium (final volume 2.4 mL) contained 100 mM Tris-HCl at pH 7.8, 240 mM L-aspartate (Asp), 0.11 mM pyridoxal phosphate, 0.16 mM NADH, 0.93 kU malate dehydrogenase (MDH), 0.42 kU lactate dehydrogenase (LDH), 12 mM 2-oxoglutarate and 200 μL enzyme extract [52]. The activity was expressed as unit per milligramme protein. One unit corresponds to the oxidation of 1 μmol of β-NADH per minute at 37°C.

The activity of asparagine synthetase (AS EC 6.3.5.4) was determined by incubating 10 mM glutamine (Gln), 30 mM ATP, 10 mM Asp, 10 mM $MgCl_2$, 2 mM DTT, 0.1 mM EDTA, Hepes 50 mM pH 7.75 with 400 μL of the desalted extract in a total volume of 500 μL at 30°C for 60 min. The reaction was terminated with 100 μL of sulfosalicylic acid (200 mg mL^{-1}), centrifuged and an aliquot of the supernatant, adjusted to pH 7.0 with NaOH and taken for the analysis and separation of Asp and Asn by HPLC of the ortho-phthalaldehyde (OPA) reagent derivates [53]. The activity was expressed as unit per milligramme protein. One unit corresponds to the production of 1 μmol of Asn per minute at 30°C.

For the assay of citrate synthase (CS EC 2.3.3.1), the leaves were homogenized using 100 mM Tris-HCl buffer pH 8.2, containing 5 mM β-mercaptoethanol (Sigma-Aldrich, St. Louis, MO, USA), 1 mM Na_2EDTA and 10% glycerol. The leaves were filtered and centrifuged as reported [54]. All steps were performed at 4°C. The CS enzyme was assayed adding 50 mL of oxalacetic acid 0.17 mM, 50 mL acetyl-CoA 0.2 mM, and 10 mL extract, to 3 mL of Tris-HCl 0.1 M (pH 8.0). This activity was measured spectrophotometrically at 25°C, by monitoring the reduction of acetyl-coenzyme A (acetyl-CoA) to CoA, at wavelength A_{232}. The activity was expressed as unit per milligramme protein. One unit corresponds to the reduction of 1 μmol of acetyl-CoA per minute at 30°C.

To extract $NADP^+$-isocitrate dehydrogenase ($NADP^+$-IDH EC 1.1.1.42) and MDH (EC 1.1.1.37), the leaves were homogenized using 100 mM Tris-HCl buffer pH 8.2 containing 5 mM β-mercaptoethanol (Sigma), 1 mM Na_2EDTA and 10% glycerol. The extracts were filtered and centrifuged as reported [54]. All steps were performed at 4°C.

For $NADP^+$-IDH activity, 50 μL of crude extract was added to 2.85 final volume of a reaction mixture containing 88 mM imidazole buffer (pH 8.0), 3.5 mM $MgCl_2$, 0.41 mM β-NADP-Na salt and 0.55 mM isocitrate-Na salt. The assay was performed at 25°C following the formation of NADP(H) at A_{340} [55]. The activity was expressed as unit per milligramme protein. One unit corresponds to the production of 1 μmol of NADP(H) per minute.

For MDH activity, 3.17 mL of assay mixture contained 94.6 mM phosphate buffer (pH 6.7), 0.2 mM β-NADH-Na$_2$ salt, 0.5 mM oxalacetic acid, and 1.67 mM MgCl$_2$. MDH activity was assayed at 25°C, following the formation of NAD$^+$ at A$_{340}$ [54]. The activity was expressed as unit per milligramme protein. One unit corresponds to the production of 1 μmol of NAD$^+$ per minute.

Free amino acids determination

The free amino acids were analyzed, as by the method of Seebauer et al. [56]. Fifty milligrammes of homogenous dry powder was extracted for 1 h at room temperature with 1.5 mL of a 5% (w/v) TCA solution. The sample was clarified by centrifugation, and 1.5 mL of the supernatant was analyzed for free amino acids. The amino acid analysis was accomplished by precolumn OPA derivatization of the sample followed by reverse phase separation, using an Agilent 1100 HPLC (Agilent Technologies, Palo Alto, CA, USA) equipped with a thermo-controlled auto-sampler, fluorescence detector and an Agilent HP Chemstation for data elaboration. The chromatographic conditions were described previously [57].

RNA extraction and cDNA synthesis

The tissue samples were ground in liquid nitrogen. Aliquots of approximately 100 mg were used for RNA extraction using TRIzol reagent (Invitrogen, Carlsbad, CA, USA) according to the manufacturer's instructions. RNA was quantified by spectrophotometric reading, and the quality was assayed by agarose gel electrophoresis. After DNAse treatment (Promega, Milan, Italy), first-strand cDNA was synthesized from 5 μg of total RNA using 200 U of MMLV Reverse Transcriptase (Promega, Milan, Italy) and oligo(dT) as a primer in 20 μL reactions, as described by Sambrook and collaborators [58].

Semi-quantitative RT-PCR analysis

To determine the expression level of *NR* (NCBI accession number M27821), *AS* (NCBI accession number X82849), *MDH* (NCBI accession number T27564), *CS* (NCBI accession number W49861) and *IDH* (NCBI accession number W21690), semi-quantitative RT-PCR experiments with specific primers were performed as described earlier [59]. Genes analyzed and relative primers are listed in Table 1. The constitutively expressed 18S rRNA (NCBI; Accession Number U42796.1) gene was used as an internal control of RNA quantity. PCR products were separated by electrophoresis in a 1-2.5% agarose gel stained with ethidium bromide, and quantified through the ImageJ program (ImageJ 1.38 J, Wayne Rasband, National Institute of Health, Bethesda, MD, USA). PCR reactions were performed on cDNAs obtained from two different RNA extractions from independent experiments and repeated at least three times for each cDNA. PCR products

were further purified by using the QIAquick gel extraction Kit (Qiagen, Hilden, Germany) and sequenced, according to Sanger et al. [60], using the ABIPRISM original Rhodamine Terminator kit (Applied Biosystems, Branchburg, NJ, USA). The sequence comparisons were performed by using Blastx and Blastn computer programs (NCBI, National Center for Biotechnology Information).

Statistical analysis

All sets of experiments were repeated three times, and for each set three replicates were assayed. Data were the means of three replicates, and the standard deviations did not exceed 5%. The results obtained were processed statistically by the Student-Newman-Keuls test [61].

Results

Effects of HS treatments on plant growth

After growing for 12 days in a complete nutrient solution, the plants were transferred in a solution containing two different concentrations of HS and treated for 48 h. For seedlings supplied with 1 mg C L^{-1} of humic solution, no significant difference from control was found for leaves, roots length and fresh weight. Conversely, treatment with a greater humic concentration (5 mg C L^{-1}) produced length of both leaves and roots 8% shorter than the control seedlings (Figure 1a,b) and fresh weight (Figure 2a,b).

Effects of HS treatments on protein content and enzyme activities

The amount of proteins was always significantly lower in roots than in leaves of maize seedlings. Seedlings treated with 1 mg C L^{-1} of HS showed a leaf content of protein larger than control (7%). On the other hand, a significant decrease in protein content (−21%) was observed in the leaves of seedlings supplied with 5 mg C L^{-1} HS (Figure 3a). In roots, no significant differences with respect to the control were detected at 1 mg C L^{-1} of HS and the 5 mg C L^{-1} treatment even caused a 5% decrement in protein content (Figure 3b).

The humic addition to the growing medium affected the activities of NR and NiR in a concentration-dependent manner (Figure 4a,b). As for NR, the application of 1 and 5 mg C L^{-1} concentrations increased the NR specific activity by about 10% and 56%, respectively (Figure 4a). A similar behaviour was found for NiR activity. In plants supplied with 1 and 5 mg C L^{-1}, the NiR activity was increased by 12% and 41%, respectively (Figure 4b).

The GS and NADH-GOGAT enzymes showed a dose-response activity similar to those described for NR and NiR. When maize seedlings were supplied with HS at 1 and 5 mg C L^{-1}, GS and GOGAT activities increased by 11% and 30%, and by 64% and 92% respectively (Figure 5a,b).

Table 1 Specific forward and reverse primer sequences used in semi-quantitative RT-PCR

Gene	Primer forward (5'-3')	Primer reverse (5'-3')	Accession number
NR (nitrate reductase)	5'-TTCATGAACACTACCGACGTCG-3'	5'-GAGCCTGTACGGATACTCGGC-3'	M27821
AS (asparagine synthetase)	5'-CATCATTGAGCTCTCGCGCAGGTTAC-3'	5'-GGGGGAAATGTTATGAAGCGTTCACAA-3'	X82849
CS (citrate synthetase)	5'-GTTTGGTCATGGAGTTCTGCGTAA-3'	5'-GGAGGTACAACTTCATACAACTTGGACAC-3'	W49861
IDH (isocitrate dehydrogenase)	5'-AAACTCGAGGCTGCTTGCGTTGAGA-3'	5'-ATAATTAGCTTGCATCGAAACTGCGG-3'	W21690
MDH (malate dehydrogenase)	5'-GCCAGATTTCTGAGAGACTTAATGTCCA-3'	5'-TCGAGGCATGAGTAAGCAAGCGTCTT-3'	T27564
18S	5'-CCATCCCTCCGTAGTTAGCTTCT-3'	5'-CCTGTCGGCCAAGGCTATATAC-3'	U42796.1

As for AspAT, the plants supplied with 1 and 5 mg C L^{-1} of HS showed an increase in enzyme activity of 16% and 20%, respectively (Figure 6a). Conversely, the AS activity had a different trend, since 1 mg C L^{-1} caused a decrease in the AS activity (–15%) and 5 mg C L^{-1} of HS caused a significant larger stimulation (+18%) compared to the control (Figure 6b).

We evaluated the effects of HS on CS, $NADP^+$-IDH and MDH of maize seedlings because of their important role in maintaining the balance between C and N metabolism. When plants were supplied with HS at 1 and 5 mg C L^{-1}, an increase of CS activity by 56% and 98%, respectively, was observed in comparison to the control (Figure 7a). A similar trend was also observed for $NADP^+$-IDH and MDH activities (Figure 7b,c). The seedlings treated with 1 and 5 mg C L^{-1} showed an increase of both MDH and $NADP^+$-IDH activities by 14% and 20%, and 93% and 81%, respectively.

Effect of HS treatment on the amino acid content

The free amino acid content in leaves was measured to follow HS effects on nitrogen assimilation. Aspartate (Asp), threonine (Thr), isoleucine (Ile) and lysine (Lys) increased by 10% to 13% with 1 mg C L^{-1} HS, while the 5 mg C L^{-1} did not change the amino acids' content (Table 2). However, asparagine (Asn) behaved differently, as it was increased (41%) only with the addition of 5 mg C L^{-1} HS, while 1 mg C L^{-1} HS decreased (–11%) this amino acid in leaves as compared to control.

Effect of humic sample on gene expression

To gain further insight on the mechanisms of regulation of the enzyme previously described by HS, the expression of five genes encoding the key enzymes of N assimilation and organic acids synthesis was evaluated in leaves of both nitrate-supplied and nitrate-depleted maize seedlings, in response to 1 mg C L^{-1} HS for 24 h. The transcript

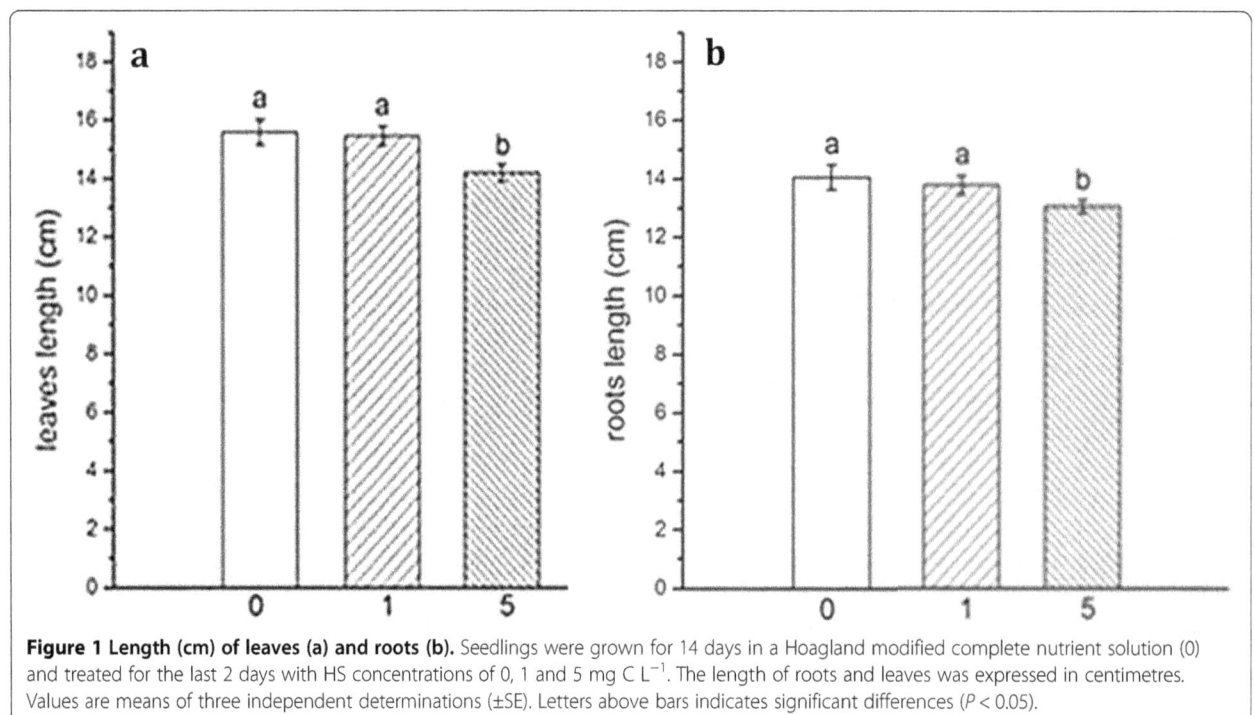

Figure 1 Length (cm) of leaves (a) and roots (b). Seedlings were grown for 14 days in a Hoagland modified complete nutrient solution (0) and treated for the last 2 days with HS concentrations of 0, 1 and 5 mg C L^{-1}. The length of roots and leaves was expressed in centimetres. Values are means of three independent determinations (±SE). Letters above bars indicates significant differences ($P < 0.05$).

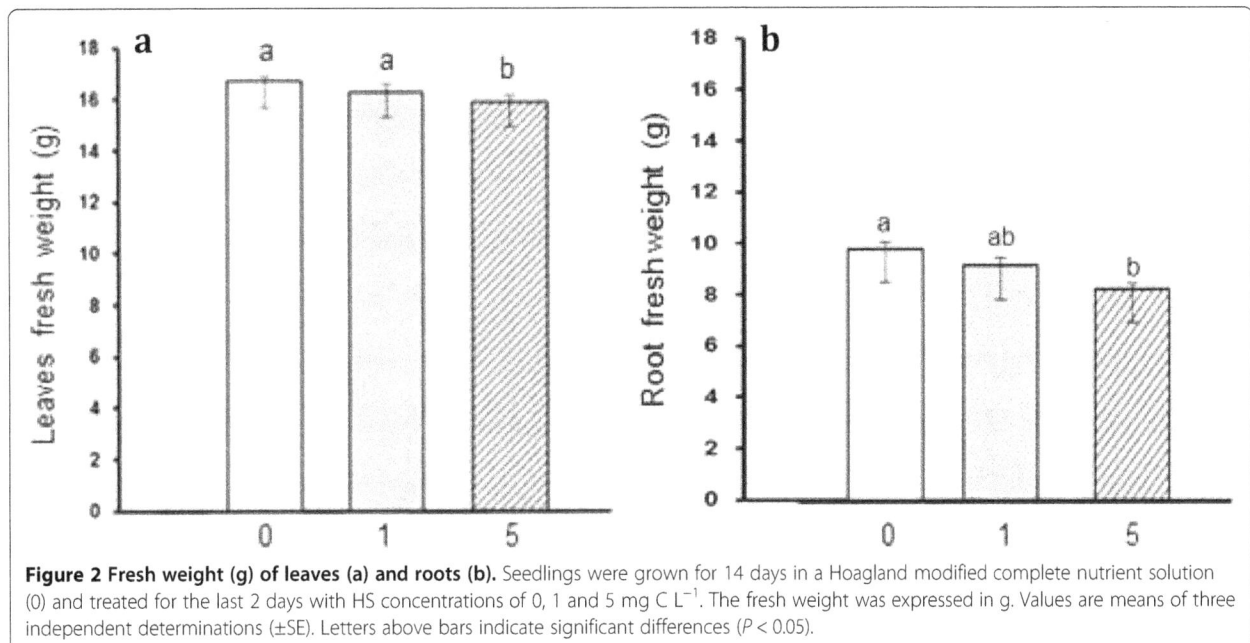

Figure 2 Fresh weight (g) of leaves (a) and roots (b). Seedlings were grown for 14 days in a Hoagland modified complete nutrient solution (0) and treated for the last 2 days with HS concentrations of 0, 1 and 5 mg C L^{-1}. The fresh weight was expressed in g. Values are means of three independent determinations (±SE). Letters above bars indicate significant differences ($P < 0.05$).

accumulation of the gene encoding a NADH-NR [59] showed a significant increase after the treatment with 1 mg C L^{-1} HS in leaves of N-supplied plants, with respect to that detected in leaves of +N control plants (Figure 8). Conversely, no NR transcript accumulation was observed in the leaves of N-deprived plants of either HS-treated or control seedlings. Similarly, humic treatment induced a substantial increase of mRNA abundance in leaves of N-supplied

seedlings, whereas no significant differences were detected when N-depleted seedlings were supplied with HS.

A similar expression pattern in response to humic substances provision was also observed for the genes encoding MDH, CS and IDH enzymes which are involved in the Krebs cycle and N organication. In the case of MDH, a slight decrease of mRNA abundance was found in N-deprived leaves after humic supply. Moreover, the transcript

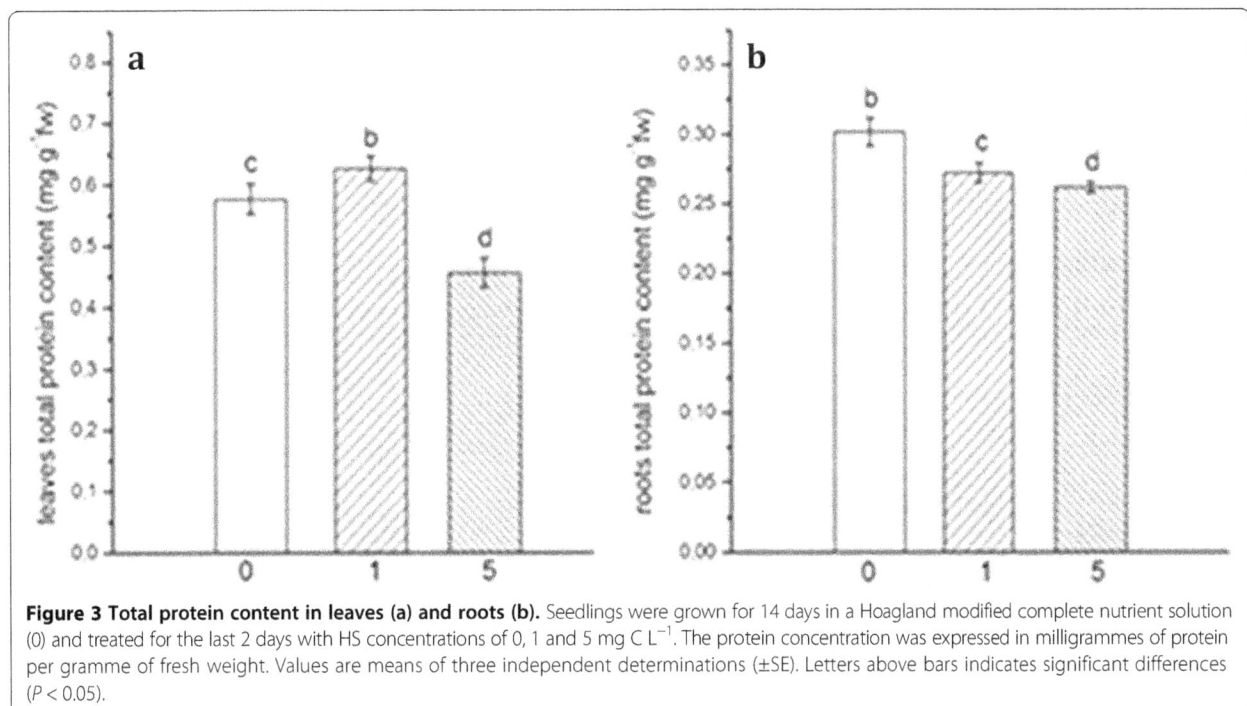

Figure 3 Total protein content in leaves (a) and roots (b). Seedlings were grown for 14 days in a Hoagland modified complete nutrient solution (0) and treated for the last 2 days with HS concentrations of 0, 1 and 5 mg C L^{-1}. The protein concentration was expressed in milligrammes of protein per gramme of fresh weight. Values are means of three independent determinations (±SE). Letters above bars indicates significant differences ($P < 0.05$).

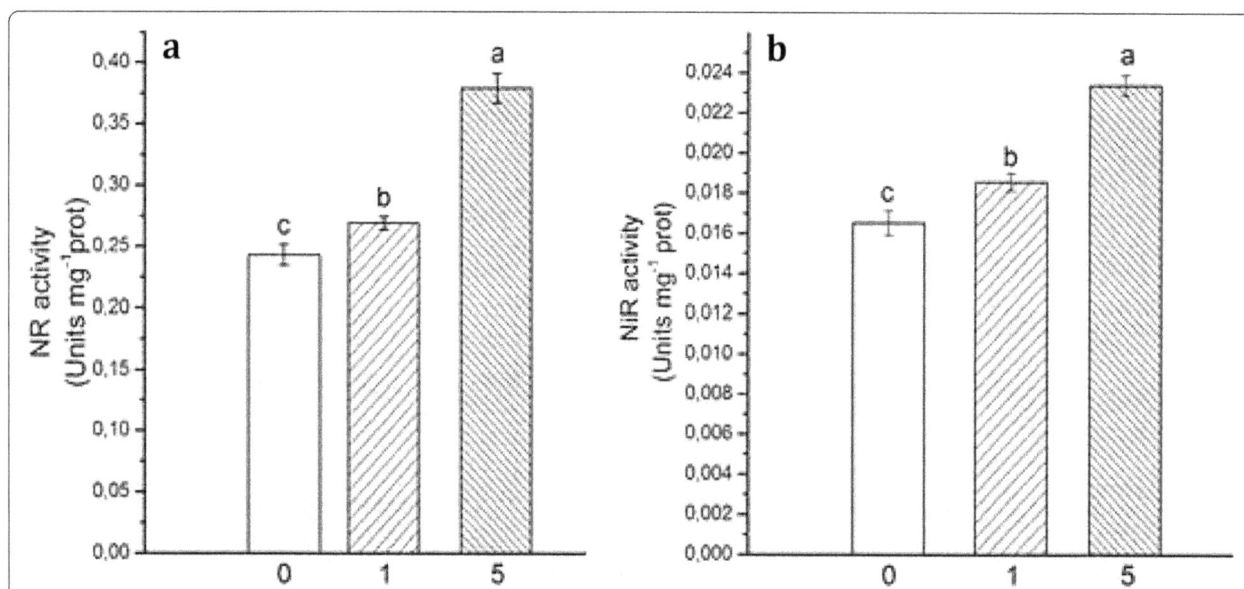

Figure 4 NR (a) and NiR (b) specific activities. Seedlings were grown for 14 days in a Hoagland modified complete nutrient solution (0) and treated for the last 2 days with HS concentrations of 0, 1 and 5 mg C L^{-1}. The specific activities were expressed in unit per milligramme protein. One unit corresponds to the production (NR) or consumption (NiR) of 1 μmol of NO$_2^-$ per minute at the assay conditions. Values are the means of three independent determinations (±SE). Letters above bars indicates significant differences ($P < 0.05$).

level of these three genes was always lower in N deficiency conditions in respect to that for N-supplied plants.

Discussion

Since the 1980s, several studies have highlighted the existence of biological effects of HS on plant physiology and metabolism [1,3,6,25,32]. However, only more recently, the attention of researchers has been focused on HS fractions characterized by an apparently low-molecular-weight (LMW), which seems to influence plant growth and physiology more than other fractions [3,63]. In fact, several works focused on effects of LMW fraction on

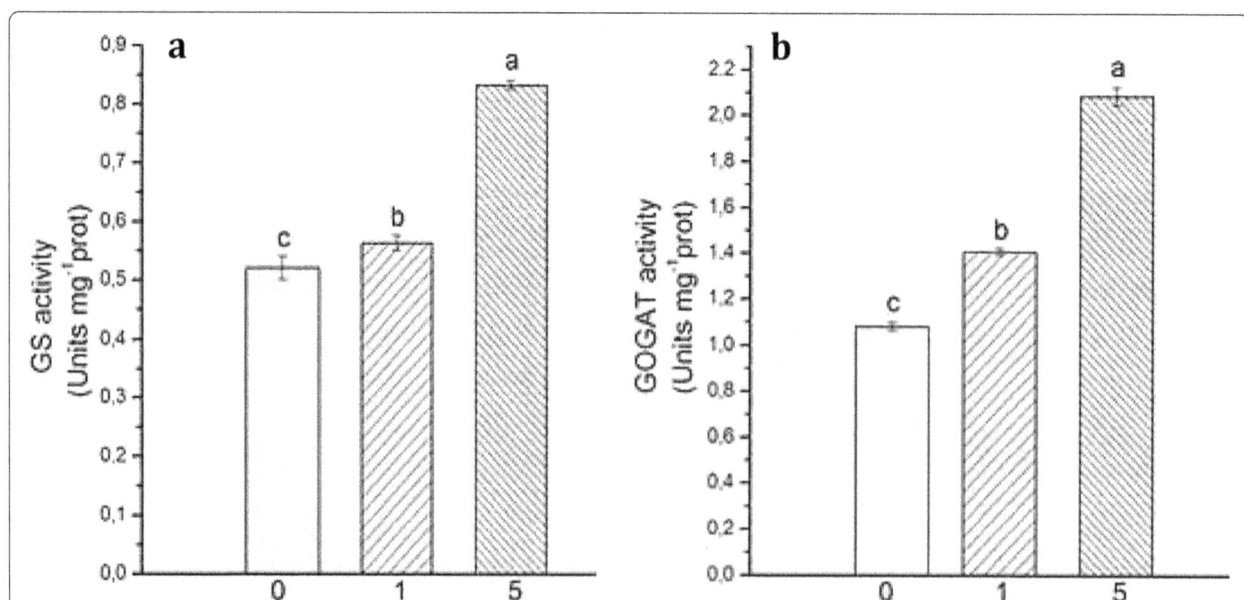

Figure 5 GS (a) and NADH-GOGAT (b) specific activities. Seedlings were grown for 14 days in a Hoagland modified complete nutrient solution (0) and treated for the last 2 days with HS concentrations of 0, 1 and 5 mg C L^{-1}. The specific activities were expressed as unit per milligramme protein. One unit corresponds to the production of 1 μmol of γ-glutamyl hydroxamate per minute in the case of GS and to the oxidation of 1 μmol of β-NADH per minute in the case of GOGAT at the assay conditions. Values are means of three independent determinations (±SE). Letters above bars indicate significant differences ($P < 0.05$).

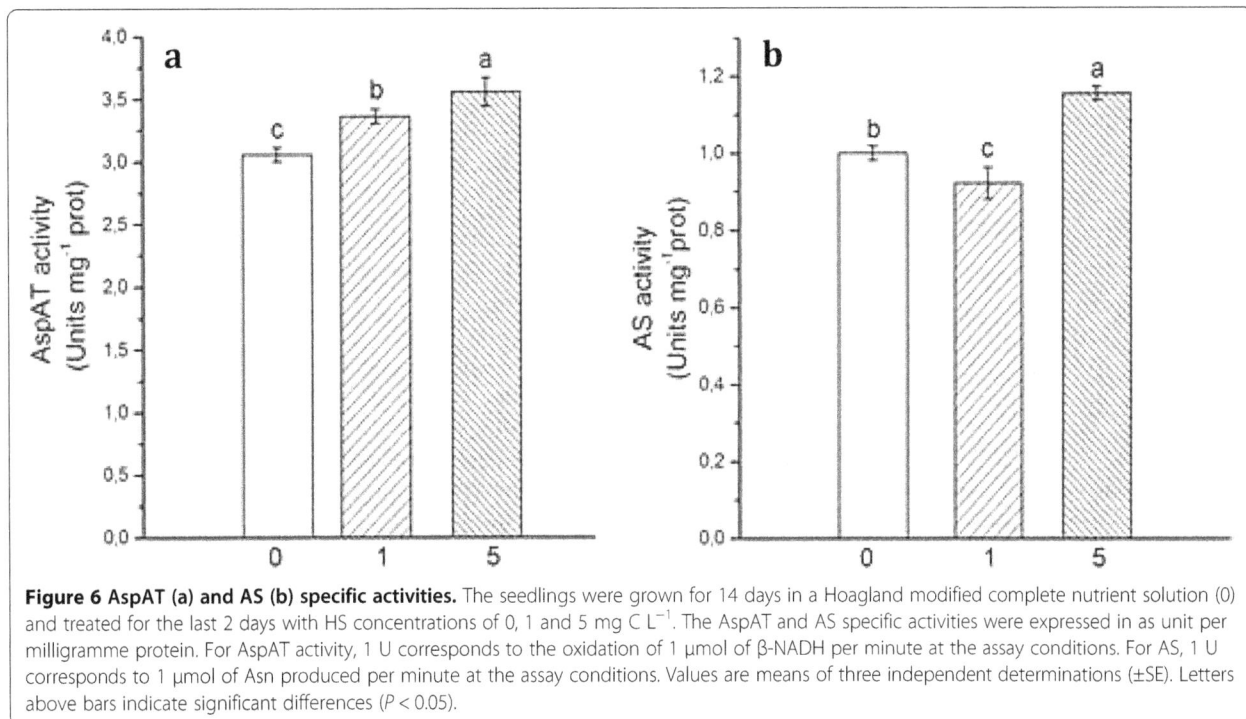

Figure 6 AspAT (a) and AS (b) specific activities. The seedlings were grown for 14 days in a Hoagland modified complete nutrient solution (0) and treated for the last 2 days with HS concentrations of 0, 1 and 5 mg C L^{-1}. The AspAT and AS specific activities were expressed in as unit per milligramme protein. For AspAT activity, 1 U corresponds to the oxidation of 1 μmol of β-NADH per minute at the assay conditions. For AS, 1 U corresponds to 1 μmol of Asn produced per minute at the assay conditions. Values are means of three independent determinations (±SE). Letters above bars indicate significant differences ($P < 0.05$).

nitrogen uptake [33,34,37,64] and assimilation [65], but the results obtained were often contradictory [66,67]. Their positive effects have mainly been attributed to a large content of carbohydratic and carboxyl-C groups [32,68].

More recently, Nardi and coworkers [36] evaluated the effect on the respiratory metabolism of maize seedlings provided by different size fractions of a soil humic acid, after their separation by preparative high-performance size-exclusion chromatography (HPSEC). The evaluation of the biological activity of size fractions showed that the size fraction with the smallest molecular size was more bioactive, followed by, in the order of, both the original bulk humic acid and the larger size fractions. Such effect was attributed to the larger flexibility of the smallest size fraction conferred by its greater content of hydrophilic components [36]. In fact, both the conformational flexibility and the hydrophilicity of this small-sized humic fraction should facilitate the release from its supramolecular structure upon the action of organic acids exuded by roots of humic molecules with plant stimulation activity. The diffusion of such bioactive molecules in solution should be less easy from humic matrices with larger and more compact conformations [37,38,69]. Here, we used the same bulk humic acid that was found bioactive and second only to its separated smallest size fractions in order to evaluate its effects on nitrogen assimilation in maize at both physiological and molecular levels. This species was chosen because of its worldwide economic and agronomic importance and also because of its importance as a model plant [70].

Our results indicate that the HS treatment influenced the activities of all these enzymes in a dose-dependent manner. In particular, the increase of enzyme activity in relation to the dose of HS supplied was exponential as a significant stimulation was observed when plants were treated with 5 mg C L^{-1}. A different trend was observed for AS, because its specific activity was strongly stimulated only in response to the 5 mg C L^{-1} humic solution, while it was slightly inhibited by the 1 mg C L^{-1} treatment.

The synthesis of Asn is in competition with that of Asp, Lys, Thr and Ile, which represent, together with the amino acids of the aspartate family [71], the main substrates for the synthesis of structural proteins and enzymes required for the optimal plant development [72]. However, Asn may also represent a transitory store of N to be later used to meet specific demands during plant growth [73]. Our results showed a significantly higher accumulation of Lys, Ile and Thr in leaves of seedlings grown with 1 mg C L^{-1} of HS in comparison to those observed in control plants that is in accordance with the higher content of soluble proteins measured in HS-treated leaves.

To deepen the understanding of molecular events underlying the HS effects on plant physiology, the expression of five genes encoding the enzymes involved in few key steps of N assimilation was evaluated. To discriminate N-dependent from N-independent molecular effects of HS, the transcript amount was measured in leaves of both nitrate-supplied and nitrate-depleted HS-treated seedlings. In this case, the choice of a 24-h treatment was aimed at evaluating the earlier molecular effects of

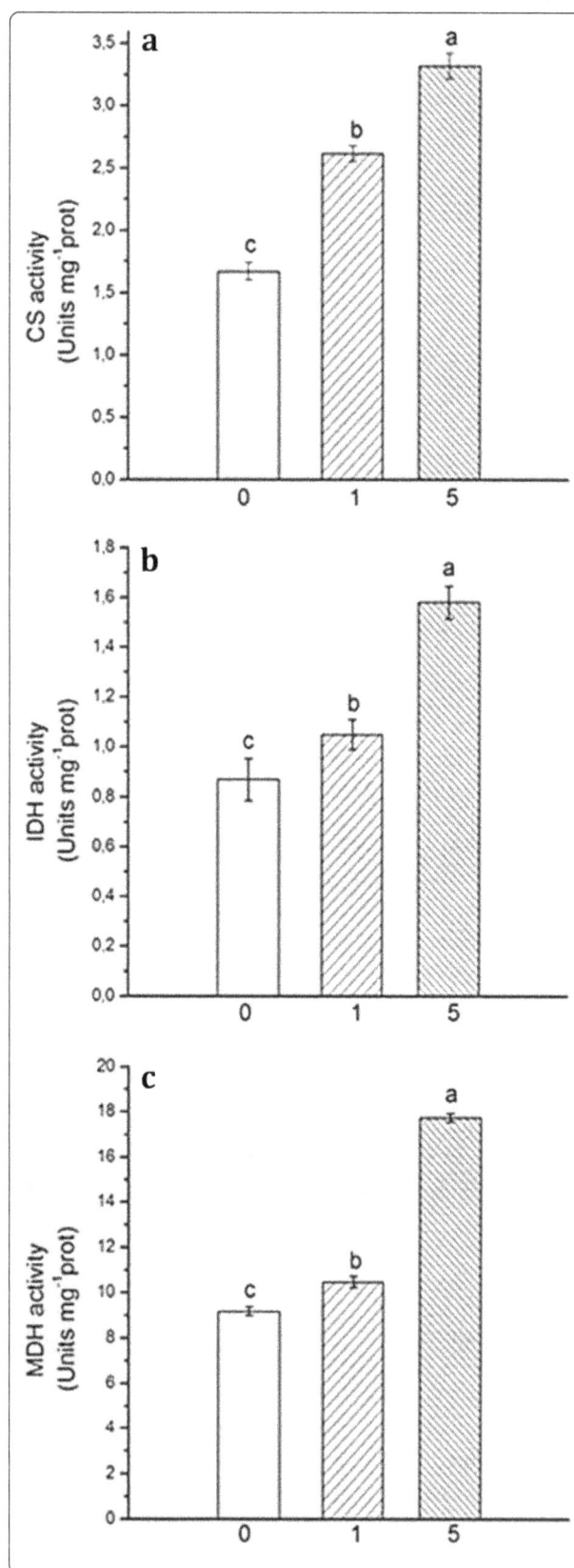

Figure 7 CS (a), NADP⁺-IDH (b) and MDH (c) specific activities. The seedlings were grown for 14 days in a Hoagland modified complete nutrient solution (0) and treated for the last 2 days with HS concentrations of 0, 1 and 5 mg C L^{-1}. The specific activities were expressed unit per milligramme protein. For CS, 1 U corresponds to the reduction of 1 μmol Ac-CoA per minute; for NADP⁺-IDH, 1 U corresponds to 1 μmol of NADPH reduced per minute and for MDH to 1 μmol of NAD⁺ formed per minute at assay conditions. Values are means of three independent determinations (±SE). Letters above bars indicate significant differences ($P < 0.05$).

humic matter. In fact, Trevisan et al. [40] showed that the gene transcription in response to HS is regulated already after a few hours of treatment.

Our findings show an increase of mRNA abundance of all genes therein studied upon HS treatment only when seedlings were grown in the presence of nitrate. On the contrary, HS did not induce any increase of transcript accumulation when the nutrient solution was depleted of nitrate. These results globally suggest that the enzyme stimulation observed after the humic supply may be generally regulated at the transcriptional level. An exception seems to be represented by the expression of the AS gene that was strongly induced after 24 h of treatment with 1 mg C L^{-1} solution, whereas the AS activity measured after 48 h was lower than for the control. This may be due either to a more rapid down-regulation of the AS activity in response to the duration of HS treatment or otherwise to the occurrence of some post-transcriptional regulatory mechanisms. Furthermore, since most of the genes analyzed are nitrate-responsive, it may be hypothesized that the HS regulation of the transcription of genes involved in nitrogen metabolism may be mediated by a nitrate-dependent signal, since HS' stimulatory effect is evident only when the nutrient solution contained nitrate.

Conclusions

Our results indicate a positive dose-dependent effect of the humic acid used here on the activities of the main enzymes involved in the reduction and assimilation of inorganic nitrogen. The enhanced bioactivity of HS may

Table 2 Amino acid contents (μmol g^{-1} fw) in leaves

Amino acid	Treatment		
	C	1	5
Aspartate	1.74 ± 0.3	1.97 ± 0.52	1.8 ± 0.45
Threonine	2.78 ± 0.92	3.06 ± 0.96	2.76 ± 0.92
Isoleucine	1.31 ± 0.27	1.46 ± 0.35	1.3 ± 0.62
Lysine	1.36 ± 0.32	1.53 ± 0.47	1.3 ± 0.31
Asparagine	1.7 ± 0.4	1.55 ± 0.18	2.4 ± 0.39
Glutamate	2.85 ± 0.90	3.19 ± 0.96	2.71 ± 0.52

Seedlings were grown for 14 days in a Hoagland modified complete nutrient solution (0) and treated for the last 2 days with HS concentrations of 0, 1 and 5 mg C L^{-1}. Mean values ± SD were obtained from five independent measurements.

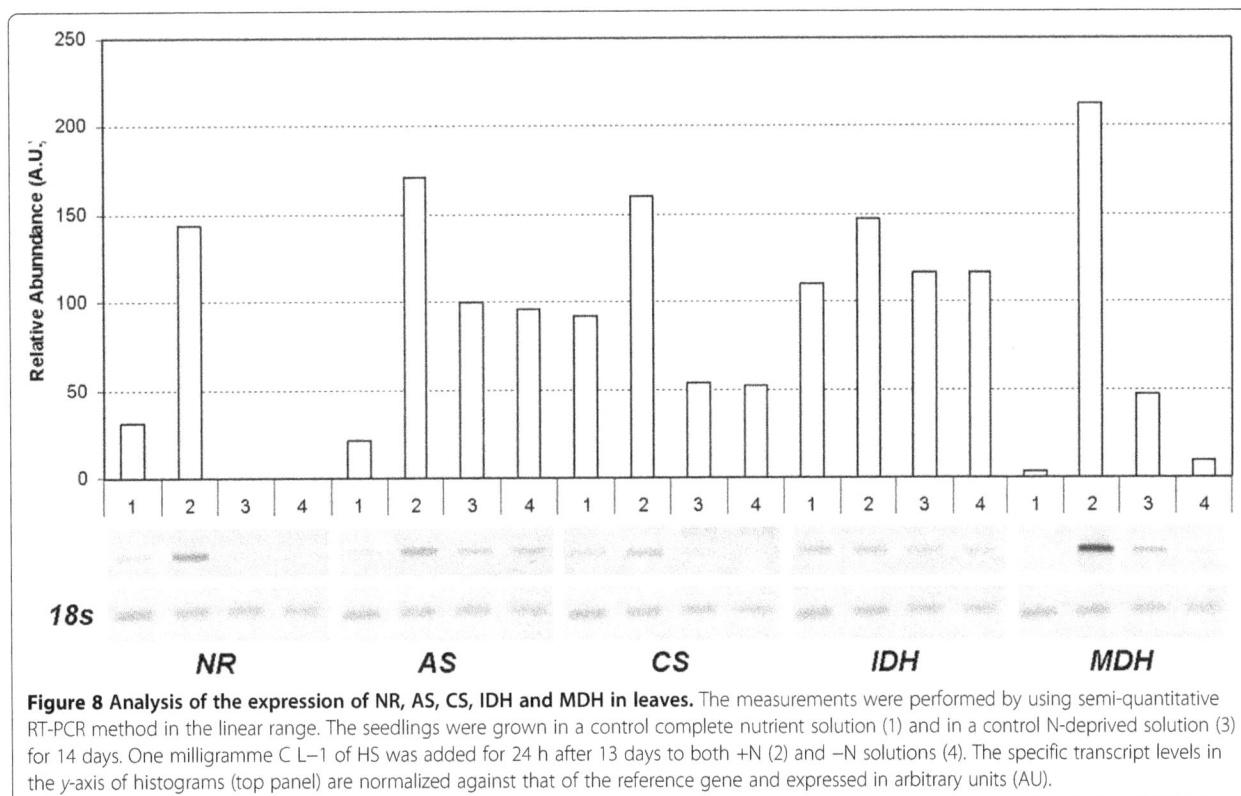

Figure 8 Analysis of the expression of NR, AS, CS, IDH and MDH in leaves. The measurements were performed by using semi-quantitative RT-PCR method in the linear range. The seedlings were grown in a control complete nutrient solution (1) and in a control N-deprived solution (3) for 14 days. One milligramme C L−1 of HS was added for 24 h after 13 days to both +N (2) and −N solutions (4). The specific transcript levels in the y-axis of histograms (top panel) are normalized against that of the reference gene and expressed in arbitrary units (AU).

be explained with the tendency to increase the size of supramolecular aggregation of humic molecules with concentration [74] and the concomitant enhanced capacity to provide specific bioactive molecules to targeted root cell membranes [6].

Among all enzymes, only the AS showed a different trend of regulation in response to HS, possibly because of a competition of its own substrate with the biosynthesis of amino acids of the Asp family [75], thus favouring the accumulation of a large amount of Asn in seedling leaves treated with the greatest HS concentration. This is not surprising considering that Asn accumulation in young leaves is an indirect result of the restriction of protein synthesis when the metabolism is subjected to stress conditions [73]. While further studies should be certainly conducted to increase our understanding on the dose-response of HS, molecular findings suggest that the biochemical response to humic treatments may principally depend on a transcriptional mechanism of regulation, as previously hypothesized for the regulation of a maize isoform of H^+-ATPase [34] and with a more holistic approach by Trevisan et al. [46]. However, the induction of a gene expression in response to HS seems to depend closely on the presence of nitrate in the growing medium because humic matter alone was not able to up-regulate the transcription of genes. These findings suggest that HS may act as an additional signal in the regulation of gene expression mediated by nitrate, by either promoting its

bioavailability or interfering with the signalling pathway that leads plants to adapt their metabolism to the nutrient availability [64,76].

Competing interests
The authors declare that they have no competing interests.

Authors' contributions
SV carried out the molecular genetic studies and participated in drafting the manuscript. AE participated in drafting the manuscript and made the corrections. AN participated in drafting the manuscript. AM, AP and SN conceived of the study, participated in its design and coordination and helped draft the manuscript. SQ carried out the molecular genetic studies and participated in drafting the manuscript. All authors read and approved the final manuscript.

Acknowledgements
This work was partially supported by the project COFIN 2003 no. 2003071271 funded by the Italian Ministry of University and Research (MIUR).

Author details
[1]Department of Agronomy, Food, Natural Resources, Animal and Environment, University of Padua, Agripolis - Viale dell'Università, 16, Legnaro, Padua 35020, Italy. [2]Centro Interdipartimentale di Ricerca sulla Risonanza Magnetica Nucleare per l'Ambiente, l'Agro-Alimentare ed i Nuovi Materiali (CERMANU), Università di Napoli Federico II, Via Università 100, Portici 80055, Italy. [3]Agriculture Department, Mediterranea University of Reggio Calabria, Feo di Vito, Reggio Calabria 89060, Italy.

References
1.   Vaughan D, Malcom RE (1985) Influence of humic substances on growth and physiological processes. In: Vaughan D, Malcom RE (ed) Soil organic matter and biological activity. Martinus Nijhoff/Junk W, Dordrecht, pp 37–76

2. Piccolo A (1996) Humic substances in terrestrial ecosystems. Elsevier, Amsterdam, pp 225–264

3. Nardi S, Carletti P, Pizzeghello D, Muscolo A (2009) Biological activities of humic substances. In: Senesi N, Xing B, Huang PM (ed) Biophysico-chemical processes involving natural nonliving organic matter in environmental systems. Part I. Fundamentals and impact of mineral-organic-biota interactions on the formation, transformation, turnover, and storage of natural nonliving organic matter (NOM). Wiley, Hoboken, pp 305–339

4. Muscolo A, Sidari M, Nardi S (2013) Humic substance: relationship between structure and activity. Deeper information suggests univocal findings. J Geochem Explor 129:57–63

5. Nardi S, Pizzeghello D, Reniero F, Rascio N (2000) Chemical and biochemical properties of humic substances isolated from forest soils and plant growth. Soil Sci Soc Am J 64:639–645

6. Canellas LP, Olivares FL (2014) Physiological responses to humic substances as plant growth promoter. Chem Biol Technol Agric 1:1–11

7. Horrigan L, Lawrence RS, Walker P (2002) How sustainable agriculture can address the environmental and human health harms of industrial agriculture. Environ Health Persp 110:445–456

8. Smith EG, Young DL (2000) The economic and environmental revolution in semi-arid cropping in North America. Annal Arid Zone 39:347–361

9. Hirel B, Le Gouis J, Ney B, Gallais A (2007) The challenge of improving nitrogen use efficiency in crop plants: towards a more central role for genetic variability and quantitative genetics within integrated approaches. J Exp Bot 58:2369–2387

10. Stevenson FJ (1994) Humus chemistry: genesis, composition, reactions, 2nd edition. Wiley, New York

11. Piccolo A (2002) The supramolecular structure of humic substances. A novel understanding of humus chemistry and implications in soil science. Adv Agron 75:57–134

12. Piccolo A, Conte P, Spaccini R, Chiarella P (2003) Effects of some dicarboxylic acids on the association of dissolved humic substances. Biol Fert Soils 37:255–259

13. Piccolo A, Conte P, Cozzolino A (2001) Chromatographic and spectrophotometric properties of dissolved humic substances compared with macromolecular polymers. Soil Sc 166:174–185

14. Orsi M (2014) Molecular dynamics simulation of humic substances. Chem Biol Technol Agric 1:1–14

15. Beiraghi A, Pourghazi K, Amoli-Diva M (2014) Mixed supramolecular hemimicelles aggregates and magnetic carrier technology for solid phase extraction of ibuprofen in environmental samples prior to its HPLC-UV determination. Chem Eng Sci 108:103–110

16. Nebbioso A, Piccolo A (2011) Basis of a humeomics science: chemical fractionation and molecular characterization of humic biosuprastructures. Biomacromolecules 12:1187–1199

17. Nebbioso A, Piccolo A (2012) Advances in humeomic: enhanced structural identification of humic molecules after size fractionation of a soil humic acid. Anal Chim Acta 720:77–90

18. Nebbioso A, Piccolo A, Lamshoft M, Spiteller M (2014) Molecular characterization of an end-residue of humeomics applied to a soil humic acid. RSC Adv 4:23658–23665

19. Niklewski B, Woiciechowski J (1937) Uber den Einfluss der wasserlöslichen Humusstoffe auf die Entwicklung einiger Kulturpflanzen. Z. Pflanzenernaehr Dueng. Bodenkd 49:294–327

20. Čatsky J (1958) Uber den Einfluss der Oxy humolithen auf die mineralische Ernährung der Pflanzen. Folia Biol 4:443–448

21. Khristeva LA, Gallushko AM, Gorovaya AI, Kolbassin AA, Shortshoi LP, Tkatshenko LK, Fot LW, Luk'Yakenko NV (1980) The main aspects of using physiologically active substances of humus nature. VI International Peat Congress, Minnesota

22. Nardi S, Concheri G, Dell'Agnola G (1996) Biological activity of humic substances. In: Piccolo A (ed) Humic substances in terrestrial ecosystems. Elsevier, Amsterdam, pp 361–406

23. Vaughan D, Linehan DJ (1976) The growth of wheat plants in humic acid solutions under axenic conditions. Plant Soil 44:445–449

24. Varanini Z, Pinton R (1995) Humic substances and plant nutrition. In: Behnke HD et al. (ed) Progress in botany, vol 56. Springer, Berlin, pp 97–117

25. Varanini Z, Pinton R (2001) Direct versus indirect effects of soil humic substances on plant growth and nutrition. In: Pinton R et al. (ed) The rizosphere. Marcel Dekker, Basel, pp 141–158

26. Nardi S, Concheri G, Pizzeghello D, Sturaro A, Rella R, Parvoli G (2000) Soil organic matter mobilization by root exudates. Chemosphere 41:653–658

27. Vaccaro S, Muscolo A, Pizzeghello D, Spaccini R, Piccolo A, Nardi S (2009) Effect of a compost and its water-soluble fractions on key enzymes of nitrogen metabolism in maize seedlings. J Agric Food Chem 57:11267–11276

28. Sladky Z (1959) The effects of extracted humus substances on growth of tomato plants. Biolol Plant 1:142–150

29. Vaughan D (1967) Effect of humic acid on the development of invertase activity in slices of beetroot tissues washed under aseptic conditions. Humus et Planta 4:268–271

30. Ferretti M, Ghisi R, Nardi S, Passera C (1991) Effect of humic substances on photosynthetic sulfate assimilation in maize seedlings. Can J Soil Sci 71:239–242

31. Nardi S, Concheri G, Dellagnola G, Scrimin P (1991) Nitrate uptake and ATPase activity in oat seedlings in the presence of 2 humic fractions. Soil Biol Biochem 23:833–836

32. Piccolo A, Nardi S, Concheri G (1992) Structural characteristics of humic substances as related to nitrate uptake and growth-regulation in plant-systems. Soil Biol Biochem 24:373–380

33. Canellas LP, Olivares FL, Okorokova-Facanha AL, Facanha AR (2002) Humic acids isolated from earthworm compost enhance root elongation, lateral root emergence, and plasma membrane H+-ATPase activity in maize roots. Plant Physiol 130:1951–1957

34. Quaggiotti S, Ruperti B, Pizzeghello D, Francioso O, Tugnoli V, Nardi S (2004) Effect of low molecular size humic substances on nitrate uptake and expression of genes involved in nitrate transport in maize (Zea mays L.). J Exp Bot 55:803–813

35. Muscolo A, Sidari M, Attinà E, Francioso O, Tugnoli V, Nardi S (2007) Biological activity of humic substances is related to their chemical structure. Soil Sci Soc Am J 71:75–85

36. Nardi S, Muscolo A, Vaccaro S, Baiano S, Spaccini R, Piccolo A (2007) Relationship between molecular characteristics of soil humic fractions and glycolytic pathway and krebs cycle in maize seedlings. Soil Biol Biochem 39:3138–3146

37. Canellas LP, Piccolo A, Dobbss LB, Spaccini R, Olivares FL, Zandonadi DB, Façanha AR (2010) Chemical composition and bioactivity properties of size-fractions separated from a vermicompost humic acid. Chemosphere 78:457–466

38. Canellas LP, Dantas DJ, Aguiar NO, Peres LEP, Zsögön A, Olivares FL, Dobbss LB, Façanha AR, Nebbioso A, Piccolo A (2011) Probing the hormonal activity of fractionated molecular humic components in tomato auxin mutants. Ann Appl Biol 159:202–211

39. Pizzeghello D, Francioso O, Ertani A, Muscolo A, Nardi S (2012) Isopentenyladenosine and cytokinin-like activity of four humic substances. J Geochem Explor 129:103–111

40. Trevisan S, Pizzeghello D, Ruperti B, Francioso O, Sassi A, Palme K, Quaggiotti S, Nardi S (2010) Humic substances induce lateral root formation and expression of the early auxin-responsive IAA19 gene and DR5 synthetic element in Arabidopsis. Plant Biol 12:604–614

41. Trevisan S, Botton A, Vaccaro S, Vezzaro A, Quaggiotti S, Nardi S (2011) Humic substances affect Arabidopsis physiology altering the expression of genes involved in primary metabolism, growth and development. EEB 74:45–55

42. Zancani M, Petrussa E, Krajňáková J, Casolo V, Spaccini R, Piccolo A, Macrì F, Vianello A (2009) Effect of humic acids on phosphate level and energetic metabolism of tobacco BY-2 suspension cell cultures. Env Exp Bot 65:287–295

43. Zancani M, Bertolini A, Petrussa E, Krajňáková J, Piccolo A, Spaccini R, Vianello A (2011) Fulvic acid affects proliferation and maturation phases in Abies cephalonica embryogenic cells. J Plant Physiol 168:1226–1233

44. Muscolo A, Cutrupi S, Nardi S (1998) IAA detection in humic substances. Soil Biol Biochem 30:1199–1201

45. Russell L, Stokes AR, Macdonald H, Muscolo A, Nardi S (2006) Stomatal responses to humic substances and auxin are sensitive to inhibitors of phospholipase A(2). Plant Soil 283:175–185

46. Piccolo A, Conte P, Trivellone E, Van Lagen B, Buurman P (2002) Reduced heterogeneity of a lignite humic acid by preparative HPSEC following interaction with an organic acid. Characterization of size-separates by PYR-GC-MS and 1H-NMR spectroscopy. Envir Sci Technol 36:76–84

47. Nardi S, Sessi E, Pizzeghello D, Sturaro A, Rella R, Parvoli G (2002) Biological activity of soil organic matter mobilized by root exudates. Chemosphere 46:1075–1081

48. Bradford MM (1976) A rapid and sensitive method for the quantitation of microgram quantities of protein utilizing the principle of protein-dye binding. Anal Biochem 72:248–254

49. Lewis OAM, James DM, Hewitt EJ (1982) Nitrogen assimilation in barley (Hordeum vulgare L. cv. Mazurka) in response to nitrate and ammonium nutrition. Ann Bot 49:39–49

50. Lillo C (1984) Diurnal variations of nitrite reductase, glutamine synthetase, glutamate synthase, alanine aminotransferase and aspartate aminotransferase in barley leaves. Physiol Plant 61:214–218

51. Cánovas FM, Cantón FR, Gallardo F, García-Gutiérrez A, De Vicente A (1991) Accumulation of glutamine synthetase during early development of maritime pine (Pinus pinaster) seedlings. Planta 185:372–378

52. Rej R, Horder M (1983) Aspartate aminotransferase (glutamate oxaloacetate transaminase). In: Methods of enzymatic analysis, 3rd edition. Bergmeyer, H. U, Weinheim

53. Lima JD, Sodek L (2003) N-stress alters aspartate and asparagine levels of xylem sap in soybean. Plant Sci 165:649–656

54. Bergmeyer HU (1986) In: Bergmeyer J, Marianne G (ed) Methods of enzymatic analysis. Weinheim, Basel

55. Goldberg DM, Ellis G (1986) Isocitrate. In: Bergmeyer HU (ed) Methods of enzymatic analysis. Academic, New York

56. Seebauer JR, Moose SP, Fabbri BJ, Crossland LD, Below FE (2004) Amino acid metabolism in maize earshoots. Implications for assimilate preconditioning and nitrogen signaling. Plant Physiol 136:4326–4334

57. Henderson JW, Ricker RD, Bidlingmeyer BA, Woodward C (2000) Rapid, accurate, sensitive and reproducible analysis of amino acids. Agilent Publication Number 5980-1193EN. Agilent Technologies, Palo Alto

58. Sambrook J, Fritsch EF, Maniatis F (1989) Molecular cloning: a laboratory manual, 2nd edition. Cold Spring Harbor Laboratory, Cold Spring Harbor, New York

59. Quaggiotti S, Ruperti B, Borsa P, Destro T, Malagoli M (2003) Expression of a putative high-affinity NO$_3^-$ transporter and of an H+-ATPase in relation to whole plant nitrate transport physiology in two maize genotypes differently responsive to low nitrogen availability. J Exp Bot 54:1023–1031

60. Sanger F, Nicklen S, Coulson AR (1977) DNA sequencing with chain-terminating inhibitors. PNAS 74:5463–5467

61. Sokal RR, Rohlf FJ (1969) Biometry, 1st edition. Freeman & Co, San Francisco

62. Stevenson FJ (1994) Humus chemistry: genesis, composition, reactions, 2nd edition. Wiley, New York

63. Nardi S, Pizzeghello D, Muscolo A, Vianello A (2002) Physiological effects of humic substances on higher plants. Soil Biol Biochem 34:1527–1536

64. Pinton R, Cesco S, Iacolettig G, Astolfi S, Varanini Z (1999) Modulation of NO$_3^-$ uptake by water-extractable humic substances: involvement of root plasma membrane H(+)ATPase. Plant Soil 215:155–161

65. Sessi E, Nardi S, Gessa C (2000) Effects of low and high molecular weight humic substances from two different soils on nitrogen assimilation pathway in maize seedlings. Humic Subst Environ 2:39–46

66. Malcolm RE, Vaughan D (1979) Humic substances and phosphatase activities in plant tissues. Soil Biol Biochem 11:253–259

67. Mato MC, Olmedo MG, Mendez J (1972) Inhibition of indoleacetic acid-oxidase by soil humic acids fractionated on sephadex. Soil Biol Biochem 4:469–473

68. Nardi S, Pizzeghello D, Gessa C, Ferrarese L, Trainotti L, Casadoro G (2000) A low molecular weight humic fraction on nitrate uptake and protein synthesis in maize seedlings. Soil Biol Biochem 32:415–419

69. Piccolo A (2001) The supramolecular structure of humic substances. Soil Sci 166:810–833

70. Hirel B, Bertin P, Quillere I, Bourdoncle W, Attagnant C, Dellay C, Gouy A, Cadiou C, Retailliau S, Falque M, Gallais A (2001) Towards a better understanding of the genetic and physiological basis for nitrogen use efficiency in maize. Plant Physiol 125:1258–1270

71. Azevedo RA, Lancien M, Lea PJ (2006) The aspartic acid metabolic pathway, an exciting and essential pathway in plants. Amino acids 30:143–162

72. Hirel B, Martin A, Terce-Laforgue T, Gonzalez-Moro MB, Estavillo JM (2005) Physiology of maize I: a comprehensive and integrated view of nitrogen metabolism in a C4 plant. Physiol Plant 124:167–177

73. Lea PJ, Sodek L, Parry MAJ, Shewry PR, Halford NG (2007) Asparagine in plants. Ann Appl Biol 150:1–26

74. Smejkalova D, Piccolo A (2008) Aggregation and disaggregation of humic supramolecular assemblies by NMR diffusion ordered spectroscopy (DOSY-NMR). Envir Sci Technol 42:699–706

75. Azevedo RA, Arruda P, Turner WL, Lea PJ (1997) The biosynthesis and metabolism of the aspartate derived amino acids in higher plants. Phytochem 46:395–419

76. Nikolic M, Cesco S, Römheld V, Varanini Z, Pinton R (2003) Uptake of iron complexed to water-extractable humic substances by sunflower leaves. J Plant Nutr 26:2243–2252

15

Effects of on-farm composted tomato residues on soil biological activity and yields in a tomato cropping system

Catello Pane[1], Giuseppe Celano[2], Alessandro Piccolo[3], Domenica Villecco[1], Riccardo Spaccini[3], Assunta M Palese[2] and Massimo Zaccardelli[1*]

Abstract

Background: The use of compost may relieve the factors that limit productivity in intensive agricultural systems, such as soil organic matter depletion and soil sickness. Concomitantly, the practice of on-farm composting allows the recycle of cropping green residues into new productive processes.

Results: We produced four vegetable composts by using tomato biomass residues in an on-farm composting plant. The tomato-based composts were assessed for their chemical, microbiological properties, and their effects on soils and plants were evaluated after their application within a tomato cropping system. Compost characteristics affected plant development and productivity through increased nutrient uptake and biostimulation functions. Soil biological activities, including basal respiration, fluorescein diacetate hydrolysis, β-glucosidase, dehydrogenase, alkaline phosphatase, arylsulphatase, and Biolog community levels of physiological profiles, were differently affected by the on-farm tomato-based composts.

Conclusions: Changes in soil activity and community structure due to compost amendments were related to classes of biomolecules such as polysaccharides and lignin-derived compounds, as revealed by nuclear magnetic resonance (NMR) spectra of compost materials. The nutrient content and fertility potential of composts were positively related to the amount of tomato residues present in the feedstock.

Keywords: Carbon structures; C-CPMAS-NMR; Soil microbial activity; Tomato yield; Vegetable compost

Background

On-farm composting is an efficient, cost-effective and environmentally safe biological process for the recycling of residual agricultural biomasses into new cropping production cycles [1]. It is a simple technology consisting of user-friendly small composting plants equipped with tools already available on a farm, where undegraded organic biomasses are transformed and stabilized through an aerobic biooxidation [2]. On-farm composting substantially contributes to solve the problem of disposing agricultural biomasses and vegetable feedstock and concomitantly provides the farmer with a self-supply of quality compost for the improvement of agricultural productivity.

Loss of soil quality is related to soil organic matter (SOM) depletion that is increased by continuous cropping without rotations, frequent soil tillage and large use of both inorganic chemical fertilizers and non-selective pesticides. Intensively exploited soils need an external supply of stabilized organic matter, such as compost, in order to counteract progressive SOM decline. Soil compost amendments contribute to the general soil quality recovery and improvement of plant growing conditions [3] by providing numerous ecosystem services, including replenishment of soil carbon stocks, increase of microbial activity and biodiversity and restoration of plant nutrition and natural soil suppressiveness [4].

In some developed horticultural areas of Southern Italy, significant amounts of agricultural wastes, such as cropping residues, unmarketable products and vegetable processing leftovers, are currently produced. They represent an

---

* Correspondence: massimo.zaccardelli@entecra.it
[1]Consiglio per la Ricerca in Agricoltura e l'Analisi dell'Economia Agraria, Centro di Ricerca per l'Orticoltura, via dei Cavalleggeri 25, I-84098 Pontecagnano, SA, Italy
Full list of author information is available at the end of the article

important source of organic matter to be composted and returned to soil. Tomato (*Solanum lycopersicum*) green wastes from greenhouse systems produce about 15 t ha^{-1} y^{-1} of fresh plant residues and are among the most abundant biomasses suitable for transformation in compost.

López-Pérez et al. [5] proposed the direct incorporation of tomato residues into soil as a green biofumigating practice, but it failed in controlling nematode *Meloidogyne incognita* infestation. Risks of plant pathogen dissemination and phytotoxicity hazards are eliminated when an effective sanitation of tomato wastes is achieved through a thermophilic composting process before amendment to soils [6,7]. Although some studies focused on tomato plant composting [8,7], little attention has been so far paid to assess the agronomic effectiveness of the produced compost. By assuming that on-farm composting of tomato plant wastes is the best sustainable practice to improve soil quality, our aim was to investigate (i) the effects of field compost amendments on tomato yields and resulting soil biological characteristics, (ii) the quality of on-farm composts from tomato plant residues in comparison with a commercial organic waste compost, and (iii) the molecular biomarkers which could differentiate tomato-based composts according to different amounts of tomato and other composted additives.

Methods
On-farm composting
Tomato plant residues were used as main compost feedstock, while escarole (*Cichorium endivia*) residues, wood chips and mature compost as starter were also added. The four composting piles had the following compositions: C_1, 17.5% tomato, 15.5% escarole residues, 65% woodchips and 2% mature compost as starter; C_2, 25% tomato residues, 13% escarole residues, 60% woodchips and 2% mature compost as starter; C_3, 37% tomato residues, 11% escarole residues, 50% woodchips and 2% mature compost as starter; and C_4, 50% tomato residues, 48% woodchips and 2% mature compost as starter. All four raw piles were set up with an initial C/N ratio of about 30 in order to hasten the composting switch-on. The mature compost starter was a 2-year-old C_{OW}, purchased at Gesenu (Perugia, Italy). The on-farm composting process was carried out in four parallel static piles of about 6 m^3 in volume, under forced aeration, through an overall 90-day cycle that included a thermophilic and a mesophilic phase, followed by a final curing period. The on-farm composting system was assembled by using currently available tools in common farms. Mechanical aeration was provided by air injection through a net of tubes connected to a blower (0.75 KW) that was periodically activated (5 min every 3 h) with an electronic timer. Pile wetting was achieved through a PVC irrigation system, manually activated on demand (when RH < 50%). Composting temperatures were measured by thermo-sensors placed in the pile core at 15 cm from the pile bottom.

Results
On-farm compost characteristics
Chemical features of feedstock and composts are reported in Tables 1 and 2. Compost samples exhibited a sub-alkaline þH value (>8.0). The levels of electrical conductivity and macronutrients, including N, P and K, increased with the amount of tomato residues used, while, in all cases, the heavy metal contents detected were below risk levels according to Italian laws.

The ^{13}C cross polarization magic angle spinning nuclear magnetic resonance (^{13}C-CPMAS-NMR) spectra of compost materials were characterized by strong signals in the O-alkyl-C (61 to 110 ppm) region, revealing a molecular composition dominated by carbohydrates (Figure 1). In fact, the signals related to O-Alkyl-C components represent most of the organic carbon, accounting for 42.5% up to 56.3% of the total area of the nuclear magnetic resonance (NMR) spectra. The different resonances in the O-alkyl-C region are currently assigned to monomeric units in oligo and polysaccharide chains of plant tissue [9]. The intense signal around 72 ppm corresponds to the overlapping resonances of carbon 2, 3 and 5 in the pyranoside structure in cellulose and some hemicelluloses, whereas the signal at 106 ppm is assigned to the anomeric carbon 1 of the glucose unit in cellulose [10]. The shoulders localized around 62 to 65 and 84 to 88 ppm results from carbon 6 and 4 of monomeric units, respectively. The low-field resonances (higher chemical shift) of each pair indicate the presence of crystalline forms of cellulose, while the high-field ones (lower chemical shift) are assigned to either amorphous

Table 1 Chemical determinations on plant residues used as composting feedstock

Residues	Chemical features											
	N (%)	P (%)	K (%)	Ca (%)	Mg (%)	Na (%)	Mn (ppm)	Cd (ppm)	Cr (ppm)	Cu (ppm)	Pb (ppm)	Zn (ppm)
Tomato	2.1	0.0062	3.04	1.46	0.20	0.57	140.94	0.253	32.33	55.81	1.06	nd
Escarole	3.8	0.0154	3.09	1.01	0.17	0.33	35.71	0.166	4.31	112.75	nd	nd
Woodchip	1.0	0.0004	0.06	0.39	0.06	0.12	7.96	0.143	6.71	5.42	nd	nd

nd, not detected.

Table 2 Main chemical quality characteristics of the composts

Composts	Chemical features													
	pH	EC (mS cm^{-1})	N (%)	P (%)	K (%)	Ca (%)	Mg (%)	Na (%)	Mn (ppm)	Cd (ppm)	Cr (ppm)	Cu (ppm)	Pb (ppm)	Zn (ppm)
C_1	8.40	2.69	1.25	0.023	1.46	3.87	1.02	0.20	328.99	0.47	34.80	40.16	4.17	64.62
C_2	8.19	4.12	1.23	0.020	1.21	5.32	1.25	0.15	297.42	0.51	34.84	55.68	4.58	108.50
C_3	8.12	5.09	1.41	0.045	1.99	4.43	1.18	0.22	415.18	0.45	57.96	52.96	4.87	140.40
C_4	8.31	8.92	1.52	0.048	1.92	4.90	1.23	0.15	260.96	0.58	17.99	45.02	3.09	57.06
C_{OW}	8.93	5.07	2.72	0.029	1.18	6.58	0.40	0.27	427.30	0.30	16.05	45.98	28.15	247.40
Legal limits										<1.5	<100	<150	<140	<500

cellulose or hemicellulose structures [11]. The various O-alkyl regions could also include signals related to carbon in the propylic side chain of lignin molecules, whose smaller resonances around 62, 72 and 82 ppm, could be masked by the predominance of polysaccharides. Besides the signals usually assigned to cellulose, the spectra of different composts revealed two additional resonances around 98 and 101 ppm. These signals may be related to the di-O-alkyl-C of, respectively, monomeric units of simple carbohydrates [10] and those of either hemicellulose or pectic polysaccharide chains contained in cell walls of tomato plants, such as α-1,5 arabinan, β-1,4 galactan and α-1,4 galacturonan [12]. The broad peak in the Alkyl-C region (0 to 45 ppm) of the NMR spectra indicated the presence of alkyl chains (-CH_2- groups) derived mainly from various lipid compounds, plant waxes and polyesters. The signal at 56 ppm is associated with either the methoxyl substituent on the aromatic rings of guaiacyl and syringyl units in lignin structures or the C-N bonds in amino acid moieties [9]. Moreover, this O-alkyl region may also include the resonances related to ether and epoxy groups of plant

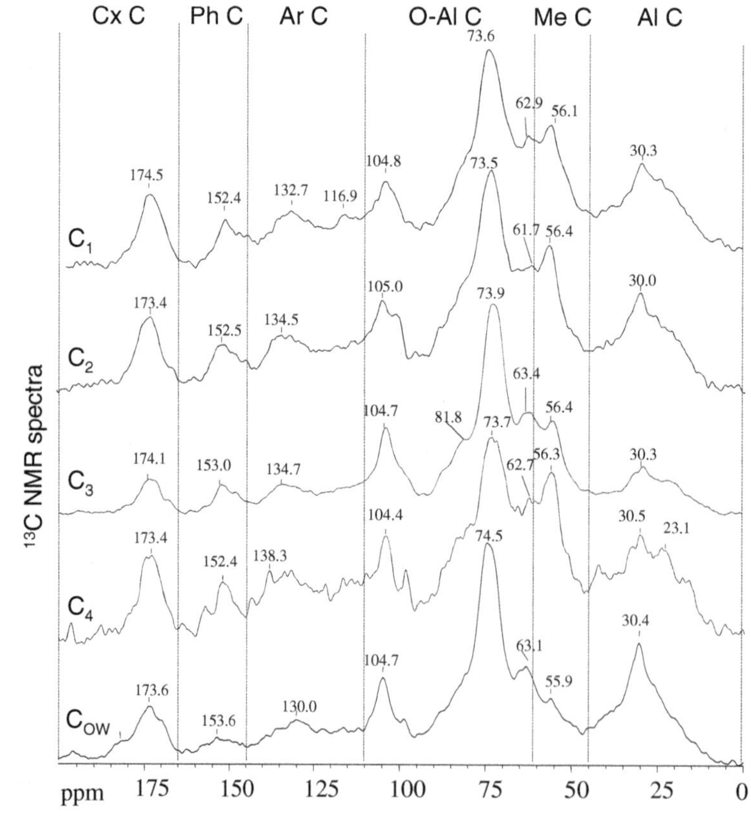

Figure 1 ^{13}C-CPMAS-NMR spectra of compost (C_1 to C_{OW}) samples. Vertical lines delimitate six different spectral regions: aliphatic and aromatic carboxyl C (Cx C, 190 to 166 ppm); oxygen-substituted aromatic C from lignin and non-hydrolyzable tannins, phenolic and O-aryl (Ph C, 165 to 146 ppm); unsubstituted and alkyl-substituted aromatic C, aryl (Ar C, 145 to 111 ppm); anomeric C and di-Oalkyl and oxidized and/or carbohydrate C, O-alkyl (O-Al C, 110 to 61 ppm); methoxyl/N-alkyl (Me C, 60 to 46 ppm); and aliphatic C, alkyl (Al C, 45 to 0 ppm).

Table 3 Relative distribution (%) of signal area over chemical shift regions (ppm) in ^{13}C-CPMAS-NMR spectra of the composts

	Carboxylic-C 195 to 166 ppm	Phenolic-C 165 to 146 ppm	Aromatic-C 145 to 111 ppm	O-Alkyl-C 110 to 61 ppm	CH$_3$O/C-N 60 to 46 ppm	Alkyl-C 45 to 0 ppm
C$_1$	8.56	4.79	13.18	43.41	11.82	8.24
C$_2$	7.46	4.62	13.12	44.82	12.08	17.90
C$_3$	5.61	4.15	11.50	52.97	11.39	14.37
C$_4$	6.25	3.75	14.55	42.41	13.05	19.99
C$_{OW}$	8.36	3.26	13.08	52.49	9.22	24.84

biopolyesters. In the aromatic/olefinic-C region (111 to 145 ppm), the different resonances around 116 and 130 ppm are related to unsubstituted and C-substituted phenyl carbon pertaining to lignin monomers of guaiacyl and syringyl units [11] as well as to the ring components of plant polyphenols. The signals shown in the specific phenolic aromatic region (146 to 165 ppm)

confirmed the presence of O-substituted ring carbon derived from different aromatic structures. In fact, the resonances included in the 148- to 155-ppm range are usually assigned to carbon 3, 4 and 5 in the aromatic ring in lignin components, carbon 3 and 5 being coupled to the corresponding methoxyl substituents. Conversely, the peaks found at 143 and 157 ppm in the NMR spectra of C$_4$ suggest the significant incorporation of polyphenol derivatives originating from tomato residues [10]. Finally, the broad signal at 173 ppm indicates the contribution of carbonyl groups of aliphatic acids and amino acid moieties in all the compost materials. The ^{13}C-CPMAS-NMR signals exhibited differences among composts (Table 3). The aliphatic alkyl C region (45 to 0 ppm) was most evident in commercial organic-waste compost (C$_{OW}$), followed by C$_1$ and C$_4$ then C$_2$ and C$_3$. The CH$_3$O/C-N region (60 to 46 ppm) was slightly variable among composts. The O-alkyl C region (110 to 61 ppm) was largely developed in C$_3$, whereas it was less noticeable in the remaining samples. Moreover, the intensity of the region associated with aromatic C (145 to 111 ppm) was large for C$_4$, decreased in the order passing from C$_1$ and C$_2$. Conversely, the spectral regions associated to phenolic C (165 to 146 ppm) was relevant in C$_2$ and limited in C$_{OW}$, while that for carboxyl C (195 to 166 ppm) was smaller in C$_3$ than for the rest of the other composts. Fluorescein diacetate hydrolytic (FDAH) activity resulted as the largest for C$_4$ and was followed, in the

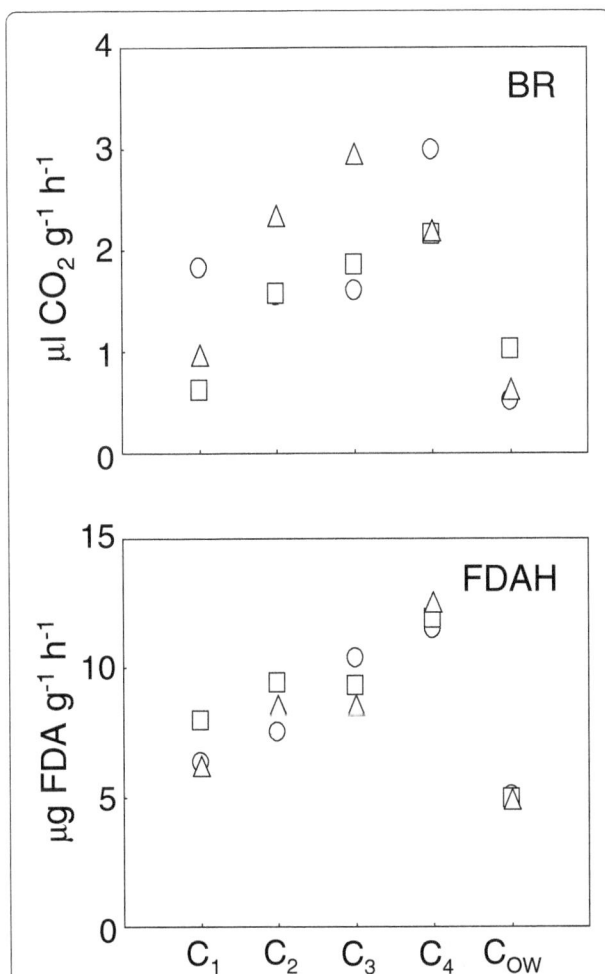

Figure 2 Biological indicators of compost stability. They were measured as basal respiration (BR) and FDA hydrolysis (FDAH) in three replicates (indicated separately with circle, triangle and square) for each compost sample.

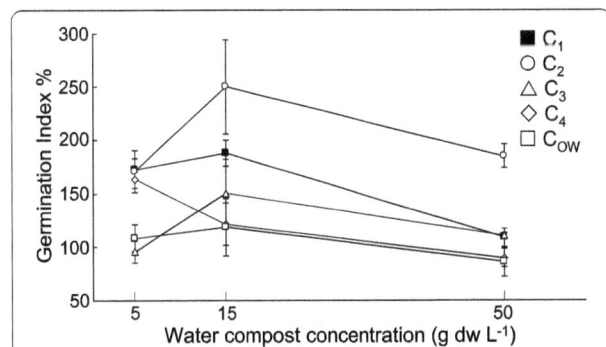

Figure 3 Evaluation of the cress germination index on water extracts of composts. C$_1$ to C$_{OW}$ samples were assayed at high, medium and low concentrations (50, 16.6 and 5 g l^{-1}, respectively).

Table 4 Effects of soil treatments on tomato cropping response

	Green biomass (Q ha^{-1})	Yield			Quality of berries		
		Total (t ha^{-1})	Marketable (t ha^{-1})	Discard (t ha^{-1})	Weight (g)	pH	Optical residue (°Bx)
C_1	98.05 bc	66.6 de	62.5 c	8.51 a	75.8 a	4.15 a	3.02 a
C_2	117.08 b	83.3 bcd	75.7 bc	10.70 a	84.3 a	4.09 a	3.37 a
C_3	128.13 b	73.6 cde	69.6 bc	8.72 a	78.4 a	4.15 a	2.98 a
C_4	149.38 a	93.1 abc	88.1 ab	13.80 a	86.9 a	4.19 a	3.71 a
C_{OW}	91.81 c	60.6 e	55.1 c	8.35 a	82.0 a	4.03 a	3.03 a
M_{NR}	144.24 a	105.8 a	101.2 a	12.1 a	74.3 a	4.31 a	3.44 a
M_{SR}	148.61 a	100.6 ab	96.7 a	12.2 a	74.9 a	4.20 a	3.56 a
CTRL	109.72 bc	73.0 cde	68.2 bc	12.4 a	81.0 a	4.17 a	2.81 a

Different letters indicate significant differences (ANOVA, Duncan's test, $P \leq 0.05$).

order, by C_3, C_2 and C_1, while C_{OW} showed the smallest value (Figure 2). Similarly, basal respiration (BR) was the greatest for the C_4 compost.

Phytotoxicity of on-farm composts

On-farm compost water extracts proved variable effects on cress germination index percentage (GI%) (Figure 3). In fact, while C_2 showed the lowest toxicity, the one observed for C_{OW} was the largest. Germination was increasingly repressed by on-farm compost extracts passing from C_1 to C_4. In the case of C_1, C_2 and C_3 extracts, the percent cress GI% showed a singular pattern, since it significantly increased at an intermediate concentration, while it dropped back down at the lowest concentration. C_4 and C_{ow} extracts exhibited a dose-dependent behaviour.

Crop response to soil amendment with on-farm composts

The commercial and total yields of C_4-treated plots were larger by about 20 t ha^{-1} than for control plots. The remaining on-farm composts (C_1, C_2 and C_3) led to increasing yields that did not differ significantly from those of untreated plots (Table 4). On-farm composts did not show any phytotoxic symptoms, whereas C_{OW} that induced the lowest yield caused a slight growth reduction in the early phases of the crop cycle. No significant differences were observed regarding the discarded production. Plant weight was significantly affected by treatments (Table 4). Similarly, berry quality (single weight, pH and optical residue) was not significantly affected by treatments, nor was plant physiologically status, which was generally observed to be at standard levels, as confirmed by chlorophyll content, likely sustained by nitrate availability throughout the tomato cycle (Figure 4). All raw compost eluates showed *in vitro* antibiosis against *Fusarium oxysporum* f. sp. *lycopersici* (data not shown).

Effects of on-farm composts on soil properties

In order to assess the impact of compost amendments on soil properties, a set of biological indicators was used. The BR analyses showed an initial burst of activity due to compost amendments that approached the control over time. Levels of BR were in the following order: C_{OW}, C_4, C_1, C_2 and C_3 (Figure 5). Soil enzymatic activities that were also significantly activated by composts showed a durable effect during the whole incubation time (Figure 5). The largest values of FDAH, β-glucosidase (βGLU) and

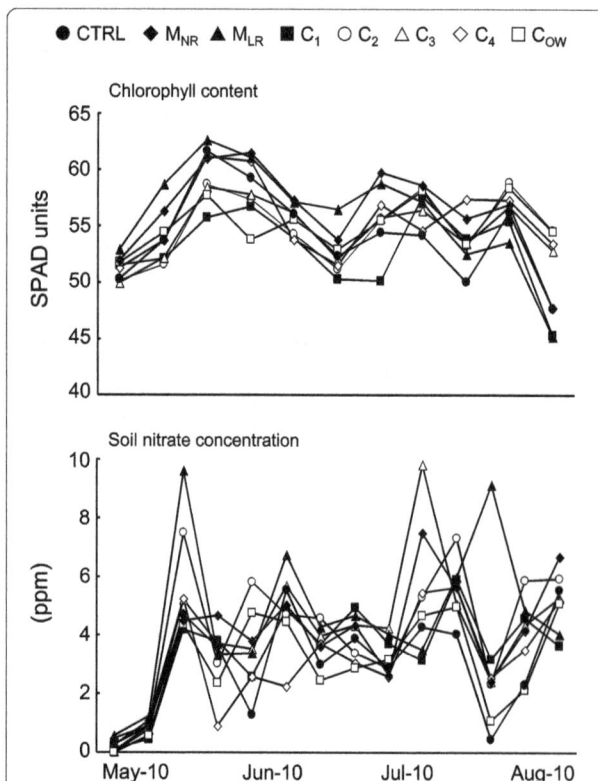

Figure 4 Plant physiological and nutritional status. It was evaluated weakly by the chlorophyll content, assessed by SPAD, and soil nitrate concentration that was available for plant nutrition.

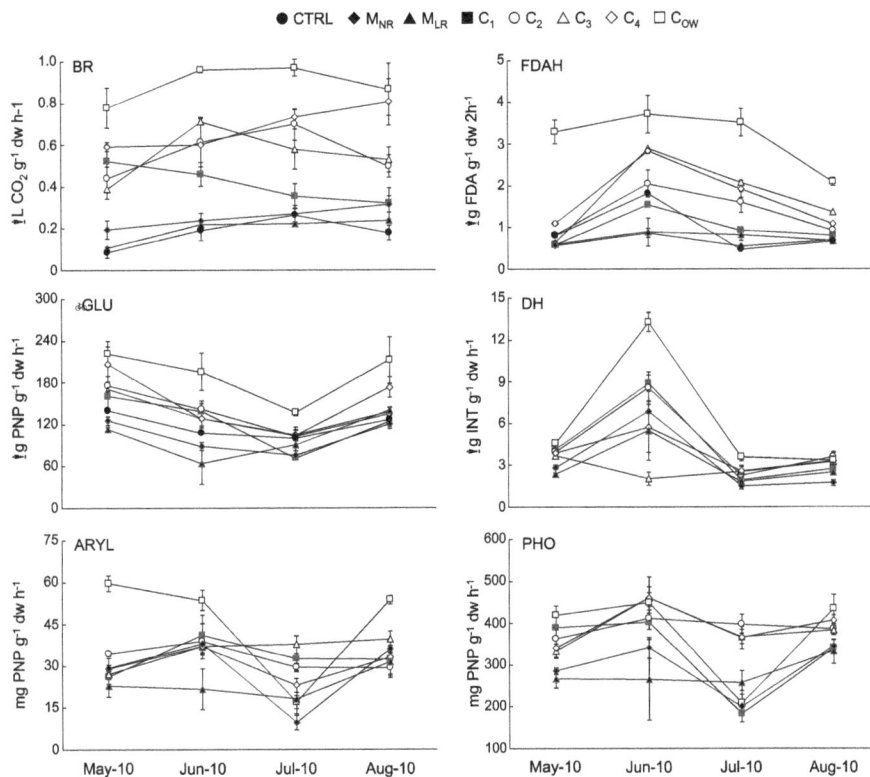

Figure 5 Changes over time in soil BR, FDAH, βGLU, DH, ARYL and PHO. They were assessed monthly in experimental plots after soil treatments.

dehydrogenase (DH) activities were observed in soils amended with the municipal waste compost. Similarly, arylsulphatase (ARYL) and alkaline phosphatase (PHO) activities that resulted in large values for soils treated with tomato-based compost also showed peaks of activity only at the later stage of incubation (Figure 5). βGLU and DH, on the other hand, showed almost constant values over the whole incubation time.

The relationships between carbon distribution in compost and soil enzymatic activities during the incubation period were elucidated by calculating the Pearson's coefficient between these two variables. In fact, coefficient profiles for cumulative regressions were generated: interestingly, it was found that polysaccharides, as well as degradation forms of lignin, produced the most significant correlations (Figure 6). Average well colour development (AWCD) and Shannon index (H′) temporal shifts showed that on-farm composts significantly increased a progressive functional diversity passing from C_1 to C_4 (Figure 7). The soil amended with municipal waste compost showed an intermediate behaviour as compared to that of C_2. However, the activity enhanced by compost treatments slightly, but substantially, regressed at the end of incubation time.

To show differences in microbial community structure, levels of carbon source catabolism were subjected

to principal component analysis (PCA). The PC1 explained 61.39% of the variance, while PC2 explained only 12.53% (Figure 7). Along the PC1 axis, the compost-amended plots clustered together and resulted different from the unamended plots, though the C_4 cluster was most distant from the rest. In general, the communities in each plots grouped closely, thus indicating a little influence from sampling time. PC1-variable correlation resulted as significant ($R > 0.60$) and negative for all carbon sources (factor loadings), with the exception of hydroxy benzoic acid, d-malic acid, l-asparagine and phenylethylamine, that were not significantly correlated. Instead, only l-arginine carbon source resulted as significantly correlated ($R - 0.67$) with PC2.

Discussion
Chemical characteristics of agricultural composts and stability
On-farm composting of crop residues is an effective method to produce highly humified organic matter from agricultural green wastes, while they are usefully recycled according to a concept of agricultural sustainability [13]. Moreover, these particular feedstocks significantly influenced compost properties and their ability to condition soil and plant response.

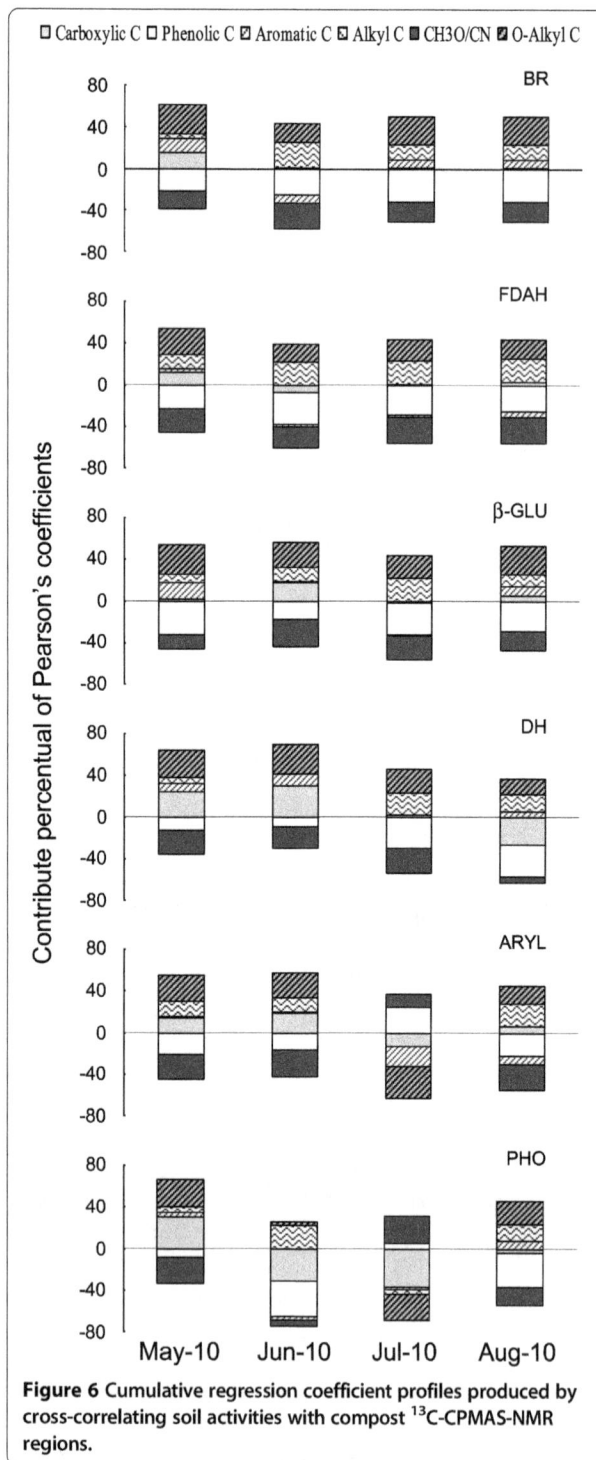

Figure 6 Cumulative regression coefficient profiles produced by cross-correlating soil activities with compost ^{13}C-CPMAS-NMR regions.

152 ppm, thereby revealing significant amounts of both alkyl components and lignin derivatives. This finding suggests that the inclusion of larger initial rates of stabilizing lignocellulosic materials, represented by wood chips, promoted the incorporation of stable and recalcitrant organic components. Conversely, the NMR spectra of the C_3 compost, showed a lower content of hydrophobic alkyl and aromatic compounds and a corresponding relative increase of more biolabile O-alkyl C components. Among the on-farm composts, evidently different characteristics were found in the C_4 sample, which was made with the initial larger amount of tomato residues. Unlike the previous composts, in addition to the peak at 30 ppm, many distinct signals were shown in the broad alkyl-C region (0 to 45 ppm), thus suggesting the simultaneous presence of different alkyl chains from linear and branched fatty acids and peptidic derivatives [14]. The inconsistency between the sharp intense peaks shown at 56 ppm, as compared to the low abundance of the O-aromatic lignin components in the 148 to 155 interval, also suggested the large contribution of peptidic moieties to the global resonance in the 46 to 60 ppm region, as also indicated by the larger N content found in the C_4 compost. Furthermore, the permanence of biolabile organic compounds was stressed by the peaks positioned at 43 and 98 ppm assigned, respectively, to Cα and Cβ of amino acids [15] and to C1 carbon of monosaccharides components [10]. Lastly, the C distribution found in the NMR spectra of the commercial C_{OW} compost was characterized by the relative predominance of carbohydrates and alkyl-C, combined with the lowest amounts of aromatic and lignin components.

Stability is the compost property, which refers to microbial degradability of organic matter [16]. Changes in biological parameters have been indicated as reasonable and informative markers of compost stability since they shall be linkable with substrate availability for microbial growth [17]. Here, our composted residues showed increasing values of FDAH and BR according to their abundance in phenolic and aromatic-C and the amount of tomato residues used in feedstock. These hydrophobic moieties could be responsible for the maintenance of an unstable carbon reservoir formed by predominating alkyl-C and lignin-deriving compounds which was still subjected to microbial breakdown. These labile carbon pools, possibly, had already been widely removed in the commercial compost.

Effects of on-farm compost on plant growth and productivity

Plant growth sustainability by compost refers to its quality and potential for agricultural applications [18]. Since this property, indicated as compost maturity, is closely linked to the loss of phytotoxicity [19], it can be directly

Although the ^{13}C-CPMAS-NMR spectra of composted materials indicated an overall similar C distribution, the analysis of specific signals exhibited clear differences in molecular composition. The C_1 and C_2 on-farm composts were characterized not only by cellulosic polysaccharides but also by prominent signals at 30, 56 and

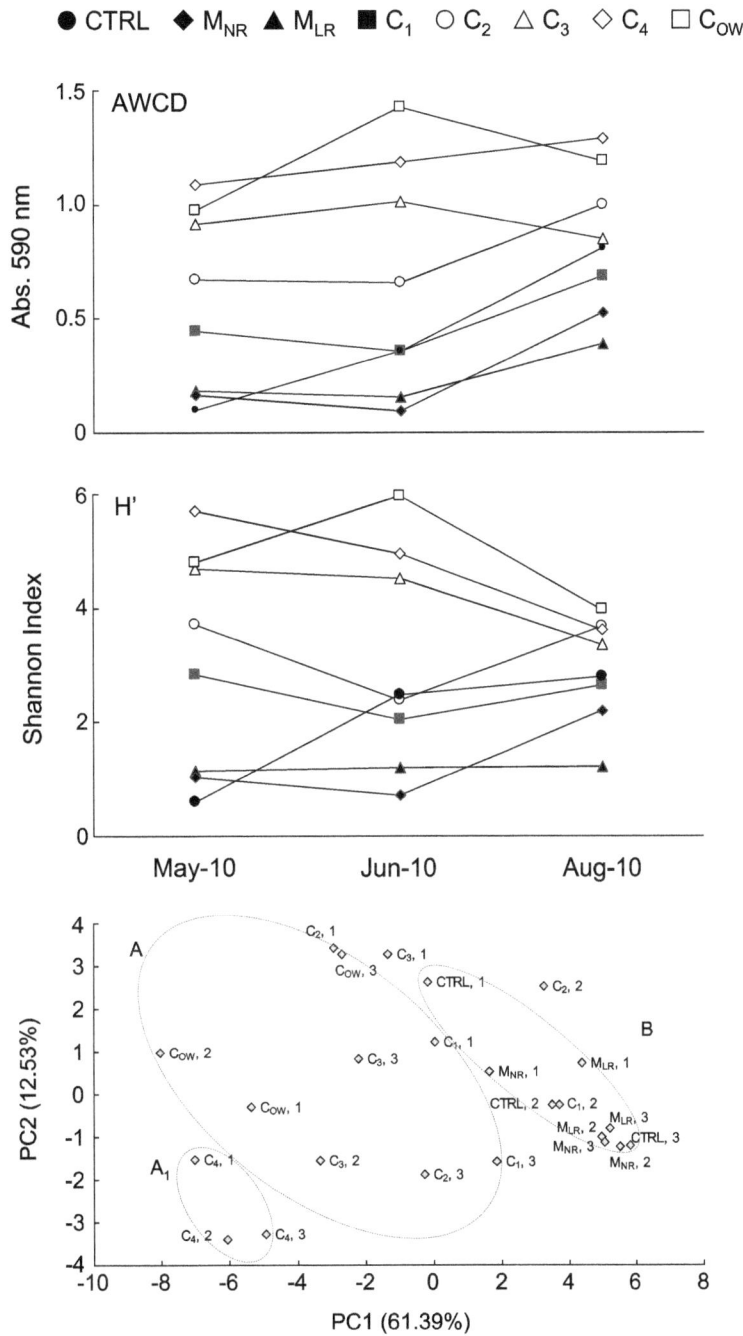

Figure 7 Changes in soil Biolog metabolic activity (AWCD), Shannon Index (H') and biplot PCA distribution of CLPPs. They were assessed in experimental plots after soil treatments at three time points of incubation.

assessed by evaluating the effects of compost eluates on seed germination [20]. In the current study, cress germination assay showed GI% levels exceeding the threshold value of 80%, which is indicative of mature and phytotoxin-free composts [21,22]. In some cases, further diluted compost extracts could have stimulate seed germination. Likely, this could be due to organic molecules

dissolved into compost water-extractable fractions, that provide seedling development promotion [22]. However, the detection and the individuation of this kind of molecules from our on-farm composts, needs of further investigations and could be an interesting future perspective of the current research. Growth stimulation effects also may occur thank to the compost humic fraction that, for example, could be responsible for root proliferation [23,24]. The productive response of the plants to composts may involve firstly nutrient supply and induction of well growth conditions till ripening. Here, tomato yield was observed significantly increased at levels of minerally fertilized plots, under C_4 amendment, as compared to the untreated soil. C_{OW}, instead, although was the most nutrient-richest compost, leads to lowest yields because of the initial detrimental impact on the plants. These findings suggest that the nutritional value alone cannot totally explain the agronomic performances of the composts. But, other factors, such as the absence of phytotoxicity [25], the activation of soil useful microorganisms [26] or disease suppressiveness mechanisms, improvement of soil physical propriety, should be totally considered. Remarkable reduction in crop yield under plant stress due to phytotoxic composts has been widely reported [27]. Furthermore, as development of this work, the extraction and characterization of humic water-extractable fraction in composted samples could contribute to clarify these aspects.

In the current study, the antifungal activity showed by composts against *F. oxysporum* f. sp. *lycopersici* indicated, in addition, their suppressive potential, which was not possible confirm in the field since, under natural conditions, tomato wilt disease did not occurred. Anyhow, a previous study showed that some of these tomato on-farm composts (C_1 and C_4) and C_{OW} significantly suppressed soil-borne diseases caused by *Rhizoctonia solani* and *Sclerotinia minor* and suppressiveness were related to their chemical characteristics and ^{13}C-CPMAS-NMR spectra [28]. Accordingly, Yogev et al. [29] reported the ability of composted tomato plants, to suppress Fusarium wilt on melon.

Effects of on-farm compost on soil biological activities

Compost amendments represent an extraordinary food event for soil microbes that play a crucial role in the turnover of all major plant nutrients. Soil biota, in fact, is involved in the SOM cycle by activating specific enzymatic pathways, through which complex carbon structures are transformed into simple organic and inorganic molecules that can be taken up by plants. External carbon supplies, such as compost additions, cause substantial shifts in SOM chemical composition and soil biological activity profiles [30]. These changes can be followed by monitoring the evolution overtime of soil quality indicators, such as microbiological and biochemical parameters [31]. The respiration rate reflects the instantaneous microbial activation induced by labile-C compounds [30]. In this study, the BR pattern was similar to those described previously by Pane et al. [26] and Cytrin et al. [32] with an initial burst of activity induced by compost, followed by a significant decrease approaching levels of the not-

amended soil. In fact, fluorescein degradation due to generalist enzymes, such as proteases, lipases and esterases [33], β-glucosidase and dehydrogenase, exhibited time-shifted trends. The kinetics of these enzymes are closely related to microbial polysaccharide breakdown with the release of low molecular weight organic compounds [31] and organic matter oxidation [34]. Biolog CLPPs showed prolonged changes in microbial structures due to compost supply and were observed over incubation time, as indicated by PCA analysis. More specific enzymatic activities, such as phosphatase, which hydrolyze organic P into phosphates and arylsulphatase, which hydrolyze aromatic sulphate esters into phenols and sulphate [35] showed, instead, different behaviours. Response of soils to amendment may be affected by the molecular quality of composts. The levels of soil activity were found positively related to NMR aromatics and polysaccharides over time, suggesting the involvement of these hydrophobic moieties in the modulation of the labile carbon flux for microbial activation. Phenolic-C passes from negative to positive correlation when sulphatase and phosphatase were strongly induced in soils. A number of previous studies highlighted the role of phenolic SOM in the regulation of soil enzymatic activities [36-38]. Jindo et al. [39] reported the up-regulation of phosphatase and other enzymes in biochar-blended composts, concomitantly to increases in lignin polyphenol oxidation. Accordingly, Grandy et al. [37] and Leinweber et al. [38] found that strong predictive character of oxidation and the depletion rates of plant-derived lignins on the intensity of microbial transformations occurred both in less degraded systems, such as forest ecosystems, as well as in secular cropped lands. Here, on-farm composts were rich in lignin-derived compounds thanks to the large contribution of plant residues. Lignin is a relatively stable constituent of SOM that can support a long-lasting broad-spectrum soil microbial activity [40]. In the present work, solid state ^{13}C-CPMAS-NMR spectroscopy revealed differences in NMR resonance signals characteristic for alkyl C (0 to 45 ppm) and aromatic C (111 to 145 ppm) among composts that may explain the differences in induced microbiological soil activities. The alkyl C spectral region includes the aliphatic macromolecular biolipids that were reported as typical biomarkers in green waste-derived composts [41]. Instead, the aromatic C type includes peaks at 152 ppm (O-substituted C in guaiacyl and syringyl units) and 130 ppm (unsubstituted C in *p*-hydroxy phenyl rings of cinnamic units in both lignin and suberin biopolymers) and indicates that the main components of these composts are just lignocellulosic-derived molecules and hydrophobic alkyl moieties [9]. Therefore, it is possible that these lignin residues can affect indirectly biological soil properties by prolonging carbon availability to microbes over time.

Chemical analyses and ^{13}C-CPMAS-NMR spectroscopy of compost samples

Total N was determined according to the Kjeldahl method. The contents of P, Ca, K, Mg, Na, Cd, Cr, Cu, Mn, Pb and Zn were determined, after compost acid digestion with a microwave oven, by ICP-OES (iCAP 6000 Series, Thermo Scientific. Waltham, MA, USA). The water content of the composts was determined after drying at 105°C for 72 h. Compost water-holding capacity was determined by measuring water content held against gravity in a filter-paper-lined funnel. Electrical Conductivity (EC) and pH were determined according to the official methods of the Italian National Society of Soil Science [42].

Molecular distribution of compost organic carbon was evaluated by ^{13}C-CPMAS-NMR spectroscopy. The ^{13}C-CPMAS-NMR spectra were obtained with a Bruker AVANCE™ 300 (Bruker BioSpin GmbH, Rheinstetten, Germany), equipped with a 4-mm wide bore MAS probe, operating at a ^{13}C resonating frequency of 75.475 MHz. Compost samples (100 to 150 mg) were packed in 4-mm zirconia rotors with Kel-F caps and spun at 13 ± 1 kHz. To account for possible inhomogeneity of the Hartmann-Hahn condition at high rotor spin rates, a 1H ramp sequence was applied in CP experiments during a contact time (CT) of 1 ms. The ^{13}C-CPMAS experiments implied a collection of 6,000 scans with 2,266 data points over an acquisition time of 25 ms and a recycle delay of 2.0 s. The Bruker Topspin 1.3 software was used to collect and process the NMR spectra. All free induction decays (FIDs) were transformed by applying a 4 k zero filling and a line broadening of 100 Hz. The areas for different ^{13}C resonances were assigned according to previous reports [4,28,43] into six integrating regions as follows: 0 to 45 ppm (alkyl C), 46 to 60 ppm (methoxyl C), 61 to 110 ppm (O-alkyl C), 111 to 145 ppm (aromatic C or aryl C), 146 to 165 ppm (phenolic C or O-aryl C) and 166 to 195 ppm (carboxylic C). The area of each spectral region was divided by the sum of all spectral areas in order to obtain a relative percentage.

Basal respiration, FDA hydrolysis and phytotoxicity of compost samples

BR and FDAH were measured with a modification of method described by Pane et al. [44]. Basal respiration was from a compost (50-g dry weight) wetted with water up to 80% of its water-holding capacity and placed in a jar (500 ml) with an airtight cap. Released CO_2 was measured using a CO_2 Analyser IRGA SBA-4 OEM (PP Systems, USA).

To evaluate fluorescein diacetate (FDA) hydrolysis, 2.5 g of compost was mixed with 15 ml of 0.2 M potassium phosphate buffered at pH 7.6, followed by the addition of 0.5 ml FDA solution (2 mg ml^{-1}). The mixture was shaken for 2 h in an orbital incubator and the hydrolysis reaction stopped by adding 15 ml CHCl$_3$/

CH$_3$OH (2:1 v/v). The reaction mixture was centrifuged (700×g) and the absorbance of the aqueous supernatant measured at 490 nm.

Composts water extracts (CWEs), prepared by vigorously shaking, were assessed for possible phytotoxicity by measuring germination and root elongation of cress (*Lepidium sativum* L. cv. Comune) under CWE treatments [45], as compared to the control H$_2$O. Experiments comprised three different CWE concentrations (50, 16.6 and 5 g l^{-1}) replicated 10 times. The number of seeds germinated and root length were recorded after 36 h following germination. GI% that was directly affected by phytotoxicity was then obtained by multiplying the number of germinated seeds by the relative mean root length, expressed as percentage of control accordingly to following formula [20]:

$$GI\% = \left(\frac{\text{No. of seeds germinated on CWEs}}{\text{No. of seeds germinated on water}}\right) \times \left(\frac{\text{Mean root length on CWEs}}{\text{Mean root length on water}}\right) \times 100$$

Assessment of *in vitro* suppressiveness of compost

In order to evaluate compost suppressiveness, we used the tomato fungal pathogen *F. oxysporum* f. sp. *lycopersici* Sacc. isolated from symptomatic plants. This fungus was maintained on a potato dextrose agar medium and stored in the fungi collection of the CRA-Centro di Ricerca per l'Orticoltura (Pontecagnano, Italy). Raw, autoclaved (122°C for 22 min) and filtered (with 0.22 mm sterilized millipore membrane, following bland centrifugation to precipitate suspended cells) CWEs, further diluted in water 1:10 vol., were used to evaluate the suppressive potential of composts against *F. oxysporum* f. sp. *lycopersici*, according to the well diffusion technique as developed by El-Masry et al. [46] and slightly modified by Pane et al. [47]. Compost pathogen suppression was assessed by measuring the CWE mycelial development inhibition as a percentage of growth reduction compared to the control plates.

Field experimental design and plant parameters

Filed experiments were carried out at the experimental farm of the CRA-Centro di Ricerca per l'Orticoltura, Battipaglia (40°35′02″ N; 14°58′50″ E), Salerno, Italy, on a clay loam soil (8.8-g organic C kg^{-1}, 1.0 g Kjeldahl N kg^{-1}, pH 7.4, 34.6% sand, 36% silt, 29.4% clay, in the top 0- to 0.40-m soil layer) and the experimental design adopted was a complete randomized block with three replicates, each consisting of a plot area of 25 m^2. Eight soil treatments were compared: on-farm composts (C$_1$, C$_2$, C$_3$ and C$_4$) and municipal organic waste compost (C$_{OW}$) amendments, mineral normal release (M$_{NR}$) and mineral low release (M_{LR}) nitrogen fertilizers; untreated plots (without any fertilizers and amendment) were used as the reference control (CTRL). Composts were applied at a rate of 30 t ha^{-1} dry weight according to previous works [48,26]. M_{NR} and M_{LR} consisted of the application of NPK synthetic fertilizers (N = 150 kg ha^{-1}; P_2O_5 = 60 kg ha^{-1}; K_2O = 50 kg ha^{-1}), in which nitrogen was ammonium nitrate and ENTEC®26 (a fertilizer containing 3,4-dimethylpyrazol phosphate, a nitrification inhibitor). Composts, PK and ENTEC®26 were incorporated into the soil, 1 week before transplanting, by rotovating, at a depth of 10 to 15 cm. Tomato plantlets (cv. Stone) were transplanted (29,000 plants ha^{-1}) in double rows. During the cultivation period, plant physiological status was evaluated by assessing foliar chlorophyll contents with Minolta Chlorophyll Meter SPAD-502 (Konica Minolta Sensing INC., Japan).

At the end of the crop cycle, total and commercial production and relative percentage of discard, as well as single-berry weight, were determined on an area of about 4 m^2 for each plot.

Soil sampling, nitrate determination and microbial activities

From May to August, after soil treatments, soil sampling was carried out by mixing 10 sub-samples that were taken from the top layer (0 to 20 cm) of each plot, sieved (2 mm), selected and stored at 4°C until biological laboratory determination. Soil moisture content was determined by measuring water content after soil drying at 50°C until constant weight. Water-holding capacity (field capacity) was determined by measuring water content held against gravity in a filter-paper-lined funnel.

Nitrate concentration in soil was analysed on weekly collected samples by colorimetric technique using Reflectoquant® strips read by a reflectometer RQflex® 10 (Merck, Darmstadt, Germany).

BR and FDAH were measured as reported above for the direct measure on composts. Soil βGLU and DH activities were determined as reported by Pane et al. [44] using 4-p-nitrophenyl-β-D-glucopyranoside (PNP) and 2-(p-iodophenyl)-3-(p-nitrophenyl)-5-phenyltetrazolium chloride (INT) substrates, respectively. βGLU was determined by adding 0.35 g of soil sample to 2 ml of 0.05 M maleate buffer, pH 5.0. The mixture was left for 5 min at 30°C, and the enzymatic reaction was started by adding 0.5 ml of 0.2 mM PNP. After incubation for 1 h at 37°C, the reaction was stopped by adding 0.5 ml of 0.5 M CaCl$_2$, 2 ml of 0.5 N NaOH and 5 ml of H$_2$O. After centrifugation at 1,500×g for 5 min at 5°C and after filtration of the aqueous phase, the absorbance of filtrates was measured at 398 nm. DH was determined by adding 0.5 g of soil sample to 1 ml of 0.2 M Tris buffer, pH 5.0. The enzymatic reaction was started by adding 0.5 ml of 0.2 mM INT. After incubation by shaking for 48 h at

37°C, the reaction was stopped by adding 10 ml of ethanol 96% N,N-dimethylformamide (1:1) and incubated in the dark by shaking for 1 h at room temperature. After centrifugation at 5,000×g for 5 min at 5°C and after filtration of the aqueous phase, the absorbance of filtrates was measured at 464 nm. Soil PHO and ARYL activities were determined using p-nitrophenyl phosphate and p-nitrophenyl sulphate as substrate, respectively [49]. A 0.5-g soil sample was added to 2 ml of maleate buffer, pH 6.5. The enzymatic reaction was started by adding 0.5 ml of 0.2 mM substrate. After incubation by shaking for 1 h at 37°C, the reaction was stopped by adding 0.5 ml of CaCl$_2$ and 5 ml H$_2$O. After centrifugation at 5,000×g for 5 min at 5°C and after filtration of the aqueous phase, the absorbance of filtrates was measured at 398 nm. All enzymatic analyses blanks, without addition of a reducing substrate, were also included to correct for background absorbance and the activity was determined against a calibration curve. Absorbance was measured by spectrophotometer model SpectroFlex 6600 (WTW, Oberbayern, Germany). Soil respiration and all enzymatic activities were determined on monthly sampled soils.

Biolog CLPPs were determined on soils sampled, after amendment, at beginning (May), at middle (Jun) and at end (Aug) of cropping cycle, as AWCD and H′ index, as previously developed by Pane et al. [44]. Aliquots of 100 μl of water-extracted soil sample, at a final dilution of 10^{-4} (w/w), were inoculated into the Eco-microplates. These were incubated at 25°C for 4 days and read, 96 h post inoculum, at 590 nm, using the Bio-Rad Microplate Reader 550 (Bio-Rad, Hercules, CA, USA).

Statistical analysis

Tomato agronomic data were analysed by ANOVA and means were separated by Duncan's test. The relationships among biological activities detected in the amended soils and relative quantities of molecular organic ^{13}C groups of composts were assessed using a regression analysis. Biolog AWCD profiles for single substrates were computed by principal component analysis, performed on OD data of the 31 carbon sources to assess distribution biplot of all community samples.

Conclusions

This study showed the great potential of on-farm technology to produce vegetable composts with peculiar characteristics that are different from commercial composted biosolids. Nutrition and biostimulation effects may be responsible for the increased productive response to agricultural compost amendments in cropping systems. NMR profiling showed that molecular composition of on-farm composts is responsible for microbial degradability in the soil and that phenolic C could play a crucial role in modulating soil biological activities.

Competing interests

The authors declare that they have no competing interests.

Authors' contributions

CP, GC and DV participated in composting activities and drafted the manuscript. CP participated to the field trial. DV carried out the microbiological studies. AP and RS carried out the NMR studies and drafted the manuscript. AMP carried out chemical analysis on composts and drafted the manuscript. MZ participated in design and coordination of the study and drafted the manuscript. All authors read and approved the final manuscript.

Author details

[1]Consiglio per la Ricerca in Agricoltura e l'Analisi dell'Economia Agraria, Centro di Ricerca per l'Orticoltura, via dei Cavalleggeri 25, I-84098 Pontecagnano, SA, Italy. [2]Dipartimento Scienze dei Sistemi Colturali, Forestali e dell'Ambiente, Università degli Studi della Basilicata, viale dell'Ateneo Lucano 10, I-85100 Potenza, Italy. [3]Centro Interdipartimentale di Ricerca sulla Risonanza Magnetica Nucleare per l'Ambiente, l'Agro-Alimentare ed i Nuovi Materiali (CERMANU), Via Università 100, I-80055 Portici, NA, Italy.

References

1. Maniadakis K, Lasaridi K, Manios Y, Kyriacou M, Manios T (2004) Integrated waste management through producers and consumers education: composting of vegetable crop residues for reuse in cultivation. J Environ Sci Health B 39:169–183

2. Christian AH, Evanylo GK, Pease JW (2009) On-Farm Composting - A Guide to Principles, Planning & Operations. VCE Publ. No. 452–232.

3. Celano G, Alluvione F, Abdel Aziz M, Spaccini R (2012) The carbon dynamics in the experimental plots. Use of ^{13}C- and ^{15}N-labelled compounds for the soil-plant balance in carbon sequestration. In: Piccolo A (ed) Carbon sequestration in agricultural soils. A multidisciplinary approach to innovative methods. Springer, Düsseldorf, pp 107–144

4. Pane C, Spaccini R, Piccolo A, Scala F, Bonanomi G (2011) Compost amendments enhance peat suppressiveness to Pythium ultimum, Rhizoctonia solani and Sclerotinia minor. Biol Cont 56:115–124

5. López-Pérez JA, Roubtsova T, Ploeg A (2005) Effect of three plant residues and chicken manure used as biofumigants at three temperatures on Meloidogyne incognita infestation of tomato in greenhouse experiments. J Nematol 37:489–494

6. Avgelis AD, Manios VI (1989) Elimination of tomato mosaic virus by composting tomato residues. Neth J Plant Pathol 95:167–170

7. Ghaly AE, Alkoaik F, Snow A (2006) Inactivation of Botrytis cinerea during thermophilic composting of greenhouse tomato plant residues. Appl Biochem Biotechnol 133:59–75

8. Alkoaik F, Ghaly AE (2006) Influence of dairy manure addition on the biological and thermal kinetics of composting of greenhouse tomato plant residues. Waste Manag 26:902–913

9. Spaccini R, Mazzei P, Squartini A, Giannattasio M, Piccolo A (2012) Molecular properties of a fermented manure preparation used as field spray in biodynamic agriculture. Environ Sci Poll Res 19:4214–4225

10. De Marco A, Spaccini R, Vittozzi P, Esposito F, Berg B, Virzo De Santo A (2012) Decomposition of black locust and black pine leaf litter in two coeval forest stands on Mount Vesuvius and dynamics of organic components assessed through proximate analysis and NMR spectroscopy. Soil Biol Biochem 51:1–15

11. Spaccini R, Piccolo A (2007) Molecular characterization of compost at increasing stages of maturity. 2. Thermochemolysis-GC-MS and ^{13}C-CPMAS-NMR spectroscopy. J Agr Food Chem 55:2303–2311

12. Rondeau-Mouro C, Crepeau M-J, Lahaye M (2003) Application of CP-MAS and liquid-like solid-state NMR experiments for the study of the ripening-associated cell wall changes in tomato. Int J Biol Macromol 31:235–244

13. Bernal-Vicente A, Ros M, Tittarelli F, Intrigliolo F, Pascual JA (2008) Citrus compost and its water extract for cultivation of melon plants in green house nurseries. Evaluation of nutriactive and biocontrol effects. Biores Technol 99:8722–8728

14. Dignac MF, Knicker H, Kogel-Knabner I (2002) Effect of N content and soil texture on the decomposition of organic matter in forest soils as revealed by solid state CPMAS NMR spectroscopy. Org Geochem 33:1715–1726

15. Kumashiro KK, Ohgo K, Niemczura WP, Onizuka AK, Asakura T (2008) Structural insights into the elastin mimetic (LGGVG)6 using solid-state ^{13}C NMR experiments and statistical analysis of the PDB. Biopol 89:668–679

16. Iannotti DA, Pang T, Toth BL, Elwell DL, Keener HM, Hoitink HAJ (1993) A quantitative respirometric method for monitoring compost stability. Compost Sci Util 1:52–65

17. Komilis D, Kontou I, Ntougias S (2011) A modified static respiration assay and ist relationship with an enzymatic test to assess compost stability and maturity. Biores Technol 102:5863–5872

18. Som MP, Lemée L, Amblès A (2009) Stability and maturity of a green waste and biowaste compost assessed on the basis of a molecular study using spectroscopy, thermal analysis, thermodesorption and thermochemolysis. Biores Technol 100:4404–4416

19. Raj D, Antil RS (2011) Evaluation of maturity and stability parameters of composts prepared from agro-industrial wastes. Biores Technol 102:2868–2873

20. Tiquia SM, Tam NFY (1998) Elimination of phytotoxicity during co-composting of spent pig-manure sawdust litter and pig sludge. Biores Technol 65:43–49

21. Zucconi F, Pera A, Forte M, de Bertoldi M (1981) Evaluating toxicity of immature compost. BioCycle 22:54–57

22. Tiquia SM, Tam NFY, Hodgkiss IJ (1996) Effects of composting on phytotoxicity of spent pig-manure sawdust litter. Environ Poll 93:249–256

23. Arancon NQ, Lee S, Edwards CA, Atiyeh RM (2003) Effects of humic acids derived from cattle, food and paper-waste vermicomposts on growth of greenhouse plants. Pedobiologia 47:741–744

24. Atyeh RM, Lee S, Edwards CA, Arancon NQ, Metzger JD (2002) The influence of humic acids derived from earthworm-processed organic wastes on plant growth. Biores Technol 84:7–14

25. Levy JS, Taylor BR (2003) Effects of pulp mill solids and three composts on early growth of tomatoes. Biores Technol 89:297–305

26. Pane C, Villecco D, Zaccardelli M (2013) Short-time response of microbial communities to waste compost amendment of an intensive cultivated soil in Southern Italy. Comm Soil Sci Plant Anal 44:2344–2352

27. Woodbury PB (1992) Trace elements in municipal solid waste composts: a review of potential detrimental effects on plants, soil biota, and water quality. Biomass Bioenerg 3:239–259

28. Pane C, Piccolo A, Spaccini R, Celano G, Villecco D, Zaccardelli M (2013) Agricultural waste-based composts exhibiting suppressivity to diseases caused by the phytopathogenic soil-borne fungi Rhizoctonia solani and Sclerotinia minor. Appl Soil Ecol 65:43–51

29. Yogev A, Raviv M, Hadar Y, Cohen R, Katan J (2006) Plant waste-based composts suppressive to diseases caused by pathogenic Fusarium oxysporum. Eur J Plant Pathol 116:267–278

30. Tejada M, Gonzales JL, García-Martínez AM, Parrado J (2008) Application of a green manure and green manure composted with beet vinasse on soil restoration: effects on soil properties. Biores Technol 99:4949–4957

31. Serra-Wittling C, Houot S, Barriuso E (1996) Soil enzymatic response to addition of municipal solid-waste compost. Biol Fert Soils 20:226–236

32. Cytryn E, Kautsky L, Ofek M, Mandelbaum RT, Minz D (2011) Short-term structure and functional changes in bacterial community composition following amendment with biosolids compost. Appl Soil Ecol 48:160–167

33. Schnürer L, Rosswall T (1982) Fluorescein diacetate hydrolysis as a measure of total microbial activity in soil and litter. Appl Environ Microbiol 43:1256–1261

34. Fernandez P, Sommer I, Cram S, Rosas I, Gutiérrez M (2005) The influence of water-soluble As(III) and As(V) on dehydrogenase activity in soils affected by mine tailings. Sci Tot Environ 348:231–243

35. Makoi JHJR, Ndakidemi PA (2008) Selected soil enzymes: examples of their potential roles in the ecosystem. Afr J Biotechnol 7:181–191

36. Fenner N, Freeman C, Reynolds B (2005) Observations of a seasonally shifting thermal optimum in peatland carbon-cycling processes; implications for the global carbon cycle and soil enzyme methodologies. Soil Biol Biochem 37:1814–1821

37. Grandy AS, Neff JC, Weintraub MN (2007) Carbon structure and enzyme activities in alpine and forest ecosystems. Soil Biol Biochem 39:2701–2711

38. Leinweber P, Jandl G, Baum C, Eckhardt KU, Kandeler E (2008) Stability and composition of soil organic matter control respiration and soil enzyme activities. Soil Biol Biochem 40:1496–1505

39. Jindo K, Suto K, Matsumoto K, García C, Sonoki T, Sanchez-Monedero MA (2012) Chemical and biochemical characterisation of biochar-blended composts prepared from poultry manure. Biores Technol 110:396–404

40. Tuomela M, Vikman M, Hatakka A, Itävaara M (2000) Biodegradation of lignin in a compost environment: a review. Biores Technol 72:169–183
41. Zmora-Nahum S, Hadar Y, Chen Y (2007) Physico-chemical properties of commercial composts varying in their source materials and country of origin. Soil Biol Biochem 39:1263–1276
42. Violante P (2000) Metodi di Analisi Chimica del Suolo, Angeli, F. (Eds.), pp. 536
43. Mathers NJ, Jalota RK, Dalal RC, Boyd SE (2007) ^{13}C NMR analysis of decomposing litter and fine roots in the semi-arid Mulga Lands of southern Queensland. Soil Biol Biochem 39:993–1006
44. Pane C, Villecco D, Pentangelo A, Lahoz E, Zaccardelli M (2012) Integration of soil solarization with *Brassica carinata* seed meals amendment in a greenhouse lettuce production system. Acta Agric Scand Sect B Soil Plant Sci 62:291–299
45. Bonanomi G, Del Sorbo G, Mazzoleni S, Scala F (2007) Autotoxicity of decaying tomato residues affects tomato susceptibility to Fusarium wilt. J Plant Pathol 89:219–226
46. El-Masry MH, Khalil AI, Hassouna MS, Ibrahim HAH (2002) *In situ* and *in vitro* suppressive effect of agricultural composts and their water extracts on some phytopathogenic fungi. World J Microb Biot 18:551–558
47. Pane C, Celano G, Villecco D, Zaccardelli M (2012) Control of *Botrytis cinerea*, *Alternaria alternata* and *Pyrenochaeta lycopersici* on tomato with whey compost-tea applications. Crop Prot 38:80–86
48. Zaccardelli M, Perrone D, Del Galdo A, Giordano I, Villari G, Bianco M (2006) Multidisciplinary approach to validate compost use in vegetable crop systems in Campania Region (Italy): effect of compost fertilization on processing tomato in field cultivation. Acta Hort 700:285–288
49. Stege PW, Messina GA, Bianchi G, Olsina RA, Raba J (2009) Determination of arylsulphatase and phosphatase enzyme activities in soil using screen-printed electrodes modified with multi-walled carbon nanotubes. Soil Biol Biochem 41:2444–2452

16

Reduced complexity of multidimensional and diffusion NMR spectra of soil humic fractions as simplified by Humeomics

Antonio Nebbioso[*], Pierluigi Mazzei and Davide Savy

Abstract

Background: Humeomics is a sequential step-wise chemical fractionation that simplifies the complex matrix of a humic acid (HA) and weakens its supramolecular interactions, thereby allowing a detailed characterization of the involved molecules. A recalcitrant residual end product of Humeomics, namely RES4, was successfully solubilized here in alkaline conditions and subjected to a semi-preparative high-performance size exclusion chromatography (HPSEC).

Results: The resulting six size fractions separated by HPSEC were analyzed by different NMR techniques. 1D ^{1}H-NMR spectra did not reveal significant molecular differences among size fractions, although all of them differed from the spectrum of the bulk RES4 especially in signal intensity for aliphatic materials, which were assigned by 2D NMR to lipidic structures. Diffusion-ordered spectroscopy (DOSY)-NMR spectra showed that the homogeneity of RES4 was significantly changed by the HPSEC separation. In fact, nominally large size fractions, rich in lipidic signals, had significantly lower and almost constant diffusivity, due to stable supramolecular associations promoted by hydrophobic interactions among alkyl chains. Conversely, diffusivity is gradually increased with the content of aromatic and hydroxyaliphatic signals, which accompanied the reduction of fractions sizes and was related to smaller superstructures.

Conclusions: This study not only confirmed the occurrence of supramolecular structures in the recalcitrant humic residue of Humeomics, but also highlighted that more homogeneous size fractions were more easily characterized by NMR spectroscopy.

Keywords: Humeomics; HPSEC; Liquid-state NMR; DOSY-NMR; Supramolecular associations; Humic acids; Recalcitrant organic matter; Diffusivity; Nuclear magnetic resonance; Molecular fractionation

Background

Humic substances (HS) are naturally occurring organic substances ubiquitously found in the environment [1], bearing properties of environmental and physical-chemical significance. Their recalcitrance, either against enzyme activity or oxidative stress, is an intrinsic property of HS and plays a pivotal role in long-term storage of organic carbon in soils and subsequently in neutralization of humic-bound pollutants. Outdated theories attributed HS recalcitrance to their nature of large macromolecules formed over a long time by unclearly defined humification processes. According to this hypothesis, the number of covalent bonds of a macromolecular structure would prevent microbial and chemical degradation, resulting in recalcitrance. Recent experimental evidence [2] overruled this theory by showing that HS were better described as small heterogeneous molecules held together by mainly hydrophobic and H-bonding forces in supramolecular associations. This novel understanding of humic chemistry attributed recalcitrance to the tight association of single small molecules into hydrophobic domains separated from water and degrading microbes and oxygen.

A chemically resistant organic matter has been estimated to reach about 11% to 18% of total carbon in some soils [3]. Characterization of refractory organic matter was attempted by ^{13}C isotope analysis [4], and chemical resistance was associated with alkyl structures. The reason for accumulation of alkyl organic matter in soils was attributed to the presence of plant-derived biopolymers cutin and suberin [5]. Other investigations indicated that

---

* Correspondence: antonio.nebbioso@unina.it

Centro Interdipartimentale di Ricerca per la Spettroscopia di Risonanza Magnetica, Nucleare per l'Ambiente, l'Agro-alimentare e i Nuovi Materiali (CERMANU), Università di Napoli Federico II, 80055 Portici, Italy

aromatic compounds with single- or polycondensed ring also play a role in the recalcitrance of organic matter in soil [6]. Despite the recalcitrance attributed to aromatic rings in polycondensed materials, both lignin derivatives [7] and charcoal [8] were found to be functionalized and, in fact, degraded over time by soil microbial activity. Based on ^{13}C isotopic dilution in organic matter extracts and their ^{13}C-cross polarization magic angle spinning (CPMAS)-NMR spectra, a preferential decrease of hydrophobicity was assessed when passing from forested to cultivated soil, thereby substantiating that accumulation of organic matter in soil proceeded by stabilization of hydrophobic molecular associations [9,10]. Moreover, Pedersen et al. [11] found evidence by ^{13}C-CPMAS-NMR that hydrophobic aliphatic and aromatic organic molecules could resist degradation in drained thaw-lake basins, thus constituting the largest part of recalcitrant organic matter.

An alternative to application of CPMAS-NMR to solid recalcitrant humic material may be based on the solubilization of humic molecules in more homogeneous and simpler fractions and, thus, favoring a more detailed evaluation by NMR in the solution state. A recently introduced chemical and chromatographic fractionation called 'Humeomics' [12,13] allowed the separation of the bulk 'Humeome' matrix into homogeneous fractions by a stepwise breaking of chemical bonds of increasing strength, except carbon-carbon bonds. An end result of this procedure was a residue composed of unextractable organic material.

The aim of this work consisted in characterizing, by mono- and bidimensional liquid-state NMR techniques, the molecular composition of such a refractory end product of a humic acid subjected to Humeomics and further separated in size fractions by preparative HPSEC chromatography.

Methods

Humeomic fractionation

All reagents were provided by Sigma-Aldrich with 99.9% purity. A humic acid (HA) was isolated from a volcanic soil (Allic Fulvudand) at Vico, near Rome (Italy), and purified as described elsewhere [14]. This HA was then submitted to a step-wise series of solvolyses and chemical fractionations defined as Humeomics [12,13]. Briefly, the procedure consisted in multiple fractionation steps, which separated either unbound or ester-bound molecules from the humic matrix, and cleaved, in the final step, alkyl-alkyl and alkyl-aryl ether bonds with an aqueous HI treatment. The resulting solid residue (defined as RES4) [12,13] was extensively washed with Milli-Q deionized water (Millipore, Billerica, MA, USA) until iodide-free and freeze-dried. The solid freeze-dried residue was suspended in water and dissolved with a 0.50 M NaOH solution by automatically titrating under N_2 to pH 7.2 (VIT 909 21 Videotitrator, Radiometer, Copenhagen) until pH remained constant for 120 min. A

final concentration of 0.20 g L^{-1} was obtained. Possible microbial growth was prevented by adding 0.3 g L^{-1} NaN_3. This solution was then filtered through a 1 µm glass microfiber filter (Whatman GF/C, Sigma-Aldrich, St. Louis, MO, USA) and kept refrigerated under N_2 atmosphere, until subjected to preparative high-performance size exclusion chromatography (HPSEC).

Preparative HPSEC

The RES4 solution was eluted through a Phenomenex Biosep SEC-S-2000 column (Phenomenex, Torrance, CA, USA) (21.2-mm diameter × 300-mm length) and precolumn (21.2-mm diameter × 78-mm length). A Gilson 305 pump, a Gilson auto-sampler model 231 equipped with a 5.0-mL loop, a Gilson FC205 fraction collector, and a Gilson 116 UV detector set at 280 nm were used to automatically and continuously separate humic fractions (Gilson, Inc., Middleton, WI, USA). Chromatographic runs and profiles were monitored with a Gilson Unipoint software. A mobile phase consisting of aqueous (Milli-Q Millipore deionized water) 0.3 g L^{-1} NaN_3, 0.01 M AcONa, and 3.5 mM of NaH_2PO_4/Na_2HPO_4 buffer solution at pH 7.0 was eluted through columns at a flow of 1.5 mL min^{-1}. The auto-sampler was loaded with 20-mL flasks, each filled with 16 mL of RES4 and used for three injections. Approximately 50 mg of RES4 were injected and, after 25 min from injection, six fractions were collected. In detail, five fractions for 5 min each and a sixth fraction comprising a final elution for 15 min were isolated to collect 83.8% of bulk material. However, the sixth fraction was collected after a longer elution time with the aim to cumulate a sufficient amount of humic material for good-quality NMR analysis (>5 mg). For the same reason, the remaining material isolated in the size fractions obtained at larger elution times was discarded because of its relatively low amount (<3%).

Fractions were acidified to pH 2.0 with 1.0-M HCl, and the precipitated humic matter was dialyzed against Milli-Q deionized water in Spectrapore 3 membranes (cut-off 3,500 Da), and freeze-dried. Yields for the six fractions are summarized in Table 1.

NMR spectroscopy

A 400-MHz Bruker Avance spectrometer (Bruker, Billerica, MA, USA), equipped with a 5-mm Bruker BBI (Broad Band inverse) probe, working at the ^{13}C and ^{1}H frequencies of

Table 1 Absolute (mg) and relative[a] (%) yields of six size fractions (F1 to F6) recovered from the preparative HPSEC of RES4

Size fractions	F1	F2	F3	F4	F5	F6	Total
mg	4.7	8.8	7.3	7.4	6.7	7.0	41.9
%	9.4	17.6	14.6	14.8	13.4	14.0	83.8

[a]Referred to 50-mg injected RES4.

100.62 and 400.13 MHz, respectively, was employed to conduct all liquid-state NMR measurements at 298 ± 1°K. Each sample (5.0 mg) was dissolved with $D_2O/NaOD$ (0.1 M) and placed into 5.0-mm quartz tubes. ^1H NMR spectra were acquired with 2 s of thermal equilibrium delay, 90° pulse length ranging within 8.5 and 9.5 μs (–2-dB power attenuation), 32,768 time domain points, and 96 transients.

2D homo- and heteronuclear spectra were acquired with 2,048 points in F2 domain and 256 experiments in F1 dimension. Both ^1H-^1H correlation spectroscopy (COSY) and total correlation spectroscopy (TOCSY) experiments were executed with 96 scans, 2 s of thermal equilibrium delay, and 16 dummy scans, while, in case of TOCSY, a trim pulse of 2,500 μs and a mixing time of 0.08 s were applied. ^1H-^{13}C gradient-enhanced heteronuclear single quantum coherence (HSQC) experiment was performed with a trim pulse of 1 ms and by optimizing experimental parameters as a function of 145-Hz short-range $^1J_{CH}$.

^1H diffusion-ordered spectroscopy (DOSY) NMR spectra were obtained by a stimulated echo pulse sequence with bipolar gradients, combined with two spoil gradients and an eddy current delay before signal acquisition. This sequence was selected to reduce signal loss due to short spin-spin relaxation times. The acquisition (8-K points) consisted in 512 scans and was executed by using a 2.5-ms long sine-shaped gradients (δ) that linearly ranged from 0.674 to 32.030 G cm^{-1} in 32 increments and selecting a 150-ms delay (Δ) between the encoding and the decoding gradients. In all cases, the ^1H and ^{13}C spectral widths were 13 (5,201.7 Hz) and 250 ppm (30,186.1 Hz), respectively. In order to suppress the residual water signal, the on-resonance presaturation technique was adopted for ^1H spectra and homonuclear 2D experiments, whereas a Watergate 3-9-19 pulsed train sequence was preferred in case of DOSY experiments.

All spectra were baseline corrected and processed with Bruker Topspin Software (v.2.1). The free induction decays (FIDs) resulting from ^1H 1D and DOSY 2D spectra were apodized by multiplying by 1- and 5-Hz exponential factors, respectively. The distorted signal resulting from water suppression was removed in 1D spectra by applying an on-resonance Qfil algorithmic function during Fourier transform.

Results and discussion
Monodimensional ^1H NMR spectroscopy
^1H NMR spectra of RES4 and the six size fractions (F1 to F6) are shown in Figure 1. The spectrum of the bulk RES4 (Figure 1A) reveals broad and overlapped signals. This broadening of proton signal is due to intermolecular associations which, combined with the relatively high viscosity of humic solution, hinder the mobility of molecular components and consequently reduce T_2 spin-spin relaxation

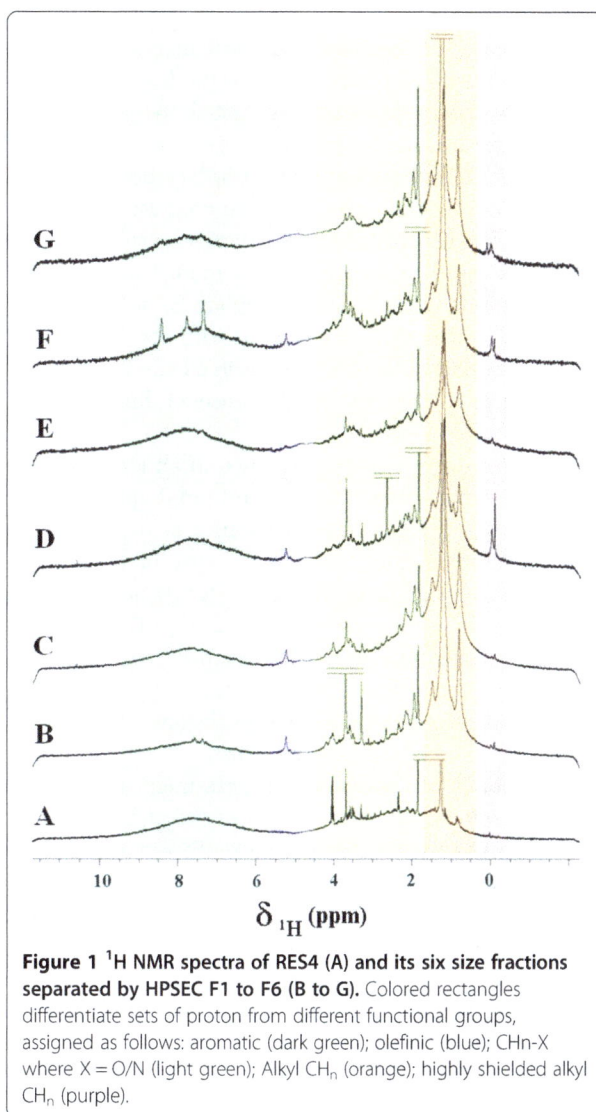

Figure 1 ^1H NMR spectra of RES4 (A) and its six size fractions separated by HPSEC F1 to F6 (B to G). Colored rectangles differentiate sets of proton from different functional groups, assigned as follows: aromatic (dark green); olefinic (blue); CHn-X where X = O/N (light green); Alkyl CHn (orange); highly shielded alkyl CHn (purple).

times [15]. Signals resonating in the region 0.5 to 4.5 ppm can be associated to alkyl, aminoalkyl, and hydroxyalkyl protons, while those signals detected between 5 and 9 ppm correspond to olefinic and aromatic protons. Signals attributable to CH-C-X (X: electron-withdrawing group) systems are detected in the 1.9- to 3.0-ppm interval, and those attributed to CH-X (X: heteroatom) groups are detected, though less abundantly, in the range between 3.0 and 4.5 ppm (Figure 1A). Similar results had been recently observed by solid-state ^{13}C-CPMAS NMR spectroscopy for the same bulk RES4 that revealed a lower intensity in the hydroxyalkyl (50 to 90 ppm) carbon region [12], following the selective Humeomic step-wise chemical fractionation of a humic matter.

A lesser extent of overlapping of proton signals was noted in spectra of size fractions (Figure 1B-G) in respect to RES4. In these spectra, most peaks appeared narrower and more pronounced than in the bulk RES4, due to the decreased molecular heterogeneity of size fractions as a result of the HPSEC separation. In particular, more intense and abundant signals were visible in the hydroxyalkyl region (3 to 5.2 ppm), which suggested the presence of hydroxy- or aminoalkyl protons CH-X (X: O/N) in the 1.50- to 2.50-ppm region next to sp^2 carbons, such as double bonds or carboxyl groups. In contrast with RES4, all size fractions revealed intense signals for alkyl protons (0.7 to 1.4 ppm), to be ascribed to methylene chains or CH_2 groups in various lipid compounds, such as plant waxes and polyesters. In agreement with previous results [12], the abundance of these groups in all size fractions indicates their role in the recalcitrance of humic supramolecular associations.

It is noteworthy that the presence in all size fractions of several signals is resonating at around 0 ppm, possibly related to highly shielded protons, such as those close to electron dense metals. Only in the case of the F5 size fraction, three significantly intense signals at 8.42, 7.78, and 7.36 ppm were detected (Figure 1F). The first two resonate at significantly greater chemical shift than for aromatic protons, thereby suggesting strong deshielding effects probably due to electron-withdrawing ring substituents or condensed ring structures.

The visibility of these signals in size-fraction spectra is to be accounted to either their undetectability under more intense resonance in the bulk RES4 or to the conformational simplification in size fractions brought about by the HPSEC separation. In fact, it has been already shown that size exclusion chromatography affects the architecture of humic supramolecular associations [2] by rearranging heterogeneous molecules into less complex assemblies and enhancing their analytical visibility. This effect was already noted for the fractions obtained by applying the Humeomic procedure to size fractions separated by HPSEC from a bulk HA [13]. As in the case of the NMR spectra of size fractions reported here, the previously reported analytical larger visibility of molecular components in less complex size fractions in respect to bulk HA was a consequence of the weaker intermolecular associations obtained after HPSEC separation.

Bidimensional NMR spectroscopy

2D NMR experiments, such as COSY, TOCSY, and HSQC, are useful tools for the identification of the molecular components in humic matter that would be hard to single out by only 1D spectra. COSY and HSQC spectra for the bulk RES4 are reported in Figure 2. The relatively low number of correlations detected in these 2D spectra reflects the poor signal visibility in the 1D ^{1}H spectrum of this material (Figure 1A). Nevertheless, a significant correlation between hydroxyalkyl and alkyl protons may account for a large presence of hydroxylated compounds. This appears to be conformed by the

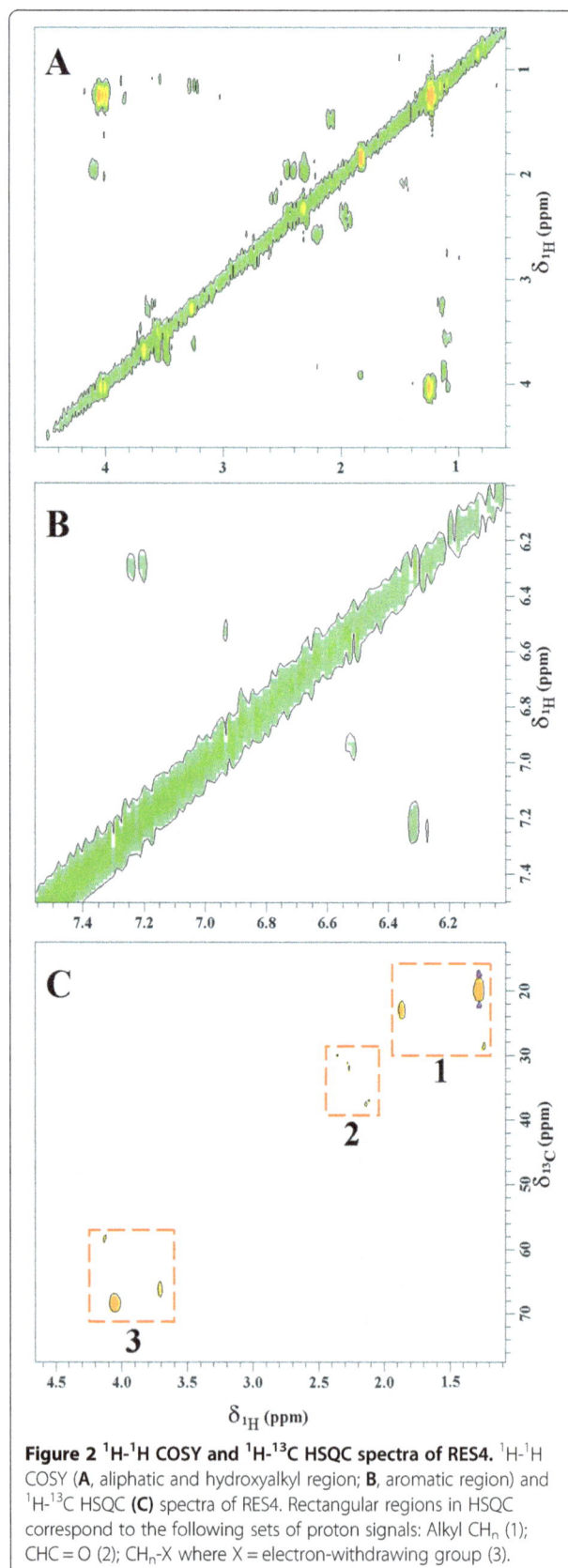

Figure 2 ^{1}H-^{1}H COSY and ^{1}H-^{13}C HSQC spectra of RES4. ^{1}H-^{1}H COSY (**A**, aliphatic and hydroxyalkyl region; **B**, aromatic region) and ^{1}H-^{13}C HSQC (**C**) spectra of RES4. Rectangular regions in HSQC correspond to the following sets of proton signals: Alkyl CH_n (1); CHC = O (2); CH_n-X where X = electron-withdrawing group (3).

most intense signals detected in HSQC spectrum for the alkyl (20 to 40 ppm) (Figure 2C) and hydroxyalkyl (60 to 70 ppm) (Figure 2C) carbon regions. As for the aromatic protons (6.5 to 8.5 ppm), the ^{1}H spectrum of RES4 (Figure 1A) showed only a very broad poorly intense resonance, whereas COSY (Figure 2B) revealed two neat proton cross peaks, representing the most abundant aromatic components in the sample. Although these cross peaks (6.52 to 6.88 ppm and 7.23 to 6.21 ppm) were not related to each other, as implied by the lack of cross correlation in COSY (Figure 2B) and TOCSY (data not shown) spectra, their chemical shifts suggest a similarity due to electron-donating substituent (such as –OH, –OCH$_3$, or OR) in the aromatic ring. Contrary to 1D and 2D COSY spectra, no clear aromatic signals were detected by HSQC experiments (Figure 2C), mainly because of the noted broadening of aromatic resonances. A reason for this may well reside in the strong intermolecular associations occurring among aromatic molecules and in the consequent shorter T$_2$ relaxation times for both ^{1}H and ^{13}C nuclei.

In line with observations in all proton spectra (Figure 1), 2D NMR of size fractions confirmed a relevant similarity among the six size fractions, except for the few intense aromatic signals in the F5 size fraction (Figure 1F). The most intense and resolved signals detected by COSY (Figure 3A,B) and HSQC (Figure 3C) spectra in the F1 size fraction were also visible in spectra of all other size fractions, except for the F5 sample which revealed a larger number of signals. Due to these similarities, only the 2D spectra of the first F1 size fraction are reported here (Figure 3A-C).

Most of 2D signals observed in the size fractions have been already noted for RES4, although size fractions produced more homo- and heterocorrelations than for the bulk RES4. This further confirms that the HPSEC separation provided simpler humic material and, consequently, less signal overlapping in size fractions. Again, several correlations, equally detected in all size fractions, predominantly involved hydroxyalkyl and alkyl protons. These signals may be attributed to hydrophobic chains deriving from different types of lipid structures, including unsaturated, mono- or polyhydroxylated chains. In particular, the most intense signal at 1.19 ppm was bound to a carbon resonating at 29.1 ppm, as revealed by HSQC (Figure 3C), and, being within a methylene proton population, it implies the presence of a hydrophobic backbone of carbon chains. In COSY spectrum (Figure 3A), the 1.19-ppm signal is correlated with the terminal methyl at 0.78 ppm that, in the HSQC spectrum, is also correlated to ^{13}C at 13.65 ppm (Figure 3C). The presence of unsaturated lipids is also suggested by the signal at 5.21 ppm (^{13}C δ = 129.3 ppm, Figure 3C), attributable to protons bound to sp^2 carbons, and, in turn, correlated to –CH$_2$ protons at 1.91 ppm (^{13}C δ = 26.99 ppm, Figure 3C). The relatively low

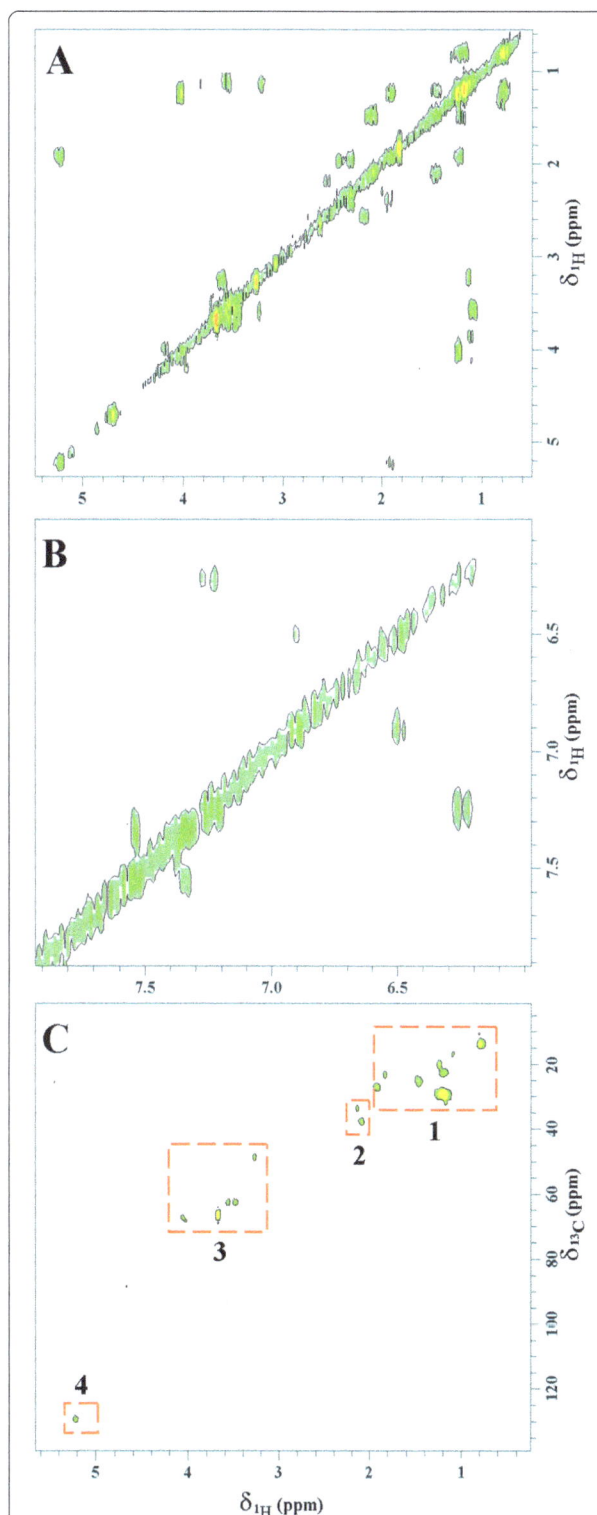

Figure 3 ^{1}H-^{1}H COSY and ^{1}H-^{13}C HSQC spectra of the F1 size fraction separated from RES4 by HPSEC. ^{1}H-^{1}H COSY (**A**, aliphatic and hydroxyalkyl region; **B**, aromatic region) and ^{1}H-^{13}C HSQC (**C**) spectra of the F1 size-fraction separated from RES4 by HPSEC. Rectangular regions (1 to 4) in HSQC correspond to the following sets of proton signals: Alkyl CH$_n$ (1); CHC=O (2); CH$_n$-X where X = electron-withdrawing group (3); CH$_n$-C=C (4).

field for these signals is due to a deshielding resulting from the adjacent double bond. The fact that these protons are also vicinally coupled to the signal at 1.18 ppm (methylene chains) strongly agrees with the assignment. The resonance at 1.18 ppm also correlates with a signal at 1.47 ppm (^{13}C δ = 25.13 ppm, Figure 3C) that is in turn related to another signal at 2.15 ppm (^{13}C δ = 33.33 ppm, Figure 3C). These signals correspond to β and α methylenes in fatty acids, respectively, whose different chemical shift varies according to the distance from a carboxyl group. The COSY spectra also reveal correlations between protons resonating at 3.98 and 4.19 ppm, which may be attributed to glycerol protons in lipids. The presence of mono- and polyhydroxylation is implied by the remaining intense correlations between the alkyl (1.1 to 1.4 ppm) and hydroxyalkyl (3.05 to 4.21 ppm) regions in COSY (Figure 3A). In fact, as suggested by HSQC in Figure 3C, such protons are adjacent to hydroxyallyl carbons resonating between 60 and 70 ppm (Figure 3C). The hypothesis that such proton and carbon frequencies could indicate possible carbohydrates structures had to be discarded, as a result of the lack of anomeric protons in 1D spectra. Moreover, the extensive treatment by hydroiodic acid in the last step of the Humeomic procedure must have resulted in cleavage of glycosidic bond. However, no typical spin systems for saccharide-like structures were detected in TOCSY spectra for the F1 size fraction (data not shown), thereby failing to substantiate their presence.

Finally, the signals that differentiate F5 from the other size fractions are resonated in the aromatic regions of both TOCSY and HSQC spectra (Figure 4). These signals at 7.3, 7.8, and 8.4 ppm (Figure 1F) were bound to carbon signals at 124, 136, and 149 ppm, respectively, as revealed by HSQC (Figure 4B). TOCSY experiment indicated that all of such peaks are part of the same spin system, whereas their deshielded chemical shifts implied the presence of an electron-withdrawing group in the substitution pattern, such as a carboxyl group. However, both COSY (Figure 4A) and TOCSY experiments (data not shown) revealed also two correlations at 7.2 and 7.7 ppm, which are evidence for aromatic rings with electron-donating substituents.

Diffusion-ordered NMR spectroscopy

^1H diffusion-ordered spectroscopy (DOSY) is a pulsed field gradient technique that enables measurement of translational diffusion of dissolved molecules and mixtures. It produces a final 2D spectral output in which each chemical shift is directly correlated to the translational diffusivity of either a single molecule or the aggregate to which the molecule is associated [16,17]. According to the Einstein-Stock diffusion theory, the larger the hydrodynamic radius of a molecular complex, the smaller is the related diffusivity [18]. Diffusion coefficients resulting from DOSY NMR experiments may thus provide direct

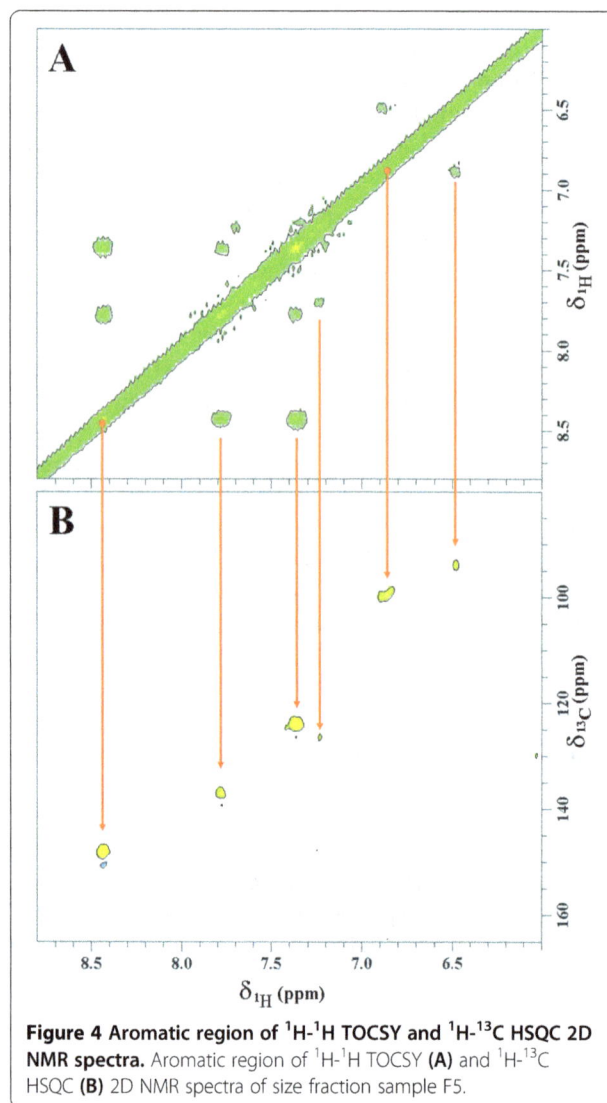

Figure 4 Aromatic region of ^1H-^1H TOCSY and ^1H-^{13}C HSQC 2D NMR spectra. Aromatic region of ^1H-^1H TOCSY **(A)** and ^1H-^{13}C HSQC **(B)** 2D NMR spectra of size fraction sample F5.

information on intermolecular interactions, state of aggregation, and conformational changes in complex mixtures [19-21].

Table 2 reports the ^1H diffusion coefficients calculated for signals in the alkyl (2 to 0.5 ppm), hydroxyalkyl (4.4 to 2.5 ppm), and aromatic (9 to 6.5 ppm) spectral regions, while 2D DOSY spectra are shown in Figure 5A-G. All RES4 protons resulted substantially aligned to the same diffusivity, since all their projections were located in a restricted logD range between –10.2 and –10.4 m^2 s^{-1} (Figure 5A). Alkyl protons resonating in the 1.1- to 0.6-ppm interval showed a small difference in respect to their projections, since they were placed in a slightly upper position (Figure 5A). These substantial similarities among the various RES4 molecular domains reflected the large sample homogeneity prior to HPSEC chromatography. In the center of aromatic region, protons with chemical shift between 7.0 and 8.0 ppm (Figure 5A) displayed slightly

Table 2 ^{1}H diffusion values (10^{-11} m^2 s^{-1} at 25°C) for signals resonating in aromatic (9.0 to 6.5 ppm), hydroxyalkyl (4.4 to 2.5 ppm), and alkyl (2 to 0.5 ppm) spectral regions

Samples	Aromatic region	Hydroxyalkyl region	Alkyl region
RES4	3.7	5.9	4.8
F1	4.1	7.0	2.1
F2	6.3	7.4	2.7
F3	7.7	9.6	2.9
F4	5.6	7.2	2.9
F5	5.2[a]	7.4	2.5
F6	18.3	25.4	2.6

[a]The reported value was obtained by excluding the contribute of signals at 7.3, 7.8, and 8.4 ppm, for which the average diffusion value resulted to 9.0.

greater diffusivity than those outside of this range. Interestingly, DOSY spectra of the fractionated RES4 of this study differed significantly from those reported in literature for bulk humic substances from oak and pine forest soil humic substances [22], which showed different values of diffusivity for aromatic, hydroxyalkyl, and alkyl protons.

Conversely, diffusion coefficients for the size fractions differed from those for the bulk RES4 (Table 2), confirming the profound changes in conformational structure induced by HPSEC separation which were already suggested by 1D spectra (Figure 1). In particular, the diffusivity for alkyl protons showed a reducing trend with decreasing fraction size, whereas that for hydroxyalkyl and aromatic protons invariably increased (Table 2). In fact, in comparison to RES4, the diffusivity values for alkyl protons were relevantly decreased for F1 and even more so for F2 and F3. For the lowest size fractions, diffusivity remained small and in the range between 2.5 and $2.9 \cdot 10^{-11}$ m^2 s^{-1} (Table 2), especially for the most shielded alkyl signals in both RES4 and size fractions (Figure 5B-G).

In the case of hydroxyalkyl and aromatic protons, the diffusion coefficients increased progressively with decreasing size of fractions (Table 2). A gradual increase of diffusivity was observed from RES4 to F3 sample (Figure 5A-D), followed by its decrease for F4 and F5 size fractions (Figure 5E,F), and a subsequent significant increase in last F6 size fraction (Figure 5G). This is also noticeable by the enhanced downward displacement of proton projections for both hydroxyalkyl (Figure 5) and aromatic (Figure 5) regions.

These variations can again be explained with an increasing molecular and conformational changes in the smaller size fractions obtained during the HPSEC separation. In fact, not only lower molecular size in the smaller F6 size fraction justifies an enhanced proton diffusivity, but also the larger amount of hydrophilic molecules in F6 may limit further aggregation size and, thus, diffusivity. In fact,

hydroxylated alkyl chains were also evident in DOSY spectra (Figure 5G), in which the projection of several alkyl protons, resonating at 1.35, 1.6, 2.1, and 2.3 ppm, were placed relatively downward (greater diffusivity) than lipid signals. It may thus be inferred that the F6 size fraction contained two populations of aggregates: the larger aggregates made of lipid components and the smaller ones constituted mainly by hydrophilic components. The presence of both kinds of small associated compounds in the same size fraction is consistent with its small nominal molecular size.

On the other hand, the diffusivity observed for hydroxyalkyl and aromatic protons in F4 and F5 (Figure 5E,F) samples may be the result of an increased supramolecular association due to aromatic interactions, driven by π-π stacking forces and/or H bondings among complementary hydroxyl groups. Furthermore, as mentioned before, the aromatic region in F5 sample showed three different signals (Figures 1 and 4) whose averaged diffusion coefficient ($9 \cdot 10^{-11}$ m^2 s^{-1}) was significantly greater than the one for the remaining aromatic protons ($5.2 \cdot 10^{-11}$ m^2 s^{-1}) (Table 2). This suggests evidence for two coexisting states for aromatic compounds in this size fraction, being held by intermolecular interactions of different strength.

Conclusions

The end product (RES4) of a step-wise Humeomic fractionation procedure applied to a humic acid from soil was successfully solubilized in alkaline solution and then separated by preparative HPSEC to provide six size fractions. High-resolution solution-state NMR spectroscopy was applied to the size fractions in order to evaluate the enhanced characterization capacity of NMR due to the size simplification provided by HPSEC separation. We confirmed that HPSEC is capable of reducing the complexity of a recalcitrant humic matrix, thereby consequently decreasing the overlapping of NMR signals and improving spectral quality and, hence, signal assignment.

1D spectra revealed that, even size fractions were not significantly different from each other, all of them differed from the bulk RES4, as the alkyl signals, assigned to lipid structures by 2D NMR spectra, were more intense and resolved. This suggests that the apparent lack of such signals in RES4 is not due to their absence but rather to their strong intermolecular associations which favor very short T$_2$ relaxation times and such a signal broadening to impair detectability. The aggregation of humic molecules in the bulk RES4 was weakened by the HPSEC separation, thereby allowing less tight molecular associations in the hydrophobic domains of the recalcitrant end product of Humeomics.

The low proton diffusivity in the alkyl region of size fractions, as shown by DOSY NMR spectra, suggests that lipid components do re-aggregate, after HPSEC separation, in

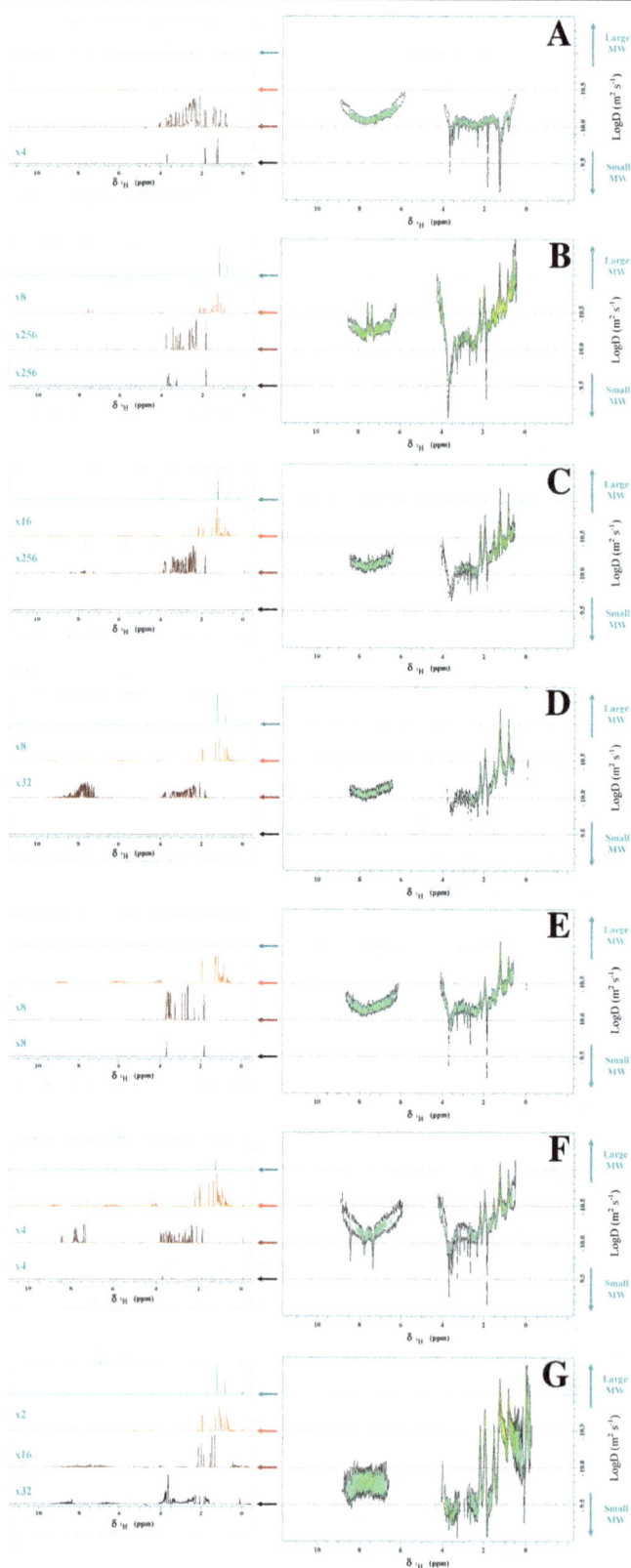

Figure 5 ^{1}H-DOSY NMR spectra of sample RES4 (A) and size fractions F1 to F6 (B-G). In left columns, proton projections for four different diffusivity values are shown (different magnification orders were applied).

newly stable superstructures. Conversely, the diffusivity for aromatic and hydroxyalkyl protons increased with decreasing size of fractions, thus indicating a molecular stabilization of molecular assemblies though H bondings and aromatic π-stacking forces. The smallest F6 size fraction showed a significantly enhanced diffusivity for these same protons that may be explained with small assemblies squeezed out from larger associations.

This work contributed to the NMR characterization of a recalcitrant end result of a Humeomic procedure applied to a terrestrial humic acid. The enhanced resolution of NMR spectra was achieved by subjecting the recalcitrant residue to a preliminary HPSEC separation in smaller size fractions, thereby confirming that an improved characterization of humic molecules can only be achieved if they are reasonably separated from the complex humic matrix.

Competing interests

The authors declare that they have no competing interests.

Authors' contributions

The experiment was conducted under the general supervision of AN. Taking equal parts, AN, DS, and PM carried out: sample preparation, Humeomic fractionation, HPSEC size fractionation, and NMR experiments. AN and PM interpreted the NMR spectra. The manuscript was written by AN. All authors read and approved the final manuscript.

Acknowledgements

The present work was carried out in partial fulfillment of AN, PM, and DS's PhD degrees funded by the host institution 'Università degli Studi di Napoli Federico II'.

References

1. Piccolo A (1996) Humus and soil conservation. In: Piccolo A (ed) Humic Substances in Terrestrial Ecosystems. Elsevier, Amsterdam, Netherlands, pp 225–264
2. Piccolo A (2002) The supramolecular structure of humic substances: A novel understanding of humus chemistry and implications in soil science. Adv Agron 75:57–134
3. Koarashi J, Iida T, Asano T (2005) Radiocarbon and stable carbon isotope compositions of chemically fractionated soil organic matter in a temperate-zone forest. J Environ Radioactiv 79:137–156
4. Wiesenberg GLB, Schwarzbauer J, Schmidt MWI, Schwark L' (2004) Source and turnover of organic matter in agricultural soils derived from n-alkane/n-carboxylic acid compositions and C-isotope signatures. Org Geochem 35:13/1–1393
5. Lorenz K, Lal R, Preston CM, Nierop KGJ (2007) Strengthening the soil organic carbon pool by increasing contributions from recalcitrant aliphatic bio(macro)molecules. Geoderma 142:1–10
6. Bayer C, Martin-Neto L, Mielniczuk J, Dieckow J, Amado TJC (2006) C and N stocks and the role of molecular recalcitrance and organomineral interaction in stabilizing soil organic matter in a subtropical Acrisol managed under no-tillage. Geoderma 133:258–268
7. Quideau SA, Anderson MA, Graham RC, Chadwick OA, Trumbore SE (2000) Soil organic matter processes: characterization by ^{13}C NMR and ^{14}C measurements. Forest Ecol Manag 138:19–27
8. Bird MI, Turney CSM, Fifield LK, Jones R, Ayliffe LK, Palmer A, Cresswell R, Robertson S (2002) Radiocarbon analysis of the early archaeological site of Nauwalabila I, Arnhem Land, Australia: implications for sample suitability and stratigraphic integrity. Quaternary Sci Rev 21:1061–1075
9. Piccolo A, Conte P, Spaccini R, Mbagwu JSC (2005) Influence of land use on the humic substances of some tropical soils of Nigeria. Eur J Soil Sci 56:343–352
10. Spaccini R, Piccolo A, Conte P, Mbagwu JSC (2006) Changes of humic substances characteristics from forested to cultivated soils in Ethiopia. Geoderma 132:9–19
11. Pedersen JA, Simpson MA, Bockheim JG, Kumar K (2011) Characterization of soil organic carbon in drained thaw-lake basins of Arctic Alaska using NMR and FTIR photoacoustic spectroscopy. Org Geochem 42:947–954
12. Nebbioso A, Piccolo A (2011) The basis for a Humeomic Science. Biomacromolecules 12:1187–1199
13. Nebbioso A, Piccolo A (2012) Advances in Humeomics. Anal Chim Acta 720:77–90
14. Nebbioso A, Piccolo A (2009) Molecular rigidity and diffusivity of Al^{3+} And Ca^{2+} humates as revealed by NMR spectroscopy. Environ Sci Technol 43:2417–2424
15. Simpson AJ (2001) Multidimensional solution state NMR of humic substances: a practical guide and review. Soil Sci 166:795–809
16. Price KE, Lucas LH, Larive CK (2004) Analytical applications of NMR diffusion measurements. Anal Bioanal Chem 378:1405–1407
17. Cobas JC, Groves P, Martin-Pastor M, Capua AD (2005) New applications, processing methods and pulse sequences using diffusion NMR. Current Anal Chem 1:289–305
18. Chapman S, Cowling TG (1990) Elementary theories of the transport Phenomena. In: John Wiley & Sons (ed) The mathematical theory of non-uniform gases, 3rd edition. Cambridge University Press, Cambridge (UK), pp 97–108
19. Brand T, Cabrita EJ, Berger S (2005) Intermolecular interaction as investigated by NOE and diffusion studies. Prog Nucl Mag Reson Spectrosc 46:159–196
20. Cohen Y, Avram L, Frish L (2005) Diffusion NMR spectroscopy. Angew Chem Int Ed 44:520–554
21. Viel S, Mannina L, Segre A (2002) Detection of a π-π complex by diffusion-order spectroscopy (DOSY). Tetrahedron Lett 43:2515–2519
22. Simpson AJ, Kingery WL, Hayes MHB, Spraul M, Humpfer E, Dvortsak P, Kerssebaum R, Godejohann M, Hofmann M (2002) Molecular structures and associations of humic substances in the terrestrial environment. Naturwissenschaften 89:84–88

17

Root exudate profiling of maize seedlings inoculated with *Herbaspirillum seropedicae* and humic acids

Lívia da Silva Lima[1], Fábio Lopes Olivares[1], Rodrigo Rodrigues de Oliveira[2], Maria Raquel Garcia Vega[2], Natália Oliveira Aguiar[1] and Luciano Pasqualoto Canellas[1*]

Abstract

Background: Co-inoculation of maize with *Herbaspirillum seropedicae* and humic substances increases the sizes of plant-associated bacterial populations and enhances grain yields under laboratory and field conditions. Root exudation is a key mechanism in the regulation of plant-bacterial interactions in the rhizosphere; humic matter supplementation is known to change the exudation of H^+ ions and organic acids from maize roots. Our starting premise was that *H. seropedicae* and humic acids would modify maize seedling exudation profiles. We postulated that a better understanding of these shifts in exudate profiles might be useful in improving the chemical environment to promote better performance of plant growth-promoting bacteria delivered as bioinoculants. Thus, root exudates of maize were collected and analyzed by gas chromatography-mass spectrometry (GC-MS) and proton nuclear magnetic resonance (^{1}H NMR).

Results: Nitrogenous compounds, fatty acids, organic acids, steroids, and terpenoid derivatives were the main structural moieties found in root exudates. Significant changes in exudation patterns occurred 14 days after the initiation of experiments. Quantities of fatty acids, phenols, and organic acids exuded by seedlings treated with humic acids alone differed from the quantities exuded in other treatments. Seedlings treated with *H. seropedicae* or *H. seropedicae* in combination with humic acids exuded a diversity of nitrogenous compounds, most of which had heterocyclic structures. Twenty-one days after initiating the experiment, seedlings treated with *H. seropedicae* alone exuded elevated quantities of steroids and terpenoid derivatives related to precursors of gibberellic acids (kaurenoic acids).

Conclusions: Changes in root exudation profiles induced by our treatments became most marked 14 and 21 days after initiation of the experiment; on those days, we observed (i) increased fatty acid exudation from seedlings treated only with humic acids and (ii) increased exudations of nitrogenated compounds and terpenes from seedlings treated only with *H. seropedicae*. Improved knowledge on the effects of bacterial inoculants and supplementation with humates on plant exudate composition may contribute substantially to improved understanding of plant metabolic responses and lead to new approaches in the use of selected compounds as additives in bioinoculant formulations that will modulate the cross-talk between bacteria and plants, thereby improving crop yields.

Keywords: Plant growth-promoting bacteria; Endophytic interaction; Humic substances

* Correspondence: lucianocanellas@gmail.com
[1]Núcleo de Desenvolvimento de Insumos Biológicos para a Agricultura (NUDIBA), Universidade Estadual do Norte Fluminense Darcy Ribeiro (UENF), Av. Alberto Lamego, 2000, Campos dos Goytacazes, 28013-602 Rio de Janeiro, Brazil
Full list of author information is available at the end of the article

Background

Root exudates function in processes of plant adaptation. They have roles in nutrient cycling in the rhizosphere and in responses to pathogens and symbiotic micro-organisms [1]. The quantities of organic compounds exuded by roots are variable, but they are frequently a significant proportion of the carbon fixed photosynthetically by plants [2]. Plant type, species, age, and environmental factors, including biotic and abiotic stressors, all affect exudation profiles [3,4]. The plant rhizosphere modulates microbial community structure and function, primarily through the release of chemical compounds [5].

The dominant organic compounds exuded by roots reflect central components of cell metabolism, including free sugars (e.g., glucose, sucrose), amino acids (e.g., glycine, glutamate), and organic acids (e.g., citrate, malate, and oxalate) [2]. Maize exudates comprise sugars (70%), phenolics (18%), organic acids (7%), and amino acids (3%) [6]. Other compounds, like fatty acids, sterols, enzymes, vitamins, and plant growth regulators (e.g., auxins, gibberellins, and cytokinins), are generally released in only very small quantities [7]. Although only small quantities of secondary metabolites are exuded by roots, their plant role in signaling to the microbial community is substantial [8].

Recently, we determined that the combined inoculation of maize with the endophytic diazotrophic bacterium *Herbaspirillum seropedicae* and humic acids increased root colonization and promoted plant growth and grain yields [9]. Canellas and Olivares [10] reviewed the use of humic substances as plant growth promoters; they considered the effects of these substances on plant metabolism and their use as carriers in procedures for the bioinoculation of plant growth-promoting bacteria in field crop systems.

Canellas et al. [11] reported changes in maize exudation profiles after humic acid application that included enhanced secretion of inorganic ions (i.e., H+ ions) and short-chain organic acids [12]. The addition of humic acids of different size fractions may have substantial effects on the quantities of bioavailable carbon deposited by maize plant roots; these additions produce significant changes in the structure and activity of soil microbial communities [13]. Hence, chemical changes induced by humic matter augmentation in the rhizosphere may enhance colonization of maize plants by inoculated endophytic diazotrophic bacteria carried in the humic substances.

H. seropedicae is a plant growth-promoting diazotrophic β-proteobacterium found mainly in association with grasses and other non-leguminous plants [14]. Roesch et al. [15] used molecular tools to assess diazotrophic bacterial diversity within rhizosphere soils, roots, and stems of field-grown maize and observed a predominance of α-proteobacteria and β-proteobacteria sequences in the rhizosphere soil and stem samples; *Herbaspirillum*

was one of the dominant genera in the interiors of maize plants but was rarer in soil. The members of this genus have been tested in the formulation of biofertilizers, with variable success in field crop trials ([16-19] and references therein [20-23]). The whole genome sequence of *H. seropedicae* has been published [24]. The species' capacity for N fixation, production of auxin and other phytohormones, and the colonization of diverse plant species has been previously demonstrated [25-27].

Root colonization is a basic first step for successful inoculation. An expansion of studies on metabolite exchange between plants and bacteria and the genetic responses of plants will fill knowledge gaps in our understanding of the colonization process. The role of root-exuded flavonoids in legume-*Rhizobium* interactions has been examined in detail. These exudations generate a finely tuned cross-molecular dialogue involving the secretion of lipochitooligosaccharides and the modulation of bacterial cell wall surface polysaccharides (extracellular polysaccharide (EPS) and lipopolysaccharide (LPS)) that result in plant root nodulation [28]. There is less information available on the role of plant metabolites in successful interactions with non-nodulating plant growth-promoting bacteria. Gough et al. [29] showed that flavonoids promote endophytic colonization of *Arabidopsis thaliana* (L.) Heynh. roots by *H. seropedicae*, and Tadra-Sfeir et al. [30] demonstrated that naringenin (a flavonoid in the flavonone class) is involved in the gene expression of cell wall components (EPS, LPS) and auxins. In addition to its role in the genetic modulation of cell wall assembly, *H. seropedicae* has an operon associated with the degradation of aromatic compounds [31]. An ability to degrade flavonoids would likely confer an important competitive advantage in rhizosphere/root colonization of the host plant by providing both a carbon source and associated detoxification mechanisms. Balsanelli et al. [32] showed that surface lipopolysaccharides produced by *H. seropedicae* strain Smr1 are required for attachment and endophytic colonization of maize plants. They [32] found that the *H. seropedicae* attachment process is partially mediated by a root lectin that specifically binds N-acetyl glucosamine residues.

Recently, Marks et al. [33] demonstrated the potential of using bacterial metabolites to enhance the performance of biofertilizers, thereby opening the possibility of chemical manipulation of carriers to benefit bacterial delivery to field crops. Furthermore, metabolites exuded by host plants may help in guiding genomic studies of plant-bacterial interactions. In the present work, we examined main changes in exudation profiles of maize seedling roots induced under laboratory conditions by (i) single applications of humic acids or (ii) *H. seropedicae* or (iii) combinations of the bacteria and humic acids.

Methods

Humic substances

Humic-like substances were extracted as described previously [11]. In brief, ten volumes of 0.5 mol L^{-1} M NaOH were mixed with one volume of earthworm compost under a N_2 atmosphere. After 12 h, the suspension was centrifuged at 5,000 × g; humic acids (HA) were extracted thrice in this manner, and the final HA pellet was de-ashed by combining it with ten volumes of a diluted mixture of HF-HCl solution (5 mL L^{-1} HCl [12 M] + 5 mL L^{-1} HF [48%, v/v]). After centrifugation (5,000 × g) for 15 min, the sample was repeatedly washed with water until a negative test against $AgNO_3$ was obtained. Subsequently, the sample was dialyzed against deionized water using a 1,000-Da cutoff membrane (Thomas Scientific, Swedesboro, NJ, USA). The dialyzate was lyophilized. We then prepared a HA solution by solubilizing HA powder in 1 mL of 0.1 M mol L^{-1} NaOH, followed by pH adjustment to 6.5 with 0.1 M HCl.

Microorganism used

H. seropedicae strain HRC 54 was originally isolated from sugarcane roots [34]. It has been used as part of the sugarcane inoculant developed by Embrapa (Brazilian Enterprise for Agricultural Research). The pre-inoculum was obtained after growth in DYGS liquid medium [35] for 24 h at 30°C on an orbital shaker rotating at 150 rpm. Subsequently, 20 μL of the suspension was transferred to JNFb liquid medium supplemented with NH_4Cl (1 g L^{-1}) and then grown for 36 h at 34°C on an orbital shaker rotating at 150 rpm. Cells were pelleted by centrifugation (4,000×g for 15 min) and resuspended in sterilized water at cell densities of 10^8 colony-forming units (cfu) mL^{-1}. The inoculant was prepared by diluting 200 mL of bacterial suspension in 800 mL of humic acid solution at pH 6.5 to produce a final humic acid concentration of 50 mg C L^{-1} and a final bacterial concentration of 2×10^7 cells mL^{-1}.

The composition of JNFb medium (per liter) was as follows: malic acid (5.0 g), K_2HPO_4 (0.6 g), KH_2PO_4 (1.8 g), $MgSO_4 \cdot 7H_2O$ (0.2 g), NaCl (0.1 g), $CaCl_2$ (0.02 g), 0.5% bromothymol blue in 0.2 N KOH (2 mL), vitamin solution (1 mL), micronutrient solution (2 mL), 1.64% Fe··EDTA solution (4 mL), and KOH (4.5 g). One-hundred milliliters of vitamin solution contained 10 mg of biotin and 20 mg of pyridoxol-HCl. The micronutrient solution contained (per liter) the following: $CuSO_4$ (0.4 g), $ZnSO_4 \cdot 7H_2O$ (0.12 g), H_3BO_3 (1.4 g), $Na_2MoO_4 \cdot 2H_2O$ (1.0 g), and $MnSO_4 \cdot H_2O$ (1.5 g); pH was adjusted to 5.8. For bacterial counts, we used the same media with a semisolid consistency obtained by adding 1.9 g L^{-1} of agar [14]. The bacterial population was determined by the most probable number technique (MPN); positive growth was recognized by the formation of a thick, white pellicle, replication was threefold, and density was expressed as the log of cell number g^{-1} root fresh mass

after growth on JNFb N-free semisolid medium (following Döbereiner et al. [35]). The presence of *H. seropedicae* was confirmed by collecting a piece of pellicle with a platinum loop, mounting it on a slide under a coverslip, and making observations under phase contrast microscopy to determine cell shape and movement and colony appearance in JNFb solid medium, as described by Döbereiner et al. [35]. When cell shape in the pellicle material differed from that of *H. seropedicae*, we identified the microbes as native bacteria associated with maize roots.

Experimental

Treatment of plants

Maize seeds (*Zea mays* L. var. Dekalb 7815) were surface-sterilized by soaking in 0.5% NaClO for 30 min, followed by rinsing and then soaking in water for 6 h. Afterward, the seeds were sown on wet filter paper and germinated in the dark at 28°C. Four days after germination, 30 maize seedlings with root length approximately 0.5 cm were transferred into 2.2-L vessels previously filled with 2 L of one-fourth-strength Furlani nutrient solution (containing 3.527 μM Ca, 2.310 μM K, 855 μM Mg, 45 μM P, 587 μM S, 25 μM B, 77 Fe, 9.1 μM Mn, 0.63 μM Cu, 0.83 μM Mo, 2.29 μM Zn, 1.74 μM Na, and 75 μM EDTA) with inorganic N content adjusted to a low concentration (100 μmol L^{-1} [$NO_3^- + NH_4^+$]). These low levels of N and P were used to simulate the low availability in highly weathered tropical soils and to avoid the inhibition of the diazotrophic bacteria. The seedlings were fixed into a perforated Teflon support with holes of 15-mm diameter in which seeds have been fitted. The system was continuously aerated by a low flux pump normally used in aquarium systems. Four treatments (*n* = 3 pots per treatment) were prepared by supplementing the nutrient solution with the following: 1. HA (50 mg C L^{-1}), 2. plant growth-promoting bacteria (PGPB) *H. seropedicae* strain HRC 54 (final bacterial suspension of 2×10^7 cells mL^{-1}), 3. humic acids plus *H. seropedicae* (HA + PGPB), and 4. control (C), without any additions. Seedlings were collected 7, 14, and 21 days after inoculation. After 1 week, and each week thereafter, one half of the nutrient solution in each pot was replaced with fresh nutrient solution through the end of the experiment. The experiment was repeated thrice independently.

Exudate collection

Maize seedlings were removed from the pots; their roots were immersed in glass tubes filled with 50 mL of 0.01 mol L^{-1} KOH for 5 min to remove organic anions adhering to the root surfaces. We then thoroughly washed the roots with tap water followed by a final rinse in distilled water. Complete root systems of seedlings from a single pot were inserted in a glass tube (6.5-cm inner diameter (i.d.) × 15-cm tall) filled with 80 mL of ultrapure water in which to collect the root exudates.

After 2 h, we collected the suspensions containing root exudates and filtered them through 0.22-μm filter membranes to remove root detritus and microbial cells. The filtered samples were kept frozen until we concentrated them by liquid chromatography using 10 cm of reverse phase (RP) C18 LiChroprep® RP-18 (15 to 25 μm; Merck Millipore, Billerica, MA, USA) as the stationary phase in an open glass column (2.5-cm i.d. × 20-cm tall). The aqueous suspension of exudate was forced through the column under low pressure provided by an aquarium pump. Compounds were eluted from the column with methanol under gravity, and the solvent was removed under low temperature (4°C) under vacuum (Rocket Evaporator System, Genevac, Stone Ridge, NY, USA). We drove our exudate capture to exclude sugars and amino acids and collected mainly products of the secondary metabolism using the RP C18 column.

NMR sample preparation and data collection

For nuclear magnetic resonance (NMR) analysis, we dissolved exudate extracts in DMSO-d6 (700 μL). All spectra were recorded at room temperature on a Bruker Avance DRX 500 spectrometer equipped with a 5-mm inverse detection probe (Bruker GmbH, Rheinstetten, Germany) operating at 500.13 MHz for 1 h. For each sample, we recorded 360 scans (FIDs) with the following parameter settings: 64 k data points, pulse width 8.5 μs (90°), spectral width of 4,401 Hz, acquisition time of 7.4 s, and a relaxation delay of 1.0 s. For spectrum processing, we used 64 k points and applied an exponential multiplication associated with a line broadening of 0.3 Hz. Spectra were referenced to tetramethylsilane (TMS) at 0.0 ppm. To obtain exudate profiles for each of the treatments in the study, we pooled extracts from treatment replicates and dissolved the dried methanolic extracts in 700 μL of DMSO-d6. Dissolved extracts were transferred to a 5-mm NMR tube for analysis.

Principal component analysis

The proton nuclear magnetic resonance (^{1}H NMR) spectra were reduced to ASCII files using OACD software; the resulting data matrix was imported into The Unscrambler 10.1 software (www.camo.com). Signals corresponding to the solvent, TMS, and noise from water suppression were removed from the data set prior to statistical analysis. Principal component analysis (PCA) was performed by auto scaling the variables using normalization and calculation of the first derivative as a transformation procedure.

Gas chromatography-mass spectrometry

After NMR analysis, we analyzed the exudates by gas chromatography-mass spectrometry (GC-MS). GC separations were performed on a GCMS QP2010 Plus instrument (Shimadzu, Tokyo, Japan) equipped with an Rtx-5MS WCOT capillary column (30 m × 0.25 mm; film thickness, 0.25 μm) (Restek, Bellefonte, PA, USA). The exudates were derivatized by refluxing 0.30 mg of sample for 1 h at 70°C with an excess of MeOH and acetyl chloride, dried under a stream of N_2, followed by silylation with 100 μL of N,N-bis[trimethylsilyl]trifluoroacetamide/1% trimethylchlorosilane (Superchrom, Milan, Italy) in closed vials at 60°C for 30 min. Chromatographic separation was achieved under the following temperature regimen: 60°C for 1 min (isothermal), rising by 7°C min^{-1} to 100°C, and then by 4°C min^{-1} to 320°C, followed by 10 min at 320°C (isothermal). Helium was the carrier gas supplied at 1.90 mL min^{-1}, the injector temperature was 250°C, and the split injection mode had a split flow of 30 mL min^{-1}. Mass spectra were obtained in EI mode (70 eV) scanning in the range of m/z 45 to 850 with a cycle time of 1 s. Compound identification was based on comparisons of mass spectra with the NIST library database (http://www.nist.gov/srd/nist1a.cfm), published spectra, and real standards. Due to the large variety of compounds with different chromatographic responses that we detected, external calibration curves for quantitative analysis were built by mixing methyl esters and/or methyl ethers of the following molecular standards: tridecanoic acid, octadecanol, 16-hydroxyhexadecanoic acid, docosandioic acid, β-sitosterol, and cinnamic acid. Increasing quantities of standard mixtures were loaded into a quartz boat and moistened with 0.5 mL of tetramethylammonium hydroxide (TMAH) solution (25% in methanol).

Results

Root tissue colonization by *H. seropedicae* strain HRC54

We examined the population dynamics of *H. seropedicae* strain HRC 54 associated with maize roots 7, 14, 21, and 30 days after inoculation (Figure 1). For all inoculation treatments (PGPB and HA + PGPB), the root-associated bacterial numbers were higher than those of uninoculated plants (controls). Cell shape and colony appearance confirmed the presence of *Herbaspirillum* in the pellicle harvested from the highest dilution, thereby indicating the effectiveness of inoculation.

Maize plants treated with only HA had higher bacterial numbers associated with roots than control plants. Even after seed surface disinfection, diazotrophic bacteria were recoverable from treated plants; these microbial populations were naturally occurring N fixers associated with maize seeds. They were clearly different from *H. seropedicae* (under phase contrast microscopy) in cell shape and colony form when grown in JNFb solid medium (data not shown).

We compared treatments PGPB and HA + PGPB, observing higher numbers of root-associated viable *H. seropedicae* cells in the latter. This result is qualitatively similar

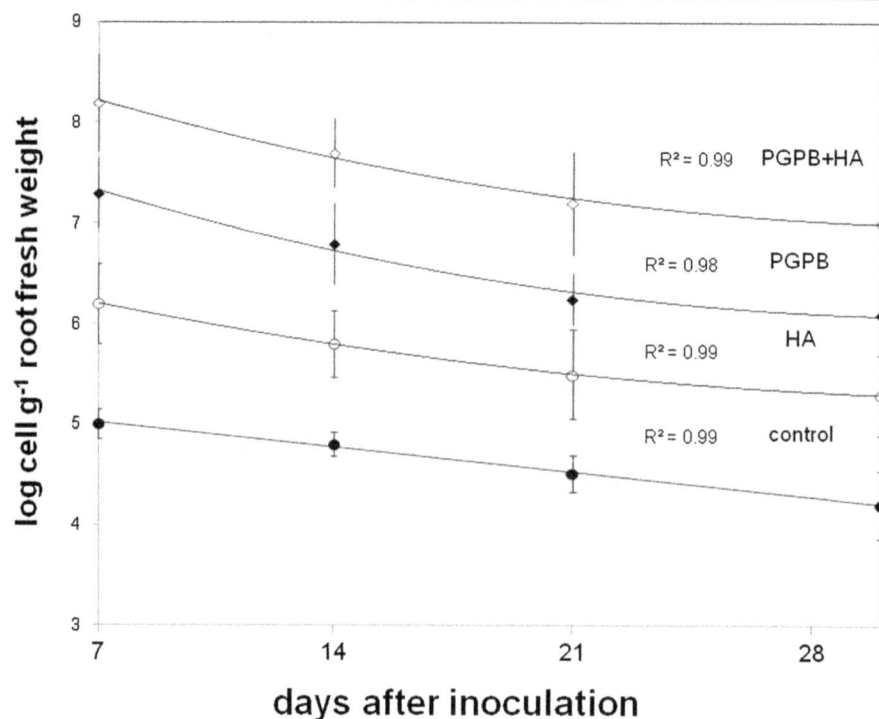

Figure 1 Number of bacterial cells (log cells g^{-1} fresh tissue) on roots of maize seedlings during different growth times. Treatments: control plants, log 10^9 cells mL^{-1} of *Herbaspirillum seropedicae* strain HRC 54, humic acids isolated from vermicompost (50 mg L^{-1}), and bacteria plus humic acids. The values represent the mean ± standard deviation.

to those obtained previously [9], indicating that HA *per se* increase the numbers of *H. seropedicae* cells colonizing root tissues and help maintain large populations in inoculated plants over protracted time periods (Figure 1). We identified similar tendencies for natural diazotrophs, whose population sizes were enhanced by HA application (in comparison with populations in control plants).

Mass of root exudates

The yields of exudates retained by the procedures used are depicted in Figure 2. Over the course of the experiment, the quantities of exudates collected across sampling occasions and treatments ranged broadly between 0.25 and 4.00 mg g^{-1} root dry weight, making it difficult to identify any treatment effects.

^{1}H NMR and PCA

Nuclear magnetic resonance (^{1}H NMR) coupled with multivariate analysis (PCA) enabled rapid discrimination among root exudate samples. The ^{1}H NMR spectra of the exudate compounds eluted from the RP C18 column with methanol on days 7, 14, and 21 are depicted in Figures 3,4,5, respectively; the main ^{1}H chemical shifts are detailed in Table 1. Visual inspection of the ^{1}H NMR spectra revealed a predominance of signals in the carbohydrate region (2.5 to 4.5 ppm) followed in rank order by signals in the aliphatic/organic acid (0.0 to 3.0 ppm)

and aromatic (5.0 to 8.0 ppm) regions. NMR spectra revealed different exudate profiles between treated and control plants. On day 7, the control spectrum contained several chemical shifts in the aliphatic region, with a small signal at 0.96 ppm and a short, intense signal at 1.24 ppm. The sugar region had a main signal at 3.71 ppm and other signals of low intensity at 3.41 and 3.93 ppm. In the aromatic region, it was possible to

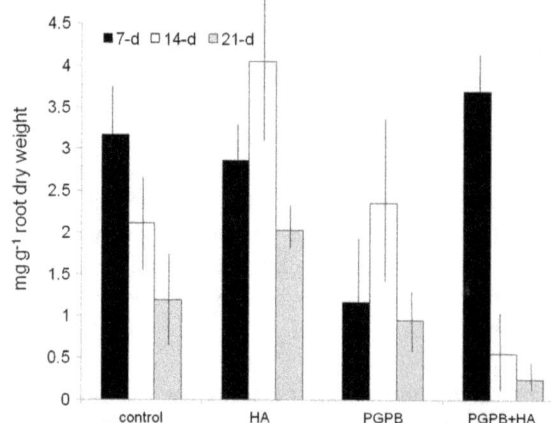

Figure 2 The yield of exudates at 7, 14, and 21 days after treatments. The values represent the mean ± standard deviation.

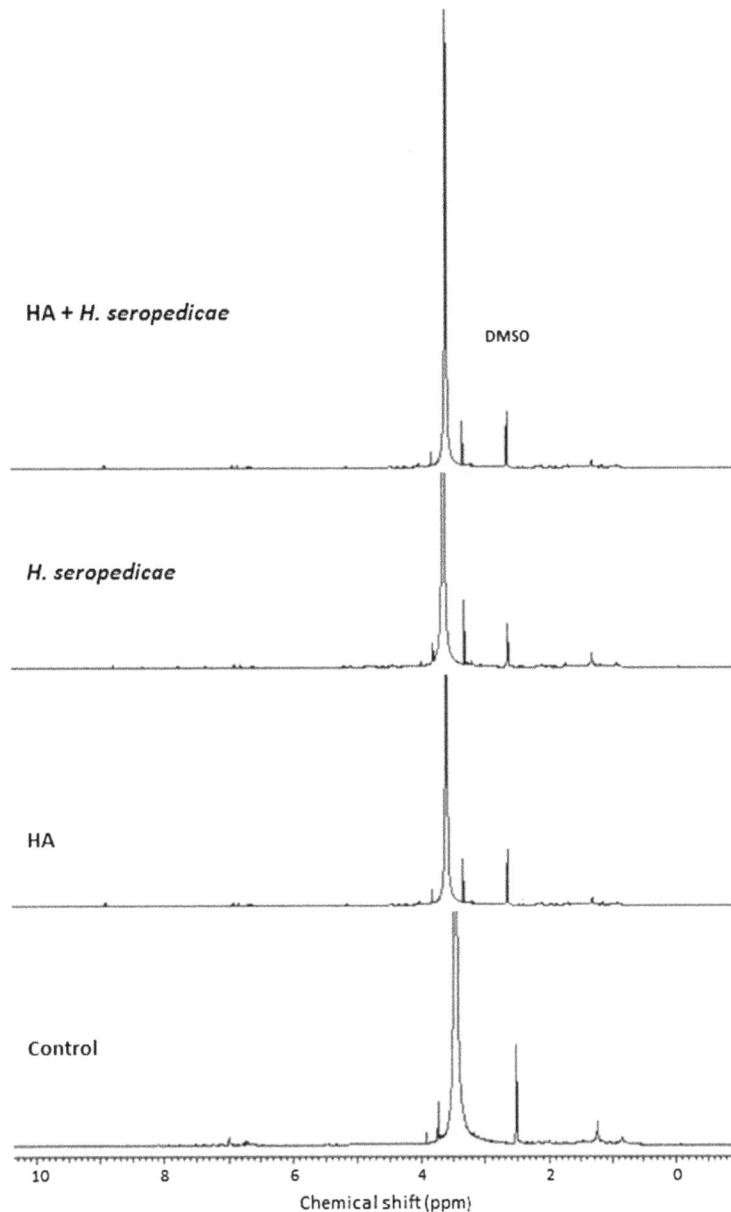

Figure 3 Spectra of root exudate at 7 days.

observe small signals in the 6.69 to 6.74 ppm region and at 7.0 ppm. The signals at 1.24 ppm may have been related to the presence of CH$_3$ compounds of aliphatic acids and βCH$_3$ in amino acids of maize extracts. The strong absorption at 3.71 ppm may be attributed to the presence of glycosylated compounds (β-Glc and α-Glc). The signals at 6.7 ppm are typical of hydroxybenzoic acids, such as cinnamic and protocatechuic acids. Tryptophan, histidine, and gallic acids have signals near 7.0 ppm. Treatment of maize seedlings with HA changed the region of aliphatic absorption, and additional signals were observed at 0.84, 0.88, and 1.10 ppm. These signals were also recorded for exudates from plants treated with

H. seropedicae (treatment PGPB), which had an additional signal at 1.64 ppm that was absent in control exudates. Additional signals were present at 1.08, 1.20, and 2.0 ppm in exudates from treatment HA + PGPB. The exudate spectrum from seedlings treated with HA alone had an additional absorption at 3.93 ppm attributable to sucrose (3.48, 3.84, 3.90, 4.22, and 5.42 ppm) and/or lysine (since a typical signal was also observed at 1.64 ppm [δCH$_2$]). Exudates from seedlings in treatment HA + PGPB had additional though very small signals in the aromatic region in the range 6.34 to 6.62 ppm and a very small signal at 8.52 ppm. Compounds similar to niacin have signals in this region.

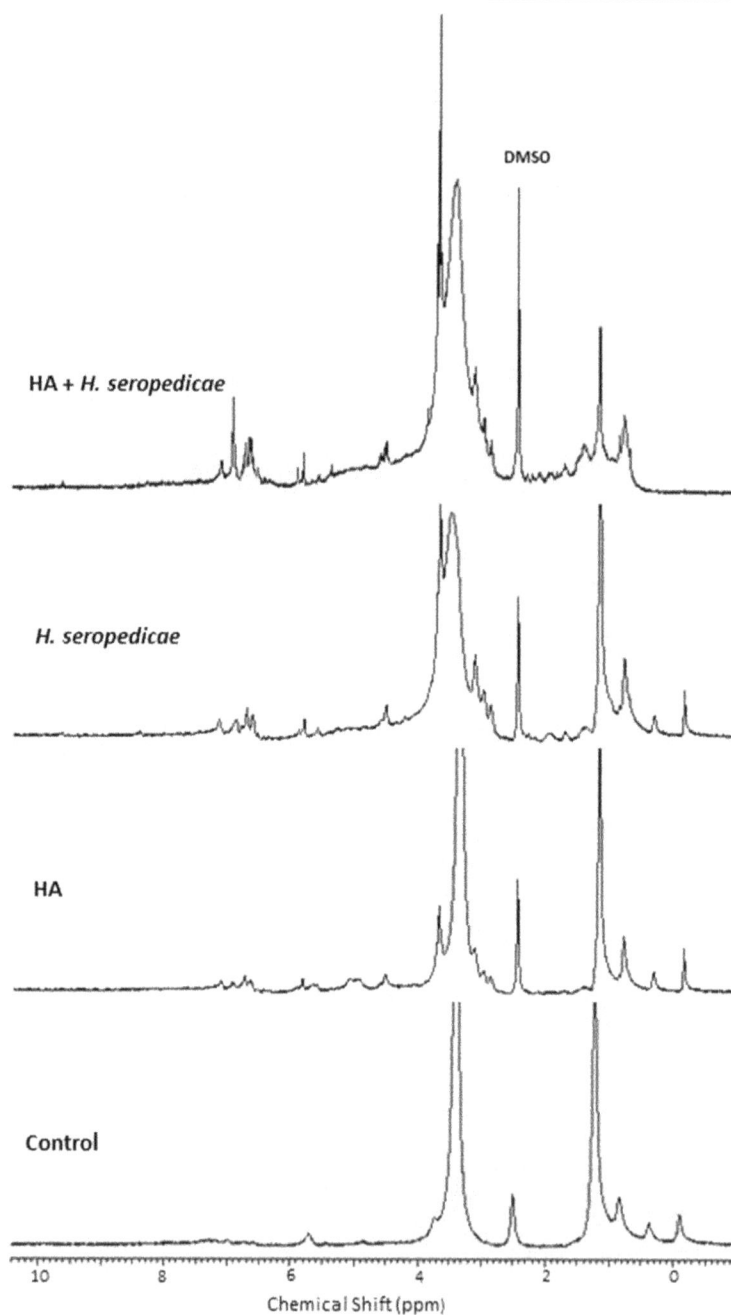

Figure 4 Spectra of root exudate at 14 days.

The profiles of exudates collected on day 14 were very different from those collected on day 7. The intensity of the signal at 1.23 ppm was enhanced in all spectra, with the highest intensity in the exudates collected from seedlings in treatment PGPB. We observed very weak signals related to aromatic compounds in the control exudate spectrum on day 14, with just one visible signal at 5.70 ppm. In contrast, signals for this region were very strong in treatment HA + PGPB exudates. In treatment HA, there was a diversity of signal shoulders in regions 4.99 to 5.13, 5.65 to 5.71, and 6.69 to 6.80 ppm and individual signals at 6.99 and 7.18 ppm. Treatment PGPB exudates had clearer signals in this region than treatment HA; treatment PGPB had well-resolved signals at 5.64, 5.85, 6.07, 6.77, 6.93, and 7.20 ppm. This spectrum was marked by weak main signals at 3.54 and 3.72 ppm. There were additional signals in the aliphatic region at 1.48, 1.77, and 2.03, which were not present in the spectra from the controls and treatment HA. The characteristics of the exudate spectra were similar between

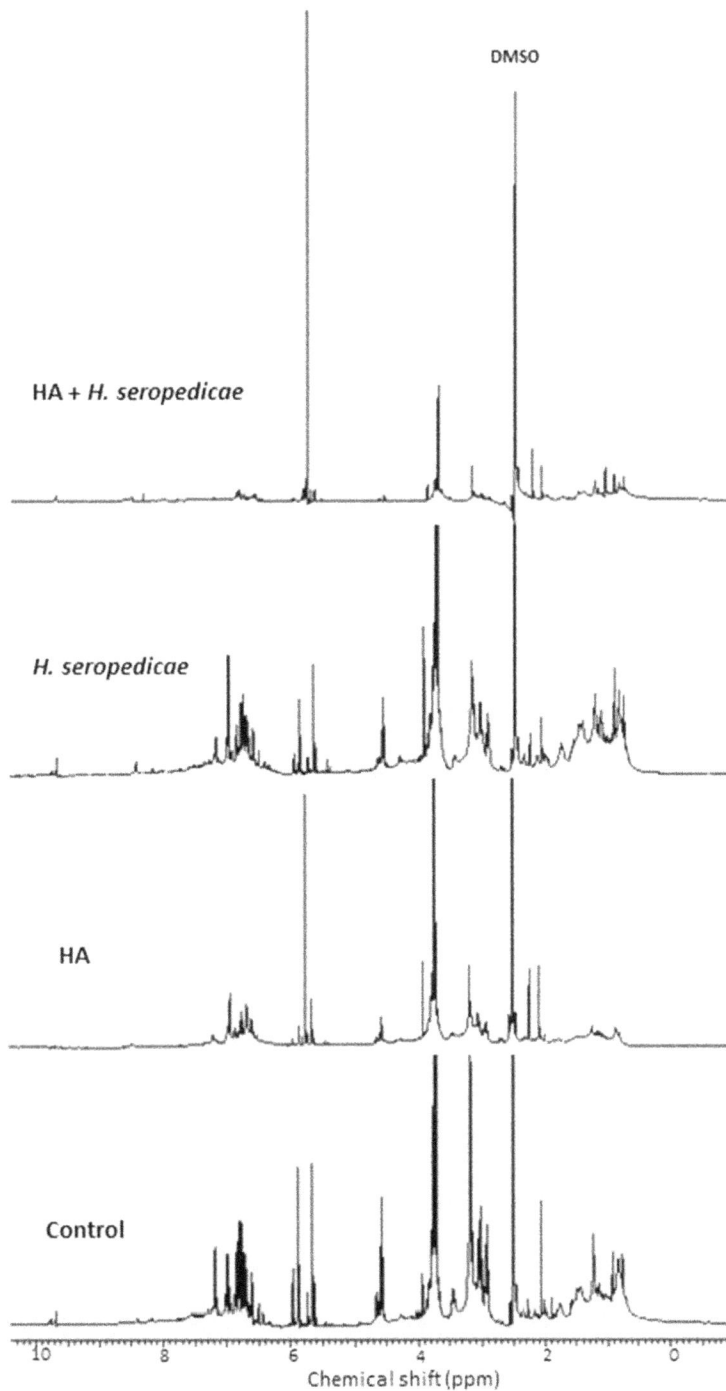

Figure 5 Spectra of root exudate at 21 days.

treatments PGPB and HA + PGPB, i.e., a strong modification in the aromatic region including a sharp signal at 6.98 ppm. However, in treatment HA + PGPB, there were changes in the sugar region including an increased intensity in the signal at 3.72 ppm.

On day 21 after inoculation, seedlings were less dependent on seed reserves and the exudate profiles were modified by the treatments (Figure 2, Table 1). The spectra were more complex than those in earlier exudates; they were characterized by an abundance of signals. Control exudates had a wide range of small signals in the aliphatic region. In the control spectra from days 7 and 14, the main signal in this region was at 2.08 ppm. Additional signals occurred in the sugar region between

Table 1 1H chemical shift

	Control			HA			PGPB			HA + PGPB		
	7	14	21	7	14	21	7	14	21	7	14	21
Aliphatic	0.96	0.10	0.10 to 0.72	0.84	0.38	0.68	0.84	0.38	0.01 to 0.59	1.08	0.77	0.78
	1.24	0.38	0.77	0.93	0.85	0.70	0.93	0.85	0.62	1.10	0.85	0.82 to 0.89
		0.85	0.80 to 1.39	1.10	1.23	0.71	1.10	1.23	0.64	1.20	1.23	0.94
		1.23	1.44	1.24		0.73 to 1.00	1.24	1.48	0.65	1.24	1.49	0.95
			1.47 to 1.50	1.64		1.03 to 1.22	1.64	1.77	0.70	2.0	2.29	1.07 to 1.10
			1.54 to 1.68			1.25		2.03	0.73	3.05	2.35	1.17 to 1.28
			1.76			1.28 to 1.54			0.77	3.18	2.92 to 3.17	1.41 to 1.54
			1.78			1.57			0.80 to 0.86			2.09
			1.81			1.60 to 2.06			0.90 to 1.24			2.24
			1.84			2.09			1.34			2.46 to 2.57
			1.86 to 2.39			2.10			1.35			2.93
			2.42			2.14 to 2.22			1.39			2.95
			2.43			2.25			1.44			2.99 to 3.28
			2.47			2.28			1.48 to 1.51			
			2.52			2.30			1.54			
			2.56			2.34 to 2.42			1.57			
			2.57			2.47			1.60 to 1.69			
			2.69			2.52			1.76			
			2.75 to 3.36			2.57			1.77			
						2.60 to 2.69			1.80			
						2.72 to 3.23			1.84			
									1.87 to 2.42			
									2.47			
									2.52			
									2.56 to 3.36			
Sugars	3.47	3.39	3.38 to 4.72		3.40	3.27 to 3.61	3.50		3.40 to 3.56	3.44	3.46	3.40
	3.71	3.73	4.92	3.74	4.56	3.66		3.54	3.66 to 5.00	3.65	3.73	3.42
	3.93			3.93	4.99 to 5.13	3.68		3.72		4.26	4.56	3.53
						3.71 to 4.65		4.56			4.58	3.56 to 3.82
						4.67 to 4.99		4.65			4.64	3.88
											4.66	
Aromatic	6.69 to 6.74	5.70	5.45	6.71	5.65 to 5.71	5.00 to 6.16	6.51	5.64	5.01 to 6.80	6.34 to 6.62	5.65	5.66
	7.0		5.61	7.01	5.88	6.19 to 6.54	6.65	5.85	6.85	8.52 (very low)	5.87	5.72
			5.63		6.69 to 6.80	6.57 to 6.90		5.94	6.87		5.96	5.77
			5.66		6.99	6.94 to 7.61		6.67	6.89 to 7.04		6.0	5.82
			5.68		7.18	7.71		6.77	7.07		6.70	6.82
			5.70 to 5.72			7.73		6.93	7.10 to 9.49		6.73	6.83
			5.75			7.87		7.20	9.64		6.98	8.34
			5.83			8.00			9.69 to 9.99		7.16	
			5.88 to 5.93			8.05						
			5.97			8.34						
			5.98			8.45						
			6.34			8.49						
			6.42 to 8.60			9.74						
			9.69 to 9.78									

3.03 and 3.45 ppm, including a sharp signal at 3.18 ppm. The main absorption in this region occurred at 3.73 ppm, as in earlier exudates. Signals between 4.55 and 4.67 ppm were very minor or absent in exudates from days 7 and 14, but much higher and with greater definition by day 21, which was also the case for signals in the aromatic region (6.50 to 7.19 ppm). Additional small signals occurred at 9.69, 9.76, and 9.78 ppm. The exudate profile from treatment HA was generally quite similar to that of the control, but significant differences were apparent. For example, there was no sharp, high-intensity signal at 2.25 ppm in the control, but this was observable in treatment HA exudates, which also had a very intense signal at 5.57 ppm (this was the main signal in treatment HA + PGPB spectra but a very minor one in treatment PGPB and in the control). On day 21, the spectra of exudates were closely similar between treatment PGPB and the control.

Similarities and dissimilarities among treatments and experimental days were examined by PCA analysis (Figure 6). PC1 (89%) and PC2 (10%) accounted for 99% of the total variance in the ^{1}H NMR data set. Symbols in the PCA plot are identified by treatment codes and days of exudate collection. Three groups of symbols are apparent in the Cartesian two-dimensional space of axes PC1 and PC2; these correspond to the three occasions of exudate collection. The day 7 group is located in the positive sector of the PC1 axis and in the negative sector of the PC2 axis. Positive values of PC1 and the loading coefficient responsible for this distinction are large for the carbohydrate ^{1}H NMR region (3 to 4 ppm). HA + PGPB symbols for the day 14 fall in the negative sector of PC2, but other treatment symbols for this collection day fall in the positive sector of this axis. HA and PGPB symbols cluster together; the control treatment had high values of aliphatic moieties (0 to 2 ppm). On day 21, the HA + PGPB and HA treatments had very different exudate compositions with elevated aromatic and aliphatic contents, respectively. The control and PGPB treatment clustered together.

Overall, the PCA analysis aided interpretation and comparisons in the data set. Thus, on day 7, exudate compositions were not significantly different among treatments, but significant differences were observed on days 14 and 21. This trend was corroborated by GC-MS analysis. The chromatograms are available in the supplementary data (Additional files 1, 2, and 3). Main exudates retained in the RP C18 column and identified by GC-MS are indicated in Table 2. On day 7, the chromatograms were closely similar; the main compounds were nitrogenous carbohydrate derivatives and fatty acids (as methyl esters). Nevertheless, there were some differences among treatments on day 7. For example, exudation of carbohydrate compounds by root seedlings in treatment HA + PGPB exceeds those in other treatments (Table 2). The chromatogram peaks of these carbohydrate compounds for

treatments PGPB and HA were twice as high as that for the control; we identified galacto, gluco, and mannose pyranoside structures in these peaks. D-glucopyranoside moieties associated with aromatic rings were also found in control exudates, but 1,6-β-glucose was found only in exudates from plants in treatment HA. The quantities and diversities of nitrogenous compounds on day 7 were also greater in treatment HA and PGPB exudates than in the controls (Table 2). Benzilamines, polyamines, pyrrol derivatives, amino acid complexes, and nucleotide derivatives were included in this group of compounds. Benzilamines and polyamines were exuded only in treatments HA and HA + PGPB. Nitrogenous compounds from the phenylalanine pathway were present; these included pyrimidinedione, pyrrole-3-carboxylic acid and a derivative from p-coumaric hydrolysis with a retention time at 28.91 min (identified as caffeine). Structures derived from adenosine were found in all treatments, but those from purine were found only in treatment PGPB. Overall, treatment HA + PGPB induced the largest and most diverse exudation of nitrogenous compounds.

Fatty acids comprised the second largest class of compounds exuded on day 7. Unlike nitrogenous compounds, treatments PGPB and HA + PGPB induced exudations of only small quantities of fatty acids (in comparison with treatment HA and the control). Chromatogram peak areas for fatty acids were elevated by 52% in treatment HA in comparison with the control on day 7. The main fatty acid in the HA exudate was hexadecanoic acid (palmitic acid). Stearic acid (octadecenoic) was exuded only by control plants, while tridecanoic acid and eicosanoic acid (also known as arachidic acid), which are minor constituents of corn oil, were found only in treatment HA exudates. Palmitic and stearic acids were found in treatment PGPB exudates. Although GC-MS is not particularly appropriate for short-chain organic acid identification, we did observe the presence of propenoic acid associated with long-chain carbon compounds in control and treatment HA exudates; we also detected succinic acid associated with aromatic moieties in treatment HA + PGPB. Benzenedicarboxylic acid was the main aromatic exudate. The steroid isosteviol was found in all exudates. Other steroids, such as cyclolanostane and androstane, were found in exudates produced in treatments HA, PGPB, and HA + PGPB.

The main classes of compounds were similar among days 7, 14, and 21 (Table 2), but the diversity of compounds diminished progressively as the seedlings grew. There were differences among the treatments: on day 14, nitrogenous compounds were still abundant in the HA, PGPB, and HA + PGPB treatment exudates; the concentrations were threefold to fourfold higher than those in the control. Nucleotide derivatives were observed on day 7, but not later. Products from phenylalanine and pyridines were exuded in treatments PGPB and HA + PGPB,

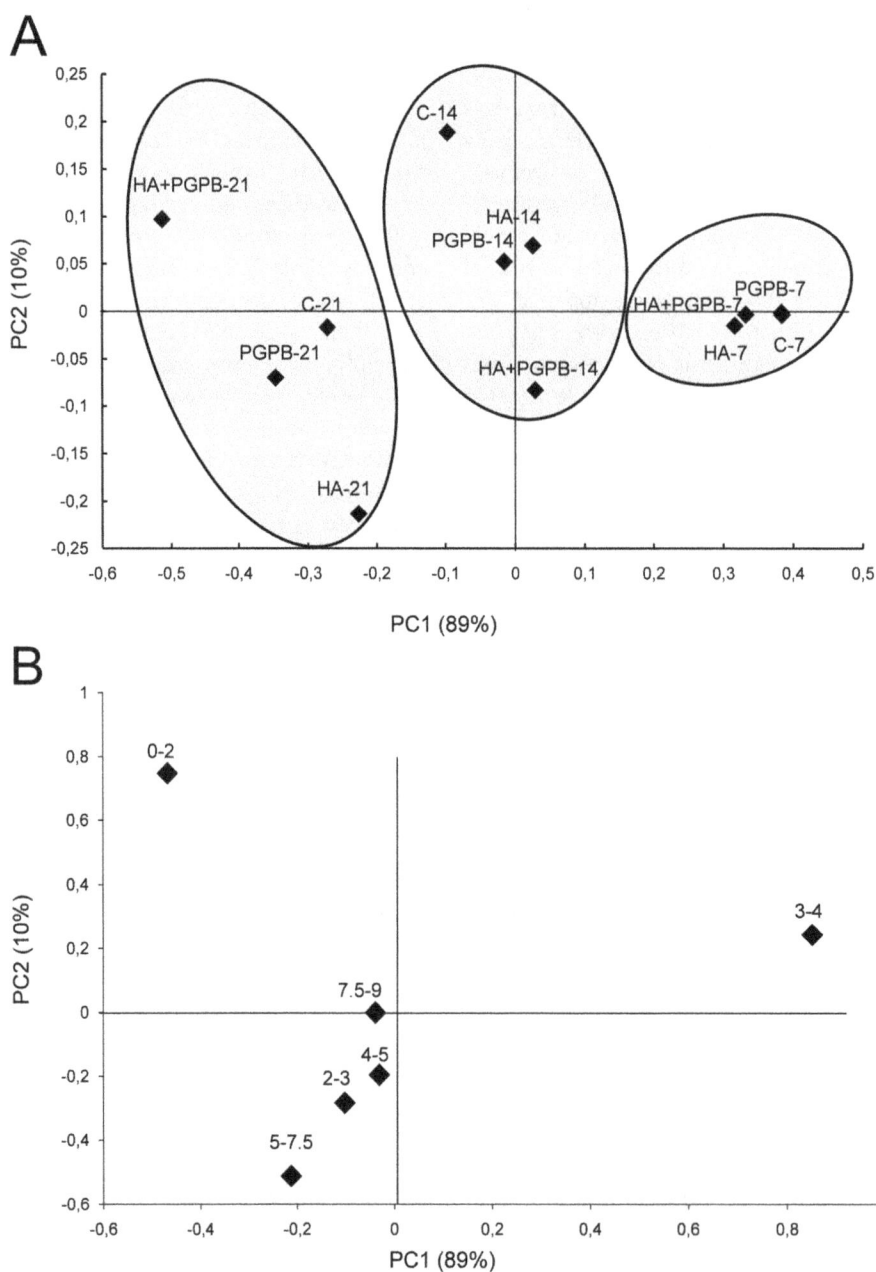

Figure 6 PCA scores and PCA loading plot. (A) PCA scores, indicating good separation in different groups according to treatments and days of treatments. **(B)** PCA loading plot. Position of the variables along the PCs indicates their importance for that PC.

while phenyl cyanate and pyrrol derivatives were found in treatment HA exudates. The compounds present in exudates from treatments PGPB and HA + PGPB were closely similar (Table 2), although benzene, 1-isocyanato-4-methoxy, was exuded only in treatment PGPB and carbonic acid, monoamine, N-(2,4-dimethoxyphenyl)-, and butyl ester only in treatment HA + PGPB. Exudations of carbohydrates were elevated in treatments HA and PGPB. These exudates contained compounds we had identified on day 7. However, quantities of carbohydrate derivatives

were reduced in treatment HA + PGPB and absent in control seedling exudates.

Hexadecanoic and octadecanoic acids were the main fatty acids occurring as methyl esters in control exudates. These compounds also occurred in other treatments, but the diversity was greatest in PGPB and HA + PGPB treatments. However, on day 7, the quantities and diversity were highest in treatment HA. As on day 7, products of the isosteviol reaction occurred in all exudates; another steroid (androstane derivative) was also

Table 2 Compounds identified by GC-MS analysis

Retention time (min)	Compound class	Control			HA			H. seropedicae			HA + H. seropedicae		
		7	14	21	7	14	21	7	14	21	7	14	21
	Nitrogenous compounds												
9.575	3-Butenoic acid, 3-methoxy-4-nitro-, methyl ester, (E)	nd	nd	nd	nd	3.82	nd	nd	nd	nd	nd	nd	nd
14.754	Benzene, 1-isocyanato-4-methoxy	nd	nd	nd	nd	0.82	nd	nd	0.83	nd	0	nd	nd
20.618	3,4-Dimethoxy-6-amino toluene	8.54	2.75	0	8.93	4.78	6.08	7.41	5.00	9.70	12.14	5.43	3.35
21.787	1H-Pyrrole-3-carboxylic acid, 5-formyl-1-(methoxymethyl)-, methyl ester	4.22	nd	nd	nd	1.21	nd	nd	nd	2.65	nd	nd	nd
21.795	3,4,5-Trimethoxybenzylamine	nd	nd	nd	5.11	nd	nd	3.47	1.18	1.18	4.56	1.21	nd
24.873	2,5-Dimethoxyphenyl isocyanate	11.41	7.41	0	16.78	13.18	24.30	14.36	10.64	48.74	18.82	24.19	16.91
25.952	2,4-Dimethoxyphenylamine, N-ethoxycarbonyl-	0.48	nd	nd	0	nd	nd	nd	0.26	0.57	0.68	0.70	nd
26.019	9H-Purin-6-amine, N,N,9-trimethyl	2.15	nd	nd	1.30	nd	nd	1.97	nd	nd	1.84	nd	nd
26.982	4-(Diethylamino)salicylaldehyde	nd	nd	nd	nd	nd	nd	nd	nd	2.87	nd	nd	nd
27.134	2,3,4-Trimethoxyphenylacetonitrile	nd	nd	nd	nd	nd	nd	nd	nd	0.94	nd	nd	nd
28.503	Bicyclo[2.2.1]heptane-2-acetamide, N-(1,3-dihydro-5,6-dimethoxy-3-oxo-4-isobenzofuranyl)	3.04	nd	nd	4.69	nd	5.30	3.51	nd	13.17	5.26	nd	3.32
28.703	3,4,5-Trimethoxyphenyl cyanate	nd	nd	nd	nd	2.23	nd	nd	nd	nd	nd	nd	nd
28.709	Tris(2,4,6-dimethylamino)pyrimidine	nd	nd	nd	nd	nd	nd	nd	1.75	nd	nd	3.34	nd
28.915	Caffeine	nd	nd	nd	0.48	nd	nd	nd	nd	nd	nd	nd	nd
29.340	Acetamide, N-[2-(3,4-dimethoxyphenyl)ethyl]	nd	nd	nd	nd	nd	nd	nd	0.89	4.06	nd	nd	nd
29.346	3-Pyridinecarboxamide, 4-(methoxymethyl)-6-methyl-2-propoxy-	nd	nd	nd	nd	nd	0.53	nd	nd	nd	nd	nd	1.49
29.537	1,3-Benzenediamine, N,N'-diethyl	nd	nd	nd	nd	0.41	nd	nd	nd	nd	nd	nd	nd
30.534	Carbonic acid, monoamide, N-(2,4-dimethoxyphenyl)-, butyl ester	nd	nd	9.55	nd	nd	nd	nd	nd	nd	nd	nd	nd
30.774	3-(2-Ethoxycarbonyl-ethyl)-4-methyl-1H-pyrrole-2-carboxylic acid, ethyl ester	nd	nd	nd	nd	12.62	nd	nd	nd	nd	nd	nd	nd
30.775	Carbonic acid, monoamide, N-(2,4-dimethoxyphenyl)-, butyl ester	nd	nd	nd	nd	nd	nd	nd	nd	nd	nd	6.93	nd
30.801	Propanamide, 3-amino-N-(2,5-dimethoxyphenyl)-3-(hydroxyimino)-	nd	nd	nd	nd	nd	nd	nd	14.99	nd	nd	nd	nd
26.134	Phenol, 4-(aminomethyl)-2-methoxy	nd	nd	nd	nd	0.33	nd	nd	nd	nd	nd	nd	nd
30.741	1,5-Bis[veratrylaminomethyl]-2,6-dihydroxynaphthalene	nd	1.25	nd	nd	nd	nd	nd	nd	nd	nd	nd	nd
30.952	3H-Purin-6-amine, N,N,3-trimethyl	nd	nd	nd	nd	nd	nd	0.61	nd	nd	0.58	nd	nd
31.834	Benzeneethanamine, N-acetyl-2,3,4,5-tetramethoxy	nd	nd	nd	nd	nd	nd	nd	2.03	nd	nd	nd	nd
32.132	Acetamide, N-(3,4,5-trimethoxyphenethyl)	nd	nd	nd	nd	nd	nd	nd	nd	0.91	nd	nd	0.40
35.077	Uridine, 2'-deoxy-3-methyl-3',5'-di-O-methyl-	nd	nd	nd	nd	nd	nd	nd	nd	nd	0.19	nd	nd
35.926	Cytidine, N-methyl-, 2',3',5'-trimethyl ether	0.62	nd	nd	1.44	nd	nd	1.52	nd	nd	0.55	nd	nd
36.019	Thymidine, 2'-deoxy-N,O,O-trimethyl-	0.91	nd	nd	1.62	nd	nd	2.24	nd	nd	3.37	nd	nd
37.792	N-[12-Aminododecyl]aziridine	0.52	nd	nd	0.58	nd	nd	0	nd	nd	0.95	nd	nd
41.125	Adenosine, 2'-deoxy-N,N,O,O-tetramethyl-	nd	nd	nd	0.65	nd	nd	1.60	nd	nd	2.11	nd	nd
42.007	Adenosine, 3'-amino-3'-deoxy-N,N-dimethyl	nd	nd	nd	0.20	nd	nd	0.46	nd	nd	0.66	nd	nd
		nd	nd	nd	nd	nd	nd	nd	nd	nd	nd	nd	nd

Table 2 Compounds identified by GC-MS analysis *(Continued)*

	Carbohydrate compounds												
17.168	1,6-Anhydro-β-D-glucose, trimethyl ether	nd	nd	nd	1.35	nd	nd	nd	0.77	nd	0.83	nd	nd
18	α-D-Galactopyranoside, methyl 2,3,4,6-tetra-O-methyl	1.74	nd	nd	3.17	nd	nd	0.73	nd	nd	4.24	nd	nd
17.688	3,4,6-Tri-O-methyl-D-glucose	nd	nd	nd	nd	0.54	nd	nd	0.83	nd	nd	nd	nd
18.1280	α-Methyl 4-methylmannoside	nd	nd	nd	nd	nd	nd	nd	0.55	nd	nd	nd	nd
21.355	2,4,5,6,7-Pentamethoxyheptanoic acid, methyl ester	nd	nd	nd	nd	0.33	nd	nd	nd	nd	nd	nd	nd
20.206	β-D-Glucopyranose, 1,6-anhydro	nd	nd	nd	nd	nd	nd	nd	nd	1.02	nd	nd	nd
28.814	1,2,3,4-Tetramethylmannose	nd	nd	nd	nd	1.85	nd	nd	1.75	nd	nd	0.58	nd
28.631	α-D-Mannopyranoside, methyl 2,3,4,6-tetra-O-methyl	nd	nd	3.33	nd	nd	nd	0	nd	nd	nd	nd	0.53
37.425	α-D-Glucopyranoside, phenyl 2, 3,4,6-tetra-O-methyl	nd	nd	nd	nd	41.61	31.93	nd	22.57	nd	nd	5.62	nd
	Fatty acids as methyl ester												
26.320	Tridecanoic acid, 12-methyl-, methyl ester	0.77	nd	nd	0	nd	nd	1.06	1.26	nd	0.80	1.62	nd
26.323	Methyl tetradecanoate	nd	nd	nd	1.22	nd	nd	nd	nd	nd	0	nd	nd
27.419	2-Hexadecenoic acid, methyl ester, (E)	nd	nd	nd	0.32	nd	nd	nd	nd	nd	0.37	nd	nd
27.901	Tridecanoic acid, 12-methyl-, methyl ester	nd	nd	nd	0.28	nd	nd	nd	nd	nd	0	nd	nd
30.133	Methyl hexadec-9-enoate	1.06	nd	nd	2.94	nd	nd	0.84	nd	nd	0.47	nd	nd
30.181	9-Octadecenoic acid (Z)-, methyl ester	0	nd	nd	0	nd	nd	nd	nd	nd	0.13	nd	nd
30.625	Hexadecanoic acid, methyl ester	4.22	2.26	0	7.28	nd	nd	4.05	nd	1.04	3.22	3.31	1.96
30.327	6-Octadecenoic acid, methyl ester	nd	nd	nd	nd	nd	nd	nd	nd	nd	nd	0.82	nd
34.053	9-Octadecenoic acid, methyl ester, (E)	1.27	nd	nd	nd	nd	nd	nd	0.73	nd	nd	0.84	nd
34.164	8-Octadecenoic acid, methyl ester, (E)	0.67	nd	nd	nd	nd	nd	nd	nd	nd	nd	nd	nd
34.548	Octadecanoic acid, methyl ester	0.95	1.11	0	1.36	0.57	nd	2.15	1.61	0.28	1.40	1.60	0.67
38.161	Eicosanoic acid, methyl ester	nd	nd	nd	0.16	nd	nd	nd	nd	nd	nd	nd	nd
	Organic acids												
26.507	Benzoic acid, 4,5-dimethoxy-2-(4-methoxybenzenesulfonylamino)-	nd	nd	nd	nd	1.31	nd	nd	nd	nd	nd	nd	nd
40.052	Succinic acid, di(5-methoxy-3-methylpent-2-yl) ester	nd	nd	nd	nd	nd	nd	nd	nd	nd	0.36	nd	nd
43.743	Cyclopropanecarboxylic acid, 2,2-dimethyl-3-(2-methyl-1-propenyl)-, 2-methyl-4-oxo-3-(2-propenyl)-2-cyclopenten-1-yl ester	nd	nd	nd	nd	nd	nd	nd	0.70	nd	nd	nd	nd
	Aromatics and phenol derivatives												
15.817	2-Methoxy-4-vinylphenol	nd	nd	nd	nd	nd	nd	nd	nd	0.3	0	nd	nd
19.337	Phenol, 4,6-di(1,1-dimethylethyl)-2-methyl	nd	nd	nd	0.24	nd	nd	nd	nd	nd	0.23	nd	nd
24.653	Isoelemicin	nd	nd	nd	nd	nd	nd	nd	nd	0.49	0	nd	nd
37.592	Benzene, (1-methyldodecyl)	0.31	nd	nd	nd	nd	nd	1.98	nd	nd	0.80	nd	nd
41.779	1,2-Benzenedicarboxylic acid, diisooctyl ester	nd	nd	69.71	nd	nd	nd	1.49	nd	nd	0.44	nd	nd
41.784	1,2-Benzenedicarboxylic acid, mono(2-ethylhexyl) ester	1.55	nd	nd	nd	nd	2.37	nd	nd	nd	0	nd	7.49
21.512	Phenol, 2,6-dimethoxy-4-(2-propenyl)	nd	nd	nd	nd	0.25	nd	nd	nd	nd	nd	nd	nd
35.007	Phenol, 5-methoxy-2,3,4-trimethyl	nd	0	nd	nd	0.22	nd	nd	nd	nd	nd	nd	nd
37.803	Benzene, (1-methylnonadecyl)	nd	0.78	nd	nd	nd	nd	nd	1.15	nd	nd	1.36	nd

Table 2 Compounds identified by GC-MS analysis *(Continued)*

22.015	Benzoic acid, 4-hydroxy-3,5-dimethoxy	nd	nd	nd	nd	0.21	nd	nd	nd	nd	nd	nd	nd
22.016	1,2,3,4-Tetramethoxybenzene	nd	nd	nd	nd	nd	nd	nd	nd	nd	nd	0.27	nd
23.399	Benzoic acid, 3,4-dimethoxy-, methyl ester	nd	nd	nd	nd	0.72	1.41	nd	nd	nd	nd	nd	1.65
25.152	2-Propenoic acid, 3-(4-methoxyphenyl)-, methyl ester	nd	nd	nd	1.17	1.03	nd	nd	0.65	nd	nd	0.66	nd
26.277	Benzenepropanoic acid, 3,4-dimethoxy-, methyl ester	nd	nd	nd	nd	nd	5.14	nd	nd	nd	nd	nd	3.27
26.328	Benzoic acid, 3,4,5-trimethoxy-, methyl ester	nd	nd	nd	0	nd	nd	nd	nd	2.70	nd	nd	nd
29.738	2-Propenoic acid, 3-(3,4-dimethoxyphenyl)-, methyl ester	0.49	nd	nd	0	2.61	5.02	nd	0.75	4.24	0.23	1.24	3.21
36.154	2-Propenoic acid, 2-methyl-, dodecyl ester	0	nd	nd	0	nd	nd	nd	nd	nd	0.12	nd	nd
	Terpenoids and steroids												
43.074	17β-Hydroxy-6α-pentyl-4-oxa-5β-androstan-3-one	nd	nd	nd	0.74	nd	nd	0.93	nd	nd	nd	nd	nd
43.412	Dihydroxyisosteviol	3.88	nd	nd	3.94	nd	nd	5.45	3.95	nd	3.69	0.73	nd
42.582	Dihydro-isosteviol methyl ester	nd	2.04	nd	nd	0	9.42	nd	nd	nd	nd	nd	17.03
43.332	Methyl dihydroisosteviol	nd	nd	nd	nd	1.47	nd	nd	nd	nd	nd	nd	nd
43.665	Isosteviol methyl ester	nd	2.88	nd	nd	4.81	nd	nd	4.90	17.18	nd	4.68	nd
39.849	5-Cholene, 3,24-dihydroxy-	nd	nd	nd	nd	nd	nd	nd	nd	nd	0.12	nd	nd
43.430	Patchoulene	nd	nd	nd	0.36	nd	nd	nd	nd	nd	nd	nd	nd
45.538	Androstan-17-one, 3-ethyl-3-hydroxy-, (5α)	nd	0.78	nd	nd	2.89	nd	nd	1.03	0.50	nd	2.30	nd
42.354	5α,6α-Epoxy-17-oxo-6β-pentyl-4-nor-3, 5-secoandrostan-3-oicacid, methyl ester	nd	nd	nd	nd	nd	nd	nd	nd	nd	nd	nd	0.31
43.091	5β-Pregnan-17α,21-diol-3,20-dione	nd	nd	nd	nd	nd	nd	nd	nd	nd	nd	nd	25.88
43.434	Pregnan-20-one, (5α,17α)	nd	nd	2.48	nd	nd	nd	nd	nd	nd	nd	nd	nd
	Alcohols												
3.974	Ethanol, 2-(2-methoxyethoxy)	nd	nd	nd	nd	nd	nd	nd	nd	1.18	0.33	nd	nd
6.715	1-Hexanol, 2-ethyl	1.65		7.48	2.71	0.39	1.13	0.86	nd	nd	3.39	nd	1.19
12.925	Ethanol, 2-phenoxy	nd	nd	nd	nd	nd	nd	nd	nd	0.63	nd	nd	nd
42.010	9-t-Butyltricyclo[4.2.1.1(2,5)]decane-9,10-diol	nd	nd	nd	nd	nd	nd	nd	nd	nd	nd	1.03	nd
	Unidentified												
30.542		nd	nd	nd	nd	nd	nd	nd	nd	nd	nd	nd	3.14
31.625		nd	nd	nd	nd	nd	nd	nd	nd	nd	nd	nd	0.74
31.825		nd	nd	nd	nd	nd	nd	nd	nd	nd	nd	nd	nd
32.184		nd	nd	nd	nd	0.79	nd	nd	nd	nd	nd	nd	nd
35.248		nd	nd	nd	nd	0.77	nd	nd	nd	nd	nd	nd	nd
38.362		nd	nd	nd	nd	nd	nd	nd	nd	nd	0.42	nd	nd
39.253		nd	nd	nd	nd	nd	nd	nd	nd	nd	0.09	nd	nd
40.033		nd	nd	nd	nd	nd	nd	nd	nd	nd	nd	0.43	nd
40.230		nd	nd	nd	0.20	nd	nd	nd	nd	nd	0.37	nd	nd
40.833		nd	nd	nd	nd	nd	nd	nd	nd	nd	nd	nd	0.18
40.964		nd	nd	nd	nd	nd	nd	nd	nd	nd	0.31	nd	nd
41.908		nd	nd	nd	0.49	nd	nd	nd	nd	nd	0.41	nd	nd
42.157		nd	nd	nd	nd	nd	nd	nd	nd	nd	0.24	nd	nd
43.746		nd	nd	nd	nd	nd	nd	nd	0.43	nd	0	nd	nd
44.305		nd	nd	nd	nd	nd	nd	nd	nd	nd	0.27	nd	nd

Table 2 Compounds identified by GC-MS analysis *(Continued)*

44.532	nd	nd	nd	0	nd	nd	nd	nd	nd	nd	nd	0.45
45.275	nd	nd	nd	0.74	nd	nd	0.35	nd	nd	0.89	nd	nd
46.026	nd	nd	nd	nd	nd	nd	0	nd	nd	0.11	nd	nd
46.277	nd	nd	nd	nd	nd	nd	0	nd	nd	nd	nd	1.72
46.606	nd	nd	nd	nd	nd	0.53	0	nd	nd	0	nd	nd

Compound exudates from maize seedlings at different growth stages (7, 14, and 21 days) induced or not (control) by humic acids (HA), *Herbaspirillum seropedicae* (*H. seropedicae*), and HA and *H. seropedicae*. nd, not detected.

found in the exudates, though in only small quantities in the control.

Compounds identified on day 21 are shown in Table 2. Exudates collected on that day contained a smaller diversity of compounds than earlier collections. Hexanol was found in control exudates and ethanol phenoxy products in treatment PGPB, likely due to carbohydrate hydrolysis reactions. Nitrogenous compounds were the main exudate products retained by the RP C18 column; carbonic acid, monoamide, *N*-(2,4-dimethoxyphenyl)-, and butyl ester were found only in control exudates on day 21. 2,5-Dimethoxyphenyl isocyanate occurred in elevated quantities in exudates of treatments HA, PGPB, and HA + PGPB; a derivative from acetamide and dimethoxy amino toluene was also abundant. A nitrogenous compound identified as 4-(diethylamino)salicylaldehyde was found in treatment PGPB exudates on day 21, but not on days 7 and 14. Finally, products from acetamide were typical of exudates in treatments PGPB and HA + PGPB; they were found on day 21 at 29.34 and 32.13 min.

Discussion

Our aim in the present work was to determine the effects of single and combined applications of humic acid and the endophytic diazotrophic bacterium *H. seropedicae* on exudation profiles of maize roots. This was a first step in identifying compounds that may be involved in rhizosphere interactions and modulation of root colonization and endophytic establishment of bacteria.

We demonstrated enhanced root colonization by *H. seropedicae* in the presence of humic acids (Figure 1). Previous work showed that the elevated bacterial population associated with maize roots may be explainable in part by humic acid sorption at the plant cell wall surface, which is associated with increased attachment and endophytic colonization by *H. seropedicae* [9,10]. Moreover, the main infection sites for *H. seropedicae* in grasses are the points of lateral root emergence, which are induced by humic acid and may contribute to enhanced establishment of the endophytic population [11].

Canellas et al. [12] and Puglisi et al. [13] previously observed changes in exudation profiles of organic acids in maize seedlings treated with humic acid. However, identifying causal relationships in changed exudate profiles is complex, and results may not always be directly attributable to treatment effects on the quantities and the composition of root exudates. Juo and Storzky [36] observed decreases in protein and carbohydrate release with increasing plant age using an electrophoretic approach. They suggested that the RP C18 column used to retain the exudates may have limited their quantitative analysis. We noticed on day 21 (across all treatments) that exudate yields had declined (Figure 2) relative to earlier collections on days 7 and 14.

^{1}H NMR spectroscopy (Figures 3,4,5) demonstrated differences in exudate profiles under the influence of the treatments, and PCA analysis revealed significant changes from day 14 onward (Figure 6). Roncatto-Maccari et al. [37] observed that 3 days after maize inoculation, the population of *H. seropedicae* mainly comprises single cells attached to the root surface. They (*loc. cit.*) found that after 12 days the cells were connected and forming aggregates; they detected (by scanning electron microscopy) a halo of mucilage surrounding the bacterial colony on the roots of inoculated maize. Although plant-bacterial signaling is considered a rapid phenomenon, changes in exudate NMR spectra in our experiment occurred only after extensive colonization on the root axis (i.e., on day 14).

One group of candidate biomolecules emerging from this prospective study comprised a number of compounds that have been identified as possible inducing agents of the quorum sensing (QS) system, which is a mechanism responsible for modulating population density, biofilm formation, and the specific genetic machinery for niche persistence and colonization at the root level. These compounds include oligopeptides and substituted gamma-butyrolactones, 3-hydroxypalmitic acid methyl ester, 3,4-dihydroxy-2-heptylquinoline, and a furanosyl borate diester ([5] and references therein). Interestingly, our analyses of the overall transcriptome profile of *H. seropedicae* strain Smr1 (as part of a broad examination of gene expression in planktonic and biofilm lifestyles) have shown that biofilm maturation is coupled with differentially regulated genes involved in aromatic metabolism and multidrug transport efflux activation (data not shown). These potential quorum sensing inducers may be key molecules in the well-known rhizosphere competence of *H. seropedicae* [14,25]. Functional studies linking these compounds to chemotaxis and biofilm induction are urgently required.

Another interesting future study that may develop from our prospective explorations might be an evaluation of the ability of *H. seropedicae* to use secondary metabolites as carbon sources that are exclusively or primarily exuded from maize roots treated with this bacterium. The presence of a catabolic machinery for specific cleavage of these compounds (especially compounds with antimicrobial properties) would represent a nutritional advantage for the colonization of the host plant rhizosphere.

We identified hexadecanoic acid (palmitic acid) in exudates of all treatments on days 7 and 14, but by day 21, these compounds were detected only in treatment PGPB. Non-quinoline derivatives of other heterocyclic aromatic N compounds belonging to the pyrimidine class were collected exclusively in treatments PGPB and HA + PGPB (Table 2). The most remarkable difference in root exudates among treatments was in N compounds; these increased with time in all treatments other than the controls, in which quantities decreased. Some nitrogenous compounds, like benzenamine ($C_6H_5CONH_2$), 2,4-dimethoxy, 3,4-dimethoxy-6-amino toluene, and 2,5-dimethoxyphenyl isocyanate, were found in the exudates; they are probably derivatives of aromatic/phenolic compounds that react with TMAH used in methoxylation since the phenolic compounds, which are the main secondary metabolites synthesized by maize, react promptly with NH_2. Pyrimidinediones comprise a class of chemical compounds characterized by a pyrimidine ring substituted with two carbonyl groups. 2,4(1*H*,3*H*)-pyrimidinedione, 1,3,5-trimethyl is typical of uracil derivatives, i.e., nucleobases in the nucleic acid of RNA. Pyrrole carboxylic acid derivatives, which may originate from the L-tryptophan pathway, have been described as antibacterial agents. Finally, we found one polyamine derivative of spermine. Such derivatives have been implicated in a wide range of metabolic processes in plants, ranging from cell division and organogenesis to protection against stress; they act as plant and microbial growth stimulants and chemoattractants [5].

Stearic and palmitic acids were the main fatty acids exuded by maize seedling roots. These fatty acids are predominant in most plants; they are synthesized by acetyl-CoA carboxylase and fatty acid synthase. Quantities of these fatty acids exuded by the controls differed between days 14 and 21. Stearic and palmitic acids were exuded by maize seedlings treated with *H. seropedicae* with or without HA (Tables 2). Terpenoids derived from the kaurenoic acid pathway, including gibberellic acids, were found in all treatments. However, their diversity was highest in treatments PGPB and HA + PGPB. A significant increase in transcript levels of *ent-kaurene*

oxidase genes that are involved in gibberellin synthesis occurs in maize seedlings treated with *H. seropedicae* [38]. The agreement between gene expression in the gibberellin biosynthesis pathway (Zmko1) and the exudation profile in the same biological model supports our suggestion that the gibberellin synthesis pathway in root tissues is modulated by *H. seropedicae* colonization.

Root colonization by *H. seropedicae* in graminaceous plants requires movement of the bacteria to the root surface followed in sequence by adsorption and attachment to the cell wall surface [25,27]. Attached bacteria may persist at the root surface as aggregates or biofilms, and they occasionally trigger plant responses that induce lateral root sites, which serve as entryways for endophytic establishment in the plant host. Humic acids are heterogeneous, irregular, and amphiphilic structures with high charge densities and hydrophobic cores that facilitate adsorption phenomena and lateral root induction [39]. As *H. seropedicae* is a flagellated cell, there is a potential for induction of flagellar activity by plant-released compounds (chemotactic substances), as previously described for other bacteria [40].

From the inventory of exuded compounds, it may be possible to select candidate molecules as quorum sensing mediators of biofilm maturation, as substrates for catabolic mechanisms to supply energy, or as detoxifiers, thereby increasing the rhizosphere competence of *Herbaspirillum*. Furthermore, changes in rhizosphere hydrophobicity by fatty acid exudates may be an important mechanism for preserving cell viability. We detected relatively large amounts of gibberellic acid precursor in exudates isolated from maize seedlings treated with *H. seropedicae*; this bacterium induced changes in nitrogenous compounds as well.

Conclusion

The number of unidentified compounds was much higher in treatments PGPB and HA + PGPB than in the control. The limitations of a conventional GC-MS approach, such as low mass resolution and the need for previous volatilization of compounds, suggest that the instrument should be combined with other analytical tools, like liquid chromatography coupled to high mass definition spectroscopy. We found that root exudates from maize seedlings mostly comprised fatty acids and nitrogenated compounds. We observed changes in root exudate profiles in maize seedlings treated with HA and *H. seropedicae*. These changes were most pronounced on days 14 and 21, when there was enhanced fatty acid exudation in treatment HA and elevated nitrogenate and terpene exudation from treatment PGPB.

This is the first study to date that has evaluated the combined application of a plant growth-promoting bacteria and humic acid on maize root exudation profiles. This study was carried out on controlled conditions in a hydroponic culture. Overall, the chemical changes we observed in the rhizosphere may improve the survival, persistence, and biological activity of *H. seropedicae* in the presence of humic substances and provide a pool of candidate molecules for use as additives in inoculant technology formulation, which will contribute substantially to progress in arable agriculture. However, the findings of this hydroponic study need to be confirmed under real soil conditions.

Competing interests

The authors declare that they have no competing interests.

Authors' contributions

LSL conducted the laboratory experiments and did the first inventory of exudates. FLO and LPC wrote and discussed the manuscript. RRO performed and interpreted the NMR data. MRGV reviewed the nomenclature of compounds and the biosynthesis route. NOA was responsible for the multivariate analysis. All authors read and approved the final manuscript.

Acknowledgements

The work was supported by Conselho Nacional de Desenvolvimento Científico e Tecnológico (CNPq), Fundação de Amparo à Pesquisa do Estado do Rio de Janeiro (FAPERJ), Instituto Nacional de Ciência e Tecnologia (INCT) para a Fixação Biológica de Nitrogênio, Internacional Foundation of Science (IFS), and OCWP.

Author details

[1]Núcleo de Desenvolvimento de Insumos Biológicos para a Agricultura (NUDIBA), Universidade Estadual do Norte Fluminense Darcy Ribeiro (UENF), Av. Alberto Lamego, 2000, Campos dos Goytacazes, 28013-602 Rio de Janeiro, Brazil. [2]Laboratório de Ciências Químicas (LCQUI), Universidade Estadual do Norte Fluminense Darcy Ribeiro (UENF), Av. Alberto Lamego, 2000, Campos dos Goytacazes, 28013-602 Rio de Janeiro, Brazil.

References

1. Vivanco JM, Baluška F (2012) Secretions and exudates in biological systems: 12. Signaling and communication in plants. Springer, Berlin, p 290
2. Jones DL, Nguyen C, Finlay RD (2009) Carbon flow in the rhizosphere: carbon trading at the soil–root interface. Plant Soil 321:5–33
3. Uren NC (2007) Types, amounts, and possible functions of compounds released into the rhizosphere by soil-grown plants. In: Pinton R, Varanini Z, Nannipieri P (ed) The rhizosphere: biochemistry and organic substances at the soil–plant interface. CRC Press, Boca Raton, pp 1–21
4. Mimmo T, Hann S, Jaitz L, Cescoa S, Gessa CE, Puschenreiter M (2011) Time and substrate dependent exudation of carboxylates by *Lupinus albus* L. and *Brassica napus* L. Plant Physiol Biochem 49:1272–1278
5. Faure D, Vereecke D, Leveau JHJ (2009) Molecular communication in the rhizosphere. Plant Soil 321:279–303
6. Azaizeh HA, Marschner H, Römheld V, Wittenmayer L (1995) Effects of vesicular abrusular micorrhizal fungus and other soil microorganisms on growth, mineral nutrient acquisition and root exudation of soil-grown maize plants. Mycorrhiza 5:321–327
7. Neumann G, Römheld V (2007) The release of root exudates as affected by the plant physiological status. In: Pinton R, Varanini Z, Nannipieri P (ed) The rhizosphere: biochemistry and organic substances at the soil–plant interface. CRC Press, Boca Raton, pp 24–72
8. Hassan S, Mathesius U (2012) The role of flavonoids in root–rhizosphere signalling: opportunities and challenges for improving plant–microbe interactions. J Exp Bot 63:3429–3444
9. Canellas LP, Martínez-Balmori D, Médici LO, Aguiar NO, Campostrini E, Rosa RC, Façanha A, Olivares FL (2013) A combination of humic substances and *Herbaspirillum seropedicae* inoculation enhances the growth of maize (*Zea mays* L.). Plant Soil 366:119–132
10. Canellas LP, Olivares FL (2014) Physiological responses to humic substances as plant growth promoter. Chem Biol Technol Agr 1:3
11. Canellas LP, Olivares FL, Okorokova-Façanha AL, Façanha AR (2002) Humic acids isolated from earthworm compost enhance root elongation, lateral root emergence, and plasma membrane H$^+$-ATPase activity in maize roots. Plant Physiol 130:1951–1957
12. Canellas LP, Teixeira Junior LRL, Dobbss LB, Silva CA, Medici LO, Zandonadi DB, Façanha AR (2008) Humic acids cross interactions with root and organic acids. Ann Appl Biol 153:157–166
13. Puglisi E, Pascazio S, Suciu N, Cattani I, Fait G, Spaccini R, Crecchio C, Piccolo A, Trevisan M (2013) Rhizosphere microbial diversity as influenced by humic substance amendments and chemical composition of rhizodeposits. J Geochem Expl 129:82–94
14. Olivares FL, Baldani VLD, Reis VM, Baldani JI, Döbereiner J (1996) Occurrence of the endophytic diazotrophs *Herbaspirillum* spp. in roots, stems and leaves predominantly of Gramineae. Biol Fertil Soils 21:197–200
15. Roesch LFW, Camargo FAO, Bento FM, Triplett EW (2007) Biodiversity of diazotrophic bacteria within the soil, root and stem of field-grown maize. Plant Soil 302:91–104
16. Pan B, Bai YM, Leibovitch S, Smith DL (1999) Plant-growth-promoting rhizobacteria and kinetin as ways to promote corn growth and yield in a short-growing-season area. Eur J Agr 11:179–186
17. Riggs PJ, Chelius MK, Iniguez AL, Kaeppler SM, Triplett EW (2001) Enhanced maize productivity by inoculation with diazotrophic bacteria. Austr J Plant Physiol 28:829–836
18. Oliveira ALM, Urquiaga S, Döbereiner J, Baldani JI (2002) The effect of inoculating endophytic N2-fixing bacteria on micropropagated sugarcane plants. Plant Soil 242:205–215
19. Lucy ME, Reed E, Glick BR (2004) Applications of free-living plant growth-promoting rhizobacteria. A van Leeuw J Microb 86:1–25
20. Kennedy IR, Choudhury ATMA, Kecskés ML (2004) Non-symbiotic bacterial diazotrophs in crop-farming systems: can their potential for plant growth promotion be better exploited? Soil Biol Biochem 36:1229–1244
21. Shaharoona B, Arshad M, Zahir ZA, Khalid A (2006) Performance of *Pseudomonas* spp. containing ACC-deaminase for improving growth and yield of maize (*Zea mays* L.) in the presence of nitrogenous fertilizer. Soil Biol Biochem 38:2971–2975
22. Mehnaz S, Kowalik T, Reynolds B, Lazarovits G (2010) Growth promoting effects of corn (*Zea mays*) bacterial isolates under greenhouse and field conditions. Soil Biol Biochem 42:1848–1856
23. Montañez A, Sicardi M (2013) Effects of inoculation on growth promotion and biological nitrogen fixation in maize (*Zea mays* L.) under greenhouse and field conditions. Bas Res J Agric Sci Rev 2:102–110
24. Pedrosa FO, Monteiro RA, Wassem R, Cruz LO, Ayub RA, Colauto NB, Fernandez M, Fungaro MHP, Grisard E, Hungria M, Madeira HM, Humberto MF, Nodari RO, Osaku CA, Petzl-Erler ML, Terenzi H, Vieira LGE, Steffens MBR, Weiss VA, Pereira LFP, Almeida MIM, Alves LR, Marin A, Araujo LM, Balsanelli E, Baura VA, Chubatsu LS, Faoro H, Favetti A, Friedermann G, et al. (2011) Genome of *Herbaspirillum seropedicae* strain SmR1, a specialized diazotrophic endophyte of tropical grasses. PLoS Genet 7(5):e1002064
25. James EK, Olivares FL (1998) Infection and colonization of sugarcane and other graminaceous plants by endophytic diazotrophs. Crit Rev Pl Sci 17:77–119
26. Baldotto LEB, Olivares FL (2008) Phylloepiphytic interaction between bacteria and different plant species in a tropical agricultural system. Can J Microb 54:918–931
27. Monteiro RA, Balsanelli E, Wassem R, Anelis M, Brusamarello-Santos LCC, Schmidt MA, Tadra-Sfeir MZ, Pankievicz VCS, Cruz LM, Chubatsu LS, Pedrosa FO, Souza EM (2012) Herbaspirillum-plant interactions: microscopical, histological and molecular aspects. Plant Soil 356:175–196
28. Hungria M (1994) Sinais Moleculares envolvidos na nodulação das leguminosas por rizóbio. Rev Bras Ci Solo 18:339–364
29. Gough C, Galera C, Vasse J, Webster G, Cocking EC, Dénarié J (1997) Specific flavonoids promote intercellular root colonization of *Arabidopsis thaliana* by *Azorhizobium caulinodans* ORS571 (1997). Mol Plant–Micr Interac 10:560–570

30. Tadra-Sfeir MZ, Souza EM, Faoro H, Müller-Santos M, Baura VA, Tuleski TR, Rigo LU, Yates MG, Wassem R, Pedrosa FO, Monteiro RA (2011) Naringenin regulates expression of genes involved in cell wall synthesis in *Herbaspirillum seropedicae*. Appl Environ Microbiol 77:2180–2183

31. Marin AM, Souza EM, Pedrosa FO, Souza LM, Sassaki GL, Baura VA, Yates MG, Wassem R, Monteiro RA (2012) Naringenin degradation by the endophytic diazotroph *Herbaspirillum seropedicae* SmR1. Microbiol 159:167–75

32. Balsanelli E, Tuleski TR, de Baura VA, Yates MG, Chubatsu LS, Pedrosa FO, de Souza EM, Monteiro RA (2013) Maize root lectins mediate the interaction with *Herbaspirillum seropedicae* via N-acetyl glucosamine residues of lipopolysaccharides. PLoS One 8(10):e77001. doi:10.1371/journal.pone.0077001

33. Marks BB, Nogueira MA, Hungria M, Megias M (2013) Biotechnological potential of rhizobial metabolites to enhance the performance of *Bradyrhizobium* spp. and *Azospirillum brasilense* inoculants with soybean and maize. AMB Express 3:2

34. Baldani JI, Pot B, Kirchhof G, Falsen E, Baldani VLD, Olivares FL, Hoste B, Kersters K, Hartmann A, Gillis M, Döbereiner J (1996) Emended description of *Herbaspirillum*, inclusion of [*Pseudomonas*] *rubrisubalbicans*, a mild plant pathogen, as *Herbaspirillum rubrisubalbicans* comb. Nov., and classification of a group of clinical isolates (EF group 1) as *Herbaspirilum* species 3. Int J Syst Bact 46:802–810

35. Döbereiner J, Baldani VLD, Baldani JI (1995) Como isolar e identificar bactérias diazotróficas de plantas não leguminosas. Embrapa Agrobiologia, Seropédica

36. Juo P-S, Storkzky G (1970) Electrophoretic separation of proteins from roots and root exudates. Can J Bot 48:713–718

37. Roncato-Maccari LDB, Ramos HJO, Pedrosa FO, Alquini Y, Chubatsu LS, Yates MG, Rigo LU, Stefens MBR, Souza EM (2003) Endophytic *Herbaspirillum seropedicae* expresses nif genes in gramineous plants. FEMS Microbiol Ecol 45:39–47

38. Amaral FP, Bueno JCF, Hermes VS, Arisi ACM (2014) Gene expression analysis of maize seedlings (DKB240 variety) inoculated with plant growth promoting bacterium *Herbaspirillum seropedicae*. Symbiosis doi:10.1007/s13199-014-0270-6

39. Piccolo A (2002) The supramolecular structure of humic substances. A novel understanding of humus chemistry and implications in soil science. Adv Agr 75:57–134

40. Ashy AM, Watson MD, Shaw CH (1987) A Ti-plasmid determined function is responsible for chemotaxis of *Agrobacterium tumefaciens* towards the plant wound acetosyringone. FEMS Microbial Lett 41:189–198

18

Extraction and characterization of bio-effectors from agro-food processing by-products as plant growth promoters

Ziad Al Chami[1*], Deaa Alwanney[1], Sandra Angelica De Pascali[2], Ivana Cavoski[1] and Francesco Paolo Fanizzi[2*]

Abstract

Background: Recently, a novel concept 'bio-effectors' rose on to describe a group of products that are able to improve plant performance rather than fertilizers. Agro-food processing residues and by-products potentially represent important sources of bio-effectors but they are currently not properly taken in consideration. To fulfill this gap, in these study, three food processing by-products: (i) brewers' spent grain, (ii) fennel processing residues, and (iii) lemon processing residues were chosen as bio-effector candidates. Raw materials were chemically characterized, and green extraction methodology was optimized by using water, ethanol, and their mixture based on the extraction yields. Aqueous extracts were used for seed germination bioassays on *Lepidium sativum* seeds to evaluate their potential bioactivities. Thereafter, the extracts were chemically characterized and metabolites were detected by 1D and 2D NMR spectroscopy.

Results: Results are summarized as follows: (i) raw materials showed an interesting nutritional content; (ii) aqueous extraction resulted higher yield more than other used solvent; (iii) at high solvent extraction ratio, aqueous extracts were not phytotoxic but enhanced seed germination and root elongation; (*iv*) all aqueous extracts are differently rich in nutrients, amino-acids, sugars, and other low molecular weight molecule compounds.

Conclusions: This study confirmed that efficient and simple recovery of bioactive compounds other than nutrients from agro-food processing by-products appear to be the new frontier in their valorization.

Keywords: Bio-effectors; By-product; Barley; Fennel; Lemon; Green extraction; Nuclear magnetic resonance (NMR)

Background

Soil fertility is considered, after water, the second most limiting factor for agricultural production. The management of soil fertility is facing problems due to the limited availability of organic fertilizer from sustainable sources, together with high prices of the available one, particularly in organic and integrated agriculture. Farmer dependence and reliance on nonrenewable resource, such as phosphorus, which will be depleted within next 50 to 100 years [1], make alternative plant nutrition strategies more urgent. Numerous researcher studies were focused to find environmental-friendly and sustainable sources of nutrients, and different plant growth promoters either synthesized or naturally-derived were tested. Recently,

a novel concept of 'bio-effectors' which describes products that are able to improve plant performances other than fertilizers rose on. Bio-effectors promote plant growth through enhancing or altering biological activity in the soil/plant system [2].

The most common bio-effector present in the market with a considerable amount produced annually (15 million metric tons) is seaweeds extract, although its collection and handling is difficult [3]. Another one is humic substances that are also produced intensively from lignite coal which is a nonrenewable source [4]. This indicates the importance of searching and studying new bio-effectors extracted from sustainable sources. New candidates could be agro-food residues and by-products because they may contain bioactive compounds like proteins, sugars and lipids, and specific aromatic and aliphatic compounds. In addition, they are low-cost and abundant materials.

* Correspondence: alchami@iamb.it; fp.fanizzi@unisalento.it
[1]CIHEAM - Istituto Agronomico Mediterraneo di Bari (IAMB), Via Ceglie 9, 70010 Valenzano, BA, Italy
[2]Dipartimento di Scienze e Tecnologie Biologiche ed Ambientali, Università del Salento, Centro Ecotekne strada provinciale Lecce Monteroni, 73100 Lecce, LE, Italy

Therefore, valorization of agro-food processing residues is receiving increased attention and interest [5]. New green extraction technologies like green extraction by using nonhazardous solvents aside from minimizing energy consumption are one of the main purposes that attract attention today [6]. The design of green and nonconventional extraction methods is currently a hot research topic in the multidisciplinary area of applied chemistry, biology, and technology.

Technology platform of the European Union (EU) enterprise provided a 2025 vision of strategic research priorities for organic researches. Bio-effectors cover the eco-functional intensification principles of increasing agricultural productivity by improving existing natural processes [7]. EU regulation No. 834 authorized the use of plant derived materials, products, and by-products of plant origin (such as oilseed cake meal, cocoa husks, and malt culms) for fertilizers in organic agriculture [8]. Moreover, lack of wide-range fertilizer options and authorized compounds like biopesticides in organic agriculture are the main driving force of this research. In fact, the available commercial fertilizers in organic farming are not comparable for their efficiency, solubility, and prices with those in conventional farming.

Therefore, authorized efficient products, economically available for the farmer, able to enhance plant growth and soil nutrients availability are needed. Such products are necessary where chemicals are prohibited and when crop nutrient demand is high. In this context, to detect new sources of bio-effectors, three common agro-food waste, such as brewers spent grain (BSG), fennel processing residues (FPRs), and lemon processing residues (LPRs), were investigated. They were chosen due to their high availability in the south of Italy. Total amount of lemon produced in Italy is around 570,000 tons year^{-1} mainly used for juice and liquor productions. Lemon production is concentrated in the south of Italy, especially in Campania, Basilicata, and Sicily regions. Lemon harvest starts from February till the end of October according to the varieties. Fennel production is also concentrated in the south of Italy, and FPRs are available in large quantities from October till March. The amount of fennel produced in Italy is around 600,000 tons year^{-1} (data from the National Institute of Statistics (ISTAT) www.istat.it). Beer production in Italy reached 13 million hl year^{-1} recently. More than 150,000 tons of malt and 50,000 tons of other cereals are consumed for beer production (ASSOBIRRA, industrial association for beer and malt production, www.assobirra.it). BSG is therefore available all the year in large quantities; breweries are also distributed all over Italy. Many international breweries are located in Apulia region ad es. Birra Peroni is located in Bari. Accordingly, the main objectives of this work are: (i) to explore some promising bio-effector candidates which can be used in agriculture (namely for organic farming), (ii) to develop more environment-friendly alternatives for plant nutrition strategies such as green extraction technology applied to agro-food processing residues and by-products, and (iii) to chemically characterize the metabolic profile of the extracts for the chosen materials by using high-resolution nuclear magnetic resonance (NMR) spectroscopy.

Experimental

Chemicals

Deionized water (Elix, Millipore Corporation, Bedford, MA, USA) and ethanol (Puriss p.a., ACS Reagent, St. Louis, MO, USA, absolute alcohol, without additive, ≥ 99.8%), HNO_3, and H_2O_2 TraceSelect were purchased from Sigma Aldrich, (TraceSelect, Sigma-Aldrich, Steinheim, Germany). Ultrapure water (18.2 MΩ cm^{-1}) was obtained with a Milli-Q purification system from Millipore (Milli-Q, Millipore Corporation, Bedford, MA, USA).

Raw material preparations

FPRs were brought from JONICA BIO packaging house (Montescaglioso, MT, Italy). FPRs were composed of unmarketable bulbs, green leaves, and external leaves of bulbs of fennel. LPRs were brought from Solagri® (Solagri®, Sant'Agnello di Sorrento, NA, Italy). They consist of juicing pulp pomace, white albedo layer and full lemon fruits which cannot be peeled mechanically. The BSGs of barley were brought from Birra Peroni company (Birra Peroni, Bari, Italy).

Raw materials were chopped manually and air dried in a greenhouse. Materials were turned and mixed for daily aeration to prevent fermentation. Greenhouse temperature did not exceed 35°C during the drying period. Then, air-dried materials were grinded with a mixer mill and passed through a 1-mm sieve to obtain homogeneous particle sizes. All chemical analysis and extraction were done on the grinded and sieved air-dried materials.

Methods

Solvent choice and extraction yield determination

Three solvents (deionized water, ethanol, and a mixture of ethanol: water 1:1; v/v), and three ratios (1:10; 1:25; and 1:50: w/v) with three replications were used for the extraction from BSG, FPR, and LPR, respectively. The substrate/solvent ratios were selected on the basis of the experimental conditions required for a nearly quantitative extraction (higher solvent extraction ratio) and the expected high phytotoxicity at lower solvent extraction ratio. The latter was also confirmed by phytotoxicity tests (see below in the text).

Materials were shacked with solvent for 30 min in 250 mL polyethylene bottles. Then centrifugation took place at 6,000 rpm for 10 min followed by filtration with

Whatman No. 1 filter paper. An aliquot of the filtered extract was dried at 105°C for yield determination according to the following equation [9]:

$$Y = \frac{D}{V} \times \frac{1}{R} \, C \times 1000$$

where Y is the yield (g dry matter in extract / kg air-dried raw material), D is the dry matter weight (g) of the extract dried at 105°C, R is the extraction ratio, V is the volume of the extract used for drying at 105°C (mL), and C is the correction factor calculated by dividing the weight of air-dried raw material on the weight of 105°C dried raw material.

The extracts with the highest yield were selected for further tests and analyses.

Seed germination bioassays

Seed germination bioassays were conducted on *Lepidium sativum* seeds using aqueous extracts according to the EPA protocol [10] and to the IRSA methods for phytotoxicity bioassays of organic substances [11]. Five different extraction ratios (1:10; 1:25; 1:50; 1:100; and 1:200) for BSG and FPR aqueous extracts, whereas six ratios (1:10; 1:25; 1:50; 1:100, 1:200 and 1:400 w/v) for LPR bioassays were used. The ratio 1:400 was conducted only for LPR aqueous extract because of its high toxicity. Bioassays were performed in plastic petri dishes with four replicates. Each plastic petri dish contained ten seeds distributed on filter paper moistened with 2 mL of tested extract. Control was moistened with 2 mL of deionized water.

Petri dishes were incubated in a growth chamber for 48 h at temperature $25 \pm 5°C$. The experiment was repeated for FPR and LPR extracts as explained previously but for 144 hours. Germination index was calculated using the following equation:

$$GI(\%) = \frac{RL_{Treatment} \times GS_{Treatment}}{RL_{Control} \times GS_{Control}}$$

Where GI is the germination index, RL is the root length, and GS is the number of germinated seed. When the germination index exceeded 60%, the extract is considered not phytotoxic [11] and was selected for the further analyses.

Chemical analysis

Standard methods have been used for the chemical analyses [12]. In brief, the raw materials, residual humidity, and ash were determined successively in oven at 105°C and 550°C. On the raw materials and on the extracts, total N was determined by the Kjeldahl method [13]. Total macro- and micronutrient concentrations were determined by wet digestion (1 mL H_2O_2 and 5 mL HNO_3) using a microwave digestion system (CEM model, MARS

Xpress); the samples were then cooled, diluted with ultra-pure water in a 50-mL volumetric flask, filtered through Whatman No. 42 filter papers, and finally measured for their cation content by means of an inductively coupled plasma optical emission spectrometer (ICP-OES; Thermo Electron ICAP 6300 Series). Total P was measured colorimetrically on the mineralized samples by a spectrophotometer (Heλios α UV-vis, UNICAM, Thermo Electron Corporation) at 650 nm using modified ascorbic acid method [14].

On the extracts, pH was measured using a pH meter (Basic 20) with a standard glass electrode (Crison 5050, Barcelona, Spain). Electrical conductivity was determined by a conductometer (XS cond 510). Ash was determined on the extract after drying in the oven at 105°C, and then samples were transferred in the muffle at 550°C. Total nitrogen was determined by the Kjeldahl method.

Total phenolic contents of the extracts were assayed according to Folin-Ciocalteu method. An aliquote of 100 µl of extracts, calibration solutions, and blank were pipetted into separate test tubes, and 900 µL of distilled water was added. After, 200 µL of Folin-Ciocalteu reagent were added to each test tube. The mixture was mixed well and allowed to equilibrate. After 5 min, 1 mL of a 10% (w/v) sodium carbonate solution was added. The mixture was swirled and put in a temperature bath at 40°C for 20 min. Then, the tubes were rapidly cooled and the maximum adsorption was measured at 740 nm using a spectrophotometer (Heλios α UV-vis, UNICAM, Thermo Electron Corporation). Data were expressed as gallic acid equivalent (GAE) using gallic acid calibration curve. Spectrophotometric analysis that used 2,2-diphenyl-1-picrylhydrazyl (DPPH) was performed to determine the antioxidant activities. This assay is based on the ability of the antioxidant to scavenge the radical cation DPPH. Data were expressed as Trolox equivalent antioxidant capacity (TEAC) using Trolox calibration curve. The *in vitro* antioxidant activities of extracts were performed in the following way: 10 µL of extracts were added to 3 mL of 0.04 mM DPPH ethyl acetate solution and mixed with glass baquet. The samples were kept in the dark for 60 min at room temperature, and then decrease in absorbance at 517 nm was measured using the spectrophotometer (Heλios α UV-vis, UNICAM, Thermo Electron Corporation). Calibration curve in the range of 0.2/0.4/0.6/1.0/2.0/4.0/6.0 mmol L^{-1} were prepared for Trolox.

NMR profiling of BSG, FPR, and LPR extracts

An amount of 0.321 g of FPR, 0.270 g of BSG, and 0.141 g of LPR dried extracts were dissolved in 1 mL of 1.2 mM TSP in D_2O solution, placed in a 5 mm NMR tube, and analyzed by multinuclear (^1H and ^{13}C), multidimensional NMR spectroscopy.

All measurements were performed on a Bruker Avance III NMR spectrometer (Bruker, Karlsruhe, Germany) operating at 400.13 MHz for ^{1}H observation, equipped with a z-axis gradient coil and automatic tuning-matching (ATM).

For each sample, a one-dimensional NOESY experiment (referred to as 1D-NOESY), including a solvent signal saturation during relaxation and mixing time and a spoil gradient, was acquired using 256 free induction decays (FIDs), 64 K data points, a spectral width of 12.019 Hz, an acquisition time of 3.42 s, a relaxation delay of 4 s, and a mixing time of 10 ms.

2D ^{1}H J-resolved spectra with pre-saturation during relaxation delay were recorded with a spectral width of 4795.396 Hz on F2 and 60.020 Hz on F1, 4 K data points, 16 FIDs for 128 experiments, and 12 s repetition delay.

^{1}H COSY spectra with pre-saturation during relaxation delay were acquired with 4 K data points, a spectral width of 4795.396 Hz, 32 FIDs for 256 experiments, 2 s repetition delay, and 16 dummy scans.

^{1}H-^{13}C HSQC and ^{1}H-^{13}C HMBC NMR spectra were acquired with 4 K data points, a spectral width of 4795.396 Hz on ^{1}H and 25156.211 Hz on ^{13}C, 16 FIDs for 256 experiments, 2 s repetition delay, 16 dummy scans.

The acquisition and processing of spectra were performed using the software TopSpin 2.1 (Bruker Biospin). The FIDs were multiplied by an exponential weighting function corresponding to a line broadening of 0.3 Hz before Fourier transformation, phasing, and base line correction. All spectra were referenced to the TSP signal ($\delta = 0.00$ ppm), used as internal reference.

The metabolites were assigned on the basis of 2D NMR spectra analysis (2D ^{1}H J-res, ^{1}H COSY, ^{1}H–^{13}C HSQC, and HMBC) and comparison with published data [15,16].

Statistical analysis

Analysis of variance (ANOVA) was carried out, and separation of means was performed using LSD test at $P = 0.05$ significance level. LSD test were computed using SAS software version 9 (SAS Institute, Cary, NC). Correlation matrix at $P = 0.05$ significance level was performed using XLSTAT 7.5.2 (Addinsoft, Paris, France).

Results and discussion
Raw material characterization

Ash content, total nitrogen, macro-nutrients, micronutrient, and total Cd and Pb contents of the raw materials (BSG, FPR, and LPR) are shown in Table 1. Ash content in FPR and LPR were almost five times higher than that in BSG. The ash content observed in BSG was approximately half of what Mussatto and Roberto [17] reported.

This result could be due to the variation in raw material in terms of cultivated plant varieties and/or processing procedures (e.g., brewing procedures).

Total N in the raw materials ranged from 14.5 to 47.9 g kg^{-1} while total P ranged from 1.95 to 4.80 g kg^{-1}. In addition to the highest nitrogen content, BSG resulted to be the second higher source of phosphorus. BSG raw material is also rich in total Cu, total Fe, and total Zn. Results of the chemical characterization of BSG was consistent with Khidzir et al. [18] for Mg but differed for Ca and P. The obtained content of N and K is similar to the values previously reported by Gupta et al. [19]. High nitrogen content in the BSG raw material is related to the richness in protein as reported by Kotlar et al. [20].

The highest concentrations of Ca and K were found in LPR and FPR raw materials. FPR raw material and FPR extract were rich in K as previously reported by Bianco et al. [21]. FPR content in macro- and micronutrients was similar to the findings of Muckensturm et al. [22]. FPR raw material was rich in total P, Na, Mn, and Ni contents. Relative high Na content of FPR was estimated to be 10 times higher than LPR and 100 times higher than BSG. Fe was found to be the most abundant microelement in all materials with content ranged from 114 to 169 mg kg^{-1}.

LPR was the lowest source of almost all analyzed nutrients except for Ca. Low nutrient content in LPR was also previously reported by Su and Horvat [23] and confirmed by USDA National Nutrient Database for Standard Reference [24].

Total Cd and Pb contents resulted below the limitation of hazardous substances required by the European regulations (EU 'eco-label'; Commission Decision 2006/799/EC) [25] for soil improvers.

Solvent choice and extraction yield determination

Figure 1 shows the extraction yields for BSG, FPR, and LPR at 1:10, 1:25, and 1:50 solvent extraction ratios. Aqueous extracts gave the highest yields for FPR and LPR, while no significant differences were observed among different solvents for BSG. Generally, EtOH extract gave the lowest yield in comparison to the aqueous and EtOH-H$_2$O mixture extract. However, yield obtained by EtOH-H$_2$O mixture was significantly lower than the yield obtained by water for both FPR and LPR. In general, and independently the solvent used, when the solvent extraction ratio increased, the extraction yield did not increase significantly. Our results showed that when the content of ethanol in the solvent increases, the total extraction yield decreases for FPR and LPR while no differences in yield was found in BSG. This may be due to the higher polarity of water in comparison to ethanol [26] and to the higher polar metabolites content in FPR and LPR. In our study, aqueous extraction resulted in

Table 1 Ash content, total macro- and micronutrients, Cd, and Pb contents in raw materials (BSG, FPR, and LPR)

Raw materials	Ash	Total N	Total P	Total Ca	Total K	Total Mg	Total Na	Total Cu	Total Fe	Total Mn	Total Ni	Total Zn	Total Cd	Total Pb
	%	g kg^{-1}	g kg^{-1}	g kg^{-1}	g kg^{-1}	g kg^{-1}	g kg^{-1}	mg kg^{-1}	mg kg^{-1}	mg kg^{-1}	mg kg^{-1}	mg kg^{-1}	mg kg^{-1}	mg kg^{-1}
BSG	4.10 ± 0.1	47.9 ± 0.7	4.79 ± 0.4	4.41 ± 0.5	0.51 ± 0.0	1.63 ± 0.1	0.07 ± 0.0	14.3 ± 1.4	169 ± 13	37.3 ± 2.1	2.5 ± 0.3	97.5 ± 7.9	< 0.01	0.28 ± 0.06
FPR	20.4 ± 0.8	30.9 ± 1.0	4.80 ± 0.8	6.67 ± 0.7	33.4 ± 1.1	1.58 ± 0.1	7.00 ± 0.2	8.94 ± 1.5	137 ± 21	75.5 ± 6.7	5.1 ± 0.8	26.4 ± 2.1	< 0.01	0.38 ± 0.09
LPR	20.1 ± 0.5	14.5 ± 0.6	1.95 ± 0.3	10.3 ± 0.5	9.04 ± 0.8	1.02 ± 0.1	0.70 ± 0.0	5.52 ± 1.0	114 ± 15	5.54 ± 0.6	1.8 ± 0.1	12.8 ± 1.0	< 0.01	0.19 ± 0.04

Values reported are average of three replicates ± standard deviation.
BSG, brewers spent grain; FPR, fennel processing residues; LPR, lemon processing residues.

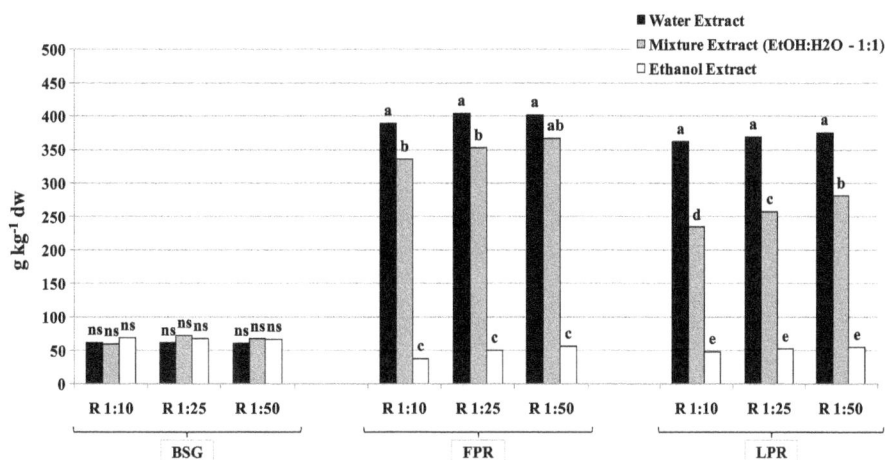

Figure 1 Extraction yield of BSG, FPR, and LPR at 1:10, 1:25, and 1:50 extraction ratios. Means with different letters within the same raw materials indicate the significant difference between values at $P < 0.05$ (LSD Fisher's test). *NS* not significant, *R* ratio.

higher yield than the mixture of EtOH-H_2O and therefore only aqueous extracts were used for further tests.

In BSG extracts, changing the ratio and/or the solvent did not affect obtained yields. In our study, applying different solvent extraction ratios did not strongly change the yields regardless the solvent or material and that was in agreement with Kalia et al. [27].

In FPR extracts, yields were not affect by changing the solvent extraction ratio, while the solvent used had significant effect. The extraction yields for the different solvents ranked as follows: EtOH < EtOH-H_2O < H_2O. The yield of the aqueous extract was almost eight times higher than for the EtOH extracts. The high extraction yield obtained from FPR could be due to the richness in high molecular weight polysaccharide as reported by Taie et al. [28]. On the other hand, Leal et al. [29] has estimated the aqueous extraction yield of FPR by 16.8%, while it exceeded 45% in this study. The different FPR yields can be explained by the differences in extraction procedures such as temperature, extraction time, and ratios. Our results varied also when compared to Il-Suk et al. [30] who obtained an extraction yield equal to 11.38%. Taie et al. [28] found significant differences in FPR extraction yield when different solvents and/or ratios were applied and these results are similar to our findings.

In LPR extract, similar results as FPR extracts were obtained. In fact, aqueous extracts gave the same yield when 1:10, 1:25, and 1:50 extraction ratios were applied. Similar results were obtained when EtOH was used. Conversely, the yield of LPR EtOH-H_2O extracts varied significantly when different ratios were used. The yield was in the order of 1:10 < 1:25 < 1:50 for LPR EtOH-H_2O extracts. The yield of LPR extraction using EtOH-H_2O extraction was equal to 135 g kg^{-1} of fresh peel as determined by Kang et al. [31]. The differences between our study and the

previous studies could be due to the higher percentage of EtOH used in the mixture. The low yield of ethanolic extraction of LPR was confirmed by Zia-ur [32]. However, the extraction yield can be improved by other assistant methods such as ultrasounding method [33].

Generally, the compounds obtained by organic solvent such as EtOH are different from those obtained by the aqueous extraction. Tsibranska et al. [34] found that the organic solvents like ethanol can give two times higher content of valuable compounds in the extract. However, aqueous extraction was adopted for many reasons other than the yield. Indeed, there are restrictions in organic solvent use for green extraction technology as well as in organic farming. Moreover, it should be underlined that (i) aqueous extraction is efficient and less expensive in comparison to other solvents, nontoxic, and environmental-friendly alternative to conventional extraction techniques; (ii) it is easy for both field application and market registration; (iii) organic solvents are phytotoxic for plants and, in addition, extra work is needed to remove them from the extract; (iv) other surfactant (e.g., tween, dimethyl sulfoxide) should be added to redissolve in water compounds extracted with other organic solvents, increasing the cost of the bio-effector; (v) the extraction yields of some bioactive components resulted better in water than ethanol at room temperature [35]; and (vi) the limitation of aqueous extraction can be reduced by increasing the temperature [36] which rises the water polarity [37].

Seed germination bioassays for optimum solvent extraction ratio determination

Figure 2 shows the germination indexes (GI) after 48 h of incubation using different aqueous extraction ratios. According to IRSA method [11], GI index below 60% is an indication of phytotoxic effect, while GI above 60%, is

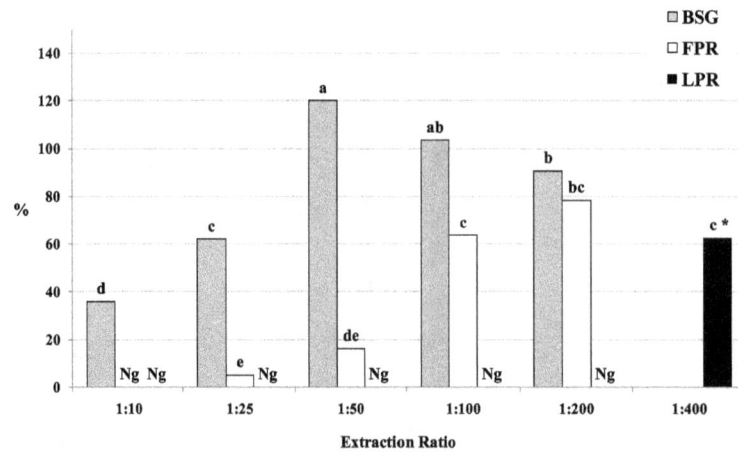

Figure 2 Germination index after 48 h. Means with different letters indicate significant difference between values at $P < 0.05$ (LSD Fisher's test). *Ng* no germination. The *asterisk* denotes that the extraction ratio 1:400 was done only for LPR.

considered not phytotoxic, and there is no harmful risk for plant growth.

The highest GI in each ratio of extraction was obtained with BSG treatment followed by FPR treatment. BSG aqueous extract enhanced seed germination and root growth. In fact, GI index in BSG aqueous extract increased by 20% in comparison to control at 1:50 extraction ratio. Seeds did not germinate in the FPR aqueous extract at 1:10 extract ratio, and GI increased when the solvent extraction ratio increased. LPR was very toxic and no seed germination was observed at 1:10, 1:25, 1:50, 1:100, or 1:200 extraction ratios while GI reached 62% at 1:400 ratio.

Due to the higher toxicity of FPR and LPR, the related experiment was repeated for a longer time (144 h). Figure 3 shows the improvement of GI for FPR and LPR aqueous extract after 144 h of seed incubations. No germination was registered for LPR extract at 1:25 and 1:50 ratios after

144 h and matched the results after 48 h, while slight increase in GI was obtained by LPR at 1:100 ratio (34%) and 1:200 ratio resulted in GI equal to 66.7%.

In order to understand the effects of the aqueous extracts on the GI, a correlation matrix was determined between pH, EC, and GI for each aqueous extract. The correlation matrices are shown in Table 2. GI is negatively correlated with EC. Salinity inhibits seed germination due to an osmotic effect or a specific ion toxicity [38]. In addition, pH influences seed germination. In fact, seed germination is inhibited when pH is below 5 and completely inhibited when the pH is below 4 [39]. The correlation between pH value and GI explains the low germination index obtained from LPR aqueous extract which had the highest acidity. The low GI obtained in FPR when low solvent extraction ratio is adopted could be due to the high EC and to the high total polyphenols which can negatively affect seed germination.

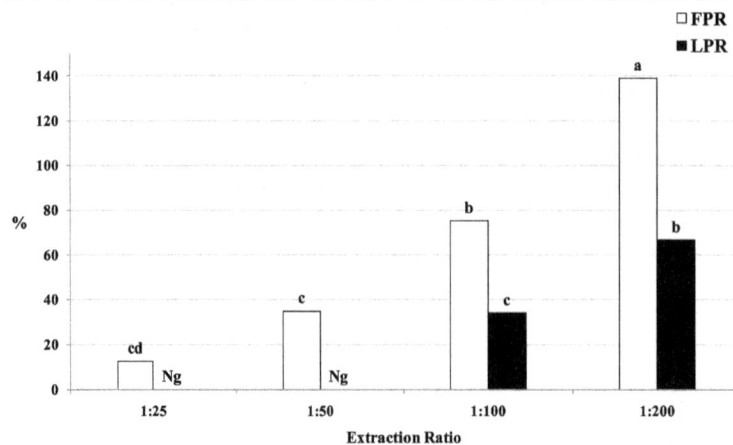

Figure 3 Germination index after 144 h. Means with different letters indicate significant difference between values at $P < 0.05$ (LSD Fisher's test). *Ng* no germination.

Table 2 Correlation matrix between pH, EC, and GI (48 and 144 h) for BSG 1:25, FPR 1:100, and LPR 1:200 aqueous extracts

		GI (48 h)	GI (144 h)
BSG	pH	0.72	Nd
	EC	**−0.86**	Nd
FPR	pH	−0.76	−0.75
	EC	**−0.85**	**−0.85**
LPR	pH	**0.92**	**0.89**
	EC	−0.40	−0.71

GI, germination index at 48 h and 144 h; BSG, brewers spent grain; FPR, fennel processing residues; LPR, lemon processing residues; Nd, not determined. Values in bold indicate significant $P = 0.05$ (two-tailed test).

Total polyphenols are relatively high in FPR and LPR extracts in comparison to BSG extract. Total polyphenols are known to have germination inhibitor effect [40].

However, when the toxicity of the extract is low, a retardation in seed germination could be observed. In fact, at higher solvent extraction ratio for FPR and LPR, an increase in germination index at 144 h test time was observed in comparison to 48 h. No seed germination was observed in LPR extract treatment at 1:200 ratio for the first 48 h while at 144 h, GI was equal to 66.7% which is considered the limit to classify the extract as nontoxic. Based on the obtained results, BSG 1:25, FPR 1:100, and LPR 1:200 (GI > 60%) were selected for the chemical characterization.

Extract characterization

Ash content, total nitrogen, macro- and micronutrients, and heavy metal contents of each aqueous extract (BSG 1: 25, FPR 1:100, and LPR 1: 200) are shown in Table 3. The results revealed that all extracts were rich in macro- and micronutrients; therefore, the use of these aqueous extracts can have an added value for plant nutrition. The highest EC (0.6 dS m^{-1}) was measured in FPR extract. The lowest pH value (3.3) was measured in LPR extract. Ash content was highest in FPR in comparison to those in LPR and BSG aqueous extracts. FPR extract was found to be almost the richest source of N, P, and K in comparison to those in LPR and BSG aqueous extracts. BSG extract was slightly higher in total P content than that in FPR extract. BSG extract showed the highest content in total Ca and Mg. We found the FPR extract to be the richest in total K content with more than 330 mg L^{-1}. Total Mg content was the highest in BSG extract, and total Fe was the highest in LPR extract. FPR content of total Na was approximately 13 times higher than those in LPR and BSG contents. The highest concentrations of K, Cu, Na, Ni, and Zn were found in FPR extract. Fe is the most abundant microelement in all three materials with content ranged from 114 to 169 mg kg^{-1}.

Table 3 pH, EC, ash content, total macro- and micronutrients, total Cd, total Pb, total polyphenols and antioxidant activity content in BSG 1:25 FPR 1:100 and LPR 1:200 aqueous extracts

Aqueous extracts		BSG 1:25	FPR 1:100	LPR 1:200
pH		5.5 ± 0.1	5.5 ± 0.2	3.3 ± 0.1
EC	dS m^{-1}	0.2 ± 0.0	0.6 ± 0.1	0.5 ± 0.0
Ash	%	18.5 ± 0.2	25.0 ± 1.1	9.3 ± 0.3
Total N		40.0 ± 0.4	157 ± 10	10.0 ± 0.2
Total P		47.2 ± 0.2	43.6 ± 0.3	3.78 ± 0.2
Total K		19.4 ± 1.8	330 ± 8.7	50.1 ± 4.5
Total Ca		58.3 ± 0.4	8.22 ± 0.55	34.7 ± 2.7
Total Mg		34.3 ± 0.40	5.41 ± 0.8	5.49 ± 0.60
Total Na		5.44 ± 1.1	72.0 ± 2.3	5.46 ± 0.48
Total Cu	mg L^{-1}	0.05 ± 0.01	0.11 ± 0.00	0.05 ± 0.00
Total Fe		0.44 ± 0.08	0.43 ± 0.10	0.55 ± 0.04
Total Mn		0.32 ± 0.01	0.17 ± 0.00	0.03 ± 0.00
Total Ni		0.03 ± 0.01	0.06 ± 0.00	0.02 ± 0.00
Total Zn		0.18 ± 0.05	0.27 ± 0.06	0.27 ± 0.05
Total Cd		<0.005	<0.005	<0.005
Total Pb		0.01 ± 0.00	0.01 ± 0.00	0.01 ± 0.00
Total polyphenols	g kg^{-1}	10.57 ± 1.96^c	57.88 ± 5.61^a	49.54 ± 3.13^b
Antioxidant activity	mM trolox L^{-1}	1.81 ± 0.31^c	5.55 ± 1.01^a	3.77 ± 0.44^b

Values reported are average of three replicates ± standard deviation. Within same row, means with different letters (a, b, c) indicate significant difference between values at P < 0.05(LSD Fisher's test).
BSG, brewers spent grain; FPR, fennel processing residues; LPR, lemon processing residues.

Total Pb content in all extract was around 0.01 mg L^{-1} which is considered very low, while total Cd content was below the instrument detection limit (<0.01 mg L^{-1}).

Total polyphenols and antioxidant activity are shown in the Table 3. Extracts of studied materials revealed different amounts of polyphenols: 10.6, 57.9 and 49.5 g L^{-1} in BSG, FPR, and LPR, respectively. Aqueous extract obtained from FPR showed high polyphenolic content and antioxidant activity. Total polyphenols content in BSG aqueous extract was significantly lower than FPR and LPR aqueous extract. FPR extract showed an important DPPH scavenging ability exhibited by the highest antioxidant activity (5.5 mM trolox L^{-1}). LPR antioxidant activity was much higher than BSG's.

The low content of total polyphenols observed in BSG extract may be due to their weak solubility in water. Both hydrolysis and saponification with NaOH are the suitable extraction techniques of BSG polyphenols [41]. Methanolic extraction was also used for quantification and extraction of polyphenolic compounds from BSG by Naczk and Shahidi [42]. The high polyphenolic compounds observed in FPR is probably the reason for its high antioxidant activity [43]. Total polyphenols content measured in this study (57.9 g kg^{-1}) is in accordance with Munir et al. [44] but was higher than the total polyphenols content observed by Taie et al. [28] (31.94 g kg^{-1}). Parejo et al. [45] referred the antioxidant activity of FPR to some compounds such as caffeoylquinic acid, rosmarinic acid, and kaempferol-3-O-glucoside. The high antioxidant activity of LPR extract was indicated by the strong scavenging ability of free radicals, and nitrites [46]. The polyphenolic content of the aqueous solution of other types of citrus peel powders seems to give different value (3.5 g kg^{-1}) [31] in comparison with our result (49.5 g kg^{-1}).

High-resolution NMR spectroscopy represents a potentially powerful tool for plant metabolite analysis [47]. This technology has been previously utilized to profile metabolites in clinical samples [48,49] but also to account for the metabolic changes that occur during fermentation and/or production processes, and to evaluate the quality of food and beverages such as oil and wine [50-52]. The main advantage of NMR spectroscopy, as well as the sample preparation, is that it is not

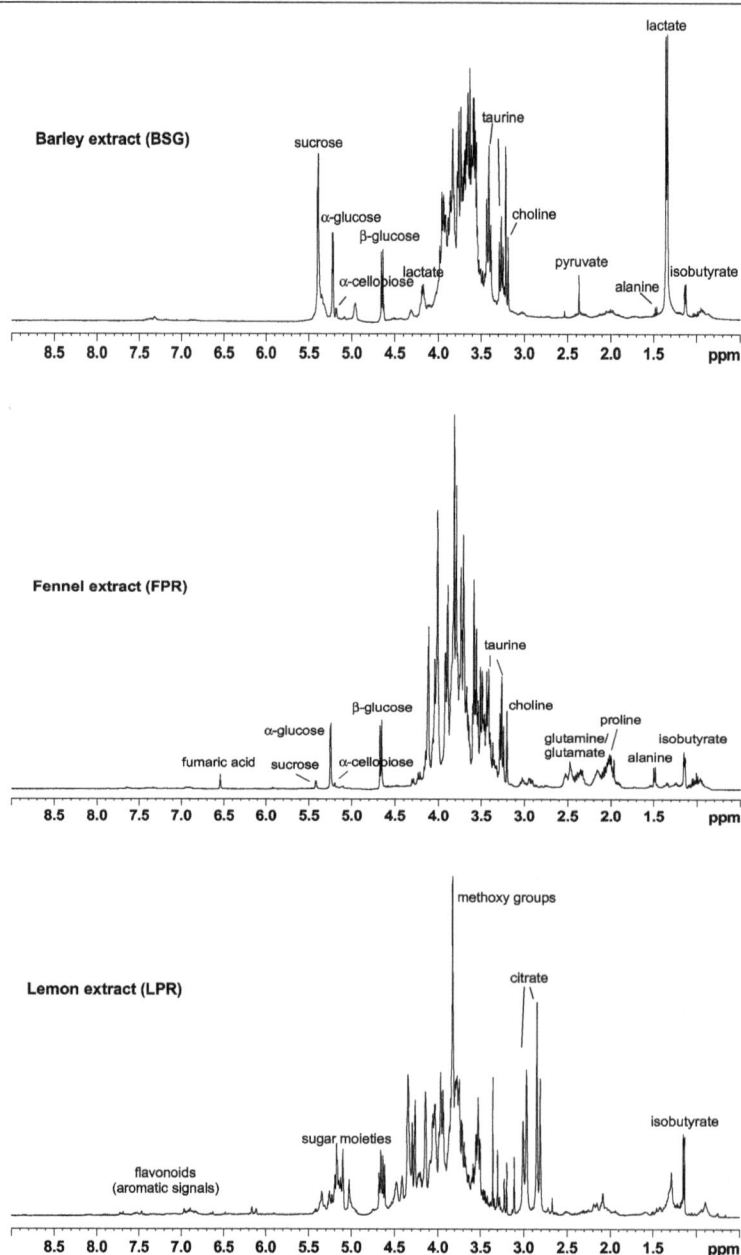

Figure 4 ^{1}H NMR spectra of FPR, BSG, and LPR extracts in D$_2$O.

discriminatory, unlike certain mass spectrometry methods that rely on the prior separation and/or derivatization of metabolites. In this way, all compounds such as carbohydrates, amino acids, organic and fatty acids, amines, esters, ethers, and lipids, which are present in a sample, could be simultaneously detected and investigated by NMR spectroscopy. Thus, NMR spectra of raw material extracts have the potential to provide a relatively unbiased fingerprint, containing the signals of the metabolites present in the solution.

^{1}H NMR spectra of extracts showed several signals attributed to amino acids, sugars, and low molecular weight compounds (Figure 4). They were assigned by comparison with published data [15] on the basis of analysis of 2D NMR spectra (2D ^{1}H J-res, ^{1}H COSY, ^{1}H-^{13}C HSQC and HMBC). The metabolites list with their chemical shifts (δ), and the presence detected in each extract are reported in Table 4. Although most metabolites are present in all extracts, there are several differences in their concentration. Interestingly, in the FPR extract, the preponderant metabolites are α and β glucose whereas sucrose and lactate are the most abundant in the BSG extract. In the LPR extract, due to the presence of high level of citrate, the pH was very acidic producing a de-shielding of most pH sensitive signals. Moreover, the lemon extract shows a very high singlet at 3.83 ppm, attributed to the methoxy moieties of flavonoids.

Concerning NMR profiling results, important differences were noticed among extracts for leucine, proline, glutamine, and methionine. The aromatic signal resulted by flavonoids was observed only in LPR extract.

Lactic acid present in all aqueous extracts is known to have positive effects on plant growth as well as on crop yield and quality, especially under stress conditions [53]. This behavior is due to the forming of stable bonds with several metal ions [54]. Furthermore, the presence of fumaric acid in BSG and FPR extracts offers an explanation for their effects. Fumaric acid is required for rapid nitrogen assimilation, also as a temporary carbon sink for photosynthate [55].

In BSG extract, α and β-glucose [56] and choline [57] were previously reported. On the contrary, methoxy group and proline mentioned in the study of Gupta et al. [19] and Huige [57], respectively, were not detected in our samples. Santos et al. [58] suggested that barley variety along with malting conditions and the type of additives in the brewing process can affect the BSG content. However, other possible explanation could be related to the sensitivity of applied NMR techniques. Moreover, BSG extract has a special characteristic. Changes occur during uncompleted germination stage of barley, during the brewing process, include increment in bioactive compounds [59]. Those changes could play an important role during seed germination. Pyruvate was detected only in the BSG extract and it is considered the direct tool by

Table 4 Metabolites observed in the fennel BSG, FPR, and LPR extracts and their ^{1}H NMR chemical shifts (δ)

Metabolites	BSG	FPR	LPR	δ
Isoleucine	x	x		0.93 [7][a] (t), 1.00 [7] (d)
Leucine	x			0.94 [7] (d), 0.96 [7] (d)
Valine	x	x		0.98 [7] (d), 1.03 [7] (d), 2.26 (m)
Isobutyrate	x	x	x	1.13 [6.5] (d)
3-Hydroxybutirate	x			1.21 [6.4] (d)
Lactate	x	x	x	1,35 [7] (d), 4.12 [7] (q)
Alanine	x	x	x	1.48 [7.2] (d), 3.78 [7.2] (q)
Proline		x		2.03 (m), 2.36 (m), 3.36 (m), 4.12 (m)
Glutamate		x		2.00 (m), 2.08 (m), 2.35 (m)
Glutamine		x		2.13 (m), 2.41 (m), 3.77 (m)
Methionine			x	2.15 (s)
Pyruvate	x			2.36 (s)
Methylamine	x		x	2.54 (s)
Dimethylamine	x		x	2.72 (s)
Trimethylamine	x		x	2.81 (s)
Asparagine		x		2.89 (m), 2.96 (m)
Choline	x	x	x	3.20 (s)
Taurine	x	x		3.26 (t), 3.41 (t)
Methoxy group		x		3.83 (s)
β-Cellobiose	x	x		4.48 [8] (d)
β-Galactose			x	4.52 [7.8] (d)
β-Glucose	x	x	x	4.65 [7.9] (d)
α-Cellobiose	x	x		5.20 [3.6] (d)
α-Glucose	x	x	x	5.25 [3.7] (d)
α-Galactose			x	5.27 [3.7] (d)
Sucrose	x	x	x	5.42 [3.8] (d)
Fumaric acid	x	x		6.56 (s)
Tyrosine	x	x	x	6.88 [8.6] (d), 7.17 [8.6] (d)
Phenylalanine	x	x		7.33 (m), 7.38 (m), 7.43 (m)
Formiate			x	8.43 (s)
Flavonoids (aromatic signals)			x	6.5 to 9.0 range

[a]Values of $J_{H,H}$ [square brackets] are given when assignable. BSG, brewers spent grain; FPR, fennel processing residues; LPR, lemon processing residues.

which gibberellin regulates the growth [60]. Moreover, Yu et al. [61] explained that pyruvate is converted in plant cell to acetyl-CoA and NADH for energy production (tricarboxylic acid cycle). Therefore the plant metabolome might be affected by such processes.

NMR profile of FPR extract showed similar results to those demonstrated by Muckensturm et al. [22]. The flavonoids observation in LPR extract was reported by Mandalari et al. [62]. LPR carbohydrates and organic acids were described also by Poli et al. [63]. Organic acids present in LPR extracts had also an important impact on plant growth. Starting from growth media, organic acids carry negative charges which allow to make cations complexation and anions displacement [64]. Organic acids can flow across lipid bilayer of hairy root cells [65]. Citric acid of LPR is an important component of the stress response and for plant growth [66].

Finally, the aqueous extracts are rich in several low molecular weight organic compounds, and can be good candidates to be tested as bio-effectors in a more detailed study.

Conclusions
Potential and practical contributions of tested raw materials aqueous extracts as bio-effector candidates were discussed. The agro-food processing residues and their extracts studied above showed an interesting content, either at nutritional or biochemical level, and could be used as plant growth promoter. They are differently rich in nutrients, amino-acids, sugars, and low molecular weight molecules as demonstrated by the chemical analysis and NMR metabolic profiling. Furthermore, aqueous extraction resulted higher yield with respect to other used solvent. Water, safe and less expensive than other organic solvents, is effective for BSG, FPR, and LPR extractions under different extraction ratios. In addition, aqueous extracts were not phytotoxic but enhanced seed germination at higher solvent extraction ratio. BSG aqueous extract enhanced seed germination and root growth. FPR aqueous extract, at low solvent extraction ratio, showed low GI which increased when the solvent extraction ratio increased. LPR was very toxic at low solvent extraction ratio; therefore, higher solvent extraction ratios are required for the use of this product.

Finally, the reuse of agro-food processing residues as a potential source of bio-effectors suggests a rethinking of plant nutritional management in a sustainable manner. Bio-effectors can be applied in both low input (organic and integrated) and high input (conventional) types of agriculture. According to our results, we suggest that aqueous extracts could be used even in organic farming. However, further studies including comparison with other commercial plant growth promoters and toxicity studies on the microbial soil community should be conducted to evaluate the effects of suggested bio-effectors on the soil/plant system.

Competing interests
The authors declare that they have no competing interests.

Authors' contributions
ACZ has made the research protocol, conception and design, analysis of macro- and micronutrients, phytotoxicity test, and extraction yields. He also participated in the other data acquisition and analysis and the whole data interpretation. He has also made the manuscript drafting. AD is the Master's Student who followed the work to obtain his Master Degree. DPSA carried

out the NMR analysis and the NMR result writings. CI has made substantial contributions to the conception and design, carried out the total polyphenols analysis and antioxidant activities, and helped to draft the manuscript and data interpretation. FFP has made substantial contributions to the conception and design, manuscript revising and has given the final approval of the version to be published. All authors read and approved the final manuscript.

Acknowledgments
The authors want to acknowledge the Mediterranean Agronomic Institute of Bari (MAIB) for supporting this research work and the Department of Biotechnology and Environmental Science, University of Salento for the NMR analysis. The authors want to thank JONICA BIO packaging house, Montescaglioso, Matera, Italy, for providing fennel processing residues; Solagri®, Sant' Agnello di Sorrento, Napoli, Italy for providing the lemon processing residues; and Birra Peroni company, Bari, Italy, for providing the brewers' spent grains.

References
1. Cordell D, Drangert JO, White S (2009) The story of phosphorus: global food security and food for thought. Global Environ Chang 19(2):292–305
2. European Commission C5068 (2011) FP7 Cooperation Work Programme Theme 2: Food, Agriculture and Fisheries, and Biotechnologies. http://ec.europa.eu/research/participants/data/ref/fp7/89419/b-wp-201201_en.pdf. Accessed 08 October 2014
3. FAO (2007) FAO yearbook of fishery statistics: aquaculture production 2005. FAO Yearbook of Fishery Statistics 100(2):202. vii
4. Chen Y, Aviad T (1990) Effects of humic substances on plant growth. In: MacCarthy P, Clapp CE, Malcolm RL, Bloom PR (ed) Humic sbstances in soil and crop sciences. Soil Science Society of America Inc, Madison, Wisconsin, USA
5. Anastas PT, Warner JC (1998) Green chemistry: theory and practice. Oxford University Press, Oxford
6. Kroyer GT (1998) Bioconversion of food processing wastes. In: Martin AM (ed) Bioconversion of waste materials to industrial products. Springer Science, Business Media, New York
7. Urs N, Anamarija S, Otto S, Niels H, Marco S (2008) Vision for an organic food and farming research agenda to 2025; organic knowledge for the future. Vision Research Agenda, European Technology Platforms (ETP). http://www.darcof.dk/research/grafik/Visions_08.pdf
8. EC (2007) Organic production and labelling of organic products and repealing regulation. 834. Official Journal of the European Union, The Council Of The European Union 91:23
9. Zhang S, Bi H, Liu C (2007) Extraction of bio-active components from Rhodiola sachalinensis under ultrahigh hydrostatic pressure. Sep Purif Technol 57(2):277–282
10. EPA (1996) Ecological effects test guidelines. Seed Germination/Root Elongation Toxicity Test, US Environmental Protection Agency. EPA. 712–C–96–154
11. IRSA-Istituto di Ricerca sulle Acque (1983) Analisi della fitotossicità della sostanza organica in decomposizione mediante bioassaggio Lepidium sativum. Metodi Analitici per i Fanghi: Parametri biochimici e biologici. Quaderno IRSA 64:8.1–8.3
12. Trinchera A, Leita L, Sequi P (2006) Metodi di Analisi per i Fertilizzanti. Istituto Sperimentale per la Nutrizione delle Piante per conto del Ministero delle politiche agricole alimentari e forestali, Roma, Italy
13. Bremner JM (1996) Nitrogen total. In: Sparks DL (ed) Methods of soil analysis. Soil Science Society of America, Madison, Wisconsin
14. Olsen SR, Sommers LE (1982) Phosphorus. In: Page AL, Miller RH (ed) Methods of Soil Analysis, Part 2. 2nd ed. Agronomy Monograph 9, ASA and SSSA, Madison, WI, USA
15. Fan TWM (1996) Metabolite profiling by one- and two-dimensional NMR analysis of complex mixtures. Prog Nucl Mag Res Sp 28(2):161–219
16. Nicholson JK, Foxall PJ, Spraul M, Farrant RD, Lindon LC (1995) 750 MHz ^{1}H and ^{1}H-^{13}C NMR spectroscopy of human blood plasma. Anal Chem 67(5):793–811
17. Mussatto SI, Roberto IC (2006) Chemical characterization and liberation of pentose sugars from brewers spent grain. J Chem Technol Biot 81(3):268–274
18. Khidzir K, Noorlidah A, Agamuthu P (2010) Brewery spent grain: chemical characteristics and utilization as an enzyme substrate. Malay J Sci 29(1):41–51

19. Gupta M, Abu-Ghannam N, Gallaghar E (2010) Barley for brewing: characteristic changes during malting, brewing and applications of its by-products. Compr Rev Food Sci Food Saf 9(3):318–328

20. Kotlar CE, Belagardi M, Roura SI (2011) Brewer's spent grain: characterization and standardization procedure for the enzymatic hydrolysis by Bacillus cereus strain. Biotechnol Appl Bioc 58(6):464–475

21. Bianco VV, Damato G, Girardi A (1994) Sowing dates, plant density and 'crown' cutting on yield and quality of florence fennel "seed". In: Quagliotti L, Belletti P (ed). Acta Hort 362:59–66

22. Muckensturm B, Foechterlen D, Reduron JP, Danton P, Hildenbrand M (1997) Phytochemical and chemotaxonomic studies of Foeniculum vulgare. Biochem Syst Ecol 25(4):353–358

23. Su HCF, Horvat R (1987) Isolation and characterization of four major components from insecticidally active lemon peel extract. J Agr Food Chem 35(4):509–511

24. USDA (2011) Nutrient lists. National Nutrient Database for Standard Reference, United States department of Agriculture. http://ndb.nal.usda.gov/ndb/foods/show?fg=9&man=&lfacet=&count=&max=25&sort=f&qlookup=&offset=175&format=Full&new=&rptfrm=nl&ndbno=09156&nutrient1=301&nutrient2=309&nutrient3=207&subset=0&totCount=336&measureby=m

25. Commission Decision 2006/799/EC (2006) Establishing revised ecological criteria and the related assessment and verification requirements for the award of the community eco-label to soil improvers. Official J L 325:28–34

26. Tian F, Li B, Ji B, Yang J, Zhang G, Chen Y, Luo Y (2009) Antioxidant and antimicrobial activities of consecutive extracts from Galla chinensis: the polarity affects the bioactivities. Food Chem 113(1):173–179

27. Kalia K, Sharma K, Singh HP, Singh B (2008) Effects of extraction methods on phenolic contents and antioxidant activity in aerial parts of Potentilla atrosanguinea Lodd. and quantification of its phenolic constituents by RP-HPLC. J Agr Food Chem 56(21):10129–10134

28. Taie HAA, Helal MMI, Helmy WA, Amer H (2013) Chemical composition and biological potentials of aqueous extracts of fennel (Foeniculum vulgare L). J Appl Sci Res 9(3):1759–1767

29. Leal PF, Almeida TS, Prado GHC, Prado JM, Meireles MAA (2011) Extraction kinetics and anethole content of fennel (Foeniculum vulgare) and anise seed (Pimpinella anisum) extracts obtained by soxhlet, ultrasound, percolation, centrifugation, and steam distillation. Separ Sci Technol 46(11):1848–1856

30. Il-Suk K, Mi-Ra Y, Ok-Hwan L, Suk-Nam K (2011) Antioxidant activities of hot water extracts from various spices. Int J Mol Sci 12(6):4120–4131

31. Kang HJ, Chawla SP, Jo C, Kwon JH, Byun MW (2006) Studies on the development of functional powder from citrus peel. Bioresource Technol 97(4):614–620

32. Zia-ur R (2006) Citrus peel extract—a natural source of antioxidant. Food Chem 99(3):450–454

33. Londoño-Londoño J, Lima VR, Lara O, Gil A, Pasa TBC, Arango GJ, Pineda JRR (2010) Clean recovery of antioxidant flavonoids from citrus peel: Optimizing an aqueous ultrasound-assisted extraction method. Food Chem 119(1):81–87

34. Tsibranska I, Tylkowski B, Kochanov R, Alipieva K (2011) Extraction of biologically active compounds from Sideritis ssp. L Food Bioprod Process 89(4):273–280

35. Kim WJ, Kim J, Veriansyah B, Kim JD, Lee YW, Oh SG, Tjandrawinata RR (2009) Extraction of bioactive components from Centella asiatica using subcritical water. J Supercrit Fluid 48(3):211–216

36. Gamiz-Gracia L, Luque de Castro MD (2000) Continuous subcritical water extraction of medicinal plant essential oil: comparison with conventional techniques. Talanta 51(6):1179–1185

37. Tubtimdee C, Shotipruk A (2011) Extraction of phenolics from Terminalia chebula Retz with water–ethanol and water–propylene glycol and sugaring-out concentration of extracts. Sep Purif Technol 77(3):339–346

38. Katembe WJ, Ungar IA, Mitchell JP (1998) Effect of Salinity on Germination and Seedling Growth of two Atriplex species (Chenopodiaceae). Ann Bot 82(2):167–175

39. Redmann RE, Abouguendia ZM (1979) Germination and seedling growth on substrates with extreme pH: laboratory evaluation of buffers. J Appl Ecol 16(3):901–907

40. Politycka B, Wójcik-Wojtkowiak D, Pudelski T (1985) Phenolic compounds as a cause of phytotoxicity in greenhouse substrates repeatedly used in cucumber growing. Acta Hort 156:89–94

41. Stalikas CD (2007) Extraction, separation, and detection methods for phenolic acids and flavonoids. J Sep Sci 30(18):3268–3295

42. Naczk M, Shahidi F (2004) Extraction and analysis of phenolics in food. J Chromatogr A 1054(1–2):95–111

43. Shahidi F, Daun JK, DeClercq DR (1997) Glucosinolates in Brassica oilseeds: processing effects and extraction. In: Antinutrients and Phytochemicals in Food. ACS Symposium Series. American Chemical Society. http://pubs.acs.org/doi/abs/10.1021/bk-1997-0662.ch009

44. Munir O, Ilhami G, I'Rfan OK (2010) Determination of in vitro antioxidant activity of fennel (Foeniculum vulgare) seed extracts. LWT-Food Sci Technol 36(2):263–271

45. Parejo I, Jauregui O, Sanchez-Rabaneda F, Viladomat F, Bastida J, Codina C (2004) Separation and characterization of phenolic compounds in fennel (Foeniculum vulgare) using liquid chromatography-negative electrospray ionization tandem mass spectrometry. J Agr Food Chem 52(12):3679–3687

46. Paari A, Naidu HK, Kanmani P, Satishkumar R, Yuvaraj N, Pattukumar V, Arul V (2012) Evaluation of irradiation and heat treatment on antioxidant properties of fruit peel extracts and its potential application during preservation of goat fish Parupenaeus indicus. Food Bioprocess Technol 5(5):1860–1870

47. Ward JL, Baker JM, Beale MH (2007) Recent applications of NMR spectroscopy in plant metabolomics. FEBS J 274(5):1126–1131

48. Nicholson JK, Wilson ID (1989) High-resolution proton magnetic resonance spectroscopy of biological fluids. Prog Nucl Magn Reson Spectrosc 21(4–5):449–501

49. Del Coco L, Assfalg M, D'Onofrio M, Sallustio F, Pesce F, Fanizzi FP, Schena FP (2013) A proton nuclear magnetic resonance-based metabolomic approach in IgA nephropathy urinary profiles. Metabolomics 9(3):740–751

50. Godelmann R, Fang F, Humpfer E, Scütz B, Bansbach M, Schäfer H, Spraul M (2013) Targeted and nontargeted wine analysis by ^{1}H NMR spectroscopy combined with multivariate statistical analysis. Differentiation of important parameters: Grape variety, geographical origin, year of vintage. J Agric Food Chem 61(23):5610–5619

51. Del Coco L, De Pascali SA, Fanizzi FP (2014) ^{1}H NMR spectroscopy and multivariate analysis of monovarietal EVOOs as a tool for modulating Coratina-based blends. Foods 3(2):238–249

52. De Pascali SA, Coletta A, Del Coco L, Basile T, Gambacorta G, Fanizzi FP (2014) Viticultural practice and winemaking effects on metabolic profile of Negroamaro. Food Chem 161:112–119

53. Bohme M (1999) Effects of lactate, humate and Bacillus subtilis on the growth of tomato plants in hydroponic systems. Acta Hort 481:231–239

54. Bohme M, Ouahid A, Shaban N (2000) Reaction of some vegetable crops to treatments with lactate as bioregulator and fertilizer. Acta Hort 514:33–40

55. Pracharoenwattana I, Zhou WX, Keech O, Francisco PB, Udomchalothorn T, Tschoep H, Stitt M, Gibon Y, Smith SM (2010) Arabidopsis has a cytosolic fumarase required for the massive allocation of photosynthate into fumaric acid and for rapid plant growth on high nitrogen. Plant J 62(5):785–795

56. Mussatto S (2009) Biotechnological potential of brewing industry by-products. In: Nigam P, Pandey A (ed) Biotechnology for agro-industrial residues utilisation. Springer, Netherlands

57. Huige N (1994) Brewery by-products and effluents. In: Wa H (ed) Handbook of brewing. Marcel Dekker, New York

58. Santos M, Jiménez JJ, Bartolomé B, Gómez-Cordovés C, del Nozal MJ (2003) Variability of brewer's spent grain within a brewery. Food Chem 80(1):17–21

59. Madhujith T, Shahidi F (2007) Antioxidative and antiproliferative properties of selected barley (Hordeum vulgarae L.) cultivars and their potential for inhibition of low-density lipoprotein (LDL) cholesterol oxidation. J Agr Food Chem 55(13):5018–5024

60. Jan A, Nakamura H, Handa H, Ichikawa H, Matsumoto H, Komatsu S (2006) Gibberellin regulates mitochondrial pyruvate dehydrogenase activity in rice. Plant Cell Physiol 47(2):244–253

61. Yu H, Du X, Zhang F, Zhang F, Hu Y, Liu S, Jiang X, Wang G, Liu D (2012) A mutation in the E2 subunit of the mitochondrial pyruvate dehydrogenase complex in Arabidopsis reduces plant organ size and enhances the accumulation of amino acids and intermediate products of the TCA Cycle. Planta 236(2):387–399

62. Mandalari G, Bennett RN, Bisignano G, Saija A, Dugo G, Lo Curto RB, Faulds CB, Waldron KW (2006) Characterization of flavonoids and Pectins from bergamot (Citrus bergamia Risso) peel, a major byproduct of essential oil extraction. J Agr Food Chem 54(1):197–203

63. Poli A, Anzelmo G, Fiorentino G, Nicolaus B, Tommonaro G, Donato PD (2011) Polysaccharides from wastes of vegetable industrial processing: new opportunities for their eco-friendly re-use. In: Elnashar M (ed) Biotechnology of Biopolymers. InTech publication. doi:10.5772/16387

64. Jones D (1998) Organic acids in the rhizosphere—a critical review. Plant Soil
     205(1):25–44
65. Dennis DT, Turpin DH, Lefebvre DD, Layzell DB (1997) Plant metabolism.
     Addison Wesley Longman Ltd, Harlow UK. Plant metabolism
66. Sun YL, Hong SK (2011) Effects of citric acid as an important component of
     the responses to saline and alkaline stress in the halophyte Leymus
     chinensis (Trin.). Plant Growth Regul 64(2):129–139

19

An analysis of Brazilian sugarcane bagasse ash behavior under thermal gasification

Catia Fredericci[1*], Gerhard Ett[1], Guilherme FB Lenz e Silva[2], João B Ferreira Neto[1], Fernando JG Landgraf[1,2], Ricardo L Indelicato[2] and Tiago R Ribeiro[1]

Abstract

Background: Ashes from sugarcane were analyzed by X-ray fluorescence, ash content, X-ray diffraction, scanning electron microscopy (SEM), and energy dispersive spectroscopy (EDS). FactSage 6.4 database software was used to estimate viscosity at high temperatures (900 - 1550°C) of them.

Results: The results showed that although ashes from sugarcane bagasse contain silica, most of its SiO_2 is from soil contamination. Higher and lower silica samples treated at 1350°C for 20 minutes showed that the fine portion of fraction of the ashes melted at this temperature.

Conclusions: The melting phase could act as sticking flux for the coarse silica particles on the gasifier bottom wall, which could compromise the gasification process.

Keywords: Biomass; Sugarcane; Bagasse; Ashes; Gasification

Background

The Institute for Technological Research (IPT) is interested in the development of the gasification process (Figure 1). Gasification produces synthetic gas or syngas ($H_2 + CO_2 + N_2 + CO_2 + H_2O + C_nH_m$) which can then be converted into energy, Fischer Tropsch biodiesel or monomers. It has higher energy efficiency when compared to other uses of bagasse and can produce important base chemicals [1]. Gasification is very flexible to raw materials, allowing for local development. The greatest challenge is to overcome the initial investment barrier, and lowering the current evaluation of US$3/W to US$1.2/W that would make it commercially competitive [2]. A project involving the construction and operation of a gasification pilot plant, one of the questions raised is the melting temperature of the ashes from bagasse and/or straw used as a feedstock fuel.

Sugarcane bagasse, as biomass, will be used in this work, due to its availability and promising potential. Brazil is the largest producer of sugarcane, with production concentrated in its South-Central region. Currently, the bagasse is used in boilers in the mills to produce energy and heat with typical 45% carbon content (dry mass) [3]. Brazil's

sugarcane industry association (UNICA) reported that the amount of crushed surgarcane in the South-Central region in 2013/14 was 596.936 Mt, about 12% up from the 532.758 Mt produced in 2012/13 [4] that generated approximately 118 Mt of wet bagasse. Even considering its utilization as fuel for energy generation (typically in steam boilers) in the production mills, Perrone et al. [5] report that there is still a 12% surplus of biomass that will increase proportionally with the increase of ethanol production, requiring solutions and innovative ideas in order to generate new economic value and opportunities for the sugar and alcohol industry. Elements including N, P, K, S, Ca, Mg, Fe, Zn, B, Cu, Mn, Cl and Si considered as macro-, micro- and functional nutrients are essentials for increasing and sustaining crop yields [6]. At high temperatures these elements in bagasse or straw are involved in reactions leading to ash formation [7].

The temperature inside the gasifier should be sufficient to melt the ash that will be deposited on the wall of the gasifier forming a liquid slag which should flow out from the bottom of the gasifier [8]. The typical temperature for highly efficient processes should be bellow than 1500°C. Entrained flow gasification works with liquid (bio-oil) or sub-millimeter solid particles (bagasse or straw particles). Entrained flow consists of a vertically placed cylindrical reactor at the top of which fuel (bio-oil or biomass) and a

* Correspondence: catiaf@ipt.br
[1]Institute for Technological Research of São Paulo State, 05508-901 São Paulo, SP, Brazil
Full list of author information is available at the end of the article

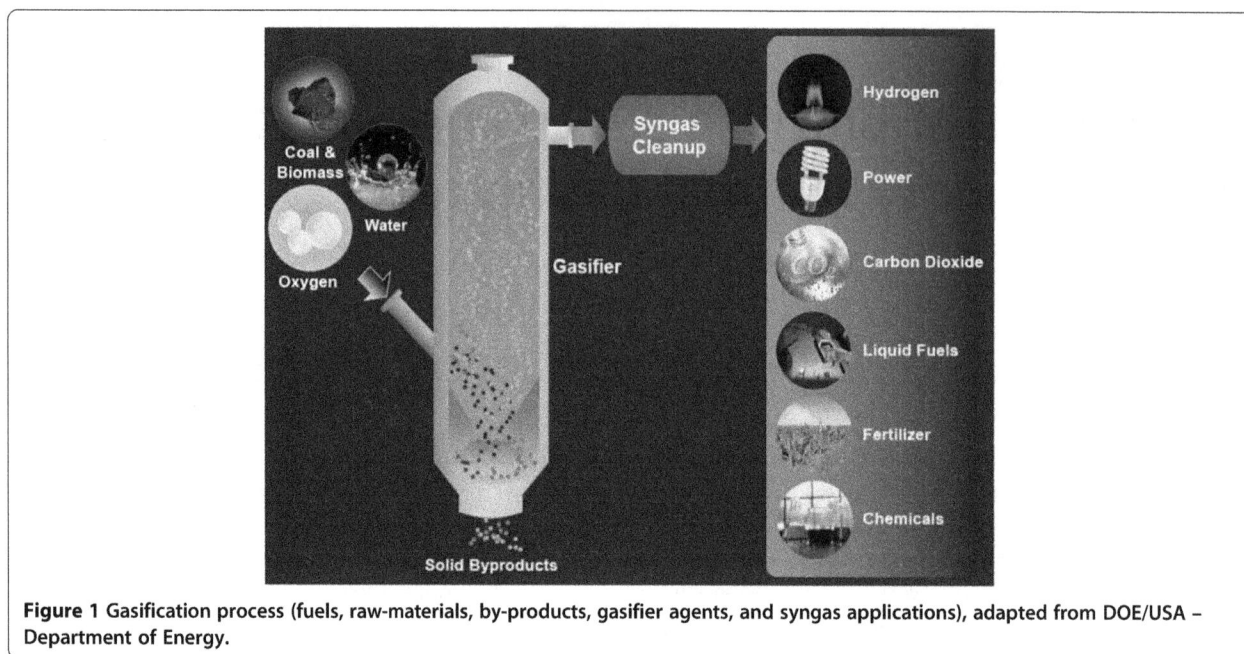

Figure 1 Gasification process (fuels, raw-materials, by-products, gasifier agents, and syngas applications), adapted from DOE/USA – Department of Energy.

gasifying agent (air or O_2) are inserted through a nozzle, usually in a swirling flow, forming a flame that carries either particles or droplets through the reactor as they undergo incomplete combustion, i.e. gasification. Residence times are small, temperature and pressure can as be high as 1500°C and 40 bar, respectively, ensuring low tar production [9].

The aim of this work is to analyze the ash from sugarcane bagasse generated from sugarcane mills in some regions of the São Paulo State – Brazil. The results obtained in this research will provide inputs for determining the preferred processing conditions for the gasifier.

Methods and Experimental

Bagasse from four different sugarcane mills in São Paulo State – Brazil were collected from the stock area of the mills, in the period of December 2011 to December 2012 and were used for the study (Table 1). Currently, the bagasse presents ~50% moisture content. They were dried at 80°C for 48 h for obtaining bagasse on a dry weight basis

(about 2% humidity). Initial scanning electron microscopy (FEI-Quanta 3D FEG-SEM) was used to evaluate the shape and morphology. A small assortment of each bagasse sample was deposited on carbon tape secured on brass stubs and then coated with a thin layer of gold. The qualitative analysis of the chemical elements was performed using an energy dispersive spectrometer (EDAX System). The secondary electron images were taken using a voltage of 5 to 15 kV.

Twenty grams of bagasse were heat treated at 600°C for 30 min to generate ashes. Each experiment was repeated 10 times for each bagasse under the same conditions and the percentage of generated ashes was given as the average of them.

Table 1 Samples of bagasse and sugarcane and their sources

Bagasse samples collected in the mill	Location of the Mill
BA	Araraquara – SP(*)
BI	Iracemapólis – SP
BRP	Rio das Pedras – SP
BS	Sorocaba – SP
Sugarcane sample collected in the field	Location of the Mill
BAL	Araraquara – SP

(*)SP – São Paulo State – South-east Brazil.

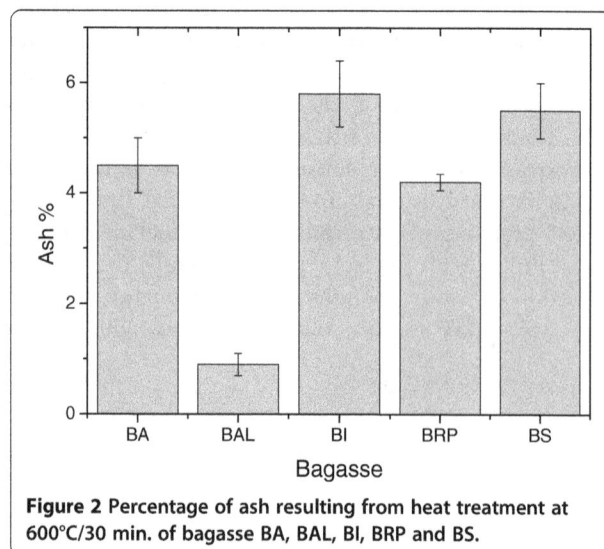

Figure 2 Percentage of ash resulting from heat treatment at 600°C/30 min. of bagasse BA, BAL, BI, BRP and BS.

Table 2 Chemical analysis of the ashes obtained by X-ray fluorescence from BA, BI, BRP, BS and BAL samples (wt%)

Oxide	BA	BI	BRP	BS	BAL
SiO_2	73.4	56.1	80.5	65.3	11.1
Al_2O_3	7.7	18.0	5.6	10.6	0.3
P_2O_5	5.1	2.0	4.7	5.5	6.9
K_2O	3.1	4.9	3.1	3.0	27.0
MgO	1.7	1.8	1.8	1.7	19.0
Fe_2O_3	5.1	7.8	1.5	8.5	0.3
CaO	1.6	5.0	1.4	1.9	8.9
SO_3	0.5	2.0	0.5	0.7	21.0
TiO_2	1.3	1.8	0.4	2.5	0.1
Cl	<0.1	<0.1	<0.1	<0.1	5.0

$Na_2O < 0.1$, $ZnO < 0.02$, $CuO < 0.02$, $MnO < 0.2$.

The samples collected from different mills were named with the following identification: BA, BI, BRP and BS. Ash samples from Araraquara region that were entirely processed in the IPT's laboratory were named BAL and they were used for comparative purposes. In this case, the BAL sample stalks were milled in the laboratory and were dried at 80°C for 48 h for obtaining bagasse on a dry weight basis (about 2% humidity). Finally, they were heat treated at 600°C for 30 min. Table 1 shows the ash sample identifications and their mill.

Ashes were analyzed by X-ray diffraction (Shimadzu XRD 6000, using Co K_α radiation), X-ray fluorescence pressed powder pellet technique (Philips, model PW 2404), scanning electron microscopy (FEI-Quanta 3D FEG-SEM) coupled to an energy dispersive spectrometer (EDAX System), and

sieving analysis for particle size distribution. The computer thermodynamic package FactSage 6.4, in the modulus Glass database, was used to calculate the viscosity of the ash as a function of temperature from 900°C to 1550°C.

About 0.2 g of ash was compacted in a stainless steel mold with a 0.5 cm diameter. The resulting cylindrical samples were put in an alumina boat crucible with 5 cm in length and were heat treated at 1350°C for 20 min in order to analyze the fusibility behavior of the ashes. The surfaces of the heat treated samples were analyzed by scanning electron microscopy using a JEOL-SEM model JSM 6300, coupled to an energy dispersive spectrometer Noran System.

Results and discussion
Ash from bagasse
The temperature of 600°C was chosen to burn out all carbon and to prevent as much as possible the loss of potassium [10]. Figure 2 shows the results of ash percentage from the bagasse samples (BA, BAL, BI, BRP and BS) on a dry weight basis. It can be observed that the amount of ash is in the range of 4 to 6 wt% and is agreement with some data reported in the literature [11,12], although one publication indicates a percentage of ash as high as 20 wt% [13]. An interesting fact was that the amount of ash resulted from the heat treatment of BAL is smaller than that obtained from the heat treated bagasse from the sugar mills, including BA bagasse. The reasons for this discrepancy will be discussed later.

Ash composition and crystalline phases
Table 2 shows the chemical composition of the ashes obtained by X-ray fluorescence (XRF). The ashes from

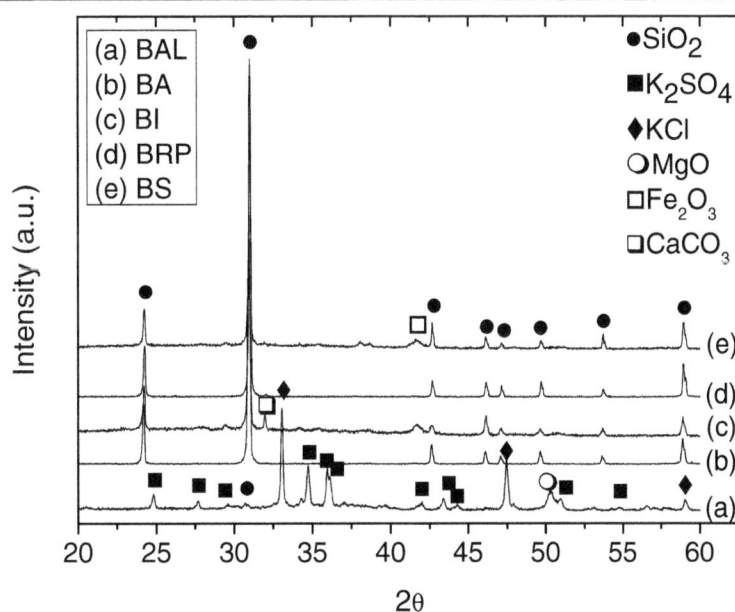

Figure 3 XRD patterns of the ashes resulted from the heat treatment at 600°C/30 min of BA, BAL, BI, BRP and BS samples.

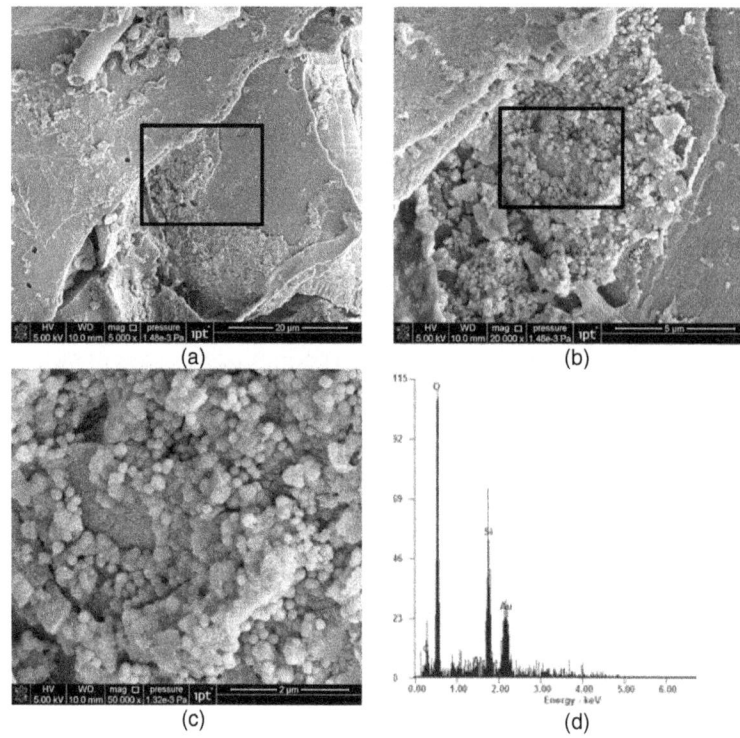

Figure 4 SEM micrographs (SE image) of bagasse BA: (a), (b) and (c). The squares represents the magnified area; (d) EDS microanalysis from area (c).

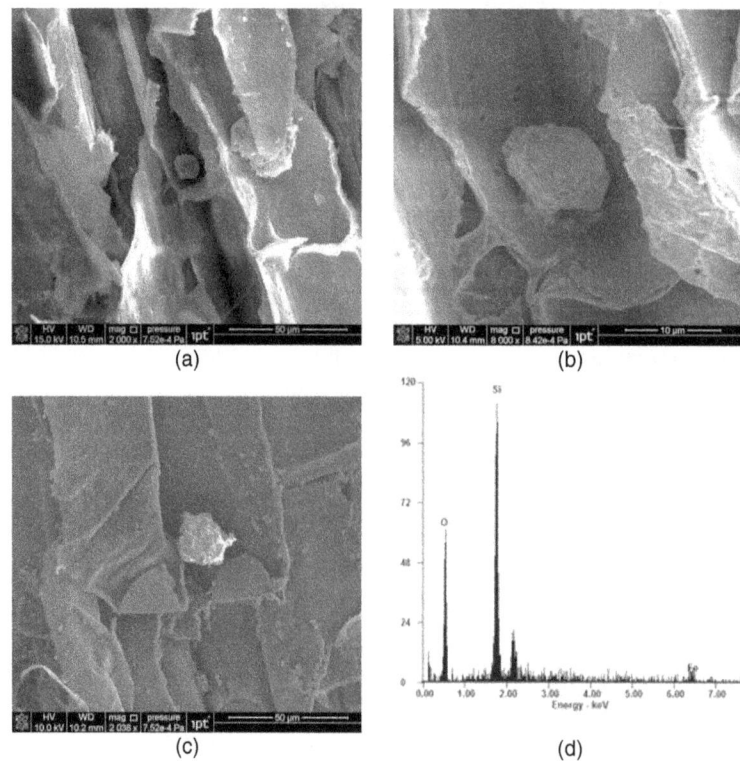

Figure 5 (a) SEM of bagasse BA, (b) magnified area from (a), (c) SEM of bagasse BA in different region of (a) and (d) EDS of the particle presented in (b).

Table 3 Granulometric size distribution by sieving for the ashes obtained through the heat treatment of BA, BI, BRP, BS and BAL samples

Sieves mesh	BA	BI	BRP	BS	BAL
+50	3.3	10.3	6.6	2.3	-
−50 + 100	22.9	13.4	20.1	15.2	-
−100 + 170	2.1	18.1	4.4	3.3	-
−170 + 270	41.7	13.7	39.2	35.8	-
−270 + 325	7.2	3.5	4.7	3.5	-
−325	21.7	40.2	24.2	39.7	100.0

bagasse collected in the sugar/ethanol mills present high amounts of SiO_2, as high as 56 wt%, and low amounts of alkaline oxide, P_2O_5 and Al_2O_3. On the other hand, the BAL bagasse presents low amounts of silica (~11 wt%) and Al_2O_3, and high amount of alkaline and earth alkaline oxides (K_2O, MgO and CaO). The contents of macro and micronutrients for plant growth Fe_2O_3, ZnO, CuO, Cl (chloride) and MnO was also determinate. Boron and Nitrogen were not determined by X-ray fluorescence since they are considered light elements.

As can be seen in Figure 3 the XRD patterns show a significant difference between the ashes from bagasse processed in the mill and in the laboratory. The ashes from bagasse BA, BRP, BI and BS present SiO_2 in the form of quartz as the mainly crystalline phase, while the primary phase for the ash from BAL is potassium chloride, followed by potassium sulfate and silica appears as a minor crystalline phase, consistent with the chemical composition showed in Table 2. It is known that silicon is considered a beneficial element to plants and is absorbed through the roots as silicic acid. Malavolta [14] reported that stalks are richer in K (109 kg/ha) than in Si (98 kg/ha), which is only in agreement with BAL bagasse.

Figure 4 shows SEM micrographs of BA bagasse and exhibits spherical nanometer scale, SiO_2 rich particles that can be attributed to absorbed silicon from the soil by sugarcane. Sugarcane harvesting in Brazil is done in

two ways: by machine and by manual process after sugarcane burning (São Paulo State law does not allow sugarcane burning from June to November due to increased air pollution). In both cases there is soil contamination of the sugarcane, especially in times of rain, as in the Brazilian summer, when bagasse samples for this study were collected. Although in some sugar and alcohol mill plants occurs the washing process of the sugarcanes, it seems not very effective in removing all their soil impurities. When the shredded sugarcane passes through several mills in the plant production of sugar and alcohol, soil particles become trapped in the fibers of the sugarcane.

The micrograph shown in Figure 5, obtained by scanning electron microscopy, shows a particle rich in Si and O, confirming the presence of soil contamination in bagasse after the milling process. Thus the SiO_2 present in the ash composition of the BA, BI, BRP and BS bagasse is a combination of silicon oxide from soil contamination and from absorbed silicon from the soil in the sugarcane. This is one of the reasons for the discrepant results of ash amount obtained from bagasse collected in plant mills [11-13].

Ash particle size distribution
Table 3 presents the particle size distribution of the ashes. Except for BAL ash, the others present particles with sizes greater than 40 μm (# 325 mesh) and significant amounts of particles in the 53 to 88 μm (+270 − 170 mesh) size range. It can be seen in Figure 6 that particles of BA ash in the 53 to 88 μm range exhibit the typical rounded morphology of quartz sand grains from soil (Figure 6b) [15].

Ash fusibility
Patterson and Hurst [16] proposed that a silica ratio of $100SiO_2/(SiO_2+ Fe_2O_3 + CaO + MgO) < 80$ is required for entrained flow gasifiers, if the SiO_2 amount is high, a flux is required for decreasing the ash viscosity, and to minimize the flux amount a SiO_2/Al_2O_3 ratio of 1.6-2.0 is optimal. Table 4 presents the relations proposed by

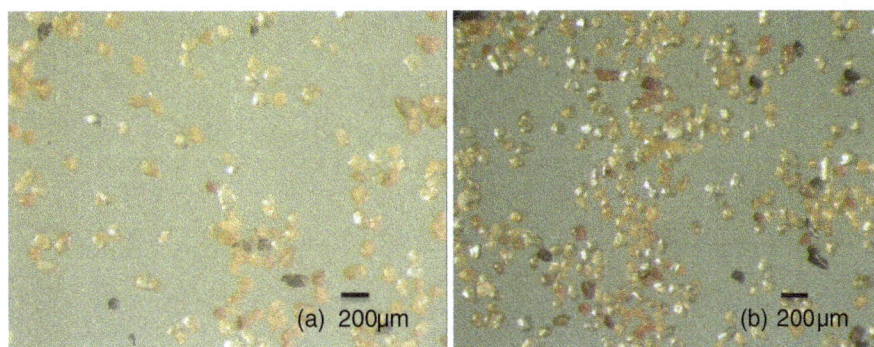

Figure 6 Reflected light microscopy of the -170+270 mesh (-88+53 μm) fraction in ash (a) BA and (b) soil from Araraquara area.

Table 4 Ash properties used to determine viscosities in the entrained flow gasifiers (Patterson and Hurst)

Ashes	$SiO_2/ (SiO_2 + Fe_2O_3 + CaO + MgO)$ (%)	SiO_2/Al_2O_3
BAL	28.2	37.0
BI	79.3	3.1
BS	84.4	6.2
BA	89.7	9.5
BRP	94.5	14.4

Patterson and Hurst for the ash from BAL, BI, BS, BA and BRP. As can be seen, only ash from BI and BAL has an index lower than 80. However, regarding the SiO_2/Al_2O_3 relation all ashes present indices higher than 2.0. According to Peterson and Hurst [16], all bagasse from the stock area of mills (BA, BRP, BI and BS) should be mixed with flux to produce an ash with optimal viscosity for an entrained flow gasifier. It is known that viscosity can only be used as an indication of slag in the gasifier, and that the melting characteristic of ash is generally lower in reducing atmosphere [16-18]. Iron is the principal fluxing element that controls ash melting under reducing conditions [17]. We have used the Fe_2O_3 and its equivalent FeO values presented in Table 2 to calculate ash viscosities using the FactSage 6.4 a database computing systems in chemical thermodynamics. The results could be a guide to evaluate bagasse quality for gasification. For software calculations, chloride and SO_3 were not considered, since they are not available on FactSage

database software. Although P_2O_5 is also not available, it was added to SiO_2, since it is a glass former.

Figure 7 presents the curve of calculated viscosity for the ash from BRP, BA, BS, BI and BAL. Higmann and Burgt [18] reported that to ensure continuous flow, a slag viscosity less than 25 Pa.s is necessary at temperatures lower than 1400°C. As can be seen in Figure 7 the calculated viscosities of the ash from BRP, BA, BS and BI at 1350°C are $2.7x10^4$ Pa.s, $8.7x10^3$ Pa.s, $1.6x10^3$ Pa.s, and $0.5x10^3$ Pa.s, which are very high. The viscosity of ash BAL is lower than 25 Pa.s (log $\eta \sim 1.4$ Pa.s) at 900°C. Figure 8 shows the ash samples heat treated at 1350°C, and it is possible to observe the difference on viscosity from it. Wang and Massoudi [8] reported that one possibility to obtain viscosity lower than 25 Pa.s is to increase the gasifier operating temperature, that however can lead to a reduction of the overall efficiency of the process. Even for temperatures as high as 1550°C, for the samples from the stock area of mills, only the ash from BI presented a viscosity near 25 Pa.s (log $\eta \sim 1.4$ Pa.s). Great differences in viscosity were not observed when using FeO instead of F_2O_3 on the chemical composition of the ashes.

Figure 9 shows that to obtain a viscosity close to 25 Pa.s, an addition of 35 wt% and 15 wt% of CaO, as flux, should be added to the BRP and BI, respectively [19]. There are some difficulties with the utilization of flux when the biomass is bagasse due the difficulty in homogenizing it with the flux. Furthermore, it is reported that with flux addition

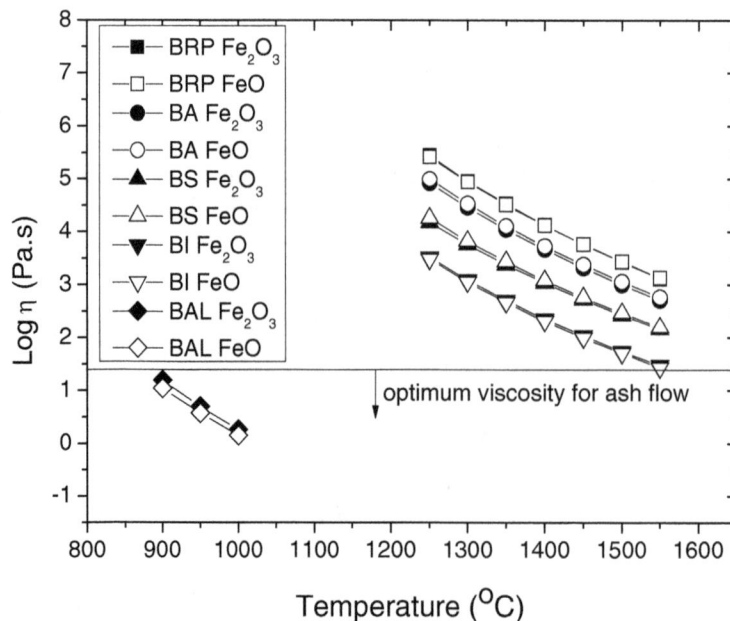

Figure 7 Ash viscosity as a function of temperature for the ashes from BRP, BA, BS BI and BAL, calculated by FactSage using the data from Table 2.

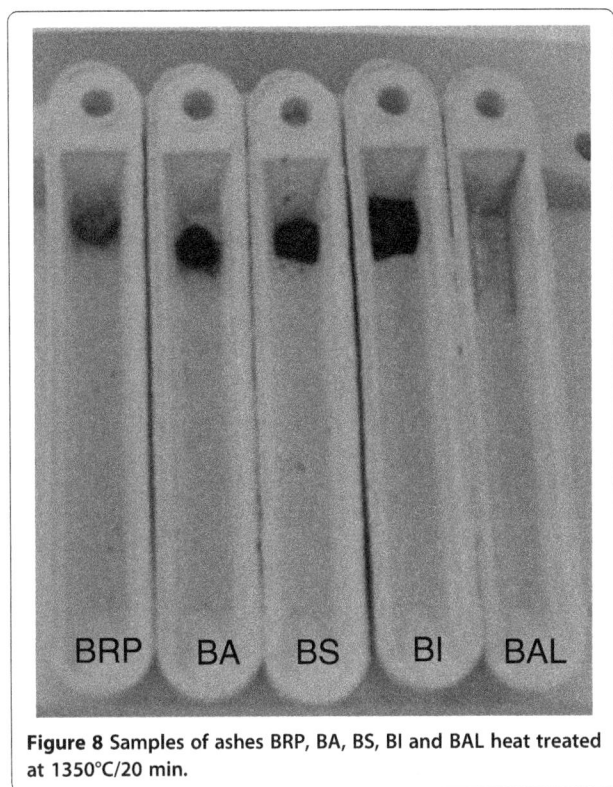

Figure 8 Samples of ashes BRP, BA, BS, BI and BAL heat treated at 1350°C/20 min.

all the ash must be melted and more heat is required for greater ash content, thereby reducing gasifier efficiency [15].

In Figure 10 and Figure 11 are shown the surfaces of the BRP and BI ash samples, heat treated at 1350°C/20 min in

an oxidizing atmosphere (air). The ashes from BRP and BI represent high and the low amounts of silica (quartz), respectively, of the bagasse from the stock area of mills. It can be seen that particles of SiO_2 are surrounded by a glass phase resulting from the reaction of the components of the fine fraction of the ash, consisting of the plant nutrients (SiO_2, Al_2O_3, K_2O, CaO, MgO, P_2O_5, and Fe_2O_3, for example).

If bagasse BA, BI, BRP and BS would be used for gasification, as received, coarse particles of SiO_2 could became stuck and tightened on the inner and bottom walls of the gasifier, which could be dangerous for the process since the agglomeration of quartz particles on the bottom wall could cause a blockage of the gasifier. This could lead to severe unscheduled shutdowns and high operation maintenance costs [20]. It is necessary to find technological and economically feasible alternatives for removing silica from bagasse before its gasification processing.

Conclusions

The ashes from sugarcane bagasse processed in the sugarcane mills from São Paulo State present quartz as the main crystalline phase from soil contamination. On the other hand, the ash from bagasse cleaned and processed in laboratory has potassium chloride and potassium sulfate as principal the crystalline phases, and SiO_2 as a minor phase. Analysis from scanning electron microscopy of the surface of ash samples from mill bagasse heat treated at 1350°C for 20 minutes showed that the fine fraction of the ashes melt at this temperature and act as flux for coarse quartz sintering. This can cause

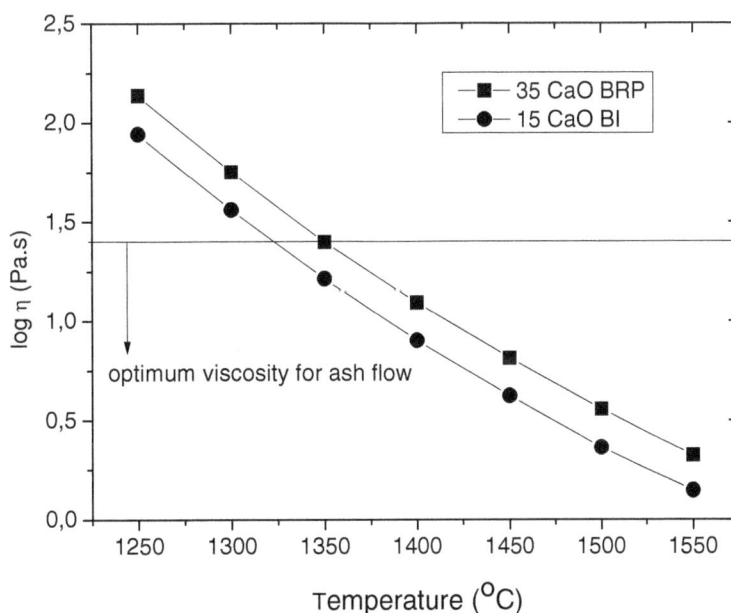

Figure 9 Ash viscosity as a function of temperature for the ash from BRP and BI, calculated by FactSage, using the data from Table 2, with addition of 35 wt% CaO in BRP and 15 wt% CaO in BI.

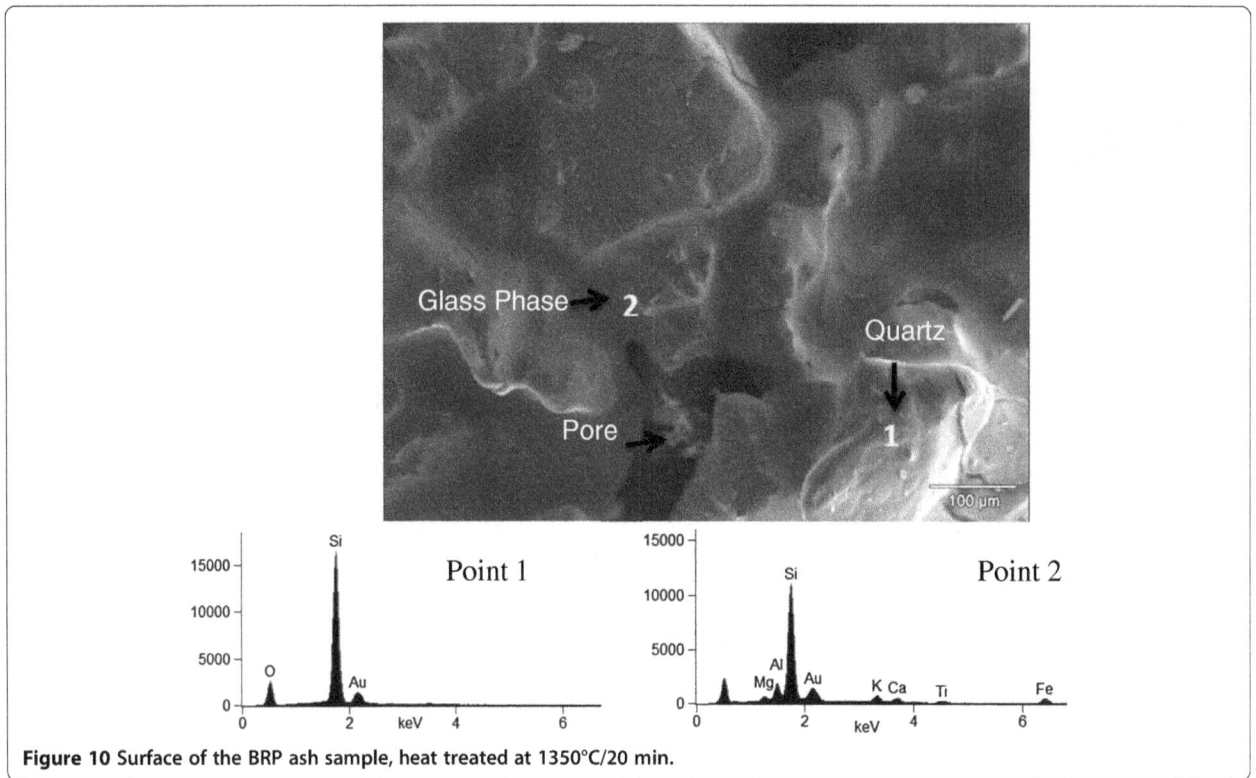

Figure 10 Surface of the BRP ash sample, heat treated at 1350°C/20 min.

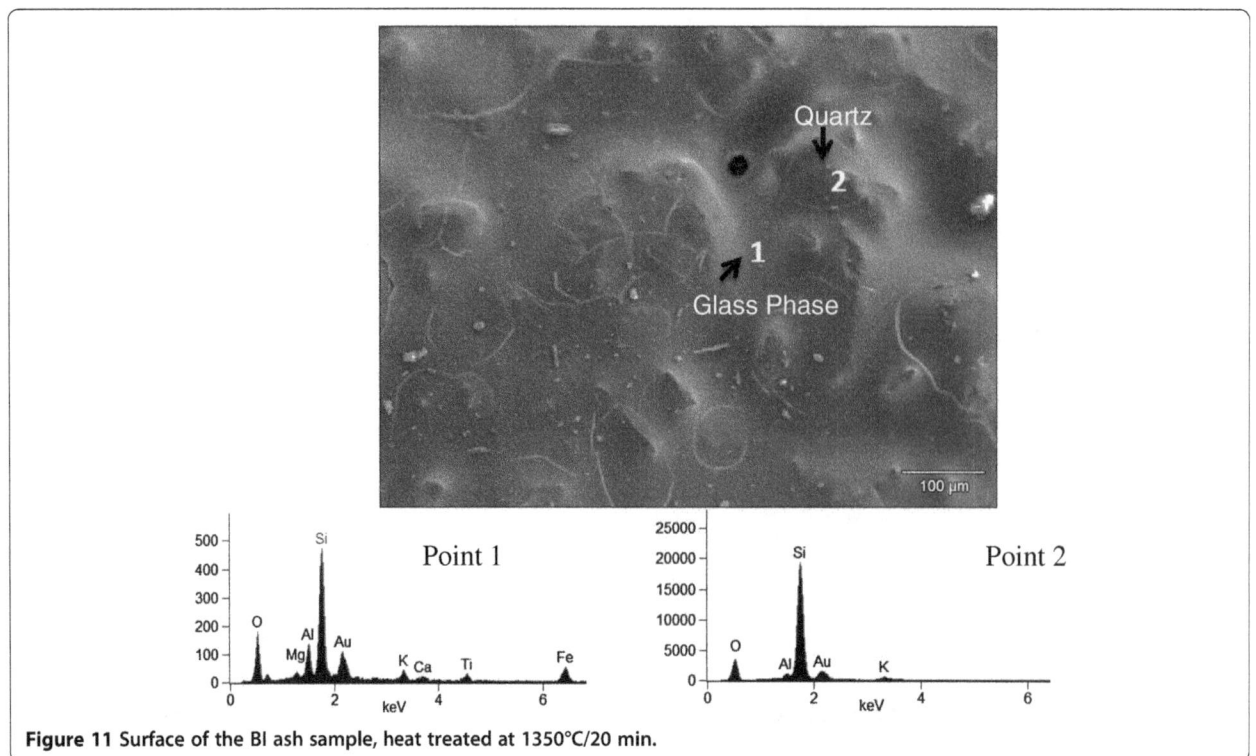

Figure 11 Surface of the BI ash sample, heat treated at 1350°C/20 min.

aggregation of SiO_2 on the bottom wall of the gasifier that can compromise the process when using sugarcane bagasse as biomass for gasification. It is very important to study possibilities for removing soil contamination before bagasse processing.

Competing interests
The authors declare that they have no competing interests.

Authors' contributions
All authors have contributed substantially to the work. They read and approved the final manuscript.

Acknowledgements
The authors kindly thank the Sugarcane Mills from São Paulo State - Brazil for donating the sugarcane bagasse, Ruben Spitz from Brown University, Claudia Maria G. de Souza and Miguel Papai Jr. of the The Center for Chemistry and Manufactured Goods – CQuim – for the X-ray fluorescence analyses.

Author details
[1]Institute for Technological Research of São Paulo State, 05508-901 São Paulo, SP, Brazil. [2]Polytechnic School, University of São Paulo, 05508-010 São, SP, Brazil.

References

1. U.S. DOE - National Energy Technology Laboratory (NETL) Advantages and efficiency of gasification. available at: http://www.netl.doe.gov/research/coal/energy-systems/gasification/gasifipedia/clean-power. Accessed 22 Jul 2014
2. Yu ASO, Landgraf FJG, Ett G, Silveira JRF (2013) IPT Bagasse Gasification Conceptual Engineering. In: International Society of Sugar Cane Technologists, 28, 2013, Proceedings. Stab and Coopersucar, São Paulo, pp 1–15
3. Paes LAD, Marian FR Carbon Storage in Sugarcane Fields of Brazilian South-Central Region, CTC Technical report 2011., available at: http://sugarcane.org/resource-library/studies/Carbon%20storage%20in%20sugarcane%20fields%20of%20Brazilian%20South-Central%20region.pdf. Accessed 22 Jul 2014
4. (2013) UNICA – União da Indústria de Cana-de-Açúcar (Sugarcane Industry Union) – UNICA Data Press Release., p 30, available at: http://www.unica-data.com.br/listagem.php?idMn=86. Accessed 22 Jul 2014
5. Perrone CC, Appel LG, Maia Lellis GL, Ferreira FM, de Sousa AM, Ferreira-Leitão VS (2010) Ethanol: an evaluation of its scientific and technological development and network of players during the period of 1995 to 2009. Waste Biomass Valor 2(1):17–32
6. Savant NK, Korndörfer GH, Datnoff LE, Snyder GH (1999) Silicon nutrition and sugarcane production: a review. J Plant Nutr 22(12):1853–1903
7. Jenkins BM, Baxter LL, Miles TR Jr, Miles TR (1998) Combustion Properties of Biomass. Fuel Process Tech 54:17–46
8. Wang P, Massoudi M (2011) Effect of Coal Properties and Operation Conditions on Flow Behavior of Coal Slag in Entrained Flow Gasifiers: A Brief Review. In: U.S. Department of Energy (DOE) National Energy Technology Laboratory (NETL)., pp 1–29, Report number: DOE/NETL-2011/1508
9. U.S. DOE National Energy Technology Laboratory (NETL)., available at. http://www.netl.doe.gov/File%20Library/Research/Coal/energy%20systems/gasification/gasifipedia/index.html. Accessed 29 Jul 2014
10. Liao Y, Yang G, Ma X (2012) Experimental study on the combustion characteristic and alkali transformation behavior of straw. Energy Fuels 26:910–916
11. Ohman M, Pommer L, Nordin A (2005) Bed agglomeration characteristics and mechanisms during gasification and combustion of biomass fuels. Energy Fuels 19(4):1742–1748
12. Batra VS, Urbonaite S, Svensson G (2008) Characterization of unburned carbon in Bagasse Fly Ash. Fuel 87:13–14, 2972–2976
13. Rezende CA, de Lima MA, Maziero P, de Azevedo ER, Garcia W, Polikarpov I (2011) Chemical and morphological characterization of sugarcane bagasse submitted to a delignification process for enhanced enzymatic digestibility. Biotech Biofuels 4(54):1–18
14. Malavolta E (1994) Fertilizing for high yield sugarcane – international potash institute Basel/Switzerland. Bull Am Meteorol Soc 14:1–102, http://ebookbrowsee.net/ipi-bulletin-14-fertilizing-for-high-yield-sugarcane-pdf-d230056405
15. Merrison JP (2012) Sand transport, erosion and granular electrification. Aeolian Res 4:1–16
16. Patterson JH, Hurst HJ (2000) Ash and slag qualities of australian bituminous coals for use in Slagging Gasifiers. Fuel 79(13):1671–1678
17. Huffman GP, Huggins FE, Dunmyre GR (1981) Investigation of the high-temperature behavior of coal ash in reducing and oxidizing atmosphere. Fuel 60(7):585–597
18. Higman C, van der Burgt M (2008) Gasification, 2nd edn. Elsevier, New York
19. Guo Z-Q, Han B-Q, Dong H (1997) Effect of coal slag on the wear rate and microstructure of the zro₂-bearing chromia refractories. Ceramics Int 23(6):489–496
20. Brooker D (1993) Chemistry of deposit formation in a coal gasification syngas cooler. Fuel 72(5):665–670

Permissions

List of Contributors

Andrea Ertani , Paolo Sambo, Carlo Nicoletto, Silvia Santagata, Michela Schiavon and Serenella Nardi
Department of Agronomy, Animals, Food, Natural Resources and Environment - DAFNAE, University of Padua, Viale dell'Università 16, 35020, Legnaro, Padova, Italy

Masakazu Aoyama
Faculty of Agriculture and Life Science, Hirosaki University, Hirosaki 036-8561, Japan

Anteneh Argaw
College of Agriculture and Environmental Sciences, School of Natural Resources Management and Environmental Sciences, Haramaya University Dire Dawa, Ethiopia

Angaw Tsigie
Ethiopian Institute of Agricultural Research, Holleta Agricultural Research Center, Holleta, Ethiopia

Alessandro Miceli, Alessandra Martorana, Giancarlo Moschetti and Luca Settanni
Department of Agricultural and Forest Science, University of Palermo, Viale delle Scienze 4, 90128 Palermo, Italy

Gennaro Roberto Abbamondi, Carmine Iodice, Barbara Nicolaus and Giuseppina Tommonaro
CNR, National Research Council of Italy, Institute of Biomolecular Chemistry, Via Campi Flegrei, 34, 80078 Pozzuoli, NA, Italy

Nele Weyens, Sofie Thijs, Wouter Sillen, Panagiotis Gkorezis, Jaco Vangronsveld and Gennaro Roberto Abbamondi
Hasselt University, Environ- mental Biology, Centre for Environmental Sciences, Agoralaan, building D, 3590 Diepenbeek, Belgium

Wesley de Melo Rangel
Laboratory of Biology, Microbiology and Soil Biological Processes, Soil Science Department, Federal University of Lavras, PO box 3037, 37200-000 Lavras, Minas Gerais, Brazil

Marina Bogicevic and Osvaldo Failla
Department of Agricultural and Environmental Sciences, University of Milan, Via Celoria 2, 20133 Milan, Italy

Vesna Maras, Milena Mugoša, Vesna Kodžulović, Jovana Raičević and Sanja Šućur
"13. Jul Plantaze" a.d., Put Radomira Ivanovica 2, 81000 Podgorica, Montenegro

Rachel L Sleighter and Hatcher
Research and Development, FBSciences, Inc, 4111 Monarch Way, Suite 408, Norfolk, VA 23508, USA
Department of Chemistry and Biochemistry, Old Dominion University, Norfolk, VA 23529, USA

Paolo Caricasole, Kristen M Richards, Terry Hanson1 and Patrick G
Research and Development, FBSciences, Inc, 4111 Monarch Way, Suite 408, Norfolk, VA 23508, USA

Zakaria Hazzoumi, Youssef Moustakime, and Khalid Amrani Joutei
Laboratory of Bioactive Molecules, Structure and Function, Faculty of Science and Technology Fez, B.P. 2202-Road of Imouzzer, Fez 30000, Morocco

El hassan Elharchli
Laboratory of microbial Biotechnology, Faculty of Science and Technology Fez, B.P. 2202-Road of Imouzzer, Fez 30000, Morocco

Vesna Dragičević, Igor Spasojević and Vesna Perić
Maize Research Institute, Slobodana Bajića 1, 11185 Zemun Polje, Serbia

Bogdan Nikolić and Sanja Đurović
Institute for Plant Protection and Environment, Teodora Drajzera 9, 11000 Belgrade, Serbia

Hadi Waisi
Institute for the Development of Water Resources, "Jaroslav Černi", Jaroslava Černog 80, 11226 Belgrade, Serbia

Milovan Stojiljković
Vinca Institute of Nuclear Sciences, 52211001 Belgrade, Serbia

Meng Li and Zhengyi Hu
College of Resources and Environment, Sino-Danish College, University of Chinese Academy of Sciences, Beijing 100049, People's Republic of China

Pierluigi Mazzei, Vincenza Cozzolino, Hiarhi Monda and Alessandro Piccolo
Interdepartmental Research Centre on Nuclear Magnetic Resonance for the Environment, Agro-Food, and New Materials (CERMANU), University of Naples Federico II, Portici 80055, Italy

Alessandro C Ramos
Laboratório de Bioquímica e Fisiologia de Microrganismos, Universidade Estadual do Norte Fluminense Darcy Ribeiro (UENF), Av. Alberto Lamego 2000, Campos dos Goytacazes 28013-602, Brazil

Leonardo B Dobbss
Laboratório de Microbiologia Ambiental e Biotecnologia, Universidade de Vila Velha (UVV), Rua Comissário José Dantas de Melo 21, Boa Vista, Vila Velha, Espírito Santo, Brazil

Leandro A Santos and Mânlio S Fernandes
Departamento de Solos da Universidade Federal Rural do Rio de Janeiro (UFRRJ), Seropédica, km 7 BR 465, Seropédica, Rio de Janeiro CEP 23851-970, Brazil

Fábio L Olivares, Natália O Aguiar and Luciano P Canellas
Universidade Estadual do Norte Fluminense Darcy Ribeiro (UENF) Núcleo de Desenvolvimento de Insumos Biológicos para Agricultura (NUDIBA), Av. Alberto Lamego 2000, Campos dos Goytacazes 28013-602, Brazil

Vesna Dragičević, Zoran Dumanović and Natalija Kravić
Maize Research Institute, Slobodana Bajića 1, 11185 Zemun Polje, Serbia

Suzana Kratovalieva
State Phytosanitary Laboratory, Ministry of Agriculture, Forestry and Water Economy, Aminta Treti 2, 1000 Skopje, Republic of Macedonia

Zoran Dimov
Faculty of Agricultural Sciences and Food, University Ss Cyril and Methodius, Aleksandar Makedonski bb, 1000 Skopje, Republic of Macedonia

Luis C Rodríguez-Zapata, Francisco L Espadas y Gil, Susana Cruz-Martínez, Carlos R Talavera-May, Fernando Contreras-Marin, Gabriela Fuentes and Jorge M Santamaría
Unidad de Biotecnología, Centro de Investigación Científica de Yucatán, Mérida 97200, México

Enrique Sauri-Duch
Instituto Tecnológico de Mérida, Mérida 97200, México

Silvia Vaccaro, Andrea Ertani, Silvia Quaggiotti and Serenella Nardi
Department of Agronomy, Food, Natural Resources, Animal and Environment, University of Padua, Agripolis - Viale dell'Università, 16, Legnaro, Padua 35020, Italy

Antonio Nebbioso and Alessandro Piccolo
Centro Interdipartimentale di Ricerca sulla Risonanza Magnetica Nucleare per l'Ambiente, l'Agro-Alimentare ed i Nuovi Materiali (CERMANU), Università di Napoli Federico II, Via Università 100, Portici 80055, Italy

Adele Muscolo
Agriculture Department, Mediterranea University of Reggio Calabria, Feo di Vito, Reggio Calabria 89060, Italy

Catello Pane, Domenica Villecco, Riccardo Spaccini and Massimo Zaccardelli
Consiglio per la Ricerca in Agricoltura e l'Analisi dell'Economia Agraria,Centro di Ricerca per l'Orticoltura, via dei Cavalleggeri 25, I-84098 Pontecagnano, SA, Italy

Giuseppe Celano and Assunta M Palese
Dipartimento Scienze dei Sistemi Colturali, Forestali e dell'Ambiente, Università degli Studi della Basilicata, viale dell'Ateneo Lucano 10, I-85100 Potenza, Italy

Alessandro Piccolo
Centro Interdipartimentale di Ricerca sulla Risonanza Magnetica Nucleare per l'Ambiente, l'Agro-Alimentare ed i Nuovi Materiali (CERMANU), Via Università 100, I-80055 Portici, NA, Italy

Antonio Nebbioso, Pierluigi Mazzei and Davide Savy
Centro Interdipartimentale di Ricerca per la Spettroscopia di Risonanza Magnetica, Nucleare per l'Ambiente, l'Agro-alimentare e i Nuovi Materiali (CERMANU), Università di Napoli Federico II, 80055 Portici, Italy

Lívia da Silva Lima , Fábio Lopes Olivares , Natália Oliveira Aguiar1 and Luciano Pasqualoto Canellas
Núcleo de Desenvolvimento de Insumos Biológicos para a Agricultura (NUDIBA), Universidade Estadual do Norte Fluminense Darcy Ribeiro (UENF), Av. Alberto Lamego, 2000, Campos dos Goytacazes, 28013-602 Rio de Janeiro, Brazil

Rodrigo Rodrigues de Oliveira and Maria Raquel Garcia Vega
Laboratório de Ciências Químicas (LCQUI), Universidade Estadual do Norte Fluminense Darcy Ribeiro (UENF), Av. Alberto Lamego, 2000, Campos dos Goytacazes, 28013-602 Rio de Janeiro, Brazil

Ziad Al Chami, Deaa Alwanney and Ivana Cavoski
CIHEAM - Istituto Agronomico Mediterraneo di Bari (IAMB), Via Ceglie 9, 70010 Valenzano, BA, Italy

Sandra Angelica De Pascali and Francesco Paolo Fanizzi
Dipartimento di Scienze e Tecnologie Biologiche ed Ambientali, Università del Salento, Centro Ecotekne strada provinciale Lecce Monteroni, 73100 Lecce, LE, Italy

Catia Fredericci, Gerhard Ett, João B Ferreira Neto, Tiago R Ribeiro and Ricardo L Indelicato
Institute for Technological Research of São Paulo State, 05508-901 São Paulo, SP, Brazil

Guilherme FB Lenz e Silva, Fernando JG Landgraf and Ricardo L Indelicato
Polytechnic School, University of São Paulo, 05508-010 São, SP, Brazil